The forced response of $m\ddot{x}+c\dot{x}+kx = F_0\sin\omega_{dr}t$ is $x(t) = A_0\sin(\omega_{dr}t+\phi)$ where the normalized amplitude A_0k/F_0 is given by $A_0k/F_0 = \dfrac{1}{\sqrt{(1-r^2)^2+(2\zeta r)^2}}$ plotted below

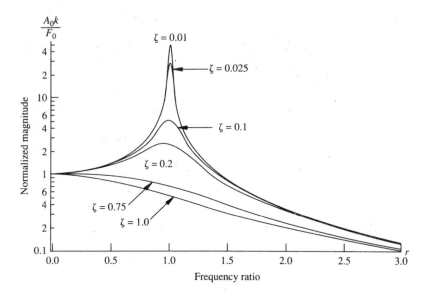

$$r = \omega_{dr}/\omega, \quad \omega = \sqrt{k/m}, \quad \zeta = \frac{c}{2\sqrt{km}}$$

and where the phase ϕ is given by $\phi = \tan^{-1}\left[\dfrac{2\zeta r}{1-r^2}\right]$ plotted below

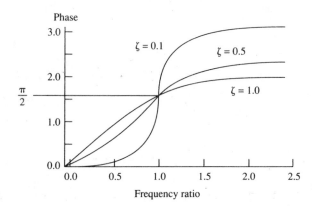

❏ Engineering Vibration ❏

DANIEL J. INMAN

Virginia Polytechnic Institute and State University

PRENTICE HALL, Englewood Cliffs, New Jersey 07632

Library of Congress Cataloging-in-Publication Data

Inman, D.J.
    Engineering vibration/ Daniel J. Inman
       p.   cm.
    Includes bibliographical references and index.
    ISBN 0-13-518531-9
    1. Vibration   I. Title.
TA355.I519   1996
620.3--dc20   95-33645
                                        CIP

Acquisitions editor: **BILL STENQUIST**
Editorial/production supervision
 and interior design: **SHARYN VITRANO**
Copy editor: **BARBARA ZEIDERS**
Cover design: **BARBARA CLAY**
Cover photo: **SOUND & VIBRATION MAGAZINE**
Manufacturing buyer: **DONNA SULLIVAN**
Editorial assistant: **MEG WEIST**

©1996 by Prentice-Hall, Inc.
Simon & Schuster/ A Viacom Company
Upper Saddle River, NJ 07458

The author and publisher of this book have used their best efforts in preparing this book. These efforts include the
development, research, and testing of the theories and programs to determine their effectiveness. The author and
publisher make no warranty of any kind, expressed or implied, with regard to these programs or the documentation
contained in this book. The author and publisher shall not be liable in any event for incidental or consequential damages
in connection with, or arising out of, the furnishing, performance, or use of these programs.

Printed in the United States of America

10 9 8 7 6 5

ISBN   0-13-518531-9

Prentice-Hall International (UK) Limited, London
Prentice-Hall of Australia Pty. Limited, Sydney
Prentice-Hall Canada Inc., Toronto
Prentice-Hall Hispanoamericana, S.A., Mexico
Prentice-Hall of India Private Limited, New Delhi
Prentice-Hall of Japan, Inc., Tokyo
Simon & Schuster Asia Pte. Ltd., Singapore
Editora Prentice-Hall do Brasil, Ltda., Rio de Janeiro

To Daniel John Inman, II

Contents

Preface

This book is intended for use in a first course in vibration for undergraduates in mechanical engineering, civil engineering, aerospace engineering or mechanics at the junior or senior level or as self study. This book contains topics normally found in undergraduate vibrations texts. In addition, new topics are introduced in the areas of design, measurement, damping and computational aspects. A goal and motivation for writing this text is the integration of traditional topics with a design emphasis, the introduction of modal analysis and the inclusion of a professional quality software package. This goal also combines the study of traditional introductory vibration with the use of vibration design, analysis and testing in engineering practice. As such this book is also intended to provide a reference for professional engineers and others in the engineering work force.

This material has grown out of lectures given to first semester seniors in mechanical, aerospace and civil engineering at the State University of New York at Buffalo. The fundamentals of introductory vibration analysis as historically defined by Den Hartog and Thomson are retained in this book. Additional topics chosen for presentation in this book reflect some of the recent advances in vibration technology, changes in ABET (Accreditation Board for Engineering and Technology) criteria and the increased importance of engineering design, as well as modal analysis (measurement).

Each instructor/teacher of undergraduate vibrations maintains their own preference for the order of presentation and grouping of topics. For instance, the geometric approach to solving for the forced response of a single degree of freedom system in Section 2.6 is considered by some to be useless because the approach cannot be used to solve any other problems, and considered by others to be very useful for helping students understand the phase relationship between the various forces. The text is written so that this topic can be omitted or covered as desired by the instructor. Another example is the inclusion of Coulomb damping (Section 2.7) in the chapter on harmonic excitation. Some instructors feel this belongs with the material in Chapter 1. It can be taught in either place, however it is included in Chapter 2 because it leads nicely into alternative forms of damping which are defined only in terms of harmonic forcing and the concept of equivalent viscous damping.

A unique feature of this book is the use of "windows," which are distributed throughout and provide reminders to the reader of essential information pertinent to the text material. The windows are placed in the text at points where such prior information is required. The windows are also used to summarize essential information.

The text is organized into ten chapters. The first chapter introduces free vibration of single degree of freedom linear systems in a rather traditional format. However, the chapter ends with some initial discussion of design, measurement and stability representing a departure from tradition texts. The second chapter discusses the response of single degree of freedom systems to harmonic inputs covering traditional aspects and also introducing alternative forms of damping not typically discussed. Chapter Three describes the response of single degree of freedom systems to general inputs with new emphasis on the response to random input for the sake of understanding vibration measurements. Stability is also discussed here. The fourth chapter combines the introduction of two degree of freedom systems with that of multiple degree of freedom systems and focuses on modal analysis. This chapter also provides a gentle on "the fly" introduction to matrix methods useful in vibration analysis, measurement and design, and essential in computational vibration.

Chapter Five represents a new organization of design related material in introductory vibration books. Presentation of traditional topics of design such as absorbers, isolators and critical speeds are augmented by information on visoelastic damping treatments and active vibration suppression methods. This design chapter runs the spectrum from using the modern notion of optimization in design to the very practical and traditional notions of design by choosing isolation materials from manufacturer's specifications. Chapter Six introduces the vibration analysis of distributed parameter systems. The procedures presented in this chapter focus on modal analysis and include damping as well as outlining traditional topics.

Chapter Seven is on experimental modal analysis, a topic not normally put forth in texts. The chapter draws on random vibration and modal analysis presented in earlier chapters and ends with a brief description of diagnostics. For readers interested in computing, the computer files for this chapter contain data from actual experiments which can be analyzed using the methods presented in Sections 7.4 and 7.5. The finite element method for vibration analysis is discussed in Chapter Eight. The focus of this chapter is the understanding of constructing finite element models of simple components. The computer files for this chapter contain a program for computing finite element models of simple structures.

Chapter Nine introduces numerical methods in the context of modern computational machines. The first few chapters present the usual topics of numerical error, influence coefficients and traditional methods of determining natural frequencies. The later sections take full advantage of modern computational algorithms and computer technology. Sections 9.5 and 9.6 on eigenvalue computation and Section 9.7 on numerical simulation are the most relevant sections in this chapter and can be covered independently at almost any convenient sequence in the book. Chapter Ten provides a brief introduction to nonlinear vibration problems and analysis. Traditional introductory topics are included along with a discussion of nonlinear damping terms, a common source of nonlinearity in practice.

The end of each chapter (except Chapter Nine) includes a brief description of the programs available for solving the problems associated with that chapter. These programs form a Vibration Toolbox and are discussed in greater detail in Chapter Nine and in the Student Edition of MATLAB® or any other recent version of MATLAB. More information on MATLAB can be obtained by using the tear out card in the inside back cover or by writing

> The Math Works, Inc.
> Cochituate Place
> 24 Prime Parkway
> Natick, Massachusetts 01760

A Macintosh version is also available (see detail card). The key factor in the software provided here is that the quality of the numerical algorithms used is the best available and the algorithms are "bug" free. The programs are also very user friendly. Every plot, graph and numerical presentation in this book is the result of a MATLAB output.

These computer exercises and programs can be ignored or used as the reader chooses. However, if used, these programs provide insight into the theory and practice of vibration and vibration measurement not before possible in books at this level. For instance, actual test data is contained on the disc for use in exercises in the modal testing chapter.

PATHS THROUGH THE TEXT

There are several possibilities for studying and teaching vibrations using this book. Whenever possible, sections have been made independent to provide maximum flexibility in organizing the order of topics to be covered. For example Chapters 1–5 have been used extensively as a 3 credit course of the first semester seniors in a curriculum designed to (successfully) be one credit of design and two credits of analysis under ABET criteria. Chapters 1, 2, 3, 4 and 7 have been used to form a 3 credit course focusing on vibration analysis and measurement. Chapters 6 through 10 have been used together as a second undergraduate vibration course. Certain sections can also be removed from each chapter such as 1.6, 1.7, 1.8, 2.6, 2.7, 2.8, 3.4, 3.5, 3.7, 3.8, 4.5, 5.1, 5.4–5.9, and 6.7, resulting in a traditional undergraduate vibration course as defined by Thompson, Rao, etc. The computer files and associated problems can be ignored or emphasized depending on the instructor's preference. They are, however, particularly useful in Chapters 4, 7, 8 and 9.

ACKNOWLEDGMENTS

The "push" for starting this project came from Doug Humphrey of Prentice Hall. Doug provided encouragement, has given me a great deal of advice and was always willing to listen to crazy ideas. The desire to write this text came from teaching students at the University at Buffalo over the last 12 years. I owe a great deal of this material to those students who sat through my lectures and taught me how to present this material. I also am indebted to the reviewers of the text, some known to me, some anonymous; some very critical, some very positive, but all very useful.

In particular I gratefully acknowledge the contributions and suggestions of Dr. Ephrahim Garcia of Vanderbilt University, who taught out of the drafts of this manuscript, and Dr. Malcolm A. Cutchins of Auburn University for helping to provide aerospace related problems and examples. I would like to thank Dr. Harley Cudney of Virginia Tech for a detailed review of chapter nine, and Professor Ladislav Starek of Bratislava Technical University in Bratislava, Slovakia, for correcting the page proofs and checking the software.

Professor Kon-Well Wang of Pennsylvania State University, Professor Noel C. Perkins of The University of Michigan, Professor Virgil W. Snyder of Michigan Technological University, Professor Hamid R. Hamidzadeh of South Dakota State University, Professor Subhash C. Sinha of Auburn University, Professor Wayne W. Walter of Rochester Institute of Technology, Professor Donald A. Grant of the University of Maine and Professor Robert G. Leonard of Virginia Polytechnic Institute and State University, all provided very helpful and detailed reviews of various drafts of this text for which I am extremely grateful. It is no small matter to review a book manuscript, and their suggestions have become as much a part of this manuscript as my own thoughts.

I had several "co-authors" in this text who deserve a great deal of credit. Chris Jerome typed the manuscript (over and over again) and provided tremendous moral support as well as skillful word processing. Sally Farmer, an undergraduate at the time, provided spirited discussions and helped me maintain a "student's" perspective in presenting the material. My graduate students, Joe Slater, Don Leo, Jeff Curtis, Marca Lam, Brett Pokines, Jeff Dosch, Andreas Kress, Tami Hamernik, Rob Carlin, Jim Lindsey, Gordon Parker, Ralph Rietz and Jim Calamita all provided endless support, proofreading, problem solving, graphics and discussion. In addition, Jeff Curtis, who was the course teaching assistant for several years, edited the Solutions Manual and had a great deal of input regarding problems. Joe Slater and Don Leo deserve special thanks for writing the Vibrations Toolbox, as does Ralph Rietz for correcting page proofs.

I would also like to thank Jennifer Wenzel, the production editor, for her outstanding help and for making the production stage of this book a pleasant experience.

A great deal of moral support as well as a sense of purpose was provided by my family. My son, Daniel J. Inman, II, my daughters, Angela W. Inman and Jennifer W. Scamacca and my wife, Catherine A. Little, encouraged this effort and shared in all the pain and fun of producing a manuscript.

Several other groups had a strong influence on my perception of what should be in this book. The American Society of Mechanical Engineers deserves some thanks for appointing me the technical editor of the *Journal of Vibrations and Acoustics*. I would also like to thank the following agencies for supporting my research in vibrations during the period that this book was written. I cannot imagine writing a book without being involved with the actual technology as well. In this regard I would like to thank Dr. W. Keith Belvin of NASA Langley Research Center, Dr. Bert Marsh of the National Science Foundation, Dr. Gary Anderson of the Army Research Office, Dr. Spencer Wu and Dr. Marc Jacobs, both of the Air Force Office of Scientific Research.

Daniel J. Inman
Blacksburg, Virginia

1 Introduction to Vibration and the Free Response

Vibration, some wanted, some unwanted. In the first case, electromechanically induced vibration of speaker cones produces music for entertainment. In the second case, vibration of a bridge caused by an earthquake results in damage which threatens lives and disrupts communities. The picture at the top is the Roger Waters' concert celebrating the demise of the Berlin Wall; the picture on the bottom is the San Francisco–Oakland Bay Bridge damaged in the 1989 California earthquake. Some of the basic concepts of vibration analysis, measurement, and design are introduced in this chapter. A complete understanding of vibration is needed by engineers involved in the analysis and design of a multitude of devices and structures.

1.1 INTRODUCTION TO FREE VIBRATION

Vibration is the study of the repetitive motion of objects relative to a stationary frame of reference or nominal position (usually equilibrium). Vibration is evident everywhere and in many cases greatly affects the nature of engineering designs. The vibrational properties of engineering devices are often limiting factors in their performance. Vibration can be harmful and should be avoided, or it can be extremely useful and desired. In either case, knowledge about vibration—how to analyze, measure and control it—is desired and forms the topic of this book.

Typical examples of vibration familiar to most are the motion of a guitar string, the quality of ride of an automobile or motorcycle, the motion of an airplane's wings and the swaying of a large building due to wind or an earthquake. In the chapters that follow, vibration is modeled mathematically based on fundamental principles, such as Newton's laws, and analyzed using results from calculus and differential equations. Information about techniques used to measure the vibrations of a system are developed. In addition, information and methods are given that are useful for designing a particular system to have a specific vibrational response.

The physical explanation of the phenomena of vibration concerns the interplay between potential energy and kinetic energy. A vibrating system must have a component that stores potential energy and releases it as kinetic energy in the form of motion (vibration) of a mass. The motion of the mass then gives up kinetic energy to the potential energy storing device. Vibration can occur in many directions and can be the result of the interaction of many objects. To simplify the introduction of the topic of vibration, only motion in a single direction of a single particle is described in the first three chapters. This chapter focuses on the response of single-degree-of-freedom systems subject to initial disturbances. In Chapter 2 the vibration response of single-degree-of-freedom systems subject to harmonic excitation forces is examined and various types of damping are introduced. The response of a single-degree-of-freedom system to general force excitations is discussed in Chapter 3. In the remaining chapters, the models of vibrating systems become more complex, as does the required analysis.

1.1.1 The Spring–Mass Model

From introductory physics and dynamics, the fundamental kinematical quantities used to describe the motion of a particle are displacement, velocity, and acceleration vectors. In addition, the laws of physics state that the motion of a mass with changing velocity is determined by the net force acting on the mass. An easy device to use in thinking about vibration is a spring (such as the one used to pull a storm door shut, or an automobile spring) with one end attached to a fixed object and a mass attached to the other end. A schematic of this arrangement is given in Figure 1.1.

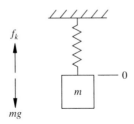

Figure 1.1   Schematic of a single-degree-of-freedom, spring–mass oscillator.

Ignoring the mass of the spring itself, the forces acting on the mass consist of the force of gravity pulling down (mg) and the force of the spring pulling back up (f_k). Note that in this case the force vectors are collinear and can easily be treated as scalars. The nature of the spring force can be deduced by performing a simple static experiment. With no mass attached, the spring stretches to the position labeled $x_0 = 0$ in Figure 1.2. As successively more mass is attached to the spring, the force of gravity causes the spring to stretch further. If the value of the mass is recorded, along with the value of the displacement of the end of the spring each time more mass is added, the plot of the force (mass, denoted by m, times the acceleration due to gravity, denoted by g), versus this displacement, denoted by x, yields a curve similar to that illustrated in Figure 1.3. Note that in the region of values for x between 0 and about 20 mm (millimeters), the curve is a straight line. This indicates that for deflections less than 20 mm and forces less than 1000 N (newtons), the force that is applied by the spring to the mass is proportional to the stretch of the spring. The constant of proportionality is the slope of the straight line between 0 and 20 mm. For the particular spring of Figure 1.3, the constant is 50 N/mm, or 5×10^4 N/m. Thus the equation that describes the force applied by the spring, denoted f_k, to the mass is

$$f_k = kx \tag{1.1}$$

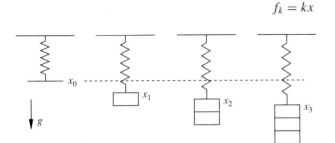

Figure 1.2   Schematic of a massless spring with no mass attached, showing its static equilibrium position, followed by increments of increasing mass, illustrating the corresponding deflections.

The value of the slope, denoted by k, is called the *stiffness* of the spring and is a property that characterizes the spring for all situations for which the displacement is less than 20 mm.

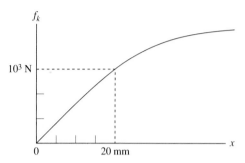

Figure 1.3 Static deflection curve for the spring of Figure 1.2.

Note that the relationship between f_k and x of equation (1.1) is *linear* (i.e., the curve is linear and f_k depends linearly on x). If the displacement of the spring is larger then 20 mm, the relationship between f_k and x becomes *nonlinear*, as indicated in Figure 1.3. Nonlinear systems are much more difficult to analyze and form the topic of Chapter 10. In this and all the other chapters, it is assumed that displacements (and forces) are limited to be in the linear range.

Next consider a free-body diagram of the mass in Figure 1.4, with the massless spring elongated from its rest or equilibrium position. As in the earlier figures, the mass of the object is taken to be m and the stiffness of the spring is taken to be k. Assuming that the mass moves on a frictionless surface along the x direction, the only force acting on the mass in the x direction is the spring force. As long as the motion of the spring does not exceed its linear range, the sum of the forces in the x direction must equal the product of mass and acceleration. For the system of Figure 1.4, this yields

$$m\ddot{x}(t) = -kx(t) \qquad \text{or} \qquad m\ddot{x}(t) + kx(t) = 0 \qquad (1.2)$$

where $\ddot{x}(t)$ denotes the second time derivative of the displacement (i.e. the acceleration). Here the displacement vector and acceleration vector are treated as scalars since the net force in the y direction is zero ($N = mg$) and the force in the x direction is collinear with the inertial force. Both the displacement and acceleration are functions of the elapsed time t, as denoted in equation (1.2). Window 1.1 illustrates three types of mechanical systems, which can be described by equation (1.2): a spring–mass system, a rotating shaft, and a swinging pendulum.

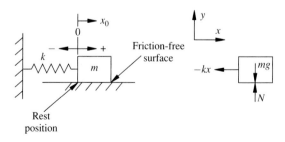

Figure 1.4 Single spring–mass system given an initial displacement x_0 from its rest or equilibrium position, and zero initial velocity.

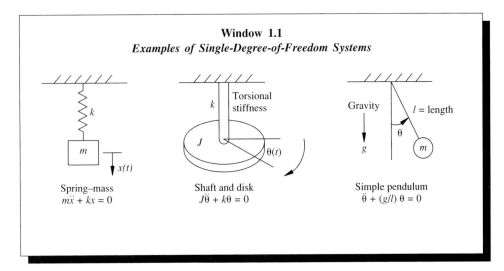

Window 1.1

Examples of Single-Degree-of-Freedom Systems

Spring–mass
$m\ddot{x} + kx = 0$

Shaft and disk
$J\ddot{\theta} + k\theta = 0$

Simple pendulum
$\ddot{\theta} + (g/l)\,\theta = 0$

One of the goals of vibration analysis is to be able to predict the response or motion of a vibrating system. Thus it is desirable to calculate the solution to equation (1.2). Fortunately, the differential equation of (1.2) is well known and is treated extensively in introductory calculus and physics texts, as well as in texts on differential equations. In fact, there are a variety of ways to calculate this solution. These are all discussed in some detail in the next section. For now it is sufficient to present a solution based on physical observation. From experience watching a spring such as the one in Figure 1.4 (or a pendulum), it is guessed that the motion is periodic, perhaps of the form

$$x(t) = A \sin(\omega t + \phi) \tag{1.3}$$

This choice is made because the sine function describes periodic phenomena. Equation (1.3) is the sine function in its most general form, where the constant A is the *amplitude* or maximum value of the function, ω, the *angular natural frequency*, determines the interval in time during which the function repeats itself; and ϕ, called the *phase*, determines the initial value of the sine function. It is standard to measure the time t in seconds (s). The phase is measured in radians and the frequency is measured in radians per second (rad/s).

To see if equation (1.3) is in fact a solution of the equation of motion, it is substituted into equation (1.2). Successive differentiation of the displacement, $x(t)$ in the form of equation (1.3), yields the velocity, $\dot{x}(t)$, given by

$$\dot{x}(t) = \omega A \cos(\omega t + \phi) \tag{1.4}$$

and the acceleration, $\ddot{x}(t)$, given by

$$\ddot{x}(t) = -\omega^2 A \sin(\omega t + \phi) \tag{1.5}$$

Substitution of equations (1.5) and (1.3) into (1.2) yields

$$-m\omega^2 A \sin(\omega t + \phi) = -kA \sin(\omega t + \phi) \tag{1.6}$$

Dividing by A and m yields the fact that this equation is satisfied if $\omega^2 = k/m$. Hence, equation (1.3) is a solution of the equation of motion. The constant ω characterizes the spring–mass system, as well as the frequency at which the motion repeats itself, and hence is called the system's *natural frequency*. A plot of the solution $x(t)$ versus time t is given in Figure 1.5. It remains to interpret the constants A and ϕ.

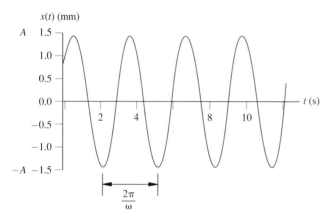

Figure 1.5 Response of a simple spring–mass system to an initial displacement of $x_0 = 0.5$ mm and an initial velocity of $v_0 = 2\sqrt{2}$ mm/s. The natural frequency is 2 rad/s and the amplitude is 1.5 mm.

Recall from differential equations that the equation of motion is of second order, so that solving equation (1.2) involves integrating twice. Thus there are two constants of integration to evaluate. These are the constants A and ϕ. The physical significance, or interpretation, of these constants is that they are determined by the initial state of motion of the spring–mass system. Again, recall Newton's laws. If no motion is imparted to the mass, it will stay at rest. If, however, the mass is displaced to a position of x_0 at time $t = 0$, the potential energy in the spring will result in motion. Also, if the mass is given an initial velocity of v_0 at time $t = 0$, motion will result. These are called *initial conditions* and when substituted into the solution (1.3) yield

$$x_0 = x(0) = A\sin(\omega 0 + \phi) = A\sin\phi \tag{1.7}$$

and

$$v_0 = \dot{x}(0) = \omega A\cos(\omega 0 + \phi) = \omega A\cos\phi \tag{1.8}$$

Solving these two simultaneous equations for the two unknowns A and ϕ yields

$$A = \frac{\sqrt{\omega^2 x_0^2 + v_0^2}}{\omega} \quad \text{and} \quad \phi = \tan^{-1}\frac{\omega x_0}{v_0} \tag{1.9}$$

Thus the solution of the equation of motion for the spring–mass system is given by

$$x(t) = \frac{\sqrt{\omega^2 x_0^2 + v_0^2}}{\omega}\sin\left(\omega t + \tan^{-1}\frac{\omega x_0}{v_0}\right) \tag{1.10}$$

and is plotted in Figure 1.5. This solution is called the *free response* of the system because no force external to the system is applied after $t = 0$. The motion of the spring–

mass system is called *simple harmonic motion* or *oscillatory motion* and is discussed in detail in the following section. The spring–mass system is also referred to as a *simple harmonic oscillator*, as well as an *undamped single-degree-of-freedom system*.

Example 1.1.1

The phase angle ϕ describes the relative shift in the sinusoidal vibration of the spring–mass system resulting from the initial displacement, x_0. Verify that equation (1.10) satisfies the initial condition $x(0) = x_0$.

Solution   Substitution of $t = 0$ in equation (1.10) yields

$$x(0) = A \sin \phi = \frac{\sqrt{\omega^2 x_0^2 + v_0^2}}{\omega} \sin\left(\tan^{-1} \frac{\omega x_0}{v_0}\right)$$

Figure 1.6 illustrates the phase angle ϕ defined by equation (1.9). This right triangle is used to define the sine and tangent of the angle ϕ. From the geometry of a right triangle, and the definitions of the sine and tangent functions, the value of $x(0)$ is computed to be

$$x(0) = \frac{\sqrt{\omega^2 x_0^2 + v_0^2}}{\omega} \frac{\omega x_0}{\sqrt{\omega^2 x_0^2 + v_0^2}} = x_0$$

which verifies that the solution given by equation (1.10) is consistent with the initial displacement condition.   □

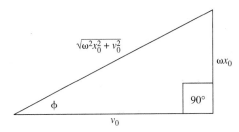

Figure 1.6   Trigometric relationship between the phase angle ϕ, the frequency and the initial conditions.

Example 1.1.2

A vehicle wheel, tire, and suspension assembly can be modeled crudely as a single degree-of-freedom spring–mass system. The mass of the assembly is measured to be about 300 kg. Its frequency of oscillation is observed to be π rad/s. What is the approximate stiffness of the tire, wheel, and suspension assembly?

Solution   The relationship between frequency, mass, and stiffness is $\omega = \sqrt{k/m}$, so that

$$k = m\omega^2 = (300 \text{ kg})(\pi \text{ rad/s})^2 = 2960 \text{ N/m}$$

This provides one simple way to estimate the stiffness of a complicated device. This stiffness could also be estimated by using a static deflection experiment similar to that suggested by Figures 1.2 and 1.3.   □

The main point of this section is summarized in Window 1.2. This illustrates harmonic motion and how the initial conditions determine the response of such a system.

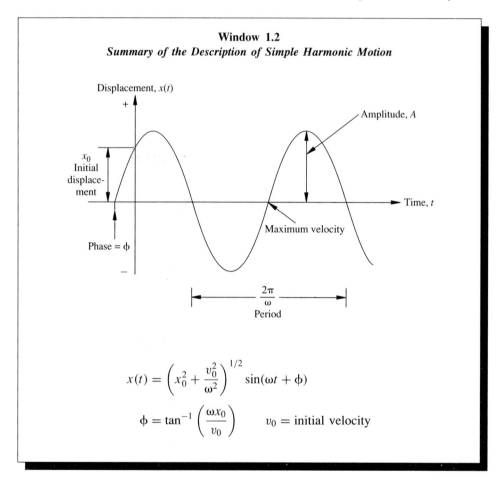

Window 1.2
Summary of the Description of Simple Harmonic Motion

$$x(t) = \left(x_0^2 + \frac{v_0^2}{\omega^2} \right)^{1/2} \sin(\omega t + \phi)$$

$$\phi = \tan^{-1} \left(\frac{\omega x_0}{v_0} \right) \qquad v_0 = \text{initial velocity}$$

1.2 HARMONIC MOTION

The fundamental kinematic properties of a particle moving in one dimension are displacement, velocity, and acceleration. For the harmonic motion of a simple spring–mass system, these are given by equations (1.3), (1.4), and (1.5), respectively. Note the different relative amplitudes of each quantity. For systems with natural frequency larger than 1 rad/s, the relative amplitude of the velocity response is larger than that of the displacement response by a factor of ω, and the acceleration response is larger by a factor of ω^2. For systems with frequency less than 1, the velocity and acceleration have smaller relative amplitudes than the displacement. Also note that the velocity is 90° (or $\pi/2$ radians) out of phase with the position [i.e., $\sin(\omega t + \pi/2 + \phi) = \cos(\omega t + \phi)$], while the acceleration is 180° out of phase with the position and 90° out of phase with the velocity. This is summarized and illustrated in Window 1.3.

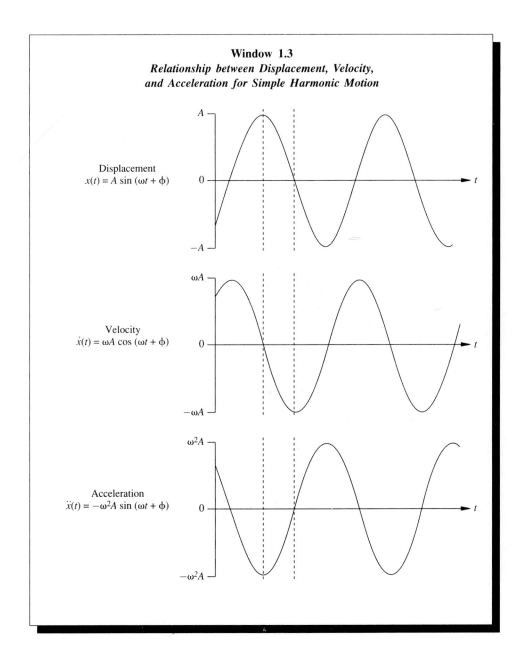

Window 1.3
*Relationship between Displacement, Velocity,
and Acceleration for Simple Harmonic Motion*

Displacement
$x(t) = A \sin(\omega t + \phi)$

Velocity
$\dot{x}(t) = \omega A \cos(\omega t + \phi)$

Acceleration
$\ddot{x}(t) = -\omega^2 A \sin(\omega t + \phi)$

The angular natural frequency, ω, used in equations (1.3) and (1.10) is measured in radians per second and describes the repetitiveness of the oscillation. As indicated in Figure 1.5, the time the cycle takes to repeat itself is the *period T*, which is related to the natural frequency by

$$T = \frac{2\pi \text{ rad}}{\omega \text{ rad/s}} = \frac{2\pi}{\omega} \text{s} \qquad (1.11)$$

This results from the elementary definition of the period of a sine function. Quite often the frequency is measured and discussed in terms of cycles per second, which is called hertz. The frequency in hertz (Hz), denoted by f, is related to the frequency in radians per second, denoted ω, by

$$f = \frac{\omega}{2\pi} = \frac{\omega \text{ rad/s}}{2\pi \text{rad/cycle}} = \frac{\omega \text{ cycles}}{2\pi \text{ s}} = \frac{\omega}{2\pi} \text{ (Hz)} \qquad (1.12)$$

Example 1.2.1

Consider a small spring about 30 mm (or $1\frac{1}{2}$ in.) long, welded to a stationary table (ground) so that it is fixed at the point of contact, with a 12-mm (or $\frac{1}{2}$-in.) bolt welded to the other end, which is free to move. The mass of this system is about 49.2×10^{-3} kg (equivalent to about 1.73 ounces). The spring stiffness can be measured using the method suggested in Figure 1.3. Such a method yields a spring constant of $k = 857.8$ N/m. Calculate the natural frequency and period. Also determine the maximum amplitude of the response if the spring is initially deflected 10 mm.

Solution   The natural frequency is

$$\omega = \sqrt{\frac{k}{m}} = \sqrt{\frac{857.8 \text{ N/m}}{49.2 \times 10^{-3} \text{ kg}}} = 132 \text{ rad/s}$$

In hertz this becomes

$$f = \frac{\omega}{2\pi} = 21 \text{ Hz}$$

The period is

$$T = \frac{2\pi}{\omega} = \frac{1}{f} = 0.0476 \text{ s}$$

To determine the maximum value of the displacement response, note from Figure 1.5 that this corresponds to the value of the constant A. Assuming that no initial velocity is given to the spring ($v_0 = 0$), equation (1.9) yields

$$x(t)_{\text{max}} = A = \frac{\sqrt{\omega^2 x_0^2 + v_0^2}}{\omega} = x_0 = 10 \text{ mm}$$

Note that the maximum value of the velocity response is ωA or $\omega x_0 = 1320$ mm/s and the acceleration response has maximum value $\omega^2 A = \omega^2 x_0 = 174.24 \times 10^3$ mm/s^2. Since $v_0 = 0$, the phase is $\phi = \tan^{-1}(\omega x_0/0) = \pi/2$, or 90°. Hence in this case the response is $x(t) = 10 \sin(132t + \pi/2) = 10 \cos(132t)$ mm.

$\square$

The solution given by equation (1.10) was developed assuming that the response should be harmonic based on physical observation. The form of the response can also be derived by a more analytical approach following the theory of elementary differential equations (see, e.g., Boyce and DiPrima, 1986). This approach is reviewed here and will be generalized in later sections and chapters to solve for the response of more complicated systems.

Assume that the solution $x(t)$ is of the form

$$x(t) = ae^{\lambda t} \qquad (1.13)$$

where a and λ are nonzero constants to be determined. Upon successive differentiation, equation (1.13) becomes $\dot{x}(t) = \lambda ae^{\lambda t}$ and $\ddot{x}(t) = \lambda^2 ae^{\lambda t}$. Substitution of the assumed exponential form into equation (1.2) yields

$$m\lambda^2 ae^{\lambda t} + kae^{\lambda t} = 0 \qquad (1.14)$$

Since the term $ae^{\lambda t}$ is never zero, expression (1.14) can be divided by $ae^{\lambda t}$ to yield

$$m\lambda^2 + k = 0 \qquad (1.15)$$

Solving this algebraically results in

$$\lambda = \pm\sqrt{-\frac{k}{m}} = \pm\sqrt{\frac{k}{m}}\,j = \pm\omega j \qquad (1.16)$$

where $j = \sqrt{-1}$ is the imaginary number and $\omega = \sqrt{k/m}$ is the natural frequency as before. Note that there are two values for λ, $\lambda = -\omega j$ and $\lambda = +\omega j$. This results because the equation for λ is of second order. This implies that there must be two solutions of equation (1.2) as well. Substitution of equation (1.16) into equation (1.13) yields that the two solutions for $x(t)$ are

$$x(t) = a_1 e^{+j\omega t} \qquad \text{and} \qquad x(t) = a_2 e^{-j\omega t} \qquad (1.17)$$

Since equation (1.2) is linear, the sum of two solutions is also a solution, hence the response $x(t)$ is of the form

$$x(t) = a_1 e^{+j\omega t} + a_2 e^{-j\omega t} \qquad (1.18)$$

where a_1 and a_2 are complex-valued constants of integration. The Euler relations for trigonometric functions states that $2j \sin\theta = (e^{\theta j} - e^{-\theta j})$, and $2\cos\theta = (e^{\theta j} + e^{-\theta j})$, where $j = \sqrt{-1}$. Using the Euler relations, equation (1.18) can be written as

$$x(t) = A \sin(\omega t + \phi) \qquad (1.19)$$

where A and ϕ are real-valued constants of integration. Note that equation (1.19) is in agreement with the physically intuitive solution given by equation (1.3). The relationships among the various constants in equations (1.18) and (1.19) are given in Window 1.4. Window 1.5 illustrates the use of Euler relations for deriving harmonic functions from exponentials for a damped case.

A precise terminology is useful in discussing an engineering problem and the subject of vibration is no exception. Since the position, velocity, and acceleration change continually with time, several other quantities are used to discuss vibration. The *peak value*, defined as the maximum displacement, or magnitude A of equation (1.9), is often used to indicate the region in space in which the object vibrates. Another quantity useful

---

Window 1.4
Three Representations of Harmonic Motion

The solution of $m\ddot{x} + kx = 0$ subject to nonzero initial conditions can be written in three equivalent ways. First, the solution can be written as

$$x(t) = a_1 e^{j\omega t} + a_2 e^{-j\omega t} \qquad \omega = \sqrt{\frac{k}{m}} \qquad j = \sqrt{-1}$$

where a_1 and a_2 are complex-valued constants: second, the solution can be written as

$$x(t) = A \sin(\omega t + \phi)$$

where A and ϕ are real-valued constants; and third, the solution can be written as

$$x(t) = A_1 \cos \omega t + A_2 \sin \omega t$$

where A_1 and A_2 are real-valued constants. Each set of two constants is determined by the initial conditions. The various constants are related by

$$A = \sqrt{A_1^2 + A_2^2} \qquad \phi = \tan^{-1}\left(\frac{A_1}{A_2}\right)$$

$$A_1 = a_1 + a_2 \qquad A_2 = (a_1 - a_2)j$$

$$a_1 = \frac{A_1 - A_2 j}{2} \qquad a_2 = \frac{A_1 + A_2 j}{2}$$

which follow from trigonometric identities and the Euler's formulas. Note that a_1 and a_2 are a complex conjugate pair, so that A_1 and A_2 are both real numbers.

---

in describing vibration is the *average value,* denoted $\bar{x}$, and defined by

$$\bar{x} = \lim_{T \to \infty} \frac{1}{T} \int_0^T x(t)\, dt \qquad (1.20)$$

Note that the average value of $x(t) = A \sin \omega t$ over one period of oscillation is zero.

Since the square of displacement is associated with a system's potential energy, the average of the displacement squared is sometimes a useful vibration property to discuss. The mean-square value of the displacement $x(t)$, denoted by $\bar{x}^2$, is defined by

$$\bar{x}^2 = \lim_{T \to \infty} \frac{1}{T} \int_0^T x^2(t)\, dt \qquad (1.21)$$

The square root of this value, called the *root mean square* (rms) value, is commonly used in specifying vibration. Because the peak value of the velocity and acceleration are multiples of the natural frequency times the displacement amplitude [i.e., equations (1.3)–(1.5)] these three basic quantities often differ in value by an order of magnitude or more. Hence logarithmic scales are often used. A common unit of measurement for vibration amplitudes and rms values is the *decibel* (db). The decibel was originally defined in terms of the base 10 logarithm of the power ratio of two electrical signals, or as the ratio of the square of the amplitudes of two signals. Following this idea, the decibel is defined as

$$dB \equiv 10 \log_{10} \left(\frac{x_1}{x_2}\right)^2 = 20 \log_{10} \frac{x_1}{x_2} \tag{1.22}$$

Using a dB scale expands or compresses vibration response information for convenience in graphical representation.

A standard way to specify all three vibration quantities—displacement, velocity, and acceleration—when discussing the response of a system, is to use a plot similar to that of Figure 1.7. In this log scale plot, the log of frequency f is plotted along the abscissa and the log of velocity along the ordinate. The lines slanting to the right of slope $+1$ are lines of constant displacement while the lines of slope -1 are lines of constant

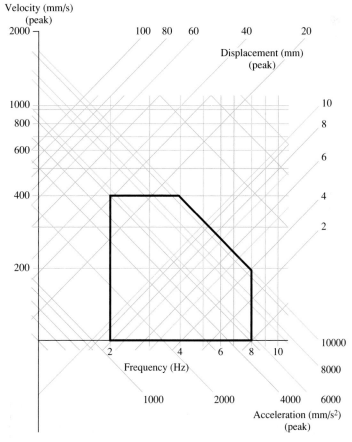

Figure 1.7   Nomograph for specifying acceptable limits of sinusoidal vibration.

acceleration. In this case peak values are plotted, but rms values are often plotted as well.

By sketching in a closed shape on this plot, such as the one indicated by the dark lines in Figure 1.7, the ranges of acceptable maximum displacement, velocities, and acceleration can easily be specified. This type of plot is called a *nomograph*. Such nomographs are useful for any type of single-degree-of-freedom system experiencing harmonic motion because the construction of the nomograph depends only on the frequency, ω, and is independent of specific values of mass or stiffness, but rather, depends only on their ratio and the displacement amplitude. A nomograph is another way to illustrate the relationship between displacement, velocity, and acceleration summarized in Window 1.3.

Example 1.2.2

In the plot of Figure 1.7, acceptable ranges of vibration for a given device are between 2 and 8 Hz. Marking off the vertical lines corresponding to 2 and 8 Hz on the nomograph allows the relationships between acceleration, velocity, and displacement to be specified. If the maximum desired acceleration is 1g (9.8 m/s^2, or 9800 mm/s^2) in a frequency range of 2 to 8 Hz, and the velocity is limited to 400 mm/s, the peak displacement will be limited to 30 mm and the motion of the resulting vibration will be contained within the borders of the region marked by the dark boundaries in Figure 1.7.

□

Example 1.2.3

A stereo turntable is modeled as a 1-kg mass supported by a spring of stiffness $k = 400$ N/m. When it oscillates it does so with a maximum deflection of 2.4 mm. With records stacked on the turntable, the mass can increase to as much as 1.4 kg with a maximum deflection of 3.4 mm. What regions in the nomogram of vibration in the nomograph of Figure 1.7 does this define?

Solution   At one extreme, the turntable oscillates at

$$\omega_1 = \sqrt{400(\text{N/m})/1(\text{kg})} = 20 \text{ rad/s} = 20\frac{\text{rad}}{\text{s}} \cdot \frac{\text{cycle}}{2\pi \text{ rad}} = 3.18 \text{ Hz}$$

with an amplitude of 2.4 mm. Thus, the response is given by equation (1.3) as

$$x(t) = 2.4(\sin 20t) \text{ mm}$$

Thus the velocity and acceleration become

$$\dot{x}(t) = 20(2.4)(\cos 20t) = 48(\cos 20t) \text{ mm/s}$$

$$\ddot{x}(t) = -(20)^2(2.4)(\sin 20t) = -960(\sin 20t) \text{ mm/s}^2$$

respectively. This defines the lines given by a frequency of 3.18 Hz, displacement amplitude of 2.4 mm, velocity amplitude of 48 mm, and acceleration amplitude of 960 mm.

The other extreme of the vibration of the stereo occurs when the mass is changed to 1.4 kg due to records stacked on the turntable. In this case the frequency becomes

$$\omega_2 = \sqrt{400 \text{ N/m}/1.4 \text{ kg}} = 16.9 \text{ rad/s} = 16.9\frac{\text{rad}}{\text{s}} \cdot \frac{\text{cycle}}{2\pi \text{ rad}} = 2.69 \text{ Hz}$$

Under the larger mass, the maximum displacement becomes 3.4 mm, so that the displacement, velocity, and acceleration become

$$x(t) = (3.4 \text{ mm})(\sin 16.9t)$$

$$\dot{x}(t) = (16.9 \text{ rad/s})(3.4 \text{ mm})(\cos 16.9t) = 57(\cos 16.9t) \text{ mm/s}$$

$$\ddot{x}(t) = (-57 \text{ mm/s})(16.9 \text{ rad/s})(\sin 16.9t) = -971 (\sin 16.9t) \text{ mm/s}^2.$$

respectively. Thus vibration at this frequency can be represented on the nomograph of Figure 1.8 corresponding to 2.69 Hz, a displacement amplitude of 3.4 mm, a velocity amplitude of 57 mm/s, and an acceleration amplitude of 971 mm/s².

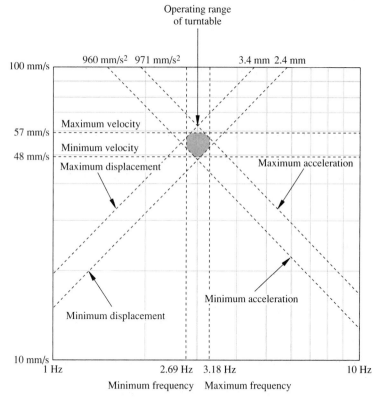

Figure 1.8 Nomograph illustrating the operating range of the turntable described in Example 1.2.3.

The area enclosed by the two groups of lines above is a reasonable representation of vibration amplitudes for frequencies ranging between 2.69 and 3.18 Hz. This corresponds to the operating range of the turntable as its operating condition changes from "empty" to fully stacked with records. On the scale of Figure 1.7, this region defined above is very small and hard to see. Thus for such applications, a nomograph with a finer scale should be used. A nomograph with the minimum and maximum values for the turntable is illustrated in Figure 1.8.

☐

Usually, the frequencies of concern in the vibration of mechanical devices range from 1 Hz in optical systems to thousands of hertz in machinery vibration. Amplitudes of vibration range from hard-to-measure micrometers up to a meter for tall buildings. Human beings can feel vibration above about 8 Hz and are more conscious of acceleration than displacement. For example, a frequency of about 30 Hz with an acceleration amplitude of about 2 m/s^2 is very disagreeable to an automobile passenger.

1.3 VISCOUS DAMPING

The response of the spring–mass model (Section 1.1) predicts that the system will oscillate indefinitely. However, everyday observation indicates that most freely oscillating systems eventually die out and reduce to zero motion. This experience suggests that the model sketched in Figure 1.4 and the corresponding mathematical model given by equation (1.2) needs to be modified to account for this decaying motion. The choice of a representative model for the observed decay in an oscillating system is based partially on physical observation and partially on mathematical convenience. The theory of differential equations suggests that adding a term to equation (1.2) of the form $c\dot{x}$, where c is a constant, will result in a solution $x(t)$ that dies out. Physical observation agrees fairly well with this model and it is used very successfully to model the damping or decay in a variety of mechanical systems. This type of damping, called *viscous damping*, is described in detail in this section.

While the spring forms a physical model for storing kinetic energy and hence causing vibration, the *dashpot*, or *damper*, forms the physical model for dissipating energy and damping the response of a mechanical system. A dashpot consists of a piston fit into a cylinder filled with oil as indicated in Figure 1.9. This piston is perforated with holes so that motion of the piston in the oil is possible. The laminar flow of the oil through the perforations as the piston moves causes a damping force on this piston.

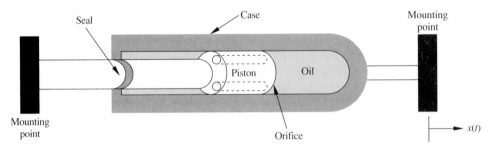

Figure 1.9 Schematic of a dashpot that produces a damping force $f_c = c\dot{x}$, where $x(t)$ is the motion of the case relative to the piston.

The force is proportional to the velocity of the piston, in a direction opposite that of the piston motion. This damping force, denoted by f_c, has the form

$$f_c = c\dot{x}(t) \tag{1.23}$$

where c is a constant of proportionality related to the oil viscosity. The constant c, called the *damping coefficient*, has units of N s/m, or kg/s.

In the case of the oil-filled dashpot, the constant c can be determined by fluid principles. However, in most cases, f_c is provided by equivalent effects occurring in the material forming the device. A good example is a block of rubber (which also provides stiffness f_k) such as an automobile motor mount, or the effects of air flowing around an oscillating mass. In all cases where the damping force f_c is proportional to velocity, the schematic of a dashpot is used to indicate the presence of this force. This is illustrated in Figure 1.10. The damping coefficient of a system cannot be measured as simply as the mass or stiffness of a system can be. This is pointed out in Section 1.6.

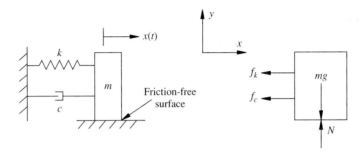

Figure 1.10   Schematic of a single-degree-of-freedom system with viscous damping indicated by a dashpot.

Using a simple force balance on the mass of Figure 1.10 in the x direction, the equation of motion for $x(t)$ becomes

$$m\ddot{x} = -f_c - f_k \tag{1.24}$$

or

$$m\ddot{x}(t) + c\dot{x}(t) + kx(t) = 0 \tag{1.25}$$

subject to the initial conditions $x(0) = x_0$ and $\dot{x}(0) = v_0$. Equation (1.25) and Figure 1.10, referred to as a *damped single-degree-of-freedom system*, form the topic of chapters 1 to 3.

To solve the damped system of equation (1.25), the same method used for solving equation (1.2) is used. In fact, this provides an additional reason to choose f_c to be of the form $c\dot{x}$. Let $x(t)$ have the form given in equation (1.13). Substitution of this form into equation (1.25) yields

$$(m\lambda^2 + c\lambda + k)ae^{\lambda t} = 0 \tag{1.26}$$

Again, $ae^{\lambda t} \neq 0$, so that this reduces to a quadratic equation in λ of the form

$$m\lambda^2 + c\lambda + k = 0 \tag{1.27}$$

called the *characteristic equation*. This is solved using the quadratic formula to yield the two solutions

$$\lambda_{1,2} = -\frac{c}{2m} \pm \frac{1}{2m}\sqrt{c^2 - 4km} \tag{1.28}$$

Examination of this expression indicates that the roots λ will be real or complex, depending on the value of the discriminant, $c^2 - 4km$. As long as m, c, and k are positive real numbers, λ_1 and λ_2 will be distinct negative real numbers if $c^2 - 4km > 0$. On the other hand, if this discriminant is negative, the roots will be a complex conjugate pair with negative real part. If the discriminant is zero, the two roots λ_1 and λ_2 are equal negative real numbers. Note that equation (1.15) represents the characteristic equation for the special undamped case (i.e., $c = 0$).

In examining these three cases, it is convenient to define the *critical damping coefficient*, c_{cr}, by

$$c_{cr} = 2m\omega = 2\sqrt{km} \tag{1.29}$$

where ω is the undamped natural frequency. Furthermore, the nondimensional number ζ, called the *damping ratio*, defined by

$$\zeta = \frac{c}{c_{cr}} = \frac{c}{2m\omega} \tag{1.30}$$

can be used to characterize the three types of solutions to the characteristic equation. Rewriting the roots given by equation (1.28) yields

$$\lambda_{1,2} = -\zeta\omega \pm \omega\sqrt{\zeta^2 - 1} \tag{1.31}$$

where it is now clear that the damping ratio ζ determines whether the roots are complex or real. This in turn determines the nature of the response of the damped single-degree-of-freedom system. For positive mass, damping, and stiffness coefficients, there are three cases, which are delineated next.

1.3.1 Underdamped Motion

In this case the damping ratio ζ is less than 1 $(0 < \zeta < 1)$ and the discriminant of equation (1.31) is negative, resulting in a complex conjugate pair of roots. These are

$$\lambda_1 = -\zeta\omega - \omega\sqrt{1 - \zeta^2}\,j \tag{1.32}$$

and

$$\lambda_2 = -\zeta\omega + \omega\sqrt{1 - \zeta^2}\,j \tag{1.33}$$

where $j = \sqrt{-1}$ and

$$\sqrt{1 - \zeta^2}\,j = \sqrt{(1 - \zeta^2)(-1)} = \sqrt{\zeta^2 - 1} \tag{1.34}$$

Following the same argument as that made for the undamped response of equation (1.18), the solution of (1.25) is then of the form

$$x(t) = e^{-\zeta\omega t}\left(a_1 e^{j\sqrt{1-\zeta^2}\,\omega t} + a_2 e^{-j\sqrt{1-\zeta^2}\,\omega t}\right) \qquad (1.35)$$

where a_1 and a_2 are arbitrary complex-valued constants of integration to be determined by the initial conditions. Using the Euler relations (see Window 1.5), this can be written as

$$x(t) = Ae^{-\zeta\omega t}\sin(\omega_d t + \phi) \qquad (1.36)$$

Window 1.5
Euler Relations and the Underdamped Solution

An underdamped solution of $m\ddot{x} + c\dot{x} + kx = 0$ to nonzero initial conditions is of the form
$$x(t) = a_1 e^{\lambda_1 t} + a_2 e^{\lambda_2 t}$$

where λ_1 and λ_2 are complex numbers of the form

$$\lambda_1 = -\zeta\omega + \omega_d j \qquad \text{and} \qquad \lambda_2 = -\zeta\omega - \omega_d j$$

where $\omega = \sqrt{k/m}$, $\zeta = c/(2m\omega)$, $\omega_d = \omega\sqrt{1-\zeta^2}$, and $j = \sqrt{-1}$. The two constants a_1 and a_2 are complex numbers and hence represent four unknown constants rather than the two constants of integration required to solve a second-order differential equation. This demands that the two complex numbers a_1 and a_2 be conjugate pairs so that $x(t)$ depends only on two undetermined constants. Substitution of the foregoing values of λ_i into the solution $x(t)$ yields

$$x(t) = e^{-\zeta\omega t}(a_1 e^{\omega_d jt} + a_2 e^{-\omega_d jt})$$

Using the Euler relations $e^{\theta j} = \cos\theta + j\sin\theta$ and $e^{-\theta j} = \cos\theta - j\sin\theta$, $x(t)$ becomes
$$x(t) = e^{-\zeta\omega t}[(a_1 + a_2)\cos\omega_d t + j(a_1 - a_2)\sin\omega_d t]$$

Choosing the real numbers $A_1 = a_1 + a_2$ and $A_2 = (a_1 - a_2)j$, this becomes

$$x(t) = e^{-\zeta\omega t}(A_1 \cos\omega_d t + A_2 \sin\omega_d t)$$

which is real-valued. Defining the constant $A = \sqrt{A_1^2 + A_2^2}$ and the angle $\theta = \tan^{-1}(A_1/A_2)$ so that $A_1 = A\sin\theta$ and $A_2 = A\cos\theta$, the form of $x(t)$ becomes [recall that $\sin a \cos b + \cos a \sin b = \sin(a+b)$]

$$x(t) = Ae^{-\zeta\omega t}\sin(\omega_d t + \theta)$$

where A and θ are the constants of integration to be determined from the initial conditions. Complex numbers are reviewed in Appendix A.

where A and ϕ are constants of integration and ω_d, called the *damped natural frequency*, is given by

$$\omega_d = \omega\sqrt{1 - \zeta^2} \tag{1.37}$$

The constants A and ϕ are evaluated using the initial conditions in exactly the same fashion as they were for the undamped system as indicated in equations (1.7) and (1.8). This yields

$$A = \sqrt{\frac{(v_0 + \zeta\omega x_0)^2 + (x_0\omega_d)^2}{\omega_d^2}}, \qquad \phi = \tan^{-1}\frac{x_0\omega_d}{v_0 + \zeta\omega x_0} \tag{1.38}$$

where x_0 and v_0 are the initial displacement and velocity. A plot of $x(t)$ versus t for this underdamped case is given in Figure 1.11. Note that the motion is oscillatory with decaying amplitude. The damping ratio ζ determines the rate of decay. The response illustrated in Figure 1.11 is exhibited in many mechanical systems and constitutes the most common case.

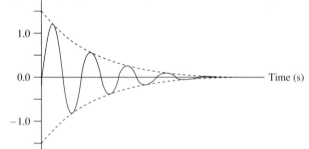

Figure 1.11 Response of an underdamped system: $0 < \zeta < 1$.

1.3.2 Overdamped Motion

In this case, the damping ratio is greater than $1(\zeta > 1)$. The discriminant of equation (1.31) is positive, resulting in a pair of distinct real roots. These are

$$\lambda_1 = -\zeta\omega - \omega\sqrt{\zeta^2 - 1} \tag{1.39}$$

and

$$\lambda_2 = -\zeta\omega + \omega\sqrt{\zeta^2 - 1} \tag{1.40}$$

The solution of equation (1.25) then becomes

$$x(t) = e^{-\zeta\omega t}(a_1 e^{-\omega\sqrt{\zeta^2-1}\,t} + a_2 e^{+\omega\sqrt{\zeta^2-1}\,t}) \tag{1.41}$$

which represents a nonoscillatory response. Again, the constants of integration a_1 and a_2 are determined by the initial conditions indicated in equations (1.7) and (1.8). In this aperiodic case, the constants of integration are real-valued and are given by

$$a_1 = \frac{-v_0 + (-\zeta + \sqrt{\zeta^2 - 1})\omega x_0}{2\omega\sqrt{\zeta^2 - 1}} \tag{1.42}$$

and

$$a_2 = \frac{v_0 + (\zeta + \sqrt{\zeta^2 - 1})\omega x_0}{2\omega\sqrt{\zeta^2 - 1}} \tag{1.43}$$

Typical responses are plotted in Figure 1.12, where it is clear that motion does not involve oscillation. An overdamped system does not oscillate, but rather, returns to its rest position exponentially.

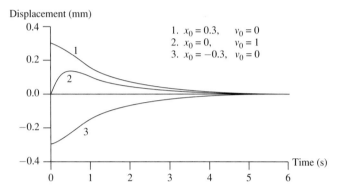

Figure 1.12   Response of an overdamped system, $\zeta > 1$, for two values of the initial displacement and zero initial velocity and one case with $x_0 = 0$ and $v_0 = 1$.

1.3.3 Critically Damped Motion

In this last case, the damping ratio is exactly 1 ($\zeta = 1$) and the discriminant of equation (1.31) is identically zero. This corresponds to the value of ζ that separates oscillatory motion from nonoscillatory motion. Since the roots are repeated, they have the value

$$\lambda_1 = \lambda_2 = -\omega \tag{1.44}$$

The solution takes the form

$$x(t) = (a_1 + a_2 t)e^{-\omega t} \tag{1.45}$$

where, again, the constants a_1 and a_2 are determined by the initial conditions. Substituting the initial displacement into equation (1.45) and the initial velocity into the derivative of equation (1.45) yields

$$a_1 = x_0, \qquad a_2 = v_0 + \omega x_0 \tag{1.46}$$

Critically damped motion is plotted in Figure 1.13 for two different values of initial conditions. It should be noted that critically damped systems can be thought of in several ways. They represent systems with the smallest value of damping rate that yields aperiodic motion. Critical damping can also be thought of as the case that separates nonoscillation from oscillation.

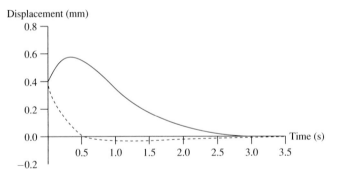

Displacement (mm)

Figure 1.13   Response of a critically damped system for two different initial velocities. The dashed line corresponds to $v_0 < 0$ and the solid line to $v_0 > 0$, both with $x_0 = 0.4$ mm.

Example 1.3.1

Recall the small spring of Example 1.2.1 (i.e., $\omega = 132$ rad/s). The damping rate of the spring is measured to be 0.11 kg/s. Calculate the damping ratio and determine if the free motion of the spring–bolt system is overdamped, underdamped, or critically damped.

Solution   From Example 1.2.1, $m = 49.2 \times 10^{-3}$ kg and $k = 857.8$ N/m. Using the definition of the critical damping coefficient of equation (1.29) and these values for m and k yields

$$c_{cr} = 2\sqrt{km} = 2\sqrt{(857.8 \text{ N/m})(49.2 \times 10^{-3} \text{ kg})}$$

$$= 12.993 \text{ kg/s}$$

If c is measured to be 0.11 kg/s, the critical damping ratio becomes

$$\zeta = \frac{c}{c_{cr}} = \frac{0.11 \text{ (kg/s)}}{12.993 \text{ (kg/s)}} = 0.0085$$

or 0.85% damping. Since ζ is less than 1, the system is underdamped. The motion resulting from giving the spring–bolt system a small displacement will be oscillatory.   □

The single-degree-of-freedom damped system of equation (1.25) is often written in a standard form. This is obtained by dividing equation (1.25) by the mass, m. This yields

$$\ddot{x} + \frac{c}{m}\dot{x} + \frac{k}{m}x = 0 \tag{1.47}$$

The coefficient of $x(t)$ is obviously ω^2, the undamped natural frequency squared. A little manipulation illustrates that the coefficient of the velocity $\dot{x}$ is $2\zeta\omega$. Thus equation (1.47) can be written as

$$\ddot{x}(t) + 2\zeta\omega\dot{x}(t) + \omega^2 x(t) = 0 \tag{1.48}$$

In this standard form, the values of the natural frequency and the damping ratio are more obvious.

1.4 MODELING AND ENERGY METHODS

Modeling is the art or process of writing down an equation, or system of equations, to describe the motion of a physical device. For example, equation (1.2) was obtained by modeling the spring–mass system of Figure 1.4. By summing the forces acting on the mass in the x direction and employing the experimental evidence of the mathematical model of the force in a spring given by Figure 1.3, equation (1.2) can be obtained. The success of this model is determined by how well the solution of equation (1.2) predicts the observed behavior of the system. This comparison between the vibration response of a device and the response predicted by the analytical model is discussed in Section 1.6. The majority of this book is devoted to the analysis of vibration models. However, two methods of modeling: Newton's law and energy methods are presented in this section. More comprehensive treatments of modeling can be found in Doebelin (1980), Shames (1980, 1989), and Cannon (1967), for example.

The force summation method is used in the previous sections and should be familiar to the reader from introductory dynamics (see, e.g., Shames, 1980). Newton's law of motion (called *Newton's second law*) states that the rate of change of the absolute momentum of the mass center is proportional to the net applied force vector and acts in a direction of the net force. For systems with constant mass (such as those considered here) moving in only one direction, the rate of change of momentum becomes the scalar relation

$$\frac{d}{dt}(m\dot{x}) = m\ddot{x}$$

which is often called the inertia force. The physical device of interest is examined by noting the forces acting on the center of mass. The forces are then summed (as vectors) to produce a dynamic equation following Newton's law. For motion in the x direction only, this becomes the scalar equation

$$\sum_i f_{xi} = m\ddot{x} \tag{1.49}$$

where f_{xi} denotes the ith force acting on the mass m in the x direction and the summation is over the number of such forces. In the first three chapters, only single-degree-of-freedom systems moving in one direction are considered; thus Newton's law takes on a scalar nature. In more practical problems, the vector nature must be considered (see Chapter 4).

For bodies that are free to rotate about a fixed axis, the sum of the torques about the rotation axis through the center of mass of the object must equal the rate of change of angular momentum of the mass. This is expressed as

$$\sum_i M_{0i} = I_0\ddot{\theta} \tag{1.50}$$

where M_{0i} are the torques acting on the object through point 0, I_0 is the moment of inertia (sometimes denoted J) about the rotation axis, and θ is the angle of rotation. This is discussed in more detail in Example 1.5.1.

If the forces and/or torques acting on an object or mechanical part are difficult to determine, an energy approach may be more efficient. In this method the differential equation of motion is established by using the principle of energy conservation. This principle is equivalent to Newton's law for conservative systems and states that the sum of the potential energy and kinetic energy of a particle remains constant at each instant of time throughout the particle's motion. Integrating Newton's law ($F = m\ddot{x}$) over an increment of displacement and identifying the work done in a conservative field as the change in potential energy yields

$$U_1 - U_2 = T_2 - T_1 \qquad (1.51)$$

where U_1 and U_2 represent the particle's potential energy at the times t_1 and t_2, respectively, and T_1 and T_2 represent the particle's kinetic energy at times t_1 and t_2, respectively. Equation (1.51) can be rearranged to yield

$$T + U = \text{constant} \qquad (1.52)$$

where T and U denote the total kinetic and potential energy, respectively.

For periodic motion, if t_1 is chosen to be the time at which the moving mass passes through its static equilibrium position, U_1 can be set to zero at that time, and if t_2 is chosen as the time at which the mass undergoes its maximum displacement so that its velocity is zero ($T_2 = 0$), equation (1.51) yields

$$T_1 = U_2 \qquad (1.53)$$

Since the reference potential energy U_1 is zero, U_2 in equation (1.53) is the maximum value of potential energy in the system. Because the energy in this system is conserved, T_2 must also be a maximum value so that equation (1.53) yields

$$T_{\max} = U_{\max} \qquad (1.54)$$

for conservative systems undergoing periodic motion. Since energy is a scalar quantity, using the conservation of energy yields a possibility of obtaining the equation of motion of a system without using vectors.

Equations (1.52), (1.53), and (1.54) are three statements of the conservation of energy. Each of these can be used to determine the equation of motion of a spring–mass system. As an illustration, consider the energy of the spring–mass system of Figure 1.14, hanging in a gravitational field of strength g. The effect of adding the mass m to the massless spring of stiffness k is to stretch the spring from its rest position at 0 to the static equilibrium position x_0. The total potential energy of the spring–mass system is the sum of the potential energy of the spring (or strain energy; see, e.g., Shames, 1989) and the gravitational potential energy. The potential energy of the spring is given by

$$U_{\text{spring}} = \tfrac{1}{2}k(x_0 + x)^2 \qquad (1.55)$$

The gravitational potential energy is

$$U_{\text{grav}} = -mgx \qquad (1.56)$$

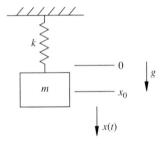

Figure 1.14 A spring–mass system hanging in a gravitational field. Here x_0 is the static equilibrium position and x is the displacement from equilibrium.

where the minus sign indicates that the mass is located below the reference point x_0. The kinetic energy of the system is

$$T = \tfrac{1}{2}m\dot{x}^2 \tag{1.57}$$

Substituting these energy expressions into equation (1.52) yields

$$\tfrac{1}{2}m\dot{x}^2 - mgx + \tfrac{1}{2}k(x_0 + x)^2 = \text{constant} \tag{1.58}$$

Differentiating this expression with respect to time yields

$$\dot{x}(m\ddot{x} + kx) + \dot{x}(kx_0 - mg) = 0 \tag{1.59}$$

Since the static force balance on the mass yields the fact that $kx_0 = mg$, this becomes

$$\dot{x}(m\ddot{x} + kx) = 0 \tag{1.60}$$

The velocity $\dot{x}$ cannot be zero for all time; otherwise, $x(t) = \text{constant}$ and no vibration would be possible. Hence equation (1.60) yields the standard equation of motion

$$m\ddot{x} + kx = 0 \tag{1.61}$$

This procedure is called the *energy method* of obtaining the equation of motion.

The energy method can also be used to obtain the frequency of vibration directly for conservative systems that are oscillatory. The maximum value of sine (and cosine) is 1. Hence, from equations (1.3) and (1.4), the maximum displacement is A and the maximum velocity is ωA. Substitution of these maximum values into the expression for $U_{\max}$ and $T_{\max}$, and using the energy equation (1.54), yields

$$\tfrac{1}{2}m(\omega A)^2 = \tfrac{1}{2}kA^2 \tag{1.62}$$

Solving this for ω yields the standard natural frequency relation $\omega = \sqrt{k/m}$.

Example 1.4.1

Figure 1.15 is a crude model of a vehicle suspension system hitting a bump. Calculate the natural frequency of oscillation using the energy method. Assume that no energy is lost during the contact.

Solution   From introductory mechanics, the rotational kinetic energy of the wheel is $T_{\text{rot}} = \tfrac{1}{2}J\dot{\theta}^2$, where J is the mass moment of inertia of the wheel and $\theta = \theta(t)$ is the angle of rotation of the wheel. This assumes that the wheel moves relative to the surface without slipping as it climbs the bump (so that no energy is lost at contact). The translational kinetic energy of the wheel is $T_T = \tfrac{1}{2}m\dot{x}^2$.

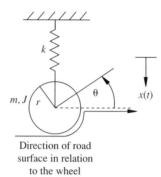

Figure 1.15   Simple model of an automobile suspension system. The rotation of the wheel relative to the horizontal as it hits a bump is given by θ. It is assumed that the wheel rolls without slipping as the car hits the bump.

The rotation θ and the translation x are related by $x = r\theta$. Thus $\dot{x} = r\dot{\theta}$ and $T_{\text{rot}} = \frac{1}{2}J\dot{x}^2/r^2$. At maximum energy $x = A$ and $\dot{x} = \omega A$, so that

$$T_{\max} = \frac{1}{2}m\dot{x}^2_{\max} + \frac{1}{2}\frac{J}{r^2}\dot{x}^2_{\max}\bigg|_{\dot{x}=\omega A} = \frac{1}{2}\left(m + J/r^2\right)\omega^2 A^2$$

and

$$U_{\max} = \frac{1}{2}kx^2_{\max}\bigg|_{x=A} = \frac{1}{2}kA^2$$

Using conservation of energy in the form of equation (1.54) yields

$$\frac{1}{2}\left(m + \frac{J}{r^2}\right)\omega^2 = \frac{1}{2}k$$

Solving this last expression for ω yields

$$\omega = \sqrt{\frac{k}{m + J/r^2}}$$

the desired frequency of oscillation of the suspension system.

The denominator in the frequency expression derived in this example is called the *effective mass* because the term $(m + J/r^2)$ has the same effect on the natural frequency as does a mass of value $(m + J/r^2)$.

□

Example 1.4.2

Determine the equation of motion of the simple pendulum shown in Window 1.1 using the energy method.

Solution   Several assumptions must first be made to ensure simple behavior (a more complicated version is considered in Example 1.4.6). The size of the mass, m, is assumed to be much smaller than the length of the pendulum l. Furthermore, the mass of the pendulum arm is considered to be negligible compared to the mass m. With these assumptions the mass moment of inertia about point 0 is

$$J_0 = ml^2$$

The angular displacement $\theta(t)$ is measured from the static equilibrium or rest position of the pendulum. The kinetic energy of the system is

$$T = \tfrac{1}{2} J_0 \dot{\theta}^2 = \tfrac{1}{2} m l^2 \dot{\theta}^2$$

The potential energy of the system is

$$U = mgl(1 - \cos\theta)$$

since $l(1-\cos\theta)$ is the geometric change in elevation of the pendulum mass. Substitution of these expressions for the kinetic and potential energy into equation (1.52) and differentiating yields

$$\frac{d}{dt}\left[\tfrac{1}{2} m l^2 \dot{\theta}^2 + mgl(1 - \cos\theta)\right] = 0$$

or

$$m l^2 \dot{\theta}\ddot{\theta} + mgl(\sin\theta)\dot{\theta} = 0$$

Factoring out $\dot{\theta}$ yields

$$\dot{\theta}(m l^2 \ddot{\theta} + mgl\sin\theta) = 0$$

Since $\dot{\theta}(t)$ cannot be zero for all time, this becomes

$$m l^2 \ddot{\theta} + mgl\sin\theta = 0$$

or

$$\ddot{\theta} + \frac{g}{l}\sin\theta = 0$$

This is a nonlinear equation in θ and is discussed in Chapter 10. However, here, since $\sin\theta$ can be approximated by θ for small angles, the linear equation of motion for the pendulum becomes

$$\ddot{\theta} + \frac{g}{l}\theta = 0$$

This corresponds to an oscillation with natural frequency $\omega = \sqrt{g/l}$ for initial conditions such that θ remains small, as defined by the approximation $\sin\theta \approx \theta$.

$\square$

Example 1.4.3

Determine the equation of motion of the shaft and disk illustrated in Window 1.1 using the energy method.

Solution   The shaft and disk of Window 1.1 are modeled as a rod stiffness in twisting, resulting in torsional motion. The shaft or rod exhibits a torque in twisting proportional to the angle of twist $\theta(t)$. The potential energy associated with the torsional spring stiffness is $U = \tfrac{1}{2} k\theta^2$, where the stiffness coefficient k is determined much like the method used to determine the spring stiffness in translation, as discussed in Section 1.1. The angle $\theta(t)$ is measured from the static equilibrium or rest position. The kinetic energy associated with the disk of mass moment of inertia J is $T = \tfrac{1}{2} J\dot{\theta}^2$. This assumes that the inertia of the rod is much smaller than that of the disk and can be neglected.

Substitution of these expressions for the kinetic and potential energy into equation (1.52) and differentiating yields

$$\frac{d}{dt}\left(\tfrac{1}{2} J\dot{\theta}^2 + \tfrac{1}{2} k\theta^2\right) = \left(J\ddot{\theta} + k\theta\right)\dot{\theta} = 0$$

so that the equation of motion becomes (because $\dot{\theta} \neq 0$)

$$J\ddot{\theta} + k\theta = 0$$

This is the equation of motion for torsional vibration of a disk on a shaft. The natural frequency of vibration is $\omega = \sqrt{k/J}$.

☐

Example 1.4.4

Model the mass of the spring of the system shown in Figure 1.16 and determine the effect of including the mass of the spring on the value of the natural frequency.

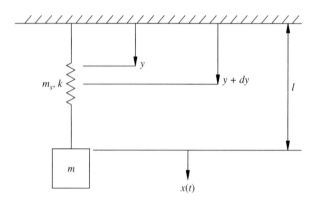

Figure 1.16 Spring–mass system with a spring of nonnegligible mass, m_s.

Solution   One approach to considering the mass of the spring in analyzing the system vibration response is to calculate the kinetic energy of the spring. Consider the kinetic energy of the element dy of the spring. If m_s is the total mass of the spring, $(m_s/l)\,dy$ is the mass of the element dy. The velocity of this element, denoted v_{dy}, may be approximated by assuming that the velocity of the tip, $\dot{x}(t)$, varies linearly over the length of the spring:

$$v_{dy} = \frac{y}{l}\dot{x}(t)$$

The total kinetic energy of the spring is the kinetic energy of the element dy integrated over the length of the spring:

$$T_{\text{spring}} = \frac{1}{2}\int_0^l \frac{m_s}{l}\left[\frac{y}{l}\dot{x}\right]^2 dy$$

$$= \frac{1}{2}\left(\frac{m_s}{3}\right)\dot{x}^2$$

From the form of this expression, the effective mass of the spring is $m_s/3$, or one-third of that of the spring. Following the energy method, the maximum kinetic energy of the system is thus

$$T_{\text{max}} = \frac{1}{2}\left(m + \frac{m_s}{3}\right)\omega^2 A^2$$

Equating this to the maximum potential energy, $\frac{1}{2}kA^2$, yields the fact that the natural frequency of the system is

$$\omega = \sqrt{\frac{k}{m + m_s/3}}$$

Thus including the effects of the mass of the spring in the system decreases the natural frequency.

☐

Example 1.4.5

Fluid systems, as well as solid systems, exhibit vibration. Calculate the natural frequency of oscillation of the fluid in the U tube manometer illustrated in Figure 1.17 using the energy method.

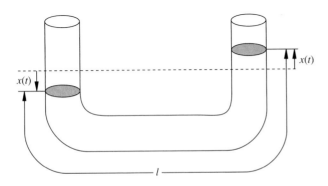

γ = weight density (volume)
A = cross-sectional area
l = length of fluid

Figure 1.17   U-tube manometer consisting of a fluid moving in a tube.

Solution   The fluid has weight density γ (i.e., the specific weight). The restoring force is provided by gravity. The potential energy of the fluid [mass × g × height = (weight)(height)] is $0.5(\gamma Ax)x$ in each column, so that the total change in potential energy is

$$U = \tfrac{1}{2}(\gamma Ax)x - \tfrac{1}{2}\gamma Ax(-x) = \gamma Ax^2$$

The change in kinetic energy is

$$T = \frac{1}{2}\frac{Al\gamma}{g}(\dot{x}^2 - 0) = \frac{1}{2}\frac{Al\gamma}{g}\dot{x}^2$$

Equating the change in potential energy to the change in kinetic energy yields

$$\frac{1}{2}\frac{Al\gamma}{g}\dot{x}^2 = \gamma Ax^2$$

Assuming an oscillating motion of the form $x(t) = X\sin(\omega t + \phi)$ and evaluating this expression for maximum velocity and position yields

$$\frac{1}{2}\frac{l}{g}\omega^2 X^2 = X^2$$

where X is used here to denote the amplitude of vibration. Solving for ω yields

$$\omega = \sqrt{\frac{2g}{l}}$$

which is the natural frequency of oscillation of the fluid in the tube. Note that it depends only on the acceleration due to gravity and the length of the fluid. Vibration of fluids inside mechanical containers (called *sloshing*) occurs in gas tanks in automobiles and airplanes as well and forms an important application of vibration analysis.

$\square$

Example 1.4.6

Consider the compound pendulum of Figure 1.18. A compound pendulum is a pendulum with a significant mass moment of inertia resulting from a distribution of mass about its length. In the figure G is the center of mass, O the pivot point, and $\theta(t)$ the angular displacement of the centerline of the pendulum of mass m and moment I measured about the z axis at point O. Point C is the *center of percussion*, which is defined as the distance q_0 along the centerline such that a simple pendulum (a massless rod pivoted at zero with mass m at its tip as in Example 1.4.2) of radius q_0 has the same period. Hence

$$q_0 = \frac{I}{mr}$$

where r is the distance from the pivot point to the center of mass. Note that the pivot point O and the center of percussion C can be interchanged to produce a pendulum with the same frequency. The *radius of gyration*, k_0, is the radius of a ring that has the same resistance to angular acceleration as the rigid body does. The radius of gyration and center of percussion are related by

$$q_0 r = k_0^2$$

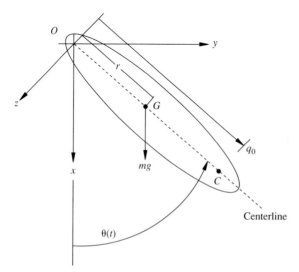

Figure 1.18 Compound pendulum pivoted to swing about point O under the influence of gravity.

Consider the equation of motion of the compound pendulum. Taking moments about its pivot point O yields

$$\Sigma M_0 = I\ddot{\theta}(t) = -mgr\sin\theta(t)$$

For small $\theta(t)$ this nonlinear equation becomes

$$I\ddot{\theta}(t) + mgr\theta(t) = 0$$

The natural frequency of oscillation becomes

$$\omega = \sqrt{\frac{mgr}{I}}$$

This frequency can be expressed in terms of the center of percussion as

$$\omega = \sqrt{\frac{g}{q_0}}$$

which is just the frequency of a simple pendulum of length q_0. This can be seen by examining the forces acting on the simple (massless) pendulum of Figure 1.19(a) or recalling the result obtained by the energy method in Example 1.4.2.

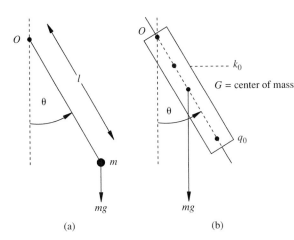

Figure 1.19 (a) Simple pendulum consisting of a massless rod pivoted at point O with a mass attached to its tip. (b) Compound pendulum consisting of a shaft with a center of mass at point G.

Summing moments about O yields

$$ml^2\ddot{\theta} = -mgl\sin\theta$$

or after approximating $\sin\theta$ with θ,

$$\ddot{\theta} + \frac{g}{l}\theta = 0$$

This yields the simple pendulum frequency of $\omega = \sqrt{g/l}$, which is equivalent to that obtained above that for the compound pendulum using $l = q_0$.

Next consider the uniformly shaped compound pendulum of Figure 1.19(b) of length l. Here it is desired to calculate the center of percussion and radius of gyration.

The mass moment of inertia about point O is I, so that summing moments about O yields

$$I\ddot{\theta} = -mg\frac{l}{2}\sin\theta$$

since the mass is assumed to be evenly distributed and the center of mass is at $r = l/2$. The moment of inertia for a slender rod about O is $I = \frac{1}{3}ml^2$; hence the equation of motion is

$$\frac{ml^2}{3}\ddot{\theta} + mg\frac{l}{2}\theta = 0$$

where $\sin\theta$ has again been approximated by θ assuming small motion. This becomes

$$\ddot{\theta} + \frac{3}{2}\frac{g}{l}\theta = 0$$

so that the natural frequency is

$$\omega = \sqrt{\frac{3}{2}\frac{g}{l}}$$

The center of percussion becomes

$$q_0 = \frac{I}{mr} = \frac{2}{3}l$$

and the radius of gyration becomes

$$k_0 = \sqrt{q_0 r} = \frac{l}{\sqrt{3}}$$

These positions are marked on Figure 1.19b.

The center of percussion and pivot point play a significant role in designing an automobile. The axle of the front wheels of an automobile is considered as the pivot point of a compound pendulum parallel to the road. If the back wheels hit a bump, the frequency of oscillation of the center of percussion will annoy passengers. Hence automobiles are designed such that the center of percussion falls over the axle and suspension system, away from passengers.

$\square$

So far, three basic systems have been modeled: rectilinear or translational motion of a spring–mass system, torsional motion of a disk–shaft system, and the pendulum motion of a suspended mass system. Each of these motions commonly experiences energy dissipation of some form. The viscous damping model of Section 1.3 developed for translational motion can be applied directly to both torsional and pendulum motion. In the case of torsional motion of the shaft, the energy dissipation is assumed to come from heating of the material and/or air resistance. Sometimes, as in the case of using the rod and disk to model an automobile crankshaft, or camshaft, the damping is assumed to come from the oil that surrounds the disk and shaft, or bearings that support the shaft.

The damping mechanism for the pendulum comes largely from friction in the joint, or point of attachment of the pendulum arm to ground, and a little from the mass pushing air out of the way as the pendulum swings. The energy dissipation or damping associated with the spring material in rectilinear motion and the rod material in torsional motion, is much larger than that associated with the pendulum.

In all three cases, the damping is modeled as proportional to velocity (i.e., $f_c = -c\dot{x}$ or $f_c = -c\dot{\theta}$). The equations of motion are then of the form indicated in Table 1.1. Both of these equations can be expressed as a damped linear oscillator given in the form of equation (1.48). Hence each of these three systems is characterized by a natural frequency and a damping ratio. Each of these three systems has a solution based on the nature of the damping ratio ζ, as discussed in Section 1.3.

TABLE 1.1   A COMPARISON OF RECTILINEAR AND ROTATIONAL SYSTEMS AND A SUMMARY OF UNITS

	Rectilinear x (m)	Torsional/pendulum θ (rad)
Spring force	kx	$k\theta$
Damping force	$c\dot{x}$	$c\dot{\theta}$
Inertia force	$m\ddot{x}$	$J\ddot{\theta}$
Equation of motion	$m\ddot{x} + c\dot{x} + kx = 0$	$J\ddot{\theta} + c\dot{\theta} + k\theta = 0$
Stiffness units	N/m	N $\cdot$ m/rad
Damping units	N $\cdot$ s/m	m $\cdot$ N $\cdot$ s/rad
Inertia units	kg	kg $\cdot$ m^2/rad
Force/torque	N $=$ kg $\cdot$ m/s^2	N $\cdot$ m $=$ kg $\cdot$ m^2/s^2

1.5 STIFFNESS

The stiffness in a spring, introduced in Section 1.1, can be related more directly to material and geometric properties of the spring. This section introduces the relationships between stiffness, elastic modulus, and geometry of various types of springs, and illustrates various situations that can lead to simple harmonic motion. A springlike behavior results from a variety of configurations, including longitudinal motion (vibration in the direction of the length), transverse motion (vibration perpendicular to the length), and torsional motion (vibration rotating around the length). Consider again the stiffness of the spring introduced in Section 1.1. A spring is generally made of an elastic material. For a slender elastic material of length l, cross-sectional area A, and elastic modulus E (or Young's modulus), the stiffness of the bar for vibration along its length is given by

$$k = \frac{EA}{l} \tag{1.63}$$

This describes the spring constant for the vibration problem illustrated in Figure 1.20, where the mass of the rod is ignored (or very small relative to the mass m in the figure). The modulus E has the units of pascal (denoted Pa), which are N/m^2. The modulus for several common materials is given in Table 1.2.

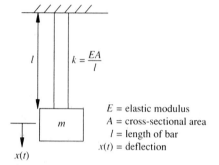

E = elastic modulus
A = cross-sectional area
l = length of bar
$x(t)$ = deflection

Figure 1.20   Stiffness associated with the longitudinal vibration of a slender bar.

TABLE 1.2   PHYSICAL CONSTANTS FOR SOME COMMON MATERIALS

Material	Young's modulus, $E\,(\text{N/m}^2)$	Density, (kg/m^3)	Shear modulus, $G\,(\text{N/m}^2)$
Steel	2.0×10^{11}	7.8×10^3	8.0×10^{10}
Aluminum	7.1×10^{10}	2.7×10^3	2.67×10^{10}
Brass	10.0×10^{10}	8.5×10^3	3.68×10^{10}
Copper	6.0×10^{10}	2.4×10^3	2.22×10^{10}
Concrete	3.8×10^9	1.3×10^3	—
Rubber	2.3×10^9	1.1×10^3	8.21×10^8
Plywood	5.4×10^9	6.0×10^2	—

Next consider a twisting motion with a similar rod as illustrated in Figure 1.21. In this case the rod possesses an area moment of inertia, J_P, and (shear) modulus of rigidity, G (see Table 1.2). For the case of a wire or shaft of diameter d, $J_P = \pi d^4/32$. The modulus of rigidity has units N/m^2. The torsional stiffness is (recall Window 1.1)

$$k = \frac{GJ_P}{l} \tag{1.64}$$

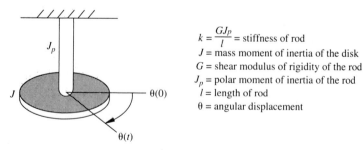

$k = \dfrac{GJ_p}{l}$ = stiffness of rod
J = mass moment of inertia of the disk
G = shear modulus of rigidity of the rod
J_p = polar moment of inertia of the rod
l = length of rod
θ = angular displacement

Figure 1.21   Stiffness associated with the torsional vibration of a shaft.

which is used to describe the vibration problem illustrated in Figure 1.21, where the mass of the shaft is ignored. In the figure, $\theta(t)$ represents the angular position of the shaft relative to its equilibrium position. The disk of radius r and rotational moment of inertia J will vibrate around the equilibrium position $\theta(0)$ with stiffness GJ_p/l.

Example 1.5.1

Calculate the natural frequency of oscillation of the torsional system given in Figure 1.21.

Solution   Using the moment equation (1.50), the equation of motion for this system is

$$J\ddot{\theta}(t) = -k\theta(t)$$

This may be written as

$$\ddot{\theta}(t) + \frac{k}{J}\theta(t) = 0$$

This agrees with the result obtained using the energy method as indicated in Example 1.4.3. This indicates an oscillatory motion with frequency

$$\omega = \sqrt{\frac{k}{J}} = \sqrt{\frac{GJ_P}{lJ}}$$

Suppose that the shaft is made of steel and is 2 m long with a diameter of 0.5 cm. If the disk has polar moment of inertia $J = 0.5$ kg $\cdot$ m^2 and considering that the shear modulus of steel is $G = 8 \times 10^{10}$ N/m^2, the frequency can be calculated by

$$\omega^2 = \frac{k}{J} = \frac{GJ_P}{lJ} = \frac{(8 \times 10^{10} \text{ N/m}^2)[\pi(0.5 \times 10^{-2} \text{ m})^4/32]}{(2 \text{ m})(0.5 \text{ kg} \cdot \text{m}^2)}$$

$$= 4.9087 \ (\text{rad}^2/\text{s}^2)$$

Thus the natural frequency is $\omega = 2.2156$ rad/s.

$\square$

Consider the helical spring of Figure 1.22. In this figure the deflection of the spring is along the axis of the coil. The stiffness is actually dependent on the "twist" of the metal rod forming this spring. The stiffness is a function of the shear modulus G, the diameter of the rod, the diameter of the coils, and the number of coils. The stiffness has

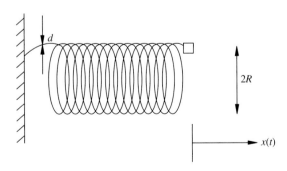

d = diameter of spring material
$2R$ = diameter of turns
n = number of turns
$x(t)$ = deflection

$$k = \frac{Gd^4}{64nR^3}$$

Figure 1.22   Stiffness associated with a helical spring.

the value

$$k = \frac{Gd^4}{64n R^3} \tag{1.65}$$

The helical-shaped spring is very common. Some examples are the spring inside a retractable ball-point pen and the spring contained in the front suspension of an automobile.

Next consider the transverse vibration of the end of a "leaf" spring illustrated in Figure 1.23. This type of spring behavior is common to the rear suspension of an automobile as well as the wings of some aircraft. In the figure, l is the length of the beam, E is the elastic (Young's) modulus of the beam, and I is the moment of inertia of the cross-sectional area. The mass m at the tip of the beam will oscillate with frequency

$$\omega = \sqrt{\frac{k}{m}} = \sqrt{\frac{3EI}{ml^3}} \tag{1.66}$$

in the direction perpendicular to the length of the beam $x(t)$.

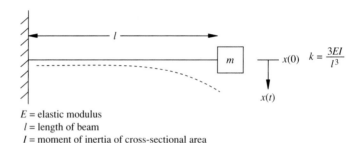

E = elastic modulus
l = length of beam
I = moment of inertia of cross-sectional area

Figure 1.23 Beam stiffness associated with the transverse vibration of the tip of a beam (Blevins, 1987).

Example 1.5.2

Consider an airplane wing with a fuel pod mounted at its tip as illustrated in Figure 1.24. The pod has a mass of 10 kg when it is empty and 1000 kg when it is full. Calculate the change in the natural frequency of vibration of the wing, modeled as in Figure 1.24 as the airplane uses up the fuel in the wing pod. The estimated physical parameters of the beam are $I = 5.2 \times 10^{-6}$ m^4, $E = 6.9 \times 10^9$ N/m^2, and $l = 2$ m.

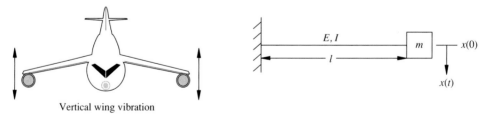

Vertical wing vibration

Figure 1.24 Simple vibration model of an airplane wing with a fuel pod mounted at its extremity.

Solution   The natural frequency of the vibration of the wings modeled as a simple massless beam with a tip mass is given by equation (1.66). The natural frequency when the fuel pod is full is

$$\omega_{full} = \sqrt{\frac{3EI}{ml^3}} = \sqrt{\frac{(3)(6.9 \times 10^9)(5.2 \times 10^{-6})}{1000(2)^3}} = 3.67 \text{ rad/s}$$

which is about 0.58 Hz (0.58 cycles per second). The natural frequency for the wing when the fuel pod is empty becomes

$$\omega_{empt} = \sqrt{\frac{3EI}{ml^3}} = \sqrt{\frac{(3)(6.9 \times 10^9)(5.2 \times 10^{-6})}{10(2)^3}} = 36.68 \text{ rad/s}$$

or 5.87 Hz. Hence the natural frequency of the airplane wing changes by a factor of 10 (i.e., becomes 10 times larger) when the fuel pod is empty.

□

If the spring of Figure 1.23 is coiled in a plane as illustrated in Figure 1.25, the stiffness of the spring is greatly affected and becomes

$$k = \frac{EI}{l} \tag{1.67}$$

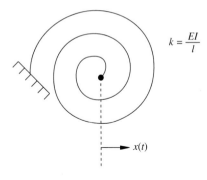

$$k = \frac{EI}{l}$$

$x(t)$

l = total length of spring
E = elastic modulus of spring
I = moment of inertia of the cross-sectional area

Figure 1.25   Stiffness associated with a coiled spring.

Several other spring arrangements and their associated stiffness values are listed in Table 1.3. Texts on solid mechanics, such as Shames (1989) should be consulted for further details.

Example 1.5.3

As another example of vibration involving fluids, consider the rolling vibration of a ship in water. Figure 1.26 illustrates a schematic of a ship rolling in water. In the figure, G is the center of gravity, C the point of intersection of the buoyant force and the centerline of the ship (called the metacenter), and h the height of the metacenter above the center of gravity.

TABLE 1.3 SAMPLE SPRING CONSTANTS

Axial stiffness of a tapered bar of length l, modulus E, and end diameters d_1 and d_2	$k = \dfrac{\pi E d_1 d_2}{4l}$
Torsional stiffness on a hollow uniform shaft of shear modulus G, length l, inside diameter d_1, and outside diameter d_2	$k = \dfrac{\pi G(d_2^4 - d_1^4)}{32l}$
Transverse stiffness of a pinned–pinned beam of modulus E, area moment of inertia I, and length l for a load applied at point a from its end	$k = \dfrac{3EIl}{a^2(l-a)^2}$
Transverse stiffness of a clamped–clamped beam of modulus, E, area moment of inertia I, and length l for a load applied at its center	$k = \dfrac{192EI}{l^3}$

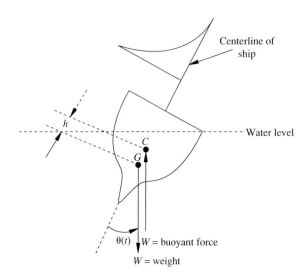

Centerline of ship

Water level

C

G

$\theta(t)$ W = buoyant force

W = weight

Figure 1.26 Dynamics of a ship rolling in water.

The angle of roll is denoted by $\theta(t)$, and the mass moment of the ship about the roll axis is denoted by J. Summing the moments about G yields

$$J\ddot{\theta}(t) = -Wh\sin\theta(t)$$

Again for small angles this nonlinear equation can be approximated by

$$J\ddot{\theta}(t) + Wh\theta(t) = 0$$

Thus the natural frequency of oscillation about the roll axis is

$$\omega = \sqrt{\frac{Wh}{J}}$$

$\square$

All of the spring types mentioned are represented schematically as indicated in Figure 1.1. If more than one spring is present in a given device, the resulting stiffness of the combined spring can be calculated by two simple rules, as given in Figure 1.27. These rules can be derived by considering the equivalent forces in the system.

Springs in series

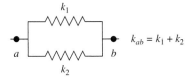

$$k_{ac} = \frac{1}{1/k_1 + 1/k_2}$$

Springs in parallel

$$k_{ab} = k_1 + k_2$$

Figure 1.27   Rules for calculating the equivalent stiffnesses for parallel and series connections of springs.

Example 1.5.4

Consider the spring mass arrangement of Figure 1.28(a) and calculate the natural frequency of the system.

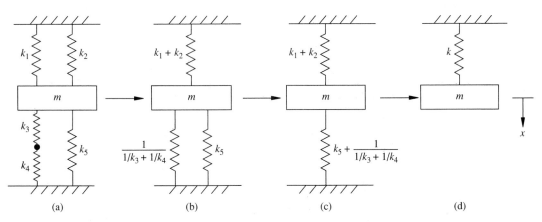

Figure 1.28   Reduction of a five-spring one-mass system to an equivalent single-spring–mass system.

Solution   To find the equivalent single stiffness representation of the five-spring system given in Figure 1.28(a), the two simple rules of Figure 1.27 are applied. First, the parallel arrangement of k_1 and k_2 is replaced by the single spring, as indicated at the top of Figure 1.28(b). Next, the series arrangement of k_3 and k_4 is replaced with a single spring of stiffness

$$\frac{1}{1/k_3 + 1/k_4}$$

as indicated in the bottom left side of Figure 1.28(b). These two parallel springs on the bottom of Figure 1.28(b) are next combined using the parallel spring formula to yield a single spring of stiffness

$$k_5 + \frac{1}{1/k_3 + 1/k_4}$$

as indicated in Figure 1.28(c). The final step is to realize that the spring acting at the top of Figure 1.28(c) and that at the bottom both attach the mass to ground and hence act in parallel. These two springs then combine to yield the single stiffness

$$k = k_1 + k_2 + k_5 + \frac{1}{1/k_3 + 1/k_4}$$

$$= k_1 + k_2 + k_5 + \frac{k_3 k_4}{k_3 + k_4} = \frac{(k_1 + k_2 + k_5)(k_3 + k_4) + k_3 k_4}{k_3 + k_4}$$

as indicated symbolically in Figure 1.28(d). Hence the natural frequency of this system is

$$\omega = \sqrt{\frac{k_1 k_3 + k_2 k_3 + k_5 k_3 + k_1 k_4 + k_2 k_4 + k_5 k_4 + k_3 k_4}{m(k_3 + k_4)}}$$

Note that even though the system of Figure 1.28 contains five springs, it consists of only one mass moving in only one (rectilinear) direction and hence is a single-degree-of-freedom system.

$\square$

Springs are usually manufactured in only certain increments of stiffness values depending on such things as the number of turns, material, and so on (recall Figure 1.22). Because mass production (and large sales) brings down the price of a product, the designer is often faced with a limited choice of spring constants when designing a system. It may thus be cheaper to use several "off-the-shelf" springs to create the stiffness value necessary than to order a special spring with specific stiffness. The rules of combining parallel and series springs given in Figure 1.27 can then be used to obtain the desired, or acceptable, stiffness and natural frequency.

Example 1.5.5

Consider the system of Figure 1.28(a) with $k_5 = 0$. Compare the stiffness and frequency of a 10-kg mass connected to ground, first by two parallel springs ($k_3 = k_4 = 0$, $k_1 = 1000$ N/m, and $k_2 = 3000$ N/m), then by two series springs ($k_1 = k_2 = 0$, $k_3 = 1000$ N/m, and $k_4 = 3000$ N/m).

Solution   First consider the case of two parallel springs so that $k_3 = k_4 = 0$, $k_1 = 1000$ N/m, and $k_2 = 3000$ N/m. Then the equivalent stiffness is given by Figure 1.27 to be the simple sum or

$$k_{eq} = 1000 \text{ N/m} + 3000 \text{ N/m} = 4000 \text{ N/m}$$

and the corresponding frequency is

$$\omega_{\text{parallel}} = \sqrt{\frac{4000 \text{ N/m}}{10 \text{ kg}}} = 20 \text{ rad/s}$$

In the case of a series connection ($k_1 = k_2 = 0$), the two springs ($k_3 = 1000$ N/m, $k_4 = 3000$ N/m) combine according to Figure 1.27 to yield

$$k_{\text{eq}} = \frac{1}{1/1000 + 1/3000} = \frac{3000}{3 + 1} = \frac{3000}{4} = 750 \text{ N/m}$$

The corresponding natural frequency becomes

$$\omega_{\text{series}} = \sqrt{\frac{750 \text{ N/m}}{10 \text{ kg}}} = 8.66 \text{ rad/s}$$

Note that using two identical sets of springs connected in the two different ways produces drastically different equivalent stiffness and resulting frequency. A series connection decreases the equivalent stiffness, while a parallel connection increases the equivalent stiffness. This is important in designing systems.

□

Example 1.5.5 illustrates that fixed values of spring constants can be used in various combinations to produce a desired value of stiffness and corresponding frequency. It is interesting to note that an identical set of physical devices can be used to create a system with drastically different frequencies simply by changing the physical arrangement of the components. This is similar to the choice of resistors in an electric circuit. The formulas of this section are intended to be aides in designing vibration systems.

1.6 MEASUREMENT

Measurements associated with vibration are used for several purposes. First, the quantities required to analyze the vibrating motion of a system all require measurement. The mathematical models proposed in previous sections all require knowledge of the mass, damping, and stiffness coefficients of the device under study. These coefficients can be measured in a variety of ways, as discussed in this section. Vibration measurements are also used to verify and improve analytical models. This is discussed in some detail in Chapter 7. Other uses of vibration testing techniques include reliability and durability studies, searching for damage, and testing for acceptability in terms of vibration parameters. These topics are also discussed briefly in Chapter 7.

In many cases, the mass of an object or device is simply determined by using a scale. In fact, mass is a relatively easy quantity to measure. However, the mass moment of inertia may require a dynamic measurement. A method of measuring the mass moment of inertia of an irregularly shaped object is to place the object on the platform of the apparatus of Figure 1.29 and measure the period of oscillation of the system, T. By using the methods of Section 1.4, it can be shown that the moment of inertia of an object, I

(about a vertical axis), placed on the disk of Figure 1.29 with its mass center aligned vertically with that of the disk, is given by

$$I = \frac{gT^2 r_0^2 (m_0 + m)}{4\pi^2 l} - I_0 \qquad (1.68)$$

Here m is the mass of the part being measured, m_0 is the mass of the disk, r_0 the radius of the disk, l the length of the wires, I_0 the moment of inertia of the disk, and g the acceleration due to gravity.

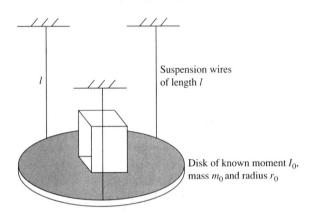

Suspension wires of length l

Disk of known moment I_0, mass m_0 and radius r_0

Figure 1.29   Trifilar suspension system for measuring the moment of inertia of irregularly shaped objects.

The stiffness of a simple spring system can be measured as suggested in Section 1.1 The elastic modulus, E, of an object can be measured in a similar fashion by performing a tensile test (see, e.g., Shames, 1989). In this method, a tensile test machine is employed which uses strain gauges (discussed in Chapter 7) to measure the strain, ϵ, in the test specimen as well as the stress, σ, both in the axial direction of the specimen. This produces a curve such as the one shown in Figure 1.30. The slope of the curve in the linear region defines the Young's modulus, or elastic modulus, for the test material. This is also known as Hooke's law.

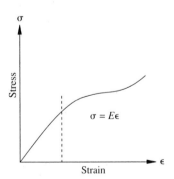

$\sigma = E\epsilon$

Strain

Figure 1.30   Stress–strain curve of a test specimen for determining the elastic modulus.

The elastic modulus can also be measured by using some of the formulas given in Section 1.5 and measurement of the vibratory response of a structure or part. For instance, consider the cantilevered arrangement of Figure 1.23. If the mass at the tip is given a small deflection, it will oscillate with frequency $\omega = \sqrt{k/m}$. If ω is measured, the modulus can be determined from equation (1.66), as illustrated in the following example.

Example 1.6.1

Consider a steel beam configuration as shown in Figure 1.23. The beam has a length $l = 1$ m and moment of inertia $I = 10^{-9}$ m^4, with a mass $m = 6$ kg attached to the tip. If the mass is given a small initial deflection in the transverse direction and oscillates with a period of $T = 0.62$ s, calculate the elastic modulus of steel.

Solution   Since $T = 2\pi/\omega$, equation (1.66) yields

$$T = 2\pi\sqrt{\frac{ml^3}{3EI}}$$

Solving for E yields

$$E = \frac{4\pi^2 ml^3}{3T^2 I} = \frac{4\pi^2 (6 \text{ kg})(1 \text{ m})^3}{3(0.62 \text{ s})^2 (10^{-9} \text{ m}^4)} = 205 \times 10^9 \text{ N/m}^2$$

The period T, and hence the frequency ω, can be measured with a stopwatch for vibrations that are large enough and last long enough to see. However, many vibrations of interest have very small amplitudes and happen very quickly. Hence several very sophisticated devices for measuring time and frequency have been developed and are discussed in Chapter 7.

The damping coefficient, or alternatively, the damping ratio, is the most difficult quantity to determine. Both mass and stiffness can be determined by static tests; however, damping requires a dynamic test to measure. A record of the displacement response of an underdamped system can be used to determine the damping ratio. One approach is to note that the decay envelope, denoted by the dashed line in Figure 1.31, for an underdamped system is $Ae^{-\zeta\omega t}$. The measured points $x(0)$, $x(t_1)$, $x(t_2)$, $x(t_3)$, and so on, can then be curve fit to A, $Ae^{-\zeta\omega t_1}$, $Ae^{-\zeta\omega t_2}$, $Ae^{-\zeta\omega t_3}$, and so on. This will yield a value for the coefficient $\zeta\omega$. If m and k are known, ζ and c can be determined from $\zeta\omega$.

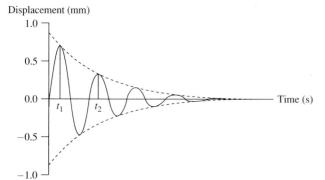

Figure 1.31   Underdamped response used to measure damping.

This approach also leads to the concept of *logarithmic decrement*, denoted by δ and defined by

$$\delta = \ln \frac{x(t)}{x(t+T)} \tag{1.69}$$

where T is the period of oscillation. Substitution of the analytical form of the underdamped response given by equation (1.36) yields

$$\delta = \ln \frac{Ae^{-\zeta\omega t} \sin(\omega_d t + \phi)}{Ae^{-\zeta\omega(t+T)} \sin(\omega_d t + \omega_d T + \phi)} \tag{1.70}$$

Since $\omega_d T = 2\pi$, the denominator becomes $e^{-\zeta\omega(t+T)} \sin(\omega_d t + \phi)$, and the expression for the decrement reduces to

$$\delta = \ln e^{\zeta\omega T} = \zeta\omega T \tag{1.71}$$

The period T in this case is the damped period so that

$$\delta = \zeta\omega \frac{2\pi}{\omega\sqrt{1-\zeta^2}} = \frac{2\pi\zeta}{\sqrt{1-\zeta^2}} \tag{1.72}$$

Solving this expression for ζ yields

$$\zeta = \frac{\delta}{\sqrt{4\pi^2 + \delta^2}} \tag{1.73}$$

which determines the damping ratio given the value of the logarithmic decrement.

Thus if the value of $x(t)$ is measured off the plot of Figure 1.31 at any two successive peaks, say $x(t_1)$ and $x(t_2)$, equation (1.69) can be used to produce a measured value of δ, and equation (1.73) can be used to determine the damping ratio. The formula for the damping ratio, equations (1.29) and (1.30), and knowledge of m and k, subsequently yield the value of the damping coefficient c. Note that peak measurements can be used over any integer multiple of the period (see Problem 41) to increase the accuracy over measurements taken at adjacent peaks.

Example 1.6.2

The free response of the system of Figure 1.10 with a mass of 2 kg is recorded to be of the form given in Figure 1.31. A static deflection test is performed and the stiffness is determined to be 1.5×10^3 N/m. The displacements at t_1 and t_2 are measured to be 9 and 1 mm, respectively. Calculate the damping coefficient.

Solution From the definition of the logarithmic decrement

$$\delta = \ln\left[\frac{x(t_1)}{x(t_2)}\right] = \ln\left[\frac{9 \text{ mm}}{1 \text{ mm}}\right] = 2.1972$$

From equation (1.73),

$$\zeta = \frac{2.1972}{\sqrt{4\pi^2 + 2.1972^2}} = 0.33 \quad \text{or} \quad 33\%$$

Also,

$$c_{cr} = 2\sqrt{km} = 2\sqrt{(1.5 \times 10^3 \text{ N/m})(2 \text{ kg})} = 1.095 \times 10^2 \text{ kg/s}$$

and from equation (1.30) the damping coefficient becomes

$$c = c_{cr}\zeta = (1.095 \times 10^2)(0.33) = 36.15 \text{ kg/s}$$

☐

Example 1.6.3

Mass and stiffness are usually measured in a straightforward manner as suggested in Section 1.6. However, there are certain circumstances that preclude using these simple methods. In these cases a measurement of the frequency of oscillation both before and after a known amount of mass is added, can be used to determine the mass and stiffness of the original system. Suppose then that the frequency of the system in Figure 1.32(a) is measured to be 2 rad/s and the frequency of Figure 1.32(b) with an added mass of 1 kg is known to be 1 rad/s. Calculate m and k.

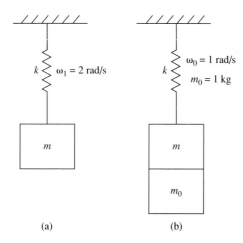

(a)   (b)

Figure 1.32 Using added mass and measured frequencies to determine m and k.

Solution   From the definition of natural frequency

$$\omega_1 = 2 = \sqrt{\frac{k}{m}} \qquad \text{and} \qquad \omega_0 = 1 = \sqrt{\frac{k}{m+1}}$$

Solving for m and k yields

$$4m = k \qquad \text{and} \qquad m + 1 = k$$

or

$$m = \frac{1}{3} \text{ kg} \qquad \text{and} \qquad k = \frac{4}{3} \text{ N/m}$$

This formulation can also be used to determine changes in mass of a system. As an example, the frequency of oscillation of a hospital patient in bed can be used to monitor the change in the patient's weight (mass) without having to move the patient from the bed. In this case the mass m_0 is considered to be the change in mass of the original system. If the original mass and frequency are known, measurement of the frequency ω_0 can be used to determine the change in mass m_0. Given that the original weight is 120 lb, the original frequency is 100.4 Hz, and the frequency of the patient bed system changes to 100 Hz, determine the change in the patient's weight.

From the two frequency relations

$$\omega_1^2 m = k$$

and

$$\omega_0^2(m + m_0) = k$$

Thus, $\omega_1^2 m = \omega_0^2(m + m_0)$. Solving for the change in mass m_0 yields

$$m_0 = m \left(\frac{\omega_1^2}{\omega_0^2} - 1 \right)$$

Multiplying by g and converting the frequency to hertz yields

$$W_0 = W \left(\frac{f_1^2}{f_0^2} - 1 \right)$$

or

$$W_0 = 120 \text{ lb} \left[\left(\frac{100.4 \text{ Hz}}{100 \text{ Hz}} \right)^2 - 1 \right]$$

$$= 0.96 \text{ lb}$$

Since the frequency decreased, the patient gained almost a pound. An increase in frequency would indicate a loss of weight. ☐

Measurement of m, c, k, ω, and ζ are used to verify the mathematical model of a system as well as for a variety of other reasons. Measurement of vibrating systems forms an important aspect of the activity in industry related to vibration technology. Chapter 7 is specifically devoted to measurement and comments on vibration measurements are mentioned throughout the remaining chapters.

1.7 DESIGN CONSIDERATIONS

Design in vibration refers to adjusting the physical parameters of a device to cause its vibration response to meet a specified shape or performance criteria. For instance, consider the response of the single-degree-of-freedom system of Figure 1.10. The shape of the response is somewhat determined by the value of the damping ratio in the sense that the response is either overdamped, underdamped, or critically damped ($\zeta > 1$, $\zeta < 1$, $\zeta = 1$, respectively). The damping ratio in turn depends on the values of m, c, and k. A designer may choose these values to produce the desired response.

Unfortunately, the values of m, c, and k have other constraints. In particular, the size and material of which the device is made determines these parameters. Hence the design procedure becomes a compromise. For example, for geometric reasons, the mass of a device may be limited to be between 2 and 3 kg, and for static displacement conditions, the stiffness may be required to be greater than 200 N/m. In this case, the natural frequency must be in the interval

$$8.16 \text{ rad/s} \leq \omega \leq 10 \text{ rad/s} \tag{1.74}$$

This severely limits the design of the vibration response, as illustrated in the following example.

Example 1.7.1

Consider the system of Figure 1.10 with mass and stiffness properties as summarized by inequality (1.74). Suppose that the system is subjected to initial velocity always less than 3 cm/s and to zero initial displacement (i.e., $x_0 = 0$, $v_0 \leq 3$ cm/s). Choose a dashpot design such that the amplitude of vibration is always less than 1 cm.

Solution For zero initial displacement, the amplitude of vibration of an underdamped system is given by equation (1.38) to be

$$A = \sqrt{\frac{v_0^2}{\omega_d^2}} = \frac{v_0}{\omega_d} = \frac{v_0}{\omega\sqrt{1 - \zeta^2}} \tag{1.75}$$

Choosing the maximum allowable amplitude, equation (1.75) becomes

$$v_0 = (0.01)\omega\sqrt{1 - \zeta^2}$$

It is desired to find ζ such that

$$(0.03) \geq v_0 = (0.01)\omega\sqrt{1 - \zeta^2}$$

Squaring this inequality and solving for ζ yields

$$\zeta \geq \sqrt{1 - \frac{9}{\omega^2}}$$

Substitution of $\omega = 10$ rad/s (the worst case) yields

$$\zeta > 0.9539$$

so that

$$c = 2\zeta\omega m > 2(0.9539)(10 \text{ rad/s})(2 \text{ kg}) = 38.156 \text{ kg/s}$$

If this number is too large to be obtained because of material restrictions, the mass and stiffness can be adjusted to the lower limit in inequality (1.74). If $\omega = 5$, then $\zeta = 0.8$ and this reduces the damping coefficient to $c = 16$ kg/s. While this is lower, it is still a relatively large value of damping that may not be available. Here a compromise must be made between available sources of damping, the requirements on the response (i.e. A), and the acceptable values of the mass m and stiffness k.   □

As another example of design, consider the problem of choosing a spring that will result in a spring–mass system having a desired or specified frequency. The formulas of Section 1.5 provide a means of designing a spring to have a specified stiffness in terms of the properties of the spring material (modulus) and its geometry. The following example illustrates this.

Example 1.7.2

Consider designing a helical spring such that when attached to a 10-kg mass, the resulting spring–mass system has a natural frequency of 10 rad/s (about 1.6 Hz).

Solution   From the definition of the natural frequency, the spring is required to have a stiffness of

$$k = \omega^2 m = (10)^2 (10) = 10^3 \text{ N/m}$$

The stiffness of a helical spring is given by equation (1.65) to be

$$k = 10^3 \text{ N/m} = \frac{Gd^4}{64nR^3} \quad \text{or} \quad 6.4 \times 10^4 = \frac{Gd^4}{nR^3}$$

This expression provides the starting point for a design. The choices available are the type of material to be used (hence various values of G); the diameter of the material, d; the radius of the coils, R; and the number of turns, n. The choices of G and d are, of course, restricted by available materials, n is restricted to be an integer, and R may have restrictions dictated by the size requirements of the device. Here it is assumed that steel of 1 cm diameter is available. The shear modulus of steel is about

$$G = 8.273 \times 10^{10} \text{ N/m}^2$$

so that the stiffness formula becomes

$$6.4 \times 10^4 \text{ N/m} = \frac{(8.273 \times 10^{10} \text{ N/m}^2)(10^{-2} \text{ m})^4}{nR^3}$$

or

$$nR^3 = 1.292 \times 10^{-2}$$

If the coil radius is chosen to be 10 cm, this yields the fact that the number of turns should be

$$n = \frac{1.29 \times 10^{-2} \text{ m}^3}{10^{-3} \text{ m}^3} = 12.9 \text{ or } 13$$

Thus if 13 turns of 1-cm-diameter steel are coiled at a radius of 10 cm, the resulting spring will have the desired stiffness and the 10-kg mass will oscillate at approximately 10 rad/s.

$\square$

In example 1.2.2, several variables were chosen to produce a desired design. In each case the design variables (such as d, R, etc.) are subject to constraints. Such constraints are considered formally in Chapter 5. Other aspects of vibration design are presented throughout the text as appropriate. There are no set rules to follow in design work. However, some organized approaches to design are presented in Chapter 5. The following example illustrates another difficulty in design, by examining what happens when conditions are changed after the design is over.

Example 1.7.3

As a last example, consider modeling the vertical suspension system of a small sports car, as a single-degree-of-freedom system of the form

$$m\ddot{x} + c\dot{x} + kx = 0$$

where m is the mass of the automobile and c and k are the equivalent damping and stiffness of the four shock absorber–spring systems. The car deflects the suspension system under its own weight 0.05 m. The suspension is chosen (designed) to be critically damped. If the car weighs 3000 lb (weight of a Porsche 968), calculate the equivalent damping and stiffness coefficients of the suspension system. If four 160-lb passengers are in the car, how does this affect the effective damping ratio?

Solution   Since the car weighs 3000 lb it has a mass of 1361 kg and its natural frequency is

$$\omega = \sqrt{\frac{k}{1361}}$$

so that

$$k = 1361\,\omega^2$$

At rest the car's springs are compressed an amount x_s, called the static deflection, by the weight of the car. Hence, from a force balance at static equilibrium, $mg = kx_s$, so that

$$k = \frac{mg}{x_s}$$

and

$$\omega = \left(\frac{k}{m}\right)^{1/2} = \left(\frac{g}{x_s}\right)^{1/2} = \left(\frac{9.8}{0.05}\right)^{1/2} = 14 \text{ rad/s}$$

The stiffness of the suspension system is thus

$$k = 1361(14)^2 = 2.668 \times 10^5 \text{ N/m}$$

For critical damping $\zeta = 1$ or $c = c_{cr}$ and equation (1.30) becomes

$$c = 2m\omega = 2(1361)(14) = 3.81 \times 10^4 \text{ kg/s}$$

Now if the four passengers are added to the car, the mass increases to $1361 + 290 = 1651$ kg. Since the stiffness and damping coefficient remains the same, the new static deflection becomes

$$x_s = \frac{mg}{k} = \frac{1651(9.8)}{2.668 \times 10^5} \approx 0.06\text{m}$$

The new frequency becomes

$$\omega = \sqrt{\frac{g}{x_s}} = \sqrt{\frac{9.8}{0.06}} = 12.7 \text{ rad/s}$$

From equations (1.29) and (1.30) the damping ratio becomes

$$\zeta = \frac{c}{c_{cr}} = \frac{3.81 \times 10^4}{2m\omega} = \frac{3.81 \times 10^4}{2(1651)(12.7)} = 0.9$$

Thus the car with four passengers is no longer critically damped and will exhibit some oscillatory motion in the vertical direction.

□

Note that this illustrates a design problem in the sense that the car cannot be exactly critically damped for all passenger situations. In this case, if critical damping is desirable, it really cannot be achieved. Designs that change dramatically when one parameter changes a small amount are said not to be *robust*. This is discussed in greater detail in Chapter 5.

1.8 STABILITY

In the preceding sections, the physical parameters m, c, and k are all considered to be positive in equation (1.25). This allows the treatment of the solutions of equation (1.25) to be classified into three groups: overdamped, underdamped, or critically damped. The case with $c = 0$ provides a fourth class, called undamped. These four solutions are all well behaved in the sense that they do not grow with time and their amplitudes are finite. There are many situations, however, in which the coefficients are not positive, and in these cases the motion is not well behaved. This situation refers to the *stability* of solutions of a system.

Recalling that the solution of the undamped case ($c = 0$) is of the form $A \sin(\omega t + \phi)$, it is easy to see that the undamped response is bounded. That is, if $|x(t)|$ denotes the absolute value of x, then

$$|x(t)| \le A|\sin(\omega t + \phi)| = A \tag{1.76}$$

for every value of t and for all choices of initial conditions. In this case the response is well behaved and said to be *stable* (sometimes called *marginally stable*). If, on the other hand, the value of k in equation (1.2) is negative and m is positive, the solutions are of the form

$$x(t) = A \sinh \omega t + B \cosh \omega t \tag{1.77}$$

which increases without bound as t does. Such solutions are called *divergent* or *unstable*. Figure 1.33 illustrates a stable response and Figure 1.34 illustrates an unstable or divergent response.

Consider the response of the damped system of equation (1.25) with positive coefficients. As illustrated in Figures 1.11, 1.12, and 1.13, it is clear that $x(t)$ approaches zero as t becomes large because of the exponential decay terms. Such solutions are called *asymptotically stable*. Again, if c or k is negative (m positive), the motion grows without bound and becomes unstable as in the undamped case. In the damped case, however, the motion may be unstable in one of two ways. Similar to overdamped solutions and underdamped solutions, the motion may grow without bound and not oscillate, or it may grow without bound and oscillate. The nonoscillatory case is called *divergent*

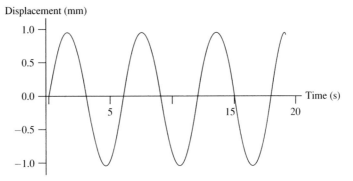

Figure 1.33   Example of a stable response.

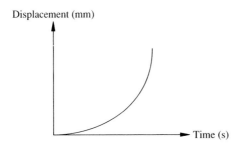

Figure 1.34   Example of an unstable or divergent response.

instability and the oscillatory case is called *flutter instability* or sometimes just *flutter*. Flutter instability is sketched in Figure 1.35. The trend of growing without bound for large *t* continues in Figures 1.34 and 1.35, even though the figure stops. These types of instability occur in a variety of situations, often called *self-excited vibrations* and require some source of energy. The following example illustrates such instabilities.

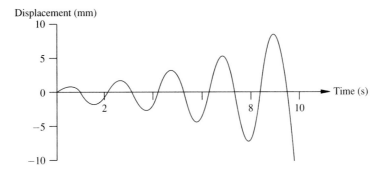

Figure 1.35   Example of flutter instability.

Example 1.8.1

Consider the inverted pendulum connected to two equal springs, shown in Figure 1.36.

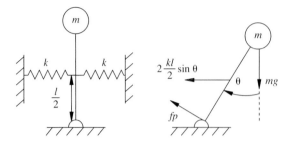

Figure 1.36 Inverted pendulum oscillator and its free-body diagram. Here f_p is the reaction force at the pin, and the pendulum is of length l.

Assume that the springs are undeflected when in the vertical position and that the mass m of the ball at the end of the pendulum rod is substantially larger than the mass of the rod itself, so that the rod is considered to be massless. If the rod is of length l and the springs are attached at the point $l/2$, the equation of motion becomes

$$ml^2\ddot{\theta} + \left(\frac{kl^2}{2}\sin\theta\right)\cos\theta - mgl\sin\theta = 0 \tag{1.78}$$

This is obtained by summing moments about the point of attachment of the pendulum to ground (hence the reaction force at the pin does not enter into the equation). For values of θ less than about $\pi/12$, $\sin\theta$ and $\cos\theta$ can be approximated by $\sin\theta \cong \theta$ and $\cos\theta \cong 1$. Applying this to equation (1.78) yields

$$ml^2\ddot{\theta} + \frac{kl^2}{2}\theta - mgl\theta = 0$$

which upon rearranging becomes

$$2ml\ddot{\theta}(t) + (kl - 2mg)\theta(t) = 0$$

where θ is now restricted to be small. If k, l, and m are all such that the effective stiffness is negative, that is, if

$$kl - 2mg < 0$$

the pendulum motion will be unstable by divergence, as illustrated in Figure 1.34. □

Example 1.8.2

The vibration of an aircraft wing can be crudely modeled as

$$m\ddot{x} + c\dot{x} + kx = \gamma\dot{x}$$

where m, c and k are the mass, damping, and stiffness values of the wing modeled as a single-degree-of-freedom system and where $\gamma\dot{x}$ is an approximate model of the aerodynamic forces on the wing ($\gamma > 0$). Rearranging this expression yields

$$m\ddot{x} + (c - \gamma)\dot{x} + kx = 0$$

If γ and c are such that $c - \gamma > 0$, the system is asymptotically stable. However, if γ is such that $c - \gamma < 0$, then $\zeta = (c - \gamma)/2m\omega < 0$ and the solutions are of the form

$$x(t) = Ae^{-\zeta\omega t}\sin(\omega_d t + \phi)$$

where $-\zeta \omega t > 0$ for all $t > 0$. Such solutions increase exponentially with time, as indicated in Figure 1.35. This is an example of flutter instability and self-excited oscillation.

□

PROBLEMS

Section 1.1

1.1. The spring of Figure 1.2 is successively loaded with mass and the corresponding (static) displacement is recorded below. Plot the data and calculate the spring's stiffness. Note that the data contain some error. Also calculate the standard deviation.

m(kg)	10	11	12	13	14	15	16
x(m)	1.14	1.25	1.37	1.48	1.59	1.71	1.82

1.2. Derive the solution of $m\ddot{x} + kx = 0$ and sketch the result for at least two periods for the case with $\omega = 2$ rad/s, $x_0 = 1$ mm, and $v_0 = \sqrt{5}$ mm/s.

1.3. Solve $m\ddot{x} + kx = 0$ for $k = 4$ N/m, $m = 1$ kg, $x_0 = 1$ mm, and $v_0 = 0$. Sketch the solution.

1.4. The amplitude of vibration of an undamped system is measured to be 1 mm. The phase shift is measured to be 2 rad and the frequency is found to be 5 rad/s. Calculate the initial conditions that caused this vibration to occur.

1.5. An undamped system vibrates with a frequency of 10 Hz and amplitude 1 mm. Calculate the maximum amplitude of the system's velocity and acceleration.

1.6. Show by calculation that $A \sin(\omega t + \phi)$ can be represented as $B \sin \omega t + C \cos \omega t$ and calculate C and B in terms of A and ϕ.

1.7. Using the solution of equation (1.2) in the form

$$x(t) = B \sin \omega t + C \cos \omega t$$

calculate the values of B and C in terms of the initial conditions x_0 and v_0.

1.8. Using Figure 1.6, verify that equation (1.10) satisfies the initial velocity condition.

1.9. A 0.5-kg mass is attached to a spring of stiffness 0.1 N/mm. Determine its natural frequency in hertz.

Section 1.2

1.10. Referring to Figure 1.7, if the maximum peak velocity of a vibrating system is 200 mm/s at 4 Hz and the maximum allowable peak acceleration is 5000 mm/s^2, what will the peak displacement be?

1.11. Show that lines of constant displacement and acceleration in Figure 1.7 have slopes of $+1$ and -1, respectively. If rms values instead of peak values are used, how does this affect the slope?

1.12. An automobile is modeled as a 1000-kg mass supported by a spring of stiffness $k = 400,000$ N/m. When it oscillates it does so with a maximum deflection of 10 cm. When loaded with passengers, the mass increases to as much as 1300 kg. Calculate the change in frequency, velocity amplitude, and acceleration amplitude if the maximum deflection remains 10 cm.

1.13. A machine oscillates in simple harmonic motion and appears to be well modeled by an undamped single-degree-of-freedom oscillation. Its acceleration is measured to have an amplitude of 10,000 mm/s^2 at 8 Hz. What is the machine's maximum displacement?

1.14. A simple undamped spring–mass system is set into motion from rest by giving it an initial velocity of 100 mm/s. It oscillates with a maximum amplitude of 10 mm. What is its natural frequency?

1.15. Repeat Example 1.2.3 for a truck of mass 2000 kg that has an unloaded maximum displacement of 10 cm and a spring stiffness of 900,000 N/m. In its fully loaded state it has a mass of 3000 kg and a maximum displacement of 4 cm. Sketch a nomograph describing its operating frequency.

1.16. An automobile traveling over a rough surface exhibits a vertical oscillating displacement of maximum amplitude 5 cm and a measured maximum acceleration of 2000 cm/s^2. Assuming that the automobile can be modeled as a single-degree-of-freedom system in the vertical direction, calculate the natural frequency of the automobile.

Section 1.3

1.17. Solve $\ddot{x} + 4\dot{x} + x = 0$ for $x_0 = 1$ mm, $v_0 = 0$ mm/s. Sketch your results and determine which root dominates.

1.18. Solve $\ddot{x} + 2\dot{x} + 2x = 0$ for $x_0 = 0$ mm, $v_0 = 1$ mm/s and sketch the response. You may wish to sketch $x(t) = e^{-t}$ and $x(t) = -e^{-t}$ first.

1.19. Derive the form of λ_1 and λ_2 given by equation (1.31) from equation (1.28) and the definition of the damping ratio.

1.20. Use the Euler formulas to derive equation (1.36) from equation (1.35) and to determine the relationships listed in Window 1.4.

1.21. Using equation (1.35) as the form of the solution of the underdamped system, calculate the values of the constants a_1 and a_2 in terms of the initial conditions x_0 and v_0.

1.22. For a damped system, m, c, and k are known to be $m = 1$ kg, $c = 2$ kg/s, $k = 10$ N/m. Calculate the values of ζ and ω. Is the system overdamped, underdamped, or critically damped?

1.23. Plot $x(t)$ for a damped system of natural frequency $\omega = 2$ and initial conditions $x_0 = 1$ mm, $v_0 = 0$, for the following values of the damping ratio: $\zeta = 0.01$, $\zeta = 0.2$, $\zeta = 0.6$, $\zeta = 0.1$, $\zeta = 0.4$, and $\zeta = 0.8$. It might be helpful to write a short program to do this or to use a standard software package (or use the one included).

1.24. Plot the response $x(t)$ of an underdamped system with $\omega = 2$ rad/s, $\zeta = 0.1$, and $v_0 = 0$ for the following initial displacements: $x_0 = 1$ mm, $x_0 = 5$ mm, $x_0 = 10$ mm, and $x_0 = 100$ mm.

1.25. Solve $\ddot{x} - \dot{x} + x = 0$ with $x_0 = 1$ and $v_0 = 0$ for $x(t)$ and sketch the response.

Section 1.4

1.26. Calculate the frequency of the compound pendulum of Figure 1.19(b) if a mass m_T is added to the tip, by using the energy method.

1.27. Calculate the total energy in a damped system with frequency 2 rad/s and damping ratio $\zeta = 0.01$ with mass 10 kg for the case $x_0 = 0.1$ and $v_0 = 0$. Plot the total energy versus time.

1.28. Use the energy method to calculate the equation of motion and natural frequency of an airplane's steering gear mechanism for the nose wheel of its landing gear. The mechanism is modeled as the single-degree-of-freedom system illustrated in Figure 1.37.

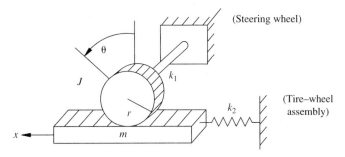

Figure 1.37   Single-degree-of-freedom model of a steering mechanism.

The steering wheel and tire assembly are modeled as being fixed at ground for this calculation. The steering rod gear system is modeled as a linear spring and mass system (m, k_2) oscillating in the x direction. The shaft–gear mechanism is modeled as the disk of inertia J and torsional stiffness k_2. The gear J turns through the angle θ such that the disk does not slip on the mass. Obtain an equation in the linear motion x.

1.29. A control pedal of an aircraft can be modeled as the single-degree-of-freedom system of Figure 1.38. Consider the lever as a massless shaft and the pedal as a lumped mass at the end of the shaft. Use the energy method to determine the equation of motion in θ and calculate the natural frequency of the system.

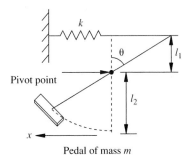

Figure 1.38   Model of a foot pedal used to operate an aircraft control surface.

1.30. To save space, two large pipes are shipped one stacked inside the other as indicated in Figure 1.39. Calculate the natural frequency of vibration of the smaller pipe (of radius R_1) rolling back and forth inside the larger pipe (of radius R). Use the energy method and assume that the inside pipe rolls without slipping and has a mass of m.

1.31. Consider the example of a simple pendulum given in Example 1.4.2. The pendulum motion is observed to decay with a damping ratio of $\zeta = 0.001$. Determine a damping coefficient and add a viscous damping term to the pendulum equation.

1.32. Determine a damping coefficient for the disk–rod system of Example 1.4.3. Assuming that the damping is due to the material properties of the rod, determine c for the rod if it is observed to have a damping ratio of $\zeta = .01$.

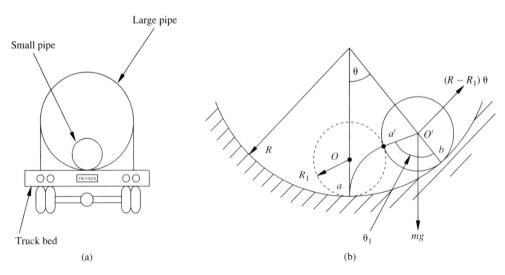

Figure 1.39 (a) Pipes stacked in a truck bed. (b) Vibration model of the inside pipe.

1.33. The rod and disk of Window 1.1 are in torsional vibration. Calculate the damped natural frequency if $J = 1000 \text{ m}^2 \cdot \text{kg}$, $c = 20 \text{ N} \cdot \text{m} \cdot \text{s/rad}$, and $k = 400 \text{ N} \cdot \text{m/rad}$.

1.34. Consider the system of Figure 1.15, which could also represent a simple model of an aircraft landing system. Suppose that a viscous damper is added to the system of damping coefficient c. What is the damped natural frequency?

1.35. Consider Problem 1.34 with $k = 400,000 \text{ N} \cdot \text{m/rad}$, $m = 1500 \text{ kg}$, $J = 10,000 \text{ m}^2 \cdot \text{kg}$, $r = 25 \text{ cm}$, and $c = 8000 \text{ N} \cdot \text{m} \cdot \text{s/rad}$. Calculate the damping ratio and the damped natural frequency. How much effect does the rotational inertia have on the undamped natural frequency?

Section 1.5

1.36. A helicopter landing gear consists of a metal framework rather than a suspension system as used in a fixed-wing aircraft. The vibration of the frame in the vertical direction can be modeled as the longitudinal vibration of a slender bar illustrated in Figure 1.20. Here $l = 0.4 \text{ m}$, $E = 20 \times 10^{10} \text{ N/m}^2$, and $m = 100 \text{ kg}$. Calculate the cross-sectional area that should be used if the natural frequency is to be $\omega = 500 \text{ Hz}$.

1.37. The frequency of oscillation of a person on a diving board can be modeled as the transverse vibration of a beam as indicated in Figure 1.23. Let m be the mass of the diver ($m = 100 \text{ kg}$) and $l = 1 \text{ m}$. If the diver wishes to oscillate at 3 Hz, what value of EI should the diving board material have?

1.38. Consider the spring system of Figure 1.28. Let $k_1 = k_5 = k_2 = 100 \text{ N/m}$, $k_3 = 50 \text{ N/m}$, and $k_4 = 1 \text{ N/m}$. What is the equivalent stiffness?

1.39. Springs are available in stiffness values of 10, 100, and 1000 N/m. Design a spring system using these values only, so that a 100-kg mass is connected to ground with frequency of about 1.5 rad/s.

1.40. Calculate the natural frequency of the system in Figure 1.28(a) if $k_1 = k_2 = 0$. Choose m and nonzero values of k_3, k_4, and k_5 so that the natural frequency is 100 rad/s.

Section 1.6

1.41. Show that the logarithmic decrement is equal to

$$\delta = \frac{1}{n} \ln \frac{x_0}{x_n}$$

where x_n is the amplitude of vibration after n cycles have elapsed.

1.42. Derive equation (1.68) for the trifalar suspension system.

1.43. A prototype composite material is formed and hence has unknown modulus. An experiment is performed consisting of forming it into a cantilevered beam of length 1 m and moment $I = 10^{-9}$ m^4 with a 6-kg mass attached at its end. The system is given an initial displacement and found to oscillate with a period of 0.5 s. Calculate the modulus E.

1.44. The free response of a 1000-kg automobile with stiffness of $k = 400{,}000$ N/m is observed to be of the form given in Figure 1.31. Modeling the automobile as a single-degree-of-freedom oscillation in the vertical direction, determine the damping coefficient if the displacement at t_1 is measured to be 2 cm and 0.22 cm at t_2.

1.45. A pendulum decays from 10 cm to 1 cm over one period. Determine its damping ratio.

1.46. The relationship between the log decrement δ and the damping ratio ζ is often approximated as $\delta = 2\pi\zeta$. For what values of ζ would you consider this a good approximation to equation (1.72)?

1.47. A damped system is modeled as illustrated in Figure 1.10. The mass of the system is measured to be 5 kg and its spring constant is measured to be 5000 N/m. It is observed that during free vibration the amplitude decays to 0.25 of its initial value after five cycles. Calculate the viscous damping coefficient, c.

Section 1.7 (see also Problem 1.39)

1.48. Choose a dashpot's viscous damping value such that when placed in parallel with the spring of Example 1.7.2 reduces the frequency of oscillation to 9 rad/s.

1.49. For an underdamped system, $x_0 = 0$ and $v_0 = 10$ mm/s. Determine m, c, and k such that the amplitude is less than 1 mm.

1.50. Repeat problem 1.49 if the mass is restricted to lie between 10 kg < m < 15 kg.

1.51. Use the formula for the torsional stiffness of a shaft from Table 1.1 to design a 1-m shaft with torsional stiffness of 10^9 N · m/rad.

1.52. Repeat Example 1.7.2 using aluminum. What difference do you note?

1.53. Try to design a bar (see Figure 1.20) that has the same stiffness as the spring of Example 1.7.2. Note that the bar must remain at least 10 times as long as it is wide in order to be modeled by the formula of Figure 1.20.

1.54. Repeat Problem 1.53 using plastic ($E = 1.40 \times 10^9$ N/m^2) and rubber ($E = 7 \times 10^6$ N/m^2). Are any of these feasible?

Section 1.8 (see also Problem 1.25)

1.55. Consider the inverted pendulum of Figure 1.36 as discussed in Example 1.8.1. Assume that a dashpot (of damping rate c) also acts on the pendulum parallel to the two springs. How does this affect the stability properties of the pendulum?

1.56. Replace the massless rod of the inverted pendulum of Figure 1.36 with a solid object compound pendulum of Figure 1.19(b). Calculate the equations of vibration and discuss values of the parameter for which the system is stable.

1.57. Consider the disk of Figure 1.40 connected to two springs. Use the energy method to calculate the system's natural frequency of oscillation for small angles $\theta(t)$.

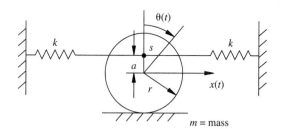

Figure 1.40 Vibration model of a rolling disk mounted against two springs, attached at point s.

MATLAB® VIBRATION TOOLBOX

MATLAB is a professional-quality software package for interactive numerical computation, data analysis, and graphics suitable for a broad range of scientific and engineering applications. MATLAB is very much like BASIC or other high-level programming languages. The program runs interactively and can be also used by creating your own files (called "M" files) for later or repeated use. These files are discussed at the end of each chapter following the problems, together with a few additional problems using MATLAB. If you do not currently have access to MATLAB, you may wish to purchase the *Student Edition of MATLAB*. This inexpensive version, which includes a user manual, is available at your bookstore or from Prentice Hall.

If MATLAB is not already installed on the machine of your choice, do so now. Macintosh users should send in the tearout card from the back of the book to receive a Macintosh compatible disk. IBM or compatible users should read the file INTRO.TXT. Open and read the text file "INTRO.TXT" which explains how to use the Vibration Toolbox. It would also be useful to run the "Demo" file in your copy of MATLAB. The folder named VTB1 contains the programs for this chapter. Each set of files is labeled VTB1_x.M to correspond to Chapter 1. There are files for each chapter (i.e., VTB2_3.M is the third file associated with Chapter 2, etc., and can be found in the folder directory named VTB2).

You may use the files contained in the *Vibration Toolbox* to help solve the problems listed above. You may also use these files to study the nature of the free response of a single-degree-of-freedom system discussed in Chapter 1 and the dependence of this response on various parameters. To become familiar with using the *Vibration Toolbox* and to build some intuition about the vibration of a single-degree-of-freedom system, try the following problems.

TOOLBOX PROBLEMS

TB1.1. Fix [your choice or use the values from Example 1.3.1 with $x(0) = 1$ mm] the values of m, c, k, and $x(0)$ and plot the responses $x(t)$ for a range of values of the initial velocity $\dot{x}(0)$ to see how the response depends on the initial velocity using VTB1_1. Remember to use numbers with consistent units.

TB1.2. Using the values from Problem TB1.1 and $\dot{x}(0) = 0$, plot the response $x(t)$ for a range of values of $x(0)$ to see how the response depends on the initial displacement.

TB1.3. Reproduce Figures 1.11, 1.12 and 1.13 using VTB1_1.

TB1.4. Consider solving Problem 1.23 and compare the time for each response to reach and stay below 0.01 mm using VTB1_1.

2 Response to Harmonic Excitation

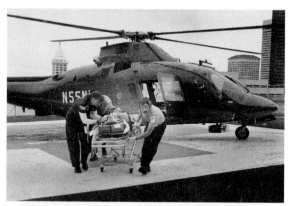

This chapter considers the vibration of simple devices when forced by harmonic excitation. Modern helicopters such as the one pictured above produce strong harmonic excitations as the blades rotate. These motions are transmitted to the cabin but are reduced to total acceptable levels by the vibration absorbers mounted at the root of the blades between the motor and cabin. Wind turbines, such as the vertical axis wind turbine pictured below are used to generate electric power. They also generate very large harmonic disturbances which need to be accounted for in designing wind power systems. Harmonic excitation is discussed in this chapter and vibration absorbers are discussed in Chapter 5.

This chapter considers the response to harmonic excitation of the single-degree-of-freedom spring–mass–damper system presented in Chapter 1. Harmonic excitation refers to a sinusoidal external force of a single frequency applied to the system. Harmonic excitations are a very common source of external force applied to machines and structures. Rotating machines such as fans, electric motors, and reciprocating engines transmit a sinusoidally varying force to adjacent components. In addition, the Fourier theorem indicates that many other forcing functions can be expressed as an infinite series of harmonic terms. Since the equations of motion considered here are linear, knowing the response to individual terms in the series allows the total response to be represented as the sum of the response to the individual terms. This is the principle of superposition. In this way knowing the response to a single harmonic input allows the calculation of the response to a variety of other input disturbances of periodic nature. General periodic disturbances are discussed in Chapter 3.

A harmonic input is also chosen for study because it can be solved mathematically with straightforward techniques. In addition, the response of a single-degree-of-freedom system to a harmonic input forms the foundation of vibration measurement, the design of devices intended to protect machines from unwanted oscillation, and the design of transducers used in measuring vibration. Harmonic excitations are simple to produce in laboratories and hence are also very useful in studying damping and stiffness properties.

2.1 HARMONIC EXCITATION OF UNDAMPED SYSTEMS

Consider the system of Figure 2.1 for the case of negligible damping ($c = 0$). For the case of harmonic excitation considered here, $F(t)$ has the form of a sine or cosine function of a single frequency. Here the driving force $F(t)$ is chosen to be of the form

$$F(t) = F_0 \cos \omega_{dr} t \tag{2.1}$$

where F_0 represents the magnitude, or maximum amplitude, of the applied force and ω_{dr} denotes the frequency of the applied force. The frequency ω_{dr} is also called the *input frequency*, or *driving frequency*, or *forcing frequency*.

The sum of the forces in the y direction yields $N = mg$, with the result of no motion in that direction. Summing forces on the mass of Figure 2.1 in the x direction for the undamped case yields the result that the displacement $x(t)$ must satisfy

$$m\ddot{x}(t) + kx(t) = F_0 \cos \omega_{dr} t \tag{2.2}$$

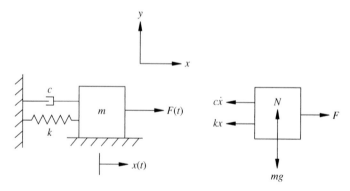

Figure 2.1   Schematic of a single-degree-of-freedom system acted on by an external force $F(t)$ and sliding on a friction-free surface. The figure on the right is a free-body diagram of the friction-free spring–mass–damper system.

Note that this expression is a linear equation in the variable $x(t)$. As in the homogeneous (unforced) case of Chapter 1, it is convenient to divide this expression by the mass, m, to yield

$$\ddot{x}(t) + \omega^2 x(t) = f_0 \cos \omega_{dr} t \qquad (2.3)$$

where $f_0 = F_0/m$. A variety of techniques can be used to solve this equation, which are commonly studied in a first course in differential equations.

First recall from differential equations that equation (2.3) is a linear nonhomogeneous equation and that its solution is therefore the sum of the homogeneous solution (i.e., the solution for the case $f_0 = 0$) and a particular solution. The particular solution can often be found by assuming that it has the same form as the forcing function. This is also consistent with observation. That is, the oscillation of a single-degree-of-freedom system excited by $f_0 \cos \omega_{dr} t$ is observed to be of the form

$$x_p = A_0 \cos \omega_{dr} t \qquad (2.4)$$

where x_p denotes the particular solution and A_0 is the amplitude of the forced response. Substitution of the assumed form of the solution (2.4) into the equation of motion (2.3) yields

$$-\omega_{dr}^2 A_0 \cos \omega_{dr} t + \omega^2 A_0 \cos \omega_{dr} t = f_0 \cos \omega_{dr} t \qquad (2.5)$$

Factoring out $\cos \omega_{dr} t (\neq 0)$, setting the coefficient to zero, and solving for A_0 yields

$$A_0 = \frac{f_0}{\omega^2 - \omega_{dr}^2} \qquad (2.6)$$

provided that $\omega \neq \omega_{dr}$. Thus as long as the driving frequency and natural frequency are different (i.e., as long as $\omega \neq \omega_{dr}$) the particular solution will be of the form

$$x_p(t) = \frac{f_0}{\omega^2 - \omega_{dr}^2} \cos \omega_{dr} t \qquad (2.7)$$

This approach, of assuming that $x_p = A_0 \cos \omega_{dr} t$, to determine the particular solution is called the *method of undetermined coefficients*.

Window 2.1
Review of the Solution of the Undamped
Homogeneous Vibration Problem from Chapter 1

$$m\ddot{x} + kx = 0 \quad \text{subject to} \quad x(0) = x_0, \qquad \dot{x}(0) = v_0$$

has solution $x(t) = A \sin(\omega t + \phi)$, which becomes after evaluating the constants A and ϕ in terms of the initial conditions:

$$x(t) = \frac{\sqrt{x_0^2 \omega^2 + v_0^2}}{\omega} \sin\left(\omega t + \tan^{-1} \frac{x_0 \omega}{v_0}\right)$$

where $\omega = \sqrt{k/m}$ is the natural frequency. Via some simple trigonometry, this solution can also be written as

$$x(t) = A_1 \sin \omega t + A_2 \cos \omega t$$

where the constants A_1, A_2, A, and ϕ are related by

$$A = \sqrt{A_1^2 + A_2^2}, \qquad \tan \phi = \frac{A_2}{A_1}$$

Since the system is linear, the total solution $x(t)$ is the sum of the particular solution of equation (2.7) plus the homogeneous solution given by equation (1.19). Recalling that $A \sin(\omega t + \phi)$ can be represented as $A_1 \sin \omega t + A_2 \cos \omega t$ (see Window 2.1), the total solution can be expressed in the form

$$x(t) = A_1 \sin \omega t + A_2 \cos \omega t + \frac{f_0}{\omega^2 - \omega_{dr}^2} \cos \omega_{dr} t \tag{2.8}$$

where it remains to determine the coefficients A_1 and A_2. These are determined by enforcing the initial conditions. Let the initial position and velocity be given by the constants x_0 and v_0 as before. Then equation (2.8) yields

$$x(0) = A_2 + \frac{f_0}{\omega^2 - \omega_{dr}^2} = x_0 \tag{2.9}$$

and

$$\dot{x}(0) = \omega A_1 = v_0 \tag{2.10}$$

Solving equations (2.9) and (2.10) for A_1 and A_2 and substituting these values into equation (2.8) yields the total response

$$x(t) = \frac{v_0}{\omega} \sin \omega t + \left(x_0 - \frac{f_0}{\omega^2 - \omega_{dr}^2}\right) \cos \omega t + \frac{f_0}{\omega^2 - \omega_{dr}^2} \cos \omega_{dr} t \tag{2.11}$$

Figure 2.2 illustrates a plot of the total response of an undamped system to a harmonic excitation and specified initial conditions.

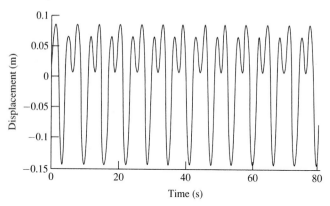

Figure 2.2   Response of an undamped system with $\omega = 1$ rad/s to harmonic excitation at $\omega_{dr} = 2$ rad/s and nonzero initial conditions of $x_0 = 0.01$ m and $v_0 = 0.01$ m/s and magnitude $f_0 = 0.1$ N/kg. The motion is the sum of two sine curves of different frequencies.

Example 2.1.1

Consider the forced vibration of a mass m connected to a spring of stiffness 2000 N/m being driven by a 20-N harmonic force at 10 Hz. The maximum amplitude of vibration is measured to be 0.1 m and the motion is assumed to have started from rest ($x_0 = v_0 = 0$). Calculate the mass of the system.

Solution   From equation (2.11) the response with $x_0 = v_0 = 0$ becomes

$$x(t) = \frac{f_0}{\omega^2 - \omega_{dr}^2}(\cos \omega_{dr} t - \cos \omega t) \tag{2.12}$$

Using simple trigonometric identities, this becomes

$$x(t) = \frac{2 f_0}{\omega^2 - \omega_{dr}^2} \sin\left(\frac{\omega - \omega_{dr}}{2} t\right) \sin\left(\frac{\omega + \omega_{dr}}{2} t\right) \tag{2.13}$$

The maximum value of the total response is evident from (2.13) so that

$$\frac{2 f_0}{\omega^2 - \omega_{dr}^2} = 0.1 \text{ m}$$

Solving this for m from $\omega^2 = k/m$ and $f_0 = F_0/m$ yields

$$m = \frac{(0.1 \text{ m})(2000 \text{ N/m}) - 2(20 \text{ N})}{(0.1 \text{ m})(10 \times 2\pi \text{ rad/s})^2} = \frac{4}{\pi^2} = 0.405 \text{ kg}$$

□

Two very important phenomenon occur when the driving frequency becomes close to the system's natural frequency. First consider the case where ($\omega - \omega_{dr}$) becomes very small. For zero initial conditions the response is given by equation (2.13), which is plotted in Figure 2.3. Since $(\omega - \omega_{dr})$ is small, $(\omega + \omega_{dr})$ is large by comparison and the term $\sin[(\omega - \omega_{dr})/2]t$ oscillates with a much longer period than does $\sin[(\omega + \omega_{dr})/2]t$. Recall that the period of oscillation T is defined as $2\pi/\omega$, or in this case $2\pi/(\omega + \omega_{dr})/2 = 4\pi/(\omega + \omega_{dr})$. The resulting motion is a rapid oscillation with slowly varying amplitude and is called a *beat*.

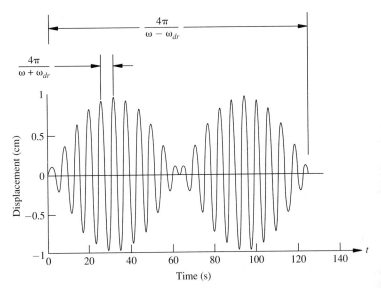

Figure 2.3   Response of an undamped system for small $\omega - \omega_{dr}$ illustrating the phenomenon of beats. In the figure, $f_0 = 0.1$ N/kg, $\omega_{dr} = 1$ rad/s, and $\omega = 1.1$ rad/s.

As ω_{dr} becomes exactly equal to the system's natural frequency, the solution given in equation (2.11) is no longer valid. In this case the choice of the function $A_0 \cos \omega_{dr} t$ for a particular solution fails because it is also a solution of the homogeneous equation. Therefore, the particular solution is of the form

$$x_p(t) = t A_0 \sin \omega_{dr} t \tag{2.14}$$

as explained, for instance, in Boyce and DiPrima (1986). Substitution of (2.14) into equation (2.3) and solving for A_0 yields

$$x_p(t) = \frac{f_0}{2\omega} t \sin \omega_{dr} t \tag{2.15}$$

Thus the total solution is now of the form

$$x(t) = A_1 \sin \omega t + A_2 \cos \omega t + \frac{f_0}{2\omega} t \sin \omega_{dr} t \tag{2.16}$$

Evaluating the initial displacement x_0 and velocity v_0 as before yields

$$x(t) = \frac{v_0}{\omega} \sin \omega t + x_0 \cos \omega t + \frac{f_0}{2\omega} t \sin \omega t \qquad (2.17)$$

where the subscripts on ω have been dropped because they are all equal. A plot of $x(t)$ is given in Figure 2.4, where it can be seen that $x(t)$ grows without bound. This defines the phenomenon of *resonance*, (i.e., that the amplitude of vibration becomes unbounded at $\omega_{dr} = \omega = \sqrt{k/m}$). This would cause the spring to fail and break. Note that equation (2.17) can also be obtained from equation (2.11) by taking the limit as $\omega_{dr} \to \omega$ using the limit theorems from calculus.

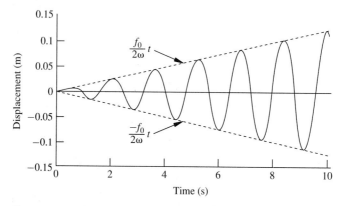

Figure 2.4 Forced response of a spring–mass system driven harmonically at its natural frequency ($\omega = \omega_{dr}$).

2.2 HARMONIC EXCITATION OF DAMPED SYSTEMS

As noted in Chapter 1, some sort of damping or energy dissipation is always present (see Window 2.2). In this section the response of a viscously damped single-degree-of-freedom system subjected to harmonic excitation is considered. Summing forces on the mass of Figure 2.1 in the x direction yields

$$m\ddot{x} + c\dot{x} + kx = F_0 \cos \omega_{dr} t \qquad (2.18)$$

Dividing by the mass m yields

$$\ddot{x} + 2\zeta\omega\dot{x} + \omega^2 x = f_0 \cos \omega_{dr} t \qquad (2.19)$$

where $\omega = \sqrt{k/m}$, $\zeta = c/(2m\omega)$, and $f_0 = F_0/m$ as before. The calculation of the particular solution for the damped case is similar to that of the undamped case and follows the method of undetermined coefficients.

From differential equations it is known that the forced response of a damped system is of the form of a harmonic function of the same frequency as the driving force with a different amplitude and phase. The phase shift is expected because of

Window 2.2
Review of the Solution of the Damped Homogeneous
Vibration Problem $(0 < \zeta < 1)$ from Chapter 1

$m\ddot{x} + c\dot{x} + kx = 0$ subject to $x(0) = x_0, \dot{x}(0) = v_0$ has the solution

$$x(t) = Ae^{-\zeta\omega t}\sin(\omega_d t + \theta)$$

where $\omega = \sqrt{k/m}$ is the undamped natural frequency

$\zeta = c/(2m\omega)$ is the damping ratio

$\omega_d = \omega\sqrt{1 - \zeta^2}$ is the damped natural frequency

and the constants A and θ are determined by the initial conditions to be

$$A = \sqrt{x_0^2 + \left(\frac{v_0 + \zeta\omega x_0}{\omega_d}\right)^2}$$

$$\theta = \tan^{-1}\frac{x_0\omega_d}{v_0 + \zeta\omega x_0}$$

the effect of the damping force. Following the method of undetermined coefficients the particular solution is assumed to be of the form

$$x_p(t) = A_0\cos(\omega_{dr}t - \phi) \tag{2.20}$$

To make the computations easy to follow, this is written in the equivalent form

$$x_p(t) = A_s\cos\omega_{dr}t + B_s\sin\omega_{dr}t \tag{2.21}$$

where the constants A_s and B_s satisfying

$$A_0 = \sqrt{A_s^2 + B_s^2} \quad \text{and} \quad \phi = \tan^{-1}\frac{B_s}{A_s} \tag{2.22}$$

are the undetermined constants.

Taking derivatives of the assumed form of the solution given by (2.21) yields

$$\dot{x}_p(t) = -\omega_{dr}A_s\sin\omega_{dr}t + \omega_{dr}B_s\cos\omega_{dr}t \tag{2.23}$$

and

$$\ddot{x}_p(t) = -\omega_{dr}^2(A_s\cos\omega_{dr}t + B_s\sin\omega_{dr}t) \tag{2.24}$$

Substitution of x_p, $\dot{x}_p$, and $\ddot{x}_p$ into the equation of motion given by equation (2.19) and grouping terms as coefficients of $\sin\omega_{dr}t$ and $\cos\omega_{dr}t$ yields

$$(-\omega_{dr}^2 A_s + 2\zeta\omega\omega_{dr}B_s + \omega^2 A_s - f_0)\cos\omega_{dr}t + (-\omega_{dr}^2 B_s - 2\zeta\omega\omega_{dr}A_s + \omega^2 B_s)\sin\omega_{dr}t = 0 \tag{2.25}$$

This equation must hold for all time, in particular for $t = 2\pi/\omega_{dr}$, so that the coefficient of $\cos \omega_{dr}t$ must vanish. Similarly, the coefficient of $\sin \omega_{dr}t$ must vanish. This yields the two equations

$$(\omega^2 - \omega_{dr}^2)A_s + (2\zeta\omega\omega_{dr})B_s = f_0$$
$$(-2\zeta\omega\omega_{dr})A_s + (\omega^2 - \omega_{dr}^2)B_s = 0 \tag{2.26}$$

in the two undetermined coefficients A_s and B_s. Solving these two equations for the undetermined coefficients, A_s and B_s, yields

$$A_s = \frac{(\omega^2 - \omega_{dr}^2)f_0}{(\omega^2 - \omega_{dr}^2)^2 + (2\,\zeta\omega\omega_{dr})^2}$$

$$B_s = \frac{2\zeta\omega\omega_{dr}\,f_0}{(\omega^2 - \omega_{dr}^2)^2 + (2\zeta\omega\omega_{dr})^2} \tag{2.27}$$

Substitution of these values into equations (2.22) and (2.20) yields that the particular solution is

$$x_p(t) = \frac{f_0}{\sqrt{(\omega^2 - \omega_{dr}^2)^2 + (2\zeta\omega\omega_{dr})^2}} \cos\left(\omega_{dr}t - \tan^{-1}\frac{2\zeta\omega\omega_{dr}}{\omega^2 - \omega_{dr}^2}\right) \tag{2.28}$$

The total solution is again the sum of the particular solution and the homogeneous solution obtained in Section 1.3. For the underdamped case $(0 < \zeta < 1)$ this becomes

$$x(t) = Ae^{-\zeta\omega t}\sin(\omega_d t + \theta) + A_0\cos(\omega_{dr}t - \phi) \tag{2.29}$$

where A_0 and ϕ are the coefficients of the particular solution as defined by equation (2.28) and A and θ (different than those of Window 2.2) are determined by initial conditions. Note that for large values of t, the first term, or homogeneous solution, approaches zero and the total solution approaches the particular solution. Thus $x_p(t)$ is called the *steady-state response* and the first term in equation (2.29) is called the *transient response*.

Note that A and θ, the constants describing the transient response in equation (2.29), will be different than calculated for the free-response case given in equation (1.38) or Window 2.2. This is because part of the transient term in equation (2.29) is due to the excitation force as well as the initial conditions. Problem 2.8 addresses this.

It is often common to ignore the transient part of the total solution given by equation (2.29) and focus only on the steady-state response: $A_0\cos(\omega_{dr}t - \phi)$. The rationale for considering only the steady-state response is based on the value of the damping ratio ζ. If the system has relatively large damping, the term $e^{-\zeta\omega t}$ causes the transient response to die out very quickly—perhaps in a fraction of a second. If on the other hand, the system is lightly damped (ζ very small), the transient part of the solution may last long enough to be significant and should not be ignored. The decision to ignore the transient part of the solution or not should also be based on the application. In fact, in some applications (such as earthquake analysis, satellite analysis) the transient response may become even more important than the steady-state response. An example is the Hubble space telescope, which originally experienced a transient vibration that lasted over 10 minutes, causing the telescope to be unusable every time it passed out of the earth's shadow, until the system was corrected.

The transient response can also be very important if it has a relatively large amplitude. Usually, devices are designed and analyzed based on the steady-state response, but the transient should always be checked to make sure that it is reasonable to ignore it, or if it should, in fact, be considered seriously.

With this caveat in mind, it is of interest to consider the magnitude, A_0, and the phase θ, of the steady-state response as a function of the driving frequency. Examining the form of equation (2.28) and comparing it to the assumed form $A_0 \cos(\omega_d t - \phi)$ yields the fact that the amplitude A_0 and the phase ϕ are

$$A_0 = \frac{f_0}{\sqrt{\left(\omega^2 - \omega_{dr}^2\right)^2 + (2\zeta\omega\omega_{dr})^2}}, \qquad \phi = \tan^{-1}\frac{2\zeta\omega\omega_{dr}}{\omega^2 - \omega_{dr}^2}$$

After some manipulation (i.e., factoring out ω^2 and dividing the magnitude by F_0/m), these expressions for the magnitude and phase can be written as

$$\frac{A_0 k}{F_0} = \frac{A_0 \omega^2}{f_0} = \frac{1}{\sqrt{(1 - r^2)^2 + (2\zeta r)^2}} \qquad \phi = \tan^{-1}\frac{2\zeta r}{1 - r^2} \qquad (2.30)$$

Here r is the frequency ratio $r = \omega_{dr}/\omega$, a dimensionless quantity. Equation (2.30) for the magnitude and phase are plotted versus the frequency ratio r in Figure 2.5 for

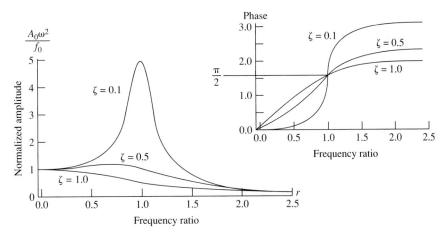

Figure 2.5 Plot of the (a) normalized magnitude ($A_0\omega^2/f_0 = A_0 k/F_0$) and (b) phase of the steady-state response of a damped system versus the frequency ratio for several different values of the damping ratio ζ as determined by equation (2.30).

several values of the damping ratio ζ. Note that as the driving frequency approaches the undamped natural frequency ($r \rightarrow 1$) the magnitude approaches a maximum value for those curves corresponding to light damping ($\zeta \le 0.1$). Also note that as the driving frequency approaches the undamped natural frequency, the phase shift crosses through $90°$. This defines *resonance* for the damped case. These two observations have important uses in both vibration design and measurement. As ω_{dr} approaches zero, the amplitude approaches f_0/ω^2, and as ω_{dr} becomes very large, the amplitude approaches zero asymptotically.

It is also important from the design point of view to note how the amplitude of steady-state vibration is affected by changing the damping ratio. This is illustrated in Figure 2.6 and Example 2.2.2. Note that as the damping ratio is increased, the peak in the magnitude curve decreases and eventually disappears. As the damping ratio decreases, however, the peak value increases and becomes sharper. In the limit as ζ goes to zero, the peak climbs to an infinite value in agreement with the undamped response at resonance (see Figure 2.4).

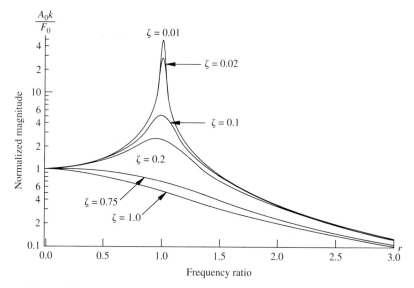

Figure 2.6   Magnitude of the steady-state response versus the frequency ratio for several values of the damping ratio ζ.

Example 2.2.1

Consider again the simple spring–mass system of Examples 1.2.1 and 1.3.1 consisting of a spring and a bolt. Calculate the value of the steady-state response if $\omega_{dr} = 132$ rad/s for $f_0 = 10$ N/kg. Calculate the change in amplitude if $\omega_{dr} = 125$ rad/s.

Solution   From Example 1.3.1, the natural frequency and damping ratio are given as $\omega = 132$ rad/s and $\zeta = 0.0085$, respectively.From the expression for A_0 above, the magnitude of $x_p(t)$ is

$$|x_p(t)| = A_0 = \frac{f_0}{\sqrt{\left(\omega^2 - \omega_{dr}^2\right)^2 + (2\zeta\omega\omega_{dr})^2}}$$

$$= \frac{10}{\left\{[(132)^2 - (132)^2]^2 + [2(0.0085)(132)(132)]^2\right\}^{1/2}}$$

$$= \frac{10}{2(0.0085)(132)^2}$$

$$= 0.034 \text{ m}$$

If the driving frequency is changed to 125 rad/s, the amplitude becomes

$$\frac{10}{\left\{[(132)^2 - (125)^2]^2 + [2(0.0085)(132)(125)]^2\right\}^{1/2}} = 0.005 \text{ m}$$

So a slight change in the driving frequency from near resonance at 132 rad/s to 125 rad/s (about 5%) causes an order-of-magnitude change in the amplitude of the steady-state response.

$\square$

It is important to note that resonance is defined to occur when $\omega_{dr} = \omega$ (i.e., when the driving frequency becomes equal to the undamped natural frequency). This also corresponds with a phase shift of $90°(\pi/2)$. Resonance does not, however, exactly correspond with the value of ω_{dr} at which the peak value of the steady-state response occurs. This can be seen by the simple calculation in the following example.

Example 2.2.2

Derive equation (2.30) for the normalized magnitude and calculate the value of $r = \omega_{dr}/\omega$ for which the amplitude of the steady-state response takes on its maximum value.

Solution   From equation (2.28) the magnitude of the steady-state response is

$$A_0 = \frac{f_0}{\sqrt{\left(\omega^2 - \omega_{dr}^2\right)^2 + (2\zeta\omega\omega_{dr})^2}} = \frac{F_0/m}{\sqrt{\left(\omega^2 - \omega_{dr}^2\right)^2 + (2\zeta\omega\omega_{dr})^2}}$$

Factoring ω^2 out of the denominator and recalling that $\omega^2 = k/m$ yields

$$A_0 = \frac{F_0/m}{\omega^2\sqrt{\left[1 - \left(\frac{\omega_{dr}}{\omega}\right)^2\right]^2 + \left[2\zeta\left(\frac{\omega_{dr}}{\omega}\right)\right]^2}} = \frac{F_0/k}{\sqrt{(1 - r^2)^2 + (2\zeta r)^2}}$$

where $r = \omega_{dr}/\omega$. Dividing both sides by F_0/k yields equation (2.30). The maximum value of A_0 will occur where the first derivative of $A_0 k/F_0$ vanishes, that is,

$$\frac{d}{dr}\left(\frac{A_0 k}{F_0}\right) = \frac{d}{dr}\left\{\left[(1 - r^2)^2 + (2\zeta r)^2\right]^{-1/2}\right\} = 0$$

Thus

$$r_{\text{peak}} = \sqrt{1 - 2\zeta^2} = \frac{\omega_{dr}}{\omega}$$

defines the value of the driving frequency at which the peak value of the magnitude occurs. This holds only for underdamped systems for which $\zeta < 1/\sqrt{2}$. Otherwise, the magnitude does not have a maximum value, or peak for any value of $\omega_{dr} > 0$ because $\sqrt{1 - 2\zeta^2}$ becomes an imaginary number for values of ζ larger than $1/\sqrt{2}$. Note also that this peak occurs a little to the left of, or before, resonance ($r = 1$) since

$$r_{\text{peak}} = \sqrt{1 - 2\zeta^2} < 1$$

This can be seen in both Figures 2.5 and 2.6. The value of the magnitude at r_{peak} is

$$\frac{A_0 k}{F_0} = \frac{1}{2\zeta\sqrt{1 - \zeta^2}}$$

which is obtained simply by substituting $r_{\text{peak}} = \sqrt{1 - 2\zeta^2}$ into the expression for the normalized magnitude $A_0 k/F_0$.

□

Note that for damped systems resonance is usually defined, as in the undamped case, by $r = 1$ or $\omega = \omega_{dr}$. However, this condition does not define precisely the peak value of the magnitude of the steady-state response as defined by equation (2.30) and as plotted in Figure 2.6. This is the point of Example 2.2.2, which illustrates that the maximum value of $A_0 k/F_0$ occurs at $r = \sqrt{1 - 2\zeta^2}$ if $0 \leq \zeta < 1/\sqrt{2}$ and at $r = 0$ if $\zeta > 1/\sqrt{2}$. For the small damping case ($\zeta < 1/\sqrt{2}$), the value of the driving frequency corresponding to the maximum value of $A_0 k/F_0$ is called the *peak frequency*, denoted ω_p, which has the value derived above:

$$\omega_p = \omega\sqrt{1 - 2\zeta^2} \qquad \text{for} \quad 0 \leq \zeta \leq \frac{1}{\sqrt{2}}$$

Note that as the damping decreases, ω_p approaches ω, resulting in the usual undamped resonance condition. As ζ increased from zero, the curves in Figure 2.6 have peaks that occur farther and farther away from the vertical line $r = 1$. Eventually, the damping ratio increases past the value $1/\sqrt{2}$ and the largest value of $A_0 k/F_0$ occurs at $r = 0$. In many applications ζ is small, so that the value $\sqrt{1 - 2\zeta^2}$ is very close to 1. Hence the undamped resonance condition $\omega_{dr} = \omega$ (i.e., $r = 1$) is often used for resonance in the (lightly) damped case as well. As an example, for $\zeta = 0.1$, a system with an undamped natural frequency of 200 Hz would have a peak value of 198 Hz, which is less than a 1% error (i.e., $r = 0.9899$ instead of $r = 1$). Hence in practice, the value of the frequency corresponding to the peak is often taken to be simply the natural frequency.

2.3 BASE EXCITATION

Often, machines or parts of machines are harmonically excited through elastic mountings, which may be modeled by springs and dashpots. For example, an automobile suspension system is excited harmonically by a road surface through a shock absorber which may be modeled by a linear spring in parallel with a viscous damper. Other examples are the rubber motor mounts which separate an automobile engine from its frame or an airplane's engine from its wing or tail section. Such systems can be modeled by considering the system to be excited by the motion of its support. This forms the *base excitation* or *support motion* problem modeled in Figure 2.7.

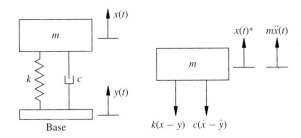

Figure 2.7 Base excitation problem models the motion of an object of mass m as being excited by a prescribed harmonic displacement acting through the spring and damper.

Summing the relevant forces on the mass m, Figure 2.7 yields (i.e., the inertial force $m\ddot{x}$ is equal to the sum of the two forces acting on m and the gravitational force is balanced against the static deflection of the spring as before).

$$m\ddot{x} + c(\dot{x} - \dot{y}) + k(x - y) = 0 \tag{2.31}$$

For the base excitation problem it is assumed that the base moves harmonically, that is, that

$$y(t) = Y \sin \omega_b t \tag{2.32}$$

where Y denotes the amplitude of the base motion and ω_b represents the frequency of the base oscillation. Substitution of $y(t)$ from equation (2.32) into the equation of motion given in (2.31) yields, after some rearrangement,

$$m\ddot{x} + c\dot{x} + kx = cY\omega_b \cos \omega_b t + kY \sin \omega_b t \tag{2.33}$$

This can be thought of as a spring–mass damper system with two harmonic inputs. The expression is very similar to the problem stated by equation (2.18) for the forced harmonic response of a damped system with $F_0 = cY\omega_b$ and $\omega_b = \omega_{dr}$, except for the "extra" forcing term $kY \sin \omega_b t$. The solution approach is to use the linearity of the equation of motion and realize that the particular solution of equation (2.33) will be the sum of the particular solution obtained by assuming an input force of $cY\omega_b \cos \omega_b t$, denoted by $x_p^{(1)}$, and the particular solution obtained by assuming an input force of $kY \sin \omega_b t$, denoted by $x_p^{(2)}$.

Calculating these particular solutions follows directly from the calculation made in Section 2.2. Dividing equation (2.33) by m and using the definitions of damping ratio and natural frequency yields

$$\ddot{x} + 2\zeta\omega\dot{x} + \omega^2 x = 2\zeta\omega\omega_b Y \cos \omega_b t + \omega^2 Y \sin \omega_b t \tag{2.34}$$

Thus substituting $f_0 = 2\zeta\omega\omega_b Y$ into equation (2.28) yields that the particular solution $x_p^{(1)}$ is

$$x_p^{(1)} = \frac{2\zeta\omega\omega_b Y}{\sqrt{(\omega^2 - \omega_b^2)^2 + (2\zeta\omega\omega_b)^2}} \cos(\omega_b t - \phi_1) \tag{2.35}$$

where

$$\phi_1 = \tan^{-1} \frac{2\zeta\omega\omega_b}{\omega^2 - \omega_b^2} \tag{2.36}$$

To calculate $x_p^{(2)}$ the method of undetermined coefficients is applied again with the harmonic input $\omega^2 Y \sin \omega_b t$. Following the procedures used to calculate equation (2.28) results in

$$x_p^{(2)} = \frac{\omega^2 Y}{\sqrt{(\omega^2 - \omega_b^2)^2 + (2\zeta\omega\omega_b)^2}} \sin(\omega_b t - \phi_1) \tag{2.37}$$

where the assumed particular solution is $x_p^{(2)} = A_0 \sin(\omega_b t - \phi_1)$. Here ϕ_1 is the same as given in equation (2.36) because it is independent of the amplitude of excitation (i.e., ζ, ω, and ω_b have not changed).

From the principle of linear superposition, the total particular solution is the sum of equations (2.35) and (2.37), (i.e., $x_p = x_p^{(1)} + x_p^{(2)}$). Adding solutions (2.35) and (2.37) yields

$$x_p(t) = \omega Y \left[\frac{\omega^2 + (2\zeta\omega_b)^2}{(\omega^2 - \omega_b^2)^2 + (2\zeta\omega\omega_b)^2} \right]^{1/2} \cos(\omega_b t - \phi_1 - \phi_2) \tag{2.38}$$

where

$$\phi_2 = \tan^{-1} \frac{\omega}{2\zeta\omega_b} \tag{2.39}$$

It is convenient to denote the magnitude of the particular solution, $x_p(t)$, by X so that

$$X = Y \left[\frac{1 + (2\zeta r)^2}{(1 - r^2)^2 + (2\zeta r)^2} \right]^{1/2} \tag{2.40}$$

where the frequency ratio $r = \omega_b/\omega$. Dividing this last expression by the magnitude of base motion, Y, yields

$$\frac{X}{Y} = \left[\frac{1 + (2\zeta r)^2}{(1 - r^2)^2 + (2\zeta r)^2} \right]^{1/2} \tag{2.41}$$

which expresses the ratio of the maximum response magnitude to the input displacement magnitude. This ratio is called the *displacement transmissibility* and is used to describe how motion is transmitted from the base to the mass as a function of the frequency ratio ω_b/ω. This ratio is plotted in Figure 2.8. Note that near $r = \omega_b/\omega = 1$, or resonance, the maximum amount of base motion is transferred to displacement of the mass.

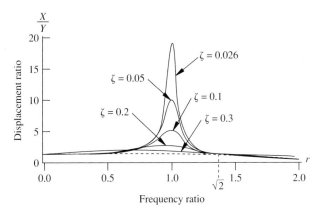

Figure 2.8 Displacement transmissibility as a function of the frequency ratio illustrating how the dimensionless deflection X/Y varies as the frequency of the base motion increases for several different damping ratios.

Note from the figure that for $r < \sqrt{2}$ the transmissibility ratio is greater than 1, indicating that for these values of the system's parameters (ω) and base frequency (ω_b), the motion of the mass is an amplification of the motion of the base. Notice also that for a given value of r the value of the damping ratio ζ determines the level of amplification. Specifically, larger ζ yields smaller transmissibility ratios. For $r > \sqrt{2}$ the transmissibility ratio is less than 1 and the motion of the mass is smaller in amplitude than that of the base for all ζ (but increases for increasing ζ).

Another quantity of interest in the base excitation problem is the force transmitted to the mass as the result of a harmonic displacement of the base. The force transmitted to the mass is done so through the spring and damper. Hence the force transmitted to the mass is the sum of the force in the spring and the force in the damper, or

$$F(t) = k(x - y) + c(\dot{x} - \dot{y}) \tag{2.42}$$

This force must balance the inertial force of the mass m; thus

$$F(t) = -m\ddot{x}(t) \tag{2.43}$$

In the steady-state, the solution for x is given by equation (2.38). Differentiating equation (2.38) twice and substituting into equation (2.43) yields

$$F(t) = m\omega_b^2 \omega Y \left[\frac{\omega^2 + (2\zeta\omega_b)^2}{\left(\omega^2 - \omega_b^2\right)^2 + (2\zeta\omega\omega_b)^2} \right]^{1/2} \cos(\omega_b t - \phi_1 - \phi_2) \tag{2.44}$$

Again using the frequency ratio r, this becomes

$$F(t) = F_T \cos(\omega_b t - \phi_1 - \phi_2) \tag{2.45}$$

where the magnitude of the transmitted force, F_T, is given by

$$F_T = kYr^2 \left[\frac{1 + (2\zeta r)^2}{(1 - r^2)^2 + (2\zeta r)^2} \right]^{1/2} \tag{2.46}$$

Equation (2.46) is used to define *force transmissibility* by forming the ratio

$$\frac{F_T}{kY} = r^2 \left[\frac{1 + (2\zeta r)^2}{(1 - r^2)^2 + (2\zeta r)^2} \right]^{1/2} \tag{2.47}$$

This force transmissibility ratio, F_T/kY, expresses a dimensionless measure of how displacement in the base of amplitude Y results in a force magnitude applied to the mass. Note from equations (2.45) and (2.38) that the force transmitted to the mass is in phase with the displacement of the mass. Figure 2.9 illustrates the force transmissibility as a function of the frequency ratio for four values of the damping ratio. Note that unlike the displacement transmissibility, the force transmitted does not necessarily fall off for $r > \sqrt{2}$. In fact as the damping increases, the force transmitted increases dramatically for $r > \sqrt{2}$.

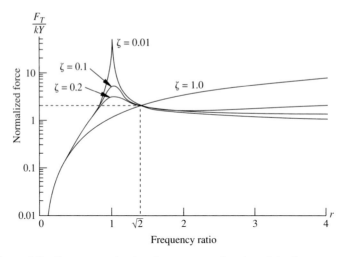

Figure 2.9   Force transmitted to the mass as a function of the frequency ratio illustrating how the dimensionless force ratio varies as the frequency of the base motion increases for four different damping ratios ($\zeta = 0.01, 0.1, 0.2$ and 1.0).

The formulas for transmissibility of force and displacement are very useful in the design of systems to provide protection from unwanted vibration. This is discussed in detail in Section 5.2 on vibration isolation, where the transmissibility ratio is derived for a *fixed* base and compared to the development here in Window 5.1. The following example illustrates some practical values of transmissibility for the base excitation problem.

Example 2.3.1

A very common example of base motion is the single-degree-of-freedom model of an automobile driving over a rough road or an airplane taxiing over a rough runway is indicated
in Figure 2.10. The road is approximated as sinusoidal in cross section providing a base
motion displacement of

$$y(t) = (0.01 \text{ m}) \sin \omega_b t$$

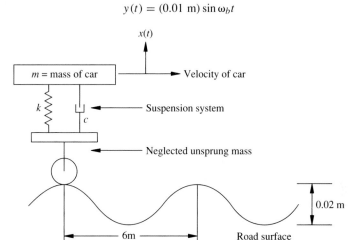

Figure 2.10   Simple model of a vehicle traveling with constant velocity on a
rough surface that is approximated as a sinusoid.

where

$$\omega_b = v \text{ (km/h)} \left(\frac{1}{0.006 \text{ km}} \right) \left(\frac{\text{hour}}{3600\text{s}} \right) \left(\frac{2\pi \text{ rad}}{\text{cycle}} \right) = 0.2909v$$

in rad/s, where v denotes the vehicle's velocity in km/h. Thus the vehicle's speed determines
the frequency of the base motion. Determine the effect of speed on the amplitude of
displacement of the automobile as well as the effect of the value of the car's mass. Assume
that the suspension system provides an equivalent stiffness of 4×10^5 N/m and damping of
20×10^3 N · s/m.

Solution   First, to determine the effect of speed on the amplitude of the vehicle's motion,
note that from the previous calculations, ω_b, and hence r, vary linearly with the car's velocity.
Hence the deflection ratio versus velocity curve will be much like the curve of Figure 2.8.
Some sample values can be calculated from equation (2.40). At 20 km/h, $\omega_b = 5.818$. If
the car is a small or sports car, its mass might be around 1007 kg (e.g., Chevrolet Beretta),
so that the natural frequency is

$$\omega = \sqrt{\frac{4 \times 10^5 \text{ N/m}}{1007 \text{ kg}}} = 19.93 \text{ rad/s}$$

so that $r = 5.818/19.93 = 0.292$ and

$$\zeta = \frac{c}{2\sqrt{km}} = \frac{20,000 \text{ N} \cdot \text{s/m}}{2\sqrt{(4 \times 10^5 \text{ N/m})(1007 \text{ kg})}} = 0.498$$

Equation (2.40) then yields that the deflection experienced by the car will be

$$X = (0.01 \text{ m})\sqrt{\frac{1 + [2(0.498)(0.292)]^2}{[1 - (0.292)^2]^2 + [2(0.498)(0.292)]^2}} = 0.0108 \text{ m}$$

This means that a 1-cm bump in the road is transmitted into a 1.1-cm "bump" experienced by the chassis and subsequently transmitted to the occupants. Hence the suspension system amplifies the rough road bumps in this circumstance.

Table 2.1 lists several different values of the vehicle displacement for two different vehicles traveling at four different speeds over the same 1-cm bump. Car 1 with frequency ratio r_1 is a 1007-kg sports car, while car 2 is a 1585 kg sedan (such as a Buick Park Avenue) with frequency ratio r_2. The same suspension system was used on both cars to illustrate the need to design suspension systems based on a given vehicle's specifications (see Chapter 5). Note that with higher speed, less vibration is experienced by the occupants of the larger car. Also, notice that the suspension system parameters chosen (k and c) work better in general for the larger car except at very low speeds.

TABLE 2.1 COMPARISON OF CAR VELOCITY, FREQUENCY, AND DISPLACEMENT FOR TWO DIFFERENT CARS

Speed (km/h)	ω_b	r_1	r_2	x_1 (cm)	x_2 (cm)
20	5.817	0.29	0.36	1.08	1.14
80	23.271	1.17	1.46	1.26	0.94
100	29.088	1.46	1.83	0.96	0.64
150	43.633	2.19	2.74	0.55	0.35

□

Example 2.3.2

A large rotating machine causes the floor of a factory to oscillate sinusoidally. A punch press is to be mounted on the same floor. The displacement of the floor at the point where the punch press is to be mounted is measured to be $y(t) = 0.1 \sin \omega_b t$ (cm). Using the base support model of this section, calculate the maximum force transmitted to the punch press at resonance if the machine is mounted on a rubber fitting of stiffness, $k = 40,000$ N/m; damping, $c = 900$ N · s/m; and mass, $m = 3000$ kg.

Solution The force transmitted to the machine is given by equation (2.47). At resonance $r = 1$, so that equation (2.47) becomes

$$\frac{F_T}{kY} = \left[\frac{1 + (2\zeta)^2}{(2\zeta)^2}\right]^{1/2}$$

or

$$F_T = \frac{kY}{2\zeta}(1 + 4\zeta^2)^{1/2}$$

From the definition of ζ and the values given above for m, c, and k,

$$\zeta = \frac{c}{2\sqrt{km}} = \frac{900}{2[(40,000)(3000)]^{1/2}} \cong 0.04$$

From the measured excitation $Y = 0.001$ m, so that

$$F_T = \frac{kY}{2\zeta}(1 + 4\zeta^2)^{1/2} = \frac{(40,000 \text{ N/m})(0.001 \text{ m})}{2(0.04)}[1 + 4(0.04)^2]^{1/2}$$

$$= 501.6 \text{ N}$$

☐

2.4 ROTATING UNBALANCE

A very common source of troublesome vibration is rotating machinery. Many machines and devices have rotating components, usually driven by electric motors. Small irregularities in the distribution of the rotating mass can cause substantial vibration. This is called a rotating unbalance. A schematic of such a rotating unbalance of mass m_0, a distance e from the center of rotation, is given in Figure 2.11. The frequency of rotation of the machine is ω_r. Assuming that the machine rotates with a constant frequency ω_r, the x component of motion of the mass m_0 is $x_r = e \sin \omega_r t$ and the reaction force, F_r, generated by the rotating mass m_0 (unbalance) has a component in the x direction of

$$F_r = m_0\ddot{x}_r = em_0\frac{d^2}{dt^2}(\sin \omega_r t) = -em_0\omega_r^2 \sin \omega_r t \tag{2.48}$$

which acts on the machine mass, m. The forces in the horizontal direction are canceled by the guides and not considered here. Dividing the machine mass into two parts and

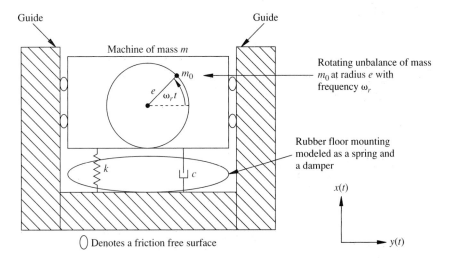

Figure 2.11   Model of a machine with rotating unbalance.

summing forces in the x direction yields

$$(m - m_0)\ddot{x} + m_0 \frac{d^2}{dt^2}(x + e \sin \omega_r t) = -kx - c\dot{x}$$

or (2.49)

$$m\ddot{x} + c\dot{x} + kx = m_0 e \omega_r^2 \sin \omega_r t$$

after rearranging the terms. Note that $(x + e \sin \omega_r t)$ is the displacement of m_0 from the static equilibrium position (i.e., the displacement of m_0 added to that of m). Equation (2.49) is similar to equation (2.18) with $F_0 = m_0 e \omega_r^2$, with the exception of the phase shift of the forcing function (i.e., $\sin \omega_r t$ instead of $\cos \omega_{dr} t$). The solution procedure is the same and results in a particular solution of the form

$$x_p(t) = X \sin(\omega_r t - \phi) \tag{2.50}$$

where $(r = \omega_r/\omega)$, as before,

$$X = \frac{m_0 e}{m} \frac{r^2}{\sqrt{(1 - r^2)^2 + (2\zeta r)^2}} \tag{2.51}$$

and

$$\phi = \tan^{-1} \frac{2\zeta r}{1 - r^2} \tag{2.52}$$

These last two expressions yield the magnitude and phase of the motion of the mass, m, due to the rotating unbalance of mass m_0.

The magnitude of the steady-state displacement, X, as a function of the rotating speed (frequency) is examined by plotting the dimensionless displacement magnitude $mX/m_0 e$ versus r as indicated in Figure 2.12 for various values of the damping ratio ζ. Note that equation (2.51) is similar to the magnitude analyzed in Example 2.2.2. From the form of the denominator, which is identical to that of Example 2.2.2, it is observed that the maximum deflection is less than or equal to 1 for any system with $\zeta > 1$. This indicates that the increase in amplification of the amplitude caused by the unbalance can be eliminated by increasing the damping in the system. However, large damping is not always practical. Note from Figure 2.12 that the magnitude of the dimensionless displacement approaches unity if r is large. Hence if the running frequency ω_r is such that $r \gg 1$, the effect of the unbalance is limited. For large values of r, all the magnitude curves for each value of ζ approach unity, so that the choice of damping coefficient for large r is not important. These results can be obtained from examining the plots of Figure 2.12 or from investigating the limit of $mX/m_0 e$ as r goes to infinity. These observations have important implications in the design of rotating machines.

The rotating unbalanced model can also be used to explain the behavior of an automobile with an out of balance wheel and tire. Here ω_r is determined by the speed of the car and e by the diameter of the wheel. The deflection x_p can be felt through the steering mechanism as shaking of the steering wheel. This usually only happens at a certain speed (near $r = 1$). As the driver increases or decreases speed, the shaking effect

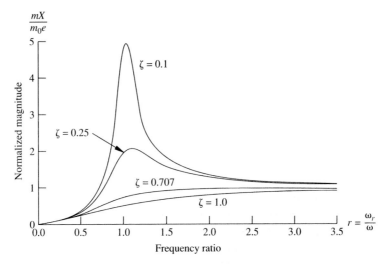

Figure 2.12   Magnitude of the dimensionless displacement versus frequency ratio caused by a rotating unbalance of mass m_0 and radius e.

in the steering wheel reduces. This change in speed is equivalent to operating conditions on either side of the peak in Figure 2.12.

Example 2.4.1

Consider a machine with rotating unbalance as described in Figure 2.11. At resonance the maximum deflection is measured to be 0.1 m. From a free decay of the system, the damping ratio is estimated to be $\zeta = 0.05$. From manufacturing data, the out-of-balance mass is estimated to be 10%. Estimate the radius e and hence the approximate location of the unbalanced mass. Also determine how much mass should be added (uniformly) to the system to reduce the deflection at resonance to 0.01 m.

Solution   At resonance, $r = 1$ so that

$$\frac{mX}{m_0\,e} = \frac{1}{2(0.05)}$$

Hence

$$(10)\frac{(0.1 \text{ m})}{e} = \frac{1}{0.1} = 10$$

so that $e = 0.1$ m Again at resonance

$$\frac{m}{m_0}\left(\frac{X}{0.1 \text{ m}}\right) = 10$$

If it is desired to change m, say by Δm, so that $X = 0.01$ m, the resonance expression above becomes

$$\frac{m + \Delta m}{m_0}\left(\frac{0.01}{0.1}\right) = 10 \qquad \frac{m + \Delta m}{m_0} = 100$$

which implies that $\Delta m = 9m$. Thus the total mass must be increased by a factor of 9 in order to reduce the deflection to a centimeter.

$\square$

Example 2.4.2

Rotating unbalance is also important in rotorcraft such as helicopters and prop planes. The tail rotor of a helicopter (the small rotor rotating in a vertical plane at the back of a helicopter used to provide yaw control and torque balance) as sketched in Figure 2.13 can be modeled as a rotating unbalance problem discussed in this section with stiffness $k = 1 \times 10^5$ N/m (provided by the tail section) and equivalent mass of 80 kg, 20 kg of which is the mass of the rotor itself. Suppose that a 500 g mass is stuck on one of the blades at a distance of 15 cm from the axis of rotation. Calculate the magnitude of the deflection of the tail section of the helicopter as the tail rotor rotates at 1500 rpm. Assume a damping ratio of 0.01. At what rotor speed is the deflection a maximum? Calculate the maximum deflection.

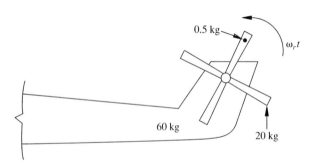

Figure 2.13   Schematic of a helicopter tail section illustrating a tail rotor. The tail rotor provides a "counterclockwise" thrust (when looking at the top of the helicopter) to counteract the "clockwise" thrust created by the main rotor, which provides lift and horizontal motion. An out-of-balance rotor can cause damaging vibrations and limit the helicopter's performance.

Solution   The natural frequency of the rotor system is (m_s is the mass of the rotor tail system providing the stiffness k)

$$\omega = \sqrt{\frac{k}{m_s}} = \sqrt{\frac{10^5 \text{ N/m}}{80 \text{ kg}}} = 35.4 \text{ rad/s}$$

The frequency of rotation is

$$\omega_r = 1500 \text{ rpm} = 1500 \frac{\text{cycles}}{\text{min}} \frac{\text{min}}{60 \text{ s}} \frac{2\pi \text{ rad}}{\text{cycles}} = 157 \text{ rad/s}$$

Hence, the frequency ratio r becomes

$$r = \frac{\omega_r}{\omega} = \frac{157 \text{ rad/s}}{35.4 \text{ rad/s}} = 4.5$$

With $r = 4.5$ and $\zeta = 0.01$, equation (2.51) yields that the magnitude of oscillation of the tail rotor is

$$X = \frac{m_0 e}{m} \frac{r^2}{\sqrt{(1 - r^2)^2 + (2\zeta r)^2}}$$

$$= \frac{(0.5 \text{ kg})(0.15 \text{ m})}{20 \text{ kg}} \frac{(4.5)^2}{\sqrt{[1 - (4.5)^2]^2 + [2(0.01)(4.5)]^2}}$$

$$\sim 3.9 \text{ mm}$$

The maximum deflection occurs at about $r = 1$, or

$$\omega_r = \omega = 35.4 \text{ rad/s} = 35.4 \frac{\text{rad}}{\text{s}} \frac{\text{cycles}}{2\pi \text{ rad}} \frac{60 \text{ s}}{\text{min}} = 338 \text{ rpm}$$

In this case the (maximum) deflection becomes

$$X = \frac{(0.5 \text{ kg})(0.15)}{20 \text{ kg}} \frac{1}{2(0.01)} = 18.75 \text{ cm}$$

which represents a large unacceptable deflection of the rotor. Thus the tail rotor should not be allowed to rotate at 338 rpm.

□

More on the special nature of rotating systems is discussed in Section 5.5.

2.5 ALTERNATIVE REPRESENTATIONS

As you may recall from the theory of differential equations there are a variety of methods useful for calculating solutions of a spring–mass–damper system excited by a harmonic force as described by equation (2.18). In Section 2.2 the method of undetermined coefficients was used. In this section, three other approaches to solve this problem are discussed: a geometric approach, a frequency response approach, and a transform approach.

The geometric approach consists of solving equation (2.18) by treating each force as a vector. Recall that x_p, $\dot{x}_p$ and $\ddot{x}_p$ will each be 90° out of phase with each other. Each of these are plotted in Figure 2.14 for the assumed solution $x_p = X \cos(\omega_{dr} t - \phi)$, $\dot{x}_p = \omega_{dr} X \cos(\omega_{dr} t - \phi + 90°)$, and $\ddot{x}_p = -\omega_{dr}^2 X \cos(\omega_{dr} t - \phi)$. Adding these three quantities as vectors indicates that X can be solved in terms of F_0 by combining the sides of the right triangle ABC to yield

$$F_0^2 = (k - m\omega_{dr}^2)^2 X^2 + (c\omega_{dr})^2 X^2 \tag{2.53}$$

or

$$X = \frac{F_0}{\sqrt{\left(k - m\omega_{dr}^2\right)^2 + (c\omega_{dr})^2}} \tag{2.54}$$

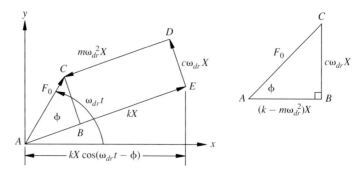

Figure 2.14   Graphical representation of the solution of equation (2.18).

and

$$\phi = \tan^{-1} \frac{c\omega_{dr}}{k - m\omega_{dr}^2} \qquad (2.55)$$

Substituting in the values for $F_0 = mf_0$ and $c = 2m\omega\zeta$ illustrates that this solution is identical with equation (2.28) derived by the method of undetermined coefficients. Note from the figure that at resonance $\omega_{dr}^2 = k/m$ and lines CD and AE become the same length. Thus the angle ϕ becomes a right angle so that at resonance the phase shift is 90°.

The graphical method of solving equation (2.18) is more illustrative than useful, as it is difficult to extend to other forcing functions or to more complicated problems. It is presented here because the method potentially helps clarify and illustrate the forced vibration of a simple single-degree-of-freedom system. An alternative method which is similar to the graphical approach is to treat the solution of equation (2.18) as a complex function. This leads to a frequency response description of forced harmonic motion and is more useful for complicated problems involving many degrees of freedom. Complex functions are reviewed in Appendix A.

Euler's formula for trigonometric functions relates the exponential function to harmonic motion by the complex relation

$$Ae^{j\omega_{dr}t} = A\cos\omega_{dr}t + (A\sin\omega_{dr}t)j \qquad (2.56)$$

where $j = \sqrt{-1}$. Thus $Ae^{j\omega_{dr}t}$ is a complex function with a real part ($A\cos\omega_{dr}t$) and an imaginary part ($A\sin\omega_{dr}t$). Appendix A reviews complex numbers and functions. With this notation in mind, $Ae^{j\omega_{dr}t}$ represents a harmonic function and can be used to discuss forced harmonic motion by rewriting equation (2.18) as the complex equation

$$m\ddot{x}(t) + c\dot{x}(t) + kx(t) = F_0 e^{j\omega_{dr}t} \qquad (2.57)$$

Here, the real part of the complex solution corresponds to the physical solution $x(t)$. This representation is extremely useful in solving multiple-degree-of-freedom systems (Chapter 4), as well as in understanding vibration measurement systems (Chapter 7).

This method proceeds by assuming that the complex particular solution of equation (2.57) is of the form

$$x_p(t) = Xe^{j\omega_{dr}t} \tag{2.58}$$

where X is again a complex-valued constant to be determined. Substitution of this into equation (2.57) yields

$$(-\omega_{dr}^2 m + cj\omega_{dr} + k)Xe^{j\omega_{dr}t} = F_0 e^{j\omega_{dr}t} \tag{2.59}$$

Since $e^{j\omega_{dr}t}$ is never zero, it can be canceled and this last expression can be rewritten as

$$X = \frac{F_0}{(k - m\omega_{dr}^2) + (c\omega_{dr})j} = H(j\omega_{dr})F_0 \tag{2.60}$$

The quantity $H(j\omega_{dr})$, defined by

$$H(j\omega_{dr}) = \frac{1}{(k - m\omega_{dr}^2) + (c\omega_{dr}j)} \tag{2.61}$$

is called the (complex) *frequency response function*. Following the rules for manipulating complex numbers (i.e., multiplying by the complex conjugate over itself and taking the modulus of the result as outlined in Appendix A) yields

$$X = \frac{F_0}{\left[(k - m\omega_{dr}^2)^2 + (c\omega_{dr})^2\right]^{1/2}} e^{-j\phi} \tag{2.62}$$

where

$$\phi = \tan^{-1} \frac{c\omega_{dr}}{(k - m\omega_{dr}^2)} \tag{2.63}$$

Substituting the value for X into equation (2.58) yields the solution

$$x_p(t) = \frac{F_0}{\left[(k - m\omega_{dr}^2)^2 + (c\omega_{dr})^2\right]^{1/2}} e^{j(\omega_{dr}t - \phi)} \tag{2.64}$$

The real part of this expression corresponds to the solution given in equation (2.28) obtained by the method of undetermined coefficients. The complex exponential approach for obtaining the forced harmonic response corresponds to the graphical approach described in Figure 2.14 by labeling the x axis as the real part of $e^{j\omega_{dr}t}$ and the y-axis as the complex part.

Next consider using the Laplace transform (see Appendix B and Section 3.4 for a review) approach to solve for the particular solution of equation (2.18). The Laplace transform method is a powerful approach that can be used for a variety of forcing functions (see Section 3.4) and can be readily applied to multiple-degree-of-freedom systems. Taking the Laplace transform of equation (2.18), assuming that the initial conditions are zero, yields

$$(ms^2 + cs + k)X(s) = \frac{F_0 s}{s^2 + \omega_{dr}^2} \tag{2.65}$$

where s is the complex transform variable and $X(s)$ denotes the Laplace transform of the unknown function $x(t)$. Solving algebraically for the unknown function $X(s)$ yields

$$X(s) = \frac{F_0 s}{(ms^2 + cs + k)(s^2 + \omega_{dr}^2)} \tag{2.66}$$

which represents the transformed solution. To calculate the inverse Laplace transform of $X(s)$, the right side of equation (2.66) can be looked up in a table of Laplace transform pairs, or the method of partial fractions can be used to reduce the right-hand side of equation (2.66) to simpler quantities for which the inverse Laplace transform is known. The solution obtained by the inversion procedure is, of course, equivalent to the solution given in (2.28) and again in (2.64). This solution technique is discussed in more detail in Section 3.4.

Of particular use is the frequency response function defined by equation (2.61). This function is related to the Laplace transform for a vibrating system. Consider equation (2.18) and its Laplace transform,

$$(ms^2 + cs + k)X(s) = F(s) \tag{2.67}$$

where $F(s)$ denotes the Laplace transform of the driving function, i.e., the right-hand side of (2.18). Manipulating equation (2.67) yields

$$\frac{X(s)}{F(s)} = \frac{1}{ms^2 + cs + k} = H(s) \tag{2.68}$$

which expresses the ratio of the Laplace transform of the output (response) to the Laplace transform of the input (driving force) for the case of zero initial conditions. This ratio, denoted $H(s)$, is called the *transfer function* of the system and provides an important tool for vibration analysis, design, and measurement as discussed in the remaining chapters.

Recall that the Laplace transform variable s is a complex number. If the value of s is restricted to lie along the imaginary axis in the complex plane (i.e., if $s = j\omega_{dr}$), the transfer function becomes

$$H(j\omega_{dr}) = \frac{1}{k - m\omega_{dr}^2 + c\omega_{dr}j} \tag{2.69}$$

which, upon comparison with equation (2.61), is the frequency response function of the system. Hence the frequency response function of the system is the transfer function of the system evaluated along $s = j\omega_{dr}$. Both the transfer function and the frequency response function are used in later chapters.

This section presented three alternative methods for calculating the particular solution for a harmonically excited system. Each was shown to yield the same result. The concepts presented in these solution techniques are generalized and used for more complicated problems in later chapters.

2.6 MEASUREMENT DEVICES

An important application of the forced harmonic vibration analysis and base excitation problem presented in Sections 2.2 and 2.3 is in the design of devices used to measure vibration. A device that changes mechanical motion into a voltage (or vice versa) is called a *transducer*. Several transducers are sketched in Figures 2.15 to 2.17. Each of these devices changes mechanical vibration into a voltage proportional to acceleration.

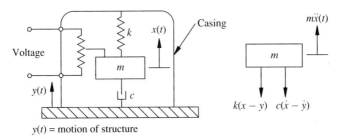

$y(t)$ = motion of structure

Figure 2.15   Schematic of an accelerometer mounted on a structure. The insert indicates the relevant forces acting on the mass m. The force $k(x - y)$ is actually parallel to the damping force because they are both connected to ground.

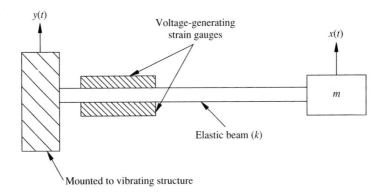

Figure 2.16   Schematic of a seismic accelerometer made of a small beam.

Referring to the accelerometer of Figure 2.15, a balance of forces on the mass m yields

$$m\ddot{x} = -c(\dot{x} - \dot{y}) - k(x - y) \tag{2.70}$$

Here it is assumed that the base which is mounted to the structure being measured undergoes a motion of $y = Y \cos \omega_b t$, (i.e., that the structure being measured is undergoing simple harmonic motion). The motion of the accelerometer mass relative to the base denoted, $z(t)$, is defined by

$$z(t) = x(t) - y(t) \tag{2.71}$$

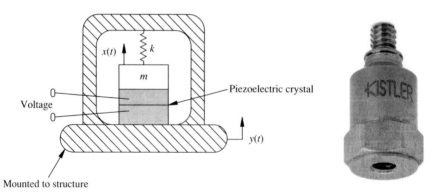

Figure 2.17 Schematic of a piezoelectric accelerometer and a photograph of a commercially available version. (Kistler Quartz K-Shear Accelerometer with internal Piezotron®Amplifier, courtesy of Kistler Instrument Corporation, Amherst, N.Y.)

Equation (2.70) can then be written in terms of the relative displacement $z(t)$. Hence equation (2.70) becomes the familiar expression

$$m\ddot{z} + c\dot{z} + kz = m\omega_b^2 Y \cos \omega_b t \qquad (2.72)$$

This expression has exactly the same form as equation (2.18). Thus the solution in steady state will be of the same form as given by equation (2.28), or

$$z(t) = \frac{\omega_b^2 Y}{\sqrt{(\omega^2 - \omega_b^2)^2 + (2\zeta\omega\omega_b)^2}} \cos\left(\omega_b t - \tan^{-1}\frac{2\zeta\omega\omega_b}{\omega^2 - \omega_b^2}\right) \qquad (2.73)$$

The difference between equations (2.28) and (2.73) is that the latter is for the relative displacement (z) and the former is for the absolute displacement (x).

Further manipulation of the magnitude of equation (2.73) yields

$$\frac{Z}{Y} = \frac{r^2}{\sqrt{(1 - r^2)^2 + (2\zeta r)^2}} \qquad (2.74)$$

for the amplitude ratio as a function of the frequency ratio $r = \omega_b/\omega$, and

$$\psi = \tan^{-1}\frac{2\zeta r}{1 - r^2} \qquad (2.75)$$

for the phase shift.

Consider the plot of the magnitude of Z/Y versus the frequency ratio as given in Figure 2.18. Note that for larger values of r (i.e., for $r \geq 3$) the magnitude ratio approaches unity, so that $Z/Y = 1$ or $Z = Y$ and the relative displacement and the displacement of the base have the same magnitude. Hence the accelerometer of Figure 2.15 can be used to measure harmonic base displacement if the frequency of the base displacement is at least three times the accelerometer's natural frequency.

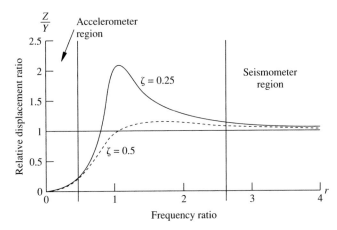

Figure 2.18 Magnitude versus frequency of the relative displacement for a transducer used to measure acceleration and for seismic measurements.

Next, consider the equation of the system in Figure 2.15 for the case when r is small. Factoring ω^2 out of the denominator, equation (2.73) can be written as

$$\omega^2 z(t) = \frac{1}{\sqrt{(1-r^2)^2 + (2\zeta r)^2}} \omega_b^2 Y \cos(\omega_b t - \phi) \tag{2.76}$$

Since $y = Y \cos(\omega_b t - \phi)$, the last term is recognized to be $-\ddot{y}(t)$, so that

$$\omega^2 z(t) = \frac{-1}{\sqrt{(1-r^2)^2 + (2\zeta r)^2}} \ddot{y}(t) \tag{2.77}$$

This expression illustrates that for small values of r, the quantity $\omega^2 z(t)$ is proportional to the base acceleration, $\ddot{y}(t)$, since

$$\lim_{r \to 0} \frac{1}{\sqrt{(1-r^2)^2 + (2\zeta r)^2}} = 1 \tag{2.78}$$

In practice this coefficient is taken as close to 1 for any value of $r < 0.5$. This indicates that for these frequencies of base motion, the relative position $z(t)$ is proportional to the base acceleration. The effect of the accelerometer internal damping, ζ, in the constant of proportionality between the relative displacement and the base acceleration is illustrated in Figure 2.19 which consists of a plot of this constant versus the frequency ratio for a variety of values of ζ, for values of $r < 1$. Note from the figure that the curve corresponding to $\zeta = 0.7$ is closest to being constant at unity over the largest range of $r < 1$. For this curve the magnitude is relatively flat for values of r between zero and about 0.2. In fact, within this region, the curve varies from 1 by less than 0.01%. This defines the useful range of operation for the accelerometer:

$$0 < \frac{\omega_b}{\omega} < 0.2 \tag{2.79}$$

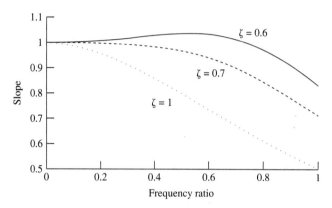

Figure 2.19 Effect of damping on the constant of proportionality between base acceleration and the relative displacement (voltage) for an accelerometer.

where ω is the natural frequency of the device. Multiplying this inequality by the device frequency yields

$$0 < \omega_b < 0.2\omega \qquad \text{or} \qquad 0 < f_m < 0.2f \qquad (2.80)$$

where f_m is the frequency to be measured by the accelerometer in hertz. For the mechanical accelerometer of Figure 2.15, the device frequency may be on the order of 100 Hz. Thus inequality (2.80) indicates that the highest frequency that can be effectively measured by the device would be 20 Hz (0.2×100).

Many structures and machines vibrate at frequencies larger than 20 Hz. The piezoelectric accelerometer design indicated in Figure 2.17 provides a device with a natural frequency of about 8×10^4 Hz (see, e.g., Kistler Instrument Company Model 8694). In this case the inequality predicts that vibration with frequency content up to 16,000 Hz can be measured. Vibration measurement is discussed in more detail in Chapter 7, where practical problems such as phase and amplitude distribution of signals measured with accelerometers is discussed.

Example 2.6.1

This example illustrates how an independent measurement of acceleration can provide a measurement of a transducer's mechanical properties. An accelerometer is used to measure the oscillation of an airplane wing caused by the plane's engine operating at 6000 rpm (628 rad/s). At this engine speed the wing is known, from other measurements, to experience 1.0 g acceleration. The accelerometer measures an acceleration of 10 m/s². If the accelerometer has a 0.01-kg moving mass and a damped natural frequency of 100 Hz (628 rad/s), the difference between the measured and the known acceleration is used to calculate the damping and stiffness parameters associated with the accelerometer.

Solution From equation (2.77), the amplitude of the measured values acceleration $|\omega^2 z(t)|$ is related to the actual values of acceleration $|\ddot{y}(t)|$ by

$$\frac{|\omega^2 z(t)|}{|\ddot{y}(t)|} = \frac{1}{\sqrt{(1 - r^2)^2 + (2\zeta r)^2}} = \frac{10 \text{ m/s}^2}{9.8 \text{ m/s}^2} = 1.02$$

Rewriting this expression yields one equation in ζ and r:

$$(1 - r^2)^2 + (2\zeta r)^2 = 0.96$$

A second expression in ζ and r can be obtained from the definition of the damped natural frequency:

$$\frac{\omega_b}{\omega_d} = \frac{\omega_b}{\omega} \frac{1}{\sqrt{1 - \zeta^2}} = r \frac{1}{\sqrt{1 - \zeta^2}} = \frac{628 \text{ rad/s}}{628 \text{ rad/s}} = 1$$

Thus $r = \sqrt{1 - \zeta^2}$, providing a second equation in ζ and r. This can be manipulated to yield $\zeta^2 = (1 - r^2)$, which when substituted with the expression above for r and ζ yields

$$\zeta^4 + 4\zeta^2(1 - \zeta^2) = 0.96$$

This is a quadratic equation in ζ^2:

$$3\zeta^4 - 4\zeta^2 + 0.96 = 0$$

This quadratic expression yields the two roots $\zeta = 0.56, 1.01$. Using $\zeta = 0.56$, the damping constant is ($\sqrt{1 - \zeta^2} = 0.83$, $\omega = \omega_d/\sqrt{1 - \zeta^2} = 758.0$ rad/s)

$$c = 2m\omega\zeta = 2(0.01)(758.0)(0.56) = 8.49 \text{ N} \cdot \text{s/m}$$

Similarly, the stiffness in the accelerometer is

$$k = m\omega^2 = (0.01)(758.0)^2 = 5745.6 \text{ N/m}$$

$\square$

2.7 COULOMB DAMPING

Damping is the term used to define the nonconservative forces acting on a spring–mass system that dissipate energy. The correct mathematical form used to model damping for a given device or material is difficult to determine. Unlike mass and stiffness, damping cannot be determined by static tests. The energy dissipation exhibited by a device may be the result of pushing air out of the way, or from electronic defects in the material of which the device is made. The model of damping force used predominantly is in the form $c\dot{x}$, largely because it simplifies solutions allowing for systematic design. This section introduces another form of damping which results from sliding friction.

A common damping mechanism occurring in machines is caused by sliding friction or dry friction and is called *Coulomb damping*. Coulomb damping is characterized by the relation

$$F_d = F_d(\dot{x}) = \begin{cases} \mu N & \dot{x} > 0 \\ 0 & \dot{x} = 0 \\ -\mu N & \dot{x} < 0 \end{cases}$$

where F_d is the dissipation force, N the normal force (see any introductory physics text), and μ the coefficient of sliding friction (or kinetic friction). Figure 2.20 is a schematic of a mass m sliding on a surface and connected to a spring of stiffness k. The frictional force F_d always opposes the direction of motion. Table 2.2 lists some measured values of the coefficient of kinetic friction for several different sliding objects. Summing forces in part (a) of Figure 2.21 in the x direction yields that (note that the mass changes direction when the velocity passes through zero)

$$m\ddot{x} + kx = \mu mg \qquad \text{for} \quad \dot{x} < 0 \tag{2.81}$$

Here the sum of the forces in the vertical direction yields the fact that the normal force N is just the weight, mg, where g is the acceleration due to gravity. In a similar fashion,

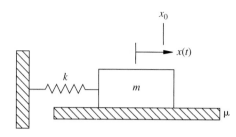

Figure 2.20   Spring–mass system sliding on a surface with kinetic coefficient of friction μ.

TABLE 2.2   APPROXIMATE COEFFICIENTS OF KINETIC FRICTION FOR VARIOUS OBJECTS SLIDING TOGETHER

Material	μ
Metal on metal (lubricated)	0.07
Wood on wood	0.2
Steel on steel (unlubricated)	0.3
Rubber on steel	1.0

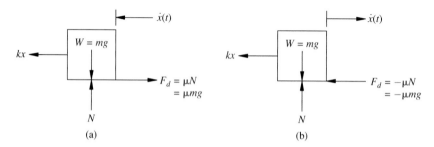

Figure 2.21   Free-body diagram of the forces acting on the sliding block system of Figure 2.20: (a) mass moving to the left ($\dot{x} < 0$); (b) mass moving to the right ($\dot{x} > 0$).

summing forces in part (b) of Figure 2.21 yields

$$m\ddot{x} + kx = -\mu mg \qquad \text{for} \quad \dot{x} > 0 \tag{2.82}$$

Since the sign of $\dot{x}$ determines the direction in which the opposing frictional force acts, equations (2.81) and (2.82) can be written as the single equation

$$m\ddot{x} + \mu mg \, \text{sgn}(\dot{x}) + kx = 0 \tag{2.83}$$

where sgn(τ) denotes the *signum function*, defined to have the value 1 for $\tau > 0$, -1 for $\tau < 0$, and 0 for $\tau = 0$. This equation cannot be solved directly using methods such as the variation of parameters or the method of undetermined coefficients. This is because equation (2.83) is a nonlinear differential equation. Rather, equation (2.83) can be solved by breaking the time intervals up into segments corresponding to the changes in direction of motion (i.e., at those time intervals separated by $\dot{x} = 0$). Alternatively, equation (2.83) can be solved numerically. A more general treatment of nonlinear vibration problems is presented in Chapter 10.

To solve equation (2.83) first assume that the initial conditions are such that the mass is moving to the left, as indicated in Figure 2.21(a). This corresponds to an initial condition of $\dot{x}(0) \leq 0$ and $x(0) = x_0$, where x_0 is to the right of the mass's equilibrium position and is large enough so that the restoring force, kx_0, is larger than the force due to static friction. Recall that the static friction coefficient is always greater than the coefficient of dynamic friction. Because of the static friction, the equilibrium position is not a single position, but rather, there are many positions for which $x(t) = 0$. The existence of multiple equilibria is also a characteristic of nonlinear systems.

With x_0 to the right of any equilibrium the mass is moving to the left, the friction force is to the right, and equation (2.81) holds. Equation (2.81) has a solution of the form

$$x(t) = A_1 \cos \omega t + B_1 \sin \omega t + \frac{\mu mg}{k} \tag{2.84}$$

where $\omega = \sqrt{k/m}$ and A_1 and B_1 are constants to be determined by the initial conditions. To that end, applying the initial conditions yields

$$x(0) = A_1 + \frac{\mu mg}{k} = x_0 \tag{2.85}$$

$$\dot{x}(0) = \omega B_1 = 0 \tag{2.86}$$

Hence $B_1 = 0$ and $A_1 = x_0 - \mu mg/k$ specifies the constants in equation (2.84). Thus when the mass starts from rest (at x_0) and moves to the left, it moves as

$$x(t) = \left(x_0 - \frac{\mu mg}{k}\right) \cos \omega t + \frac{\mu mg}{k} \tag{2.87}$$

This motion continues until the first time $\dot{x}(t) = 0$. This happens when the derivative of equation (2.87) is zero, or when

$$\dot{x}(t) = -\omega \left(x_0 - \frac{\mu mg}{k}\right) \sin \omega t_1 = 0 \tag{2.88}$$

Thus when $t_1 = \pi/\omega$, $\dot{x}(t) = 0$ and the mass starts to move to the right provided that the spring force, kx, is large enough to overcome the maximum frictional force μmg. Hence equation (2.82) now describes the motion. Solving equation (2.82) yields

$$x(t) = A_2 \cos \omega t + B_2 \sin \omega t - \frac{\mu mg}{k} \tag{2.89}$$

for $\pi/\omega < t < t_2$, where t_2 is the second time that $\dot{x}$ becomes zero. The initial conditions for equation (2.89) are calculated from the previous solution given by equation (2.87) at t_1:

$$x\left(\frac{\pi}{\omega}\right) = \left(x_0 - \frac{\mu mg}{k}\right) \cos \pi + \frac{\mu mg}{k} = \frac{2\mu mg}{k} - x_0 \tag{2.90}$$

$$\dot{x}\left(\frac{\pi}{\omega}\right) = -\omega \left(x_0 - \frac{\mu mg}{k}\right) \sin \pi = 0 \tag{2.91}$$

From equation (2.89) and its derivatives it follows that

$$A_2 = x_0 - \frac{3\mu mg}{k} \qquad B_2 = 0 \tag{2.92}$$

The solution for the second interval of time then becomes

$$x(t) = \left(x_0 - \frac{3\mu mg}{k}\right) \cos \omega t - \frac{\mu mg}{k} \qquad \frac{\pi}{\omega} < t < \frac{2\pi}{\omega} \tag{2.93}$$

This procedure is repeated until the motion stops. The motion will stop when the velocity is zero ($\dot{x} = 0$) and the spring force (kx) is insufficient to overcome the maximum frictional force (μmg). The response is plotted in Figure 2.22.

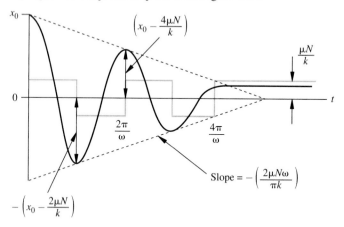

Figure 2.22   Free response of a system with Coulomb friction present.

Several things can be noted about the free response with Coulomb friction versus the free response with viscous damping. First, with Coulomb damping the amplitude decays linearly with slope

$$-\frac{2\mu mg\,\omega}{\pi k} \tag{2.94}$$

rather than exponentially as does a viscously damped system. Second, the motion under Coulomb friction comes to a complete stop, at a potentially different equilibrium position than when initially at rest, whereas a viscously damped system oscillates around a single equilibrium, $x = 0$, with infinitesimally small amplitude. Finally, the frequency of oscillation of a system with Coulomb damping is the same as the undamped frequency, whereas viscous damping alters the frequency of oscillation.

Example 2.7.1

The response of a mass oscillating on a surface is measured to be of the form indicated in Figure 2.22. The initial position is measured to be 30 mm from its zero rest position, and the final position is measured to be 3.5 mm from its zero rest position after four cycles of oscillation in 1 s. Determine the coefficient of friction.

Solution   First the frequency of motion is 4 Hz, or 25.13 rad/s , since four cycles were completed in 1 s. The slope of the line of decreasing peaks is

$$\frac{-30 + 3.5}{1} = -26.5 \text{ mm/s}$$

Therefore, from expression (2.94),

$$-26.5 \text{ mm/s} = \frac{-2\mu m g \omega}{\pi k} = \frac{-2\mu g}{\pi} \frac{\omega}{\omega^2} = \frac{-2\mu g}{\pi \omega}$$

Solving for μ yields

$$\mu = \frac{\pi (25.13 \text{ rad/s})(-26.5 \text{ mm/s})}{(-2)(9.81 \times 10^3 \text{ mm/s}^2)} = 0.107$$

This small value for μ indicates that the surface is probably very smooth or lubricated.   □

Next consider the forced harmonic response of a system with Coulomb damping. The equation of motion of interest becomes

$$m\ddot{x} + \mu m g \text{ sgn}(\dot{x}) + kx = F_0 \cos \omega_{dr} t \tag{2.95}$$

Rather than solving this equation directly, one can approximate the solution of (2.95) with the solution of a viscously damped system that dissipates an equivalent amount of energy per cycle. This is a reasonable assumption if the magnitude of the applied force is much larger than the Coulomb force ($F_0 >> \mu m g$). This approximation is accomplished by again assuming that the steady-state response will be of the form

$$x_{ss}(t) = X \sin \omega_{dr} t \tag{2.96}$$

The energy dissipated, ΔE, in a viscously damped system per cycle with viscous damping coefficient c is given by

$$\Delta E = \oint F_d \, dx = \int_0^{2\pi/\omega_{dr}} \left(c\dot{x} \frac{dx}{dt} \right) dt = \int_0^{2\pi/\omega_{dr}} \left(c\dot{x}^2 \, dt \right) \tag{2.97}$$

At steady state $x = X \sin \omega_{dr} t$, $\dot{x} = \omega_{dr} X \cos \omega_{dr} t$, and equation (2.97) becomes

$$\Delta E = c \int_0^{2\pi/\omega_{dr}} \left(\omega_{dr}^2 X^2 \cos^2 \omega_{dr} t \right) dt = \pi c \omega_{dr} X^2 \tag{2.98}$$

This is the energy dissipated per cycle by a viscous damper. On the other hand, the energy dissipated by the Coulomb friction per cycle is

$$\Delta E = \mu mg \int_0^{2\pi/\omega_{dr}} [\text{sgn}(\dot{x})\dot{x}] dt \tag{2.99}$$

Substitution of the steady-state velocity into this expression and splitting the integrations up into segments corresponding to the sign change in $\dot{x}$ yields

$$\Delta E = \mu mgX \left(\int_0^{\pi/2} \cos u \, du - \int_{\pi/2}^{3\pi/2} \cos u \, du + \int_{3\pi/2}^{2\pi} \cos u \, du \right) \tag{2.100}$$

where $u = \omega_{dr} t$. Completing the integration yields that the energy dissipated by Coulomb friction is

$$\Delta E = 4\mu mgX \tag{2.101}$$

To create a viscously damped system of equivalent energy loss, the energy loss expression for viscous damping of equation (2.98) is equated to the energy loss associated with Coulomb friction, given by equation (2.101) to yield

$$\pi c_{eq} \omega_{dr} X^2 = 4\mu mgX \tag{2.102}$$

where c_{eq} denotes the equivalent viscous damping coefficient. Solving for c_{eq} yields

$$c_{eq} = \frac{4\mu mg}{\pi \omega_{dr} X} \tag{2.103}$$

In terms of an equivalent damping ratio, ζ_{eq}, equation (2.103) must also equal $2\zeta_{eq}\omega m$, so that

$$\zeta_{eq} = \frac{2\mu g}{\pi \omega \omega_{dr} X} \tag{2.104}$$

Thus the viscously damped system described by

$$\ddot{x} + 2\zeta_{eq}\omega\dot{x} + \omega^2 x = f_0 \cos \omega_{dr} t \tag{2.105}$$

will dissipate as much energy as does the Coulomb system described by equation (2.95).

Considering (2.105) as an approximation of equation (2.95), the approximate magnitude and phase of the steady-state response of equation (2.95) can be calculated. Substitution of the equivalent viscous damping ratio given in equation (2.104) into the magnitude of equation (2.30) yields the result that the magnitude X of the steady-state response is

$$X = \frac{F_0/k}{\sqrt{(1 - r^2)^2 + (2\zeta_{eq} r)^2}} = \frac{F_0/k}{\left[(1 - r^2)^2 + (4\mu mg/\pi kX)^2 \right]^{1/2}} \tag{2.106}$$

Solving this expression for the amplitude X yields

$$X = \frac{F_0}{k} \frac{\sqrt{1 - (4\mu mg/\pi F_0)^2}}{|(1 - r^2)|} \tag{2.107}$$

with phase shift given by equation (2.29) as

$$\phi = \tan^{-1} \frac{2\zeta_{eq}r}{1 - r^2} = \tan^{-1} \frac{4\mu mg}{\pi k X (1 - r^2)} \tag{2.108}$$

The expression for the phase can be further examined by substituting for the value of X from equation (2.107). This yields

$$\phi = \tan^{-1} \frac{\pm 4\mu mg}{\pi F_0 \sqrt{1 - (4\mu mg/\pi F_0)^2}} \tag{2.109}$$

where the $\pm$ originates from the absolute value in equation (2.107). Thus ϕ is positive if $r < 1$ and negative if $r > 1$. Also note from equation (2.109) that ϕ is constant for a given F_0 and μ and is independent of the driving frequency.

Several differences in the behavior of the approximate phase and amplitude of the response with Coulomb friction compared with that of viscous friction are apparent. First at resonance, $r = 1$, the magnitude in equation (2.107) becomes infinite, unlike the viscously damped case. Second, the phase is discontinuous at resonance, rather than passing through 90° as in the viscous case. Note also from equation (2.109) that the approximation is good only if the argument in the radical of (2.107) is positive, that is, if

$$4\mu mg < \pi F_0 \tag{2.110}$$

This confirms the physical statement made at the outset (i.e., that the applied force must be larger in magnitude than the sliding friction force in order to overcome the friction to provide motion).

Example 2.7.2

Consider the system of Figure 2.20 with stiffness $k = 1.5 \times 10^4$ N/m, driving harmonically a 10-kg mass by a force of 90 N at 25 Hz. Calculate the approximate amplitude of steady-state motion assuming that both the mass and the surface are steel (unlubricated).

Solution   First note from Table 2.1 that the coefficient of friction is $\mu = 0.3$. Then from inequality (2.110),

$$4\mu mg = 4(0.3)(10 \text{ kg})(9.8 \text{ m/s}^2) = 117.6 \text{ N}$$

$$< (90 \text{ N})(3.1415) = 282.74 = \pi F_0$$

so that the approximation developed above for the steady-state response amplitude is valid. Converting 25 Hz to 157 rad/s and using equation (2.107) then yields

$$X = \frac{90 \text{ N}}{1.5 \times 10^4 \text{ N/m}} \frac{\sqrt{1 - (117.6 \text{ N}/282.74 \text{ N})^2}}{|1 - (2.467 \times 10^4/1.5 \times 10^3)|} = 3.53 \times 10^{-4} \text{ m}$$

Thus the amplitude of oscillation will be less than 1 mm.

$\square$

2.8 OTHER FORMS OF DAMPING

Several other forms of damping are available for modeling a particular mechanical device or structure in addition to viscous and Coulomb damping. It is common to study damping mechanisms by examining the energy dissipated per cycle under a harmonic loading. Often, force versus displacement curves, or stress versus strain curves, are used to measure the energy lost and hence determine a measure of the damping in the system. Recall that the energy lost per cycle was defined in equation (2.97) as

$$\Delta E = \oint F_d dx \tag{2.111}$$

where F_d is the force due to damping. This was calculated for simple viscous damping in equation (2.98) to be $\pi c \omega_{dr} X^2$. This is used to define the *specific damping capacity* as the energy lost per cycle divided by the peak potential energy, $\Delta E /U$. A more commonly used quantity is the energy lost per radian divided by the peak potential or strain energy U_{max}. This is defined to be the *loss factor* or *loss coefficient*, denoted by η, and given by

$$\eta = \frac{\Delta E}{2\pi U_{max}} \tag{2.112}$$

where U_{max} is defined as the potential energy at maximum displacement of X.

The loss factor is related to the damping ratio of a viscously damped system at resonance. To see this, substitute the value for ΔE from (2.98) into (2.112) to get

$$\eta = \frac{\pi c \omega_{dr} X^2}{2\pi \left(\frac{1}{2} k X^2\right)} \tag{2.113}$$

At resonance, $\omega_{dr} = \omega = \sqrt{k/m}$, so that (2.112) becomes

$$\eta = \frac{c}{\sqrt{km}} = 2\zeta \tag{2.114}$$

Hence, at resonance, the loss factor is twice the damping ratio.

Next, consider a force displacement curve for a system with viscous damping. The force required to displace the mass is that force required to overcome the spring and damper forces, or

$$F = kx + c\dot{x} \tag{2.115}$$

At steady state as given by equation (2.96), this becomes

$$F = kx + cX \omega \cos \omega t \tag{2.116}$$

Using a trigonometric identity $(\sin^2 \phi + \cos^2 \phi = 1)$ on the $\cos \omega t$ term yields

$$F = kx \pm c\omega X (1 - \sin^2 \omega t)^{1/2}$$

$$= kx \pm c\omega [X^2 - (X \sin \omega t)^2]^{1/2} \tag{2.117}$$

$$= kx \pm c\omega \sqrt{X^2 - x^2}$$

Squaring this expression yields, upon rearrangement,

$$F^2 + (c^2\omega^2 + k^2)x^2 - (2k)xF - c^2\omega^2 X^2 = 0 \qquad (2.118)$$

which can be recognized as the general equation for an ellipse ($c^2\omega^2 > 0$) rotated about the origin in the F-x plane (see a precalculus text). This is plotted in Figure 2.23. The ellipse in the Figure 2.23 is called a hysteresis loop and the area enclosed is the energy lost per cycle as calculated in equation (2.98) and is equal to $\pi c\omega_{dr} X^2$. Note that if $c = 0$, the ellipse of Figure 2.23 collapses to the straight line of slope k given in Figure 1.3.

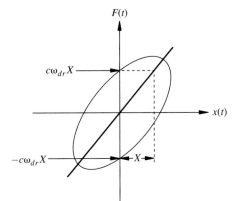

Figure 2.23 Plot of force versus displacement defining the hysteresis loop for a viscously damped system.

Materials are often tested by measuring stress (force) and strain (displacement) under carefully controlled steady-state harmonic loading. Many materials exhibit internal friction between various planes of material as the material is deformed. Such tests produce hysteresis loops of the form shown in Figure 2.24. Note that for increasing strain (loading), the path is different than for decreasing strain (unloading). This type of energy dissipation is called *hysteretic damping, solid damping*, or *structural damping*.

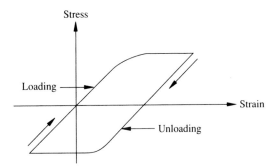

Figure 2.24 Experimental stress strain plot for one cycle of harmonically loaded material for steady-state illustrating a hysteresis loop associated with internal damping.

The area enclosed by the hysteresis loop is again equal to the energy loss. If the experiment is repeated for a number of different frequencies at constant amplitude, it is found that the area is independent of frequency and proportional to the amplitude of vibration and stiffness:

$$\Delta E = \pi k \beta X^2 \tag{2.119}$$

where k the stiffness, X the amplitude of vibration, and β is defined as the *hysteretic damping constant*. Note that some texts formulate this equation differently by defining $h = k\beta$ to be the hysteretic damping constant.

Next, recall the equivalent viscous damping concept used in Section 2.7 for Coulomb damping. If this concept is applied here, equating the energy dissipated by a viscously damped system to that of a hysteretic system is equivalent to finding the ellipse of Figure 2.23 that has the same area as the loop of Figure 2.24. Thus equating (2.119) with the energy calculated in (2.98) for the viscously damped system yields

$$\pi c_{eq} \omega_{dr} X^2 = \pi k \beta X^2 \tag{2.120}$$

Solving this expression yields that the viscously damped system dissipating the same amount of energy per cycle as the hysteretic system will have the equivalent damping constant given by

$$c_{eq} = \frac{k\beta}{\omega_{dr}} \tag{2.121}$$

where β is determined experimentally from the hysteresis loop.

The approximate steady-state response of a system with hysteretic damping can be determined from substitution of this equivalent damping expression into the equation of motion to yield

$$m\ddot{x} + \frac{\beta k}{\omega_{dr}} \dot{x} + kx = F_0 \cos \omega_{dr} t \tag{2.122}$$

Again the steady-state response is approximated by assuming the response is of the form $x_{ss}(t) = X \cos(\omega_{dr} - \phi)$. Following the procedures of Section 2.2, the magnitude of the response, X, is given by equation (2.30) to be

$$X = \frac{F_0/k}{\sqrt{(1 - r^2)^2 + (2\zeta_{eq}r)^2}} \tag{2.123}$$

where ζ_{eq} is now $c_{eq}/(2\sqrt{km})$ and $r = \omega_{dr}/\omega$. Substituting for ζ_{eq} yields

$$X = \frac{F_0/k}{\sqrt{(1 - r^2)^2 + \beta^2}} \tag{2.124}$$

Similarly, the phase becomes

$$\phi = \tan^{-1} \frac{\beta}{1 - r^2} \tag{2.125}$$

These are plotted in Figure 2.25 for several values of β.

Compared to a viscously damped system, the hysteretic system's magnitude obtains a maximum of $F_0/\beta k$. This is obtained by setting $r = 1$ in equation (2.124) so that the maximum value is obtained at the resonant frequency rather than below it as is the case for viscous damping. Examination of the phase shift shows that the response of a

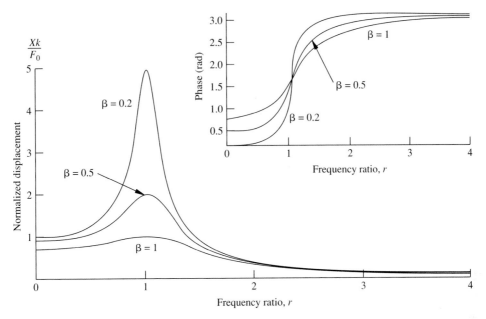

Figure 2.25   Steady-state magnitude and phase versus frequency ratio for a system with hysteretic damping coefficient β approximated by a system with viscous damping.

hysteretic system is never in phase with the applied force, which is not true for viscous damping.

In Section 2.5 the complex exponential was used to represent a harmonic input. Using equation (2.57), the equivalent hysteretic system can be written as

$$m\ddot{x} + \frac{\beta k}{\omega_{dr}}\dot{x} + kx = F_0 e^{j\omega_{dr}t} \tag{2.126}$$

Substitution of the assumed form of the solution given by $x(t) = Xe^{j\omega_{dr}t}$ for just the velocity term yields

$$m\ddot{x} + k(1 + j\beta)x = F_0 e^{j\omega_{dr}t} \tag{2.127}$$

This gives rise to the notion of *complex stiffness* or *complex modulus*. The damped problem is represented in equation (2.127) as an undamped problem with complex stiffness coefficient $k(1 + j\beta)$. This approach is very popular in the material engineering literature on damping.

Example 2.8.1

An experiment is performed on a hysteretic system with known spring stiffness of $k = 4 \times 10^4$ N/m. The system is driven at resonance, the area of the hysteresis loop is measured to be $\Delta E = 30$ N · m, and the amplitude X is measured to be $X = 0.02$ m. Calculate the magnitude of the driving force and the hysteretic damping constant.

Solution   At resonance equation (2.124) yields

$$X = \frac{F_0}{k\beta}$$

or $k\beta = F_0/X$. The area enclosed by the hysteresis loop is equal to $\pi k\beta X^2$, so that

$$30 \text{ N} \cdot \text{m} = \pi k\beta X^2 = \pi\frac{F_0}{X}X^2$$

and hence

$$F_0 = \frac{30 \text{ N} \cdot \text{m}}{\pi X} = \frac{30 \text{ N} \cdot \text{m}}{\pi(0.02 \text{ m})} = 477.5 \text{ N}$$

From the resonance expression,

$$\beta = \frac{F_0}{Xk} = \frac{477.5}{(0.02 \text{ m})(4 \times 10^4 \text{ N/m})} = 0.60$$

which is the hysteretic damping constant calculated based on the principle of equivalent viscous damping.

$\square$

Several other useful models of damping mechanisms can be analyzed by using the equivalent viscous damping approach. For instance, if an object vibrates in air (or a fluid), it often experiences a force proportional to the square of the velocity (Blevins, 1977). The equation of motion for such a vibration is

$$m\ddot{x} + \alpha \, \text{sgn}(\dot{x})\dot{x}^2 + kx = F_0 \cos \omega_{dr} t \tag{2.128}$$

The damping force is

$$F_d = \alpha \, \text{sgn}(\dot{x})\dot{x}^2 = \frac{C\rho A}{2} \text{sgn}(\dot{x})\dot{x}^2$$

which opposes the direction of motion, similar to Coulomb friction, and depends on the square of the velocity; the drag coefficient of the mass, C; the density of the fluid, ρ; and the cross-sectional area A of the mass. This type of damping is referred to as *air damping*, *quadratic damping*, or *velocity squared damping*. As in the Coulomb friction case, this is a nonlinear equation that does not have a convenient closed-form solution to analyze. While it can be solved numerically (see Chapter 9), an approximation to the behavior of the solution during steady-state harmonic excitation can be made by the equivalent viscous damping method. Assuming a solution of the form $x = X \sin \omega_{dr} t$ and computing the energy integrals following the steps taken in equation (2.100) yields that the energy dissipated per cycle is

$$\Delta E = \tfrac{8}{3}\alpha X^3 \omega_{dr}^2 \tag{2.129}$$

Again equating this to the energy dissipated by a viscously damped system given in equation (2.98) yields

$$c_{eq} = \frac{8}{3\pi}\alpha\omega_{dr}X \tag{2.130}$$

This equivalent viscous damping value can then be used in the amplitude and phase formulas for a linear viscously damped forced harmonic motion to approximate the steady-state response.

Example 2.8.2

Calculate the approximate amplitude at resonance for velocity squared damping.

Solution   Using the magnitude expression for viscous damping at resonance, equation (2.30) yields

$$X = \frac{f_0}{2\zeta\omega\omega_{dr}} = \frac{f_0}{c_{eq}\omega_{dr}}$$

Substitution of (2.130) yields

$$X = \frac{f_0}{(8/3\pi)\alpha\omega^2 X}$$

so that

$$X = \sqrt{\frac{3\pi f_0}{8\alpha\omega^2}} = \sqrt{\frac{3\pi f_0 m}{4kC\rho A}}$$

As expected, for larger values of the mass density of the fluid, the drag coefficient of the cross-sectional area produces a smaller amplitude at resonance.   □

If several forms of damping are present, one approach to examining the harmonic response is to calculate the energy lost per cycle of each form of damping present, add them up, and compare them to the energy loss from a single viscous damper. Then the formulas for magnitude and phase from equation (2.30) are used to approximate the response. For n damping mechanisms dissipating energy per cycle of ΔE_i, for the ith mechanism, the equivalent viscous damping constant is

$$c_{eq} = \frac{\sum_{i=1}^{n} \Delta E_i}{\pi\omega_{dr} X^2} \tag{2.131}$$

A study of various damping mechanisms is presented by Bandstra (1983).

PROBLEMS

Section 2.1

2.1. To familiarize yourself with the nature of the forced response, plot the solution of a forced response of equation (2.2) with $\omega_{dr} = 2$ rad/s, given by equation (2.11) for a variety of values of the initial conditions and ω/ω_{dr} as given in the following chart:

Case	x_0	v_0	f_0	ω
1	0.1	0.1	0.1	1
2	−0.1	0.1	0.1	1
3	0.1	0.1	1.0	1
4	0.1	0.1	0.1	2.1
5	1	0.1	0.1	1

2.2. Repeat the calculation made in Example 2.1.1 for the mass of a simple spring–mass system where the mass of the spring is considered and known to be 1 kg.

2.3. A spring–mass system is driven from rest harmonically such that the displacement response exhibits a beat of period of 0.2π s. The period of oscillation is measured to be 0.02π s Calculate the natural frequency and the driving frequency of the system.

2.4. An airplane wing modeled as a spring–mass system with natural frequency 40 Hz is driven harmonically by the rotation of its engines at 39.9 Hz. Calculate the period of the resulting beat.

2.5. Derive equation (2.13) from equation (2.12) using standard trigonometric identities.

Section 2.2

2.6. Calculate the constants A and θ for arbitrary initial conditions, x_0 and v_0, in the case of the forced response given by equation (2.29). Compare this solution to the transient response obtained in the case of no forcing function (i.e., $F_0 = 0$).

2.7. Show that equations (2.20) and (2.21) are equivalent by verifying equations (2.21) and (2.22).

2.8. Plot the solution of equation (2.19) for the case that $m = 1$ kg, $\zeta = 0.01$, $\omega = 2$ rad/s, $F_0 = 3$ N, and $\omega_{dr} = 10$ rad/s, with initial conditions $x_0 = 1$ m and $v_0 = 1$ m/s.

2.9. A 100 kg mass is suspended by a spring of stiffness 30×10^3 N/m with a viscous damping constant of 1000 N · s/m. The mass is initially at rest. Calculate the steady-state displacement amplitude and phase if the mass is excited by a harmonic force of 80 N at 3 Hz.

2.10. Plot the total solution of the system of Problem 2.9 including the transient.

2.11. Consider the pendulum mechanism of Figure 2.26, which is pivoted at point 0. Calculate both the damped and undamped natural frequency of the system. Assume that the mass of the rod, spring, and damper are negligible. What driving frequency will cause resonance?

2.12. Consider the pivoted mechanism of Figure 2.26 with $k = 4 \times 10^3$ N/m, $l_1 = 1.5$ m, $l_2 = 0.5$ m, $l = 1$ m, and $m = 40$ kg. The mass of the beam is 40 kg; it is pivoted at point 0 and assumed to be rigid. Design the dashpot (i.e., calculate c) so that the damping ratio of the system is 0.2. Also determine the amplitude of vibration of the steady-state response if a 10-N force is applied to the mass, as indicated in the figure, at a frequency of 10 rad/s.

2.13. In the design of Problem 2.12, the damping ratio was chosen to be 0.2 because it limits the amplitude of the forced response. If the driving frequency is shifted to 11 rad/s, calculate the change in damping coefficient needed to keep the amplitude less than calculated in Problem 12.

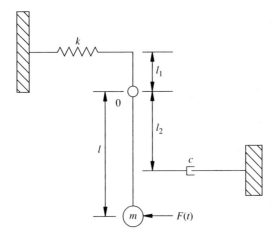

Figure 2.26 Damped pendulum mechanism (for problems 2.11 to 2.13). Such a mechanism could be used to model a brake pedal in an automobile.

Section 2.3

2.14. A machine weighing 2000 N rests on a floor. The floor deflects about 5 cm as a result of the weight of the machine. The floor is somewhat flexible and moves, because of the motion of a nearby machine, harmonically near resonance ($r = 1$) with an amplitude of 0.2 cm. Assume a damping ratio of $\zeta = 0.01$ and calculate the transmitted force and the amplitude of the transmitted displacement.

2.15. Derive equation (2.40) from (2.38).

2.16. From the equation describing Figure 2.8, show that the point ($\sqrt{2}, 1$) corresponds to the value TR > 1 (i.e., for all $r < \sqrt{2}$, TR > 1).

2.17. Consider the base excitation problem for the configuration shown in Figure 2.27. In this case the base motion is a displacement transmitted through a dashpot or pure damping element. Derive an expression for the force transmitted to the support in steady state.

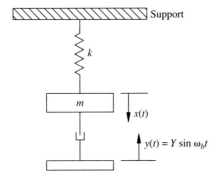

Figure 2.27 Model of displacement through a pure viscous damping element.

2.18. A fan of 45 kg has an unbalance that creates a harmonic force. A spring–damper system is designed to minimize the force transmitted to the base of the fan. A damper is used having a damping ratio of $\zeta = 0.2$. Calculate the required spring stiffness so that only 10% of the force is transmitted to the ground when the fan is running at 10,000 rpm.

2.19. A vibrating mass of 300 kg, mounted on a massless support by a spring of stiffness 40,000 N/m and a damper of unknown damping coefficient, is observed to vibrate with a 10-mm amplitude while the support vibration has a maximum amplitude of only 2.5 mm (at resonance). Calculate the damping constant and the amplitude of the force on the base.

2.20. Referring to Example 2.3.1, at what speed does car 1 experience resonance? At what speed does car 2 experience resonance? Calculate the maximum deflection of both cars at resonance.

2.21. For cars of Example 2.3.1, calculate the best choice of the damping coefficient so that the transmissability is as small as possible by comparing the magnitude of $\zeta = 0.01$, $\zeta = 0.1$ and $\zeta = 0.2$ for the case $r = 2$. What happens if the road "frequency" changes?

2.22. A system modeled by Figure 2.7 has a mass of 225 kg with a spring stiffness of 3.5×10^4 N/m. Calculate the damping coefficient given that the system has a deflection (X) of 0.7 cm when driven at its natural frequency while the base amplitude (Y) is measured to be 0.3 cm.

2.23. Consider Example 2.3.1 for car 1 if three passengers totaling 200 kg are riding in the car. Calculate the effect of the mass of the passengers on the deflection at 20, 80, 100, and 150 km/h. What is the effect of the added passenger mass on car 2?

Section 2.4

2.24. Consider Example 2.3.1. Choose values of c and k for the suspension system for car 2 (the sedan) such that the amplitude transmitted to the passenger compartment is as small as possible for the 1-cm bump at 50 km/h. Also calculate the deflection at 100 km/h for your values of c and k.

2.25. A lathe can be modeled as an electric motor mounted on a steel table. The table plus the motor have a mass of 50 kg. The rotating parts of the lathe have a mass of 10 kg at a distance 0.1 m from the center. The damping ratio of the system is measured to be $\zeta = 0.06$ (viscous damping) and its natural frequency is 7.5 Hz. Calculate the amplitude of the steady-state displacement of the motor, assuming $\omega_r = 30$ Hz.

2.26. The system of Figure 2.11 produces a forced oscillation of varying frequency. As the frequency is changed, it is noted that at resonance, the amplitude of the displacement is 10 mm. As the frequency is increased several decibels past resonance the amplitude of the displacement remains fixed at 1 mm. Estimate the damping ratio for the system.

2.27. An electric motor (Figure 2.28) has a mass eccentricity of 10 kg and is set on two identical springs ($k = 3.2$ N/mm). The motor runs at 1750 rpm, and the mass eccentricity is 100 mm from the center. The springs are mounted 250 mm apart with the motor shaft in the center. Neglect damping and determine the force transmissibility ratio for vertical vibration.

2.28. Consider a system with rotating unbalance as illustrated in Figure 2.11. The deflection at resonance is 0.05 m and the damping ratio is measured to be $\zeta = 0.1$. The out-of-balance mass is estimated to be 10%. Locate the unbalanced mass by estimating e.

Section 2.5

2.29. Referring to Figure 2.14, draw the solution for the magnitude X for the case $m = 100$ kg, $c = 4000$ N·s/m, and $k = 10,000$ N/m. Assume that the system is driven at resonance by a 10-N force.

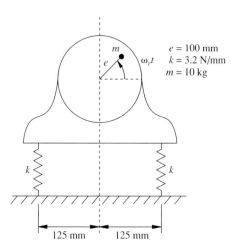

$e = 100$ mm
$k = 3.2$ N/mm
$m = 10$ kg

125 mm   125 mm

Figure 2.28   Vibration model for an electric motor with an unbalance.

2.30. Use the graphical method to compute the phase shift for the system of Problem 2.29 if $\omega_{dr} = \omega/2$ and again for the case $\omega_{dr} = 2\omega$.

2.31. A body of mass 100 kg is suspended by a spring of stiffness of 30 kN/m and a dashpot of damping constant 1000 N · s/m. Vibration is excited by a harmonic force of amplitude 80 N and frequency 3 Hz. Calculate the amplitude of the displacement for the vibration and the phase angle between the displacement and the excitation force using the graphical method.

2.32. Calculate the real part of equation (2.64) to verify that it yields equation (2.28) and hence establish the equivalence of the exponential approach to solving the damped vibration problem.

2.33. Referring to equation (2.65) and Appendix B, calculate the solution $x(t)$ by using a table of Laplace transform pairs and show that the solution obtained this way is equivalent to (2.28).

Section 2.6

2.34. Calculate damping and stiffness coefficients for the accelerometer of Figure 2.15 with moving mass of 0.04 kg such that the accelerometer is able to measure vibration between 0 and 50 Hz within 5%. (*Hint:* For an accelerometer it is desirable for $Z/Y =$ constant.)

2.35. The damping constant for a particular accelerometer of the type illustrated in Figure 2.15 is 50 N · s/m. It is desired to design the accelerometer (i.e., choose m and k) for a maximum error of 3% over the frequency range 0 to 75 Hz.

2.36. The accelerometer of Figure 2.15 has a natural frequency of 120 kHz and a damping ratio of 0.2. Calculate the error in measurement of a sinusoidal vibration at 60 kHz.

2.37. Design an accelerometer (i.e., choose m, c and k) configured as in Figure 2.15 with very small mass that will be accurate to 1% over the frequency range 0 to 50 Hz.

Section 2.7

2.38. Consider the system of Figure 2.20 with stiffness 1.2×10^4 N/m and mass 10 kg, driven harmonically by a force of 50 N at 10 Hz. Calculate the approximate amplitude of steady-state motion assuming that both the mass and the surface that it slides on are made of lubricated steel.

2.39. A 2-kg mass connected to a spring of stiffness 10^3 N/m has a dry sliding friction force (F_d) of 3 N. As the mass oscillates, its amplitude decreases 20 cm in 15 cycles. How long does this take?

2.40. Consider the system of Figure 2.20 with $m = 5$ kg and $k = 9 \times 10^3$ N/m with a friction force of magnitude 6 N. If the initial amplitude is 4 cm, determine the amplitude one cycle later as well as the damped frequency.

2.41. The system of Figure 2.20 with mass of 10 kg, stiffness of 2000 N/m, and coefficient of friction of 0.1 is driven harmonically at 10 Hz. The amplitude at steady state is 5 cm. Calculate the magnitude of the driving force.

2.42. A system of mass 10 kg and stiffness 1.5×10^4 N/m is subject to Coulomb damping. If the mass is driven harmonically by a 90-N force at 25 Hz, determine the equivalent viscous damping coefficient if the coefficient of friction is 0.1.

2.43. **a.** Plot the free response of the system of Problem 2.42 to initial conditions of $x(0) = 0$ and $\dot{x}(0) = |F_0/m| = 9$ m/s.
   b. Use the equivalent viscous damping coefficient calculated in Problem 2.42 and plot the free response of the "equivalent" viscously damped system to the same initial conditions.

2.44. Referring to the system of Example 2.7.2, calculate how large the magnitude of the driving force must be to sustain motion if the steel is lubricated. How large must this magnitude be if the lubrication is removed?

2.45. Calculate the phase shift between the driving force and the response for the system of Problem 2.44 using the equivalent viscous damping approximation.

2.46. Derive the equation of vibration for the system of Figure 2.20 assuming that a viscous dashpot of damping constant c is connected in parallel to the spring. Calculate the energy loss and determine the magnitude and phase relationships for the forced response of the equivalent viscous system.

2.47. A system of unknown damping mechanism is driven harmonically at 10 Hz with an adjustable magnitude. The magnitude is changed, and the energy lost per cycle and amplitudes are measured for five different magnitudes. The measured quantities are

ΔE (J)	0.25	0.45	0.8	1.16	3.0
X (M)	0.01	0.02	0.04	0.08	0.15

Is the damping viscous or Coulomb?

2.48. Calculate the equivalent loss factor for a system with Coulomb damping.

Section 2.8

2.49. A steel block is sliding on a steel floor with spring constant 105 N/m and mass 100 kg. Calculate the largest value of the initial conditions such that no motion occurs.

2.50. Using the initial condition calculated in Problem 2.49, calculate the position of the mass after 2 s if the mass is initially placed to the left of the central position a distance of $2x_0$.

2.51. Calculate the slope of the decay envelope of Figure 2.22 and hence derive equation (2.94).

2.52. Calculate the number of half-cycles that elapse before the system of Figure 2.21 comes to rest.

2.53. A spring–mass system ($m = 10$ kg, $k = 4 \times 10^3$ N/m) vibrates horizontally on a surface with coefficient of friction $\mu = 0.15$. When excited harmonically at 5 Hz, the steady-state displacement of the mass is 5 cm. Calculate the amplitude of the harmonic force applied.

2.54. Calculate the maximum displacement for a system with air damping using the equivalent viscous damping method.

2.55. Calculate the semimajor and semiminor axis of the ellipse of equation (2.118). Then calculate the area of the ellipse. Use $c = 10$kg/s, $\omega = 2$ rad/s and $x = 0.01$ m.

2.56. The area of a force deflection curve of Figure 2.24 is measured to be 2.5 N·m, and the maximum deflection is measured to be 8 mm. From the "slope" of the ellipse the stiffness is estimated to be 5×10^4 N/m. Calculate the hysteretic damping coefficient. What is the equivalent viscous damping if the system is driven at 10 Hz?

2.57. The area of the hysteresis loop of a hysterically damped system is measured to be 5 N·m and the maximum deflection is measured to be 1 cm. Calculate the equivalent viscous damping coefficient for a 20-Hz driving force. Plot c_{eq} versus ω_{dr} for $2\pi \le \omega_{dr} \le 100\pi$ rad/s.

2.58. Calculate the nonconservative energy of a system subject to both viscous and hysteretic damping.

2.59. Derive a formula for equivalent viscous damping for the damping force of the form, $F_d = c(\dot{x})^n$ where n is an integer.

2.60. Using the equivalent viscous damping formulation, determine an expression for the steady-state amplitude under harmonic excitation for a system with both Coulomb and viscous damping present.

2.61. Show that equation (2.84) is a solution of equation (2.81) and that equation (2.89) satisfies (2.82) by simple substitution.

M A T L A B   V I B R A T I O N   T O O L B O X

   If you did not use the *Vibration Toolbox* program from Chapter 1, refer to that section for information regarding using MATLAB files.

   The files for Chapter 2, entitled VTB2_1, VTB2_2, and so on, can be found in folder VTB2. The files in VTB2 can be used to help solve the homework problems stated above and to help gain information about the nature of the response of single-degree-of-freedom systems to harmonic inputs. Of course, the MATLAB functions themselves can be used to help you create your own vibration problem solver. The following problems are intended to help you gain some experience with the concepts in this chapter.

TOOLBOX PROBLEMS

TB2.1. Using file VTB2_1, reproduce Figure 2.2.

TB2.2. Carefully investigate the response of an undamped system near resonance by trying several values of ω_{dr} near ω for the values of Figure 2.2. Do you get the beats of Figure 2.3?

TB2.3. Using file VTB2_3 reproduce Figure 2.6. Also plot $A_0 k / f_0$ versus r for the values given in Example 2.2.1 and plot the associated time response $x_p(t)$ for a value of $r = 0.5$ using VTB2_2. Do these plots again for $\zeta = 0.01$ and $\zeta = 0.1$ and comment on how the time response changes as the damping ratio ζ changes by an order of magnitude.

TB2.4. Using file VTB2_5 for rotating unbalance, make a plot of x versus r for the helicopter of Example 2.4.2.

TB2.5. Using file VTB2_6 for damping mechanisms, compare the time response of a system (with physical parameters of $m = 10$, $k = 100$, $\alpha = 0.05$, $X = 1$) with air damping as given by equation (2.128) with initial conditions $x_0 = 10$ and $v_0 = 0$, to that of an equivalent viscously damped system using equation (2.130).

3

General Forced Response

This chapter considers ways to calculate the vibration resulting from a variety of different excitations. A major concern to manufacturers of computers is the speed and accuracy with which a disk drive head can move about a disk to read data. A disk and head such as the one pictured here undergoes vibration to a variety of excitations. The analysis presented in this chapter is useful in predicting the response of the head. The trail bike pictured below also undergoes vibration from a variety of different loadings especially shock loads and random excitation, topics discussed in this chapter. Understanding the response of such devices can lead to improved designs and better performance.

In Chapter 2 the forced response of a single-degree-of-freedom system was considered for the special case of a harmonic driving force. *Harmonic excitation* refers to an applied force that is sinusoidal of a single frequency. In this chapter the response of a system to a variety of different forces is considered, as well as a general formulation for calculating the forced response for any applied force. Since the single-degree-of-freedom system considered here is linear, the principle of superposition can be used to calculate the response to various combinations of forces based on the individual response to a specific force.

Superposition refers to the fact that for a linear equation of motion, say $\ddot{x} + \omega^2 x = 0$, if x_1 and x_2 are both solutions of the equation, then $x = a_1 x_1 + a_2 x_2$ is also a solution where a_1 and a_2 are any constants. This concept also implies that if x_1 is a particular solution to $\ddot{x} + \omega^2 x = f_1$ and x_2 is a solution to $\ddot{x} + \omega^2 x = f_2$, then $x_1 + x_2$ is a solution of $\ddot{x} + \omega^2 x = f_1 + f_2$. Thus this method of superposition can be used to construct the solution to a complicated forcing function by solving a series of simpler problems. This is effectively how the base excitation problem of Section 2.3 is solved. The principle of superposition in linear systems is very powerful and used extensively.

A variety of forces are applied to mechanical systems that result in vibration. Earthquake forces are sometimes modeled as sums of decaying periodic or harmonic forces. High winds can be a source of impulsive or step loadings to structures. Rough roads provide a variety of forcing conditions to automobiles. The ocean provides a force to ships at sea. Various manufacturing processes produce applied forces that are random, periodic, nonperiodic, or transient in nature. Air and relative motion provide forces to the wing of an aircraft that can cause it to oscillate. All of these forces can cause vibration.

Periodic forces are those that repeat in time. An example is an applied force consisting of the sum of two harmonic forces at different frequencies. A nonperiodic force is one that does not repeat itself in time. A step function is an example of a force that is a nonperiodic excitation. A transient force is one that reduces to zero after a finite, usually small, time. An impulse or a shock loading are examples of transient excitations. All of the above-mentioned classes of excitation are deterministic (i.e., they are known precisely as a function of time). On the other hand, a random excitation is one that is unpredictable in time and must be described in terms of probability and statistics. This chapter introduces a sample of these various classes of force excitations and how to calculate and analyze the resulting motion when applied to a single-degree-of-freedom spring–mass–damper system.

3.1 IMPULSE RESPONSE FUNCTION

A very common source of vibration is the sudden application of a short-duration force called an impulse. An impulse excitation is a force that is applied for a very short, or infinitesimal, length of time and represents one example of a *shock loading* and is a nonperiodic force. The response of a system to an impulse is identical to the free response of the system to certain initial conditions. This is illustrated in the following. In many useful situations the applied force $F(t)$ is impulsive in nature (i.e., acts with large magnitude for a very short period of time).

First, consider a mathematical model of an impulse excitation. A graphical time history of a model of the impulse is given in Figure 3.1. This is a rectangular pulse of very large magnitude and very small width (duration).

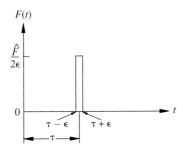

Figure 3.1 Time history of an impulse force used to model impulsive loading of large magnitude applied over a short time interval.

The rule of Figure 3.1 is stated symbolically as

$$F(t) = \begin{cases} 0 & t < \tau - \epsilon \\ \dfrac{\hat{F}}{2\epsilon} & \tau - \epsilon < t < \tau + \epsilon \\ 0 & t > \tau + \epsilon \end{cases} \qquad (3.1)$$

where ϵ is a small positive number. This simple rule, $F(t)$, can be integrated to define the *impulse*. The impulse of the force $F(t)$ is defined by the integral $I(\epsilon)$ by

$$I(\epsilon) = \int_{\tau-\epsilon}^{\tau+\epsilon} F(t)\, dt$$

which provides a measure of the strength of the forcing function $F(t)$. Since the rule $F(t)$ is zero outside the time interval from $\tau - \epsilon$ to $\tau + \epsilon$, the limits of integration on $I(\epsilon)$ can be extended to yield

$$I(\epsilon) = \int_{-\infty}^{\infty} F(t)\, dt \qquad (3.2)$$

which has the units of N·s.

In this case the integral of equation (3.2) is evaluated by calculating the area under the curve using equation (3.1), which becomes

$$I(\epsilon) = \int_{-\infty}^{\infty} F(t)\, dt = \frac{\hat{F}}{2\epsilon} 2\epsilon = \hat{F} \qquad (3.3)$$

independent of the value of ϵ as long as $\epsilon \neq 0$. In the limit as $\epsilon \rightarrow 0$ (but $\epsilon \neq 0$), the integral takes the value $I(\epsilon) = \hat{F}$. This is used to define the impulse function as the function $F(t)$ with the two properties

$$F(t - \tau) = 0 \qquad t \neq \tau \tag{3.4}$$

and

$$\int_{-\infty}^{\infty} F(t - \tau)\, dt = \hat{F} \tag{3.5}$$

If the magnitude of $\hat{F}$ is unity, this becomes the definition of the unit impulse function, denoted $\delta(t)$, also called the *Dirac delta function* (Boyce and DiPrima, 1986).

The solution for response of the single-degree-of-freedom system (see Window 3.1) to an impulsive load for the system initially at rest is calculated by recalling from physics that an impulse imparts a change in momentum to a body. For the sake of simplicity take $\tau = 0$ in the definition of an impulse. Consider the mass to be at rest just prior to the application of an impulse force. This instant of time is denoted 0^-. The initial conditions are both zero, so that $x(0^-) = \dot{x}(0^+) = 0$, since the system is initially at rest. Thus the change in momentum at impact is $m\dot{x}(0^+) - m\dot{x}(0^-) = mv_0$, so that $\hat{F} = F\Delta t = mv_0 - 0 = mv_0$, while the initial displacement remains at zero. Thus an impulse applied to a single-degree-of-freedom spring–mass system is the same as applying the initial conditions of zero displacement and an initial velocity of $v_0 = F\Delta t / m$.

Window 3.1
Review of the Response of the Single-Degree-of-Freedom System of Chapter 1

$$m\ddot{x} + c\dot{x} + kx = F(t)$$

$$x(0) = x_0 \qquad \dot{x}(0) = v_o$$

$$\ddot{x} + 2\zeta\omega\dot{x} + \omega^2 x = f(t)$$

This system has free response [i.e., $f(t) = 0$] in the underdamped case (i.e., $0 < \zeta < 1$) given by

$$x(t) = \frac{\sqrt{(v_0 + \zeta\omega x_0)^2 + (x_0\omega_d)^2}}{\omega\sqrt{1 - \zeta^2}} e^{-\zeta\omega t} \sin(\omega_d t + \phi)$$

where

$$\omega_d = \omega\sqrt{1 - \zeta^2} \qquad \text{and} \qquad \phi = \tan^{-1}\frac{x_0\omega_d}{v_0 + \zeta\omega x_0}$$

Here $\omega = \sqrt{k/m}$, $\zeta = c/(2m\omega)$, and $0 < \zeta < 1$ must hold for the solution above to be valid [from equations (1.36) to (1.38)].

For an underdamped system $(0 < \zeta < 1)$ the response to the initial conditions $x_0 = 0$, $v_0 = \hat{F}/m$ is of the form $(\hat{F} = F\Delta t$, has units of N·s)

$$x(t) = \frac{\hat{F}e^{-\zeta\omega t}}{m\omega_d} \sin \omega_d t \qquad (3.6)$$

as predicted by equations (1.36) and (1.38), repeated in Window 3.1. It is convenient to write this solution in the form

$$x(t) = \hat{F}h(t) \qquad (3.7)$$

where $h(t)$ is defined by

$$h(t) = \frac{1}{m\omega_d} e^{-\zeta\omega t} \sin \omega_d t \qquad (3.8)$$

Note that the function $h(t)$ is the response to a unit impulse applied at time $t = 0$. If applied at time $t = \tau$, $\tau \neq 0$, this can also be written as (replace t above with $t - \tau$)

$$h(t - \tau) = \frac{1}{m\omega_d} e^{-\zeta\omega(t-\tau)} \sin \omega_d(t - \tau) \qquad t > \tau \qquad (3.9)$$

and zero for the interval $0 < t < \tau$. The functions $h(t)$ and $h(t - \tau)$ are each called the *impulse response function* of the system.

While the impulse is a mathematical abstraction of an infinite force applied over an infinitesimal time, in applications it presents an excellent model of a force applied over a short period of time. The impulse response is physically interpreted as the response to an initial velocity with no initial displacement (hence no phase shift for $\tau = 0$). The impulse response function is useful for calculating the response of a system to a general applied force excitation as discussed in Section 3.2. A common occurrence that causes such an excitation is an impact. In vibration testing, a mechanical device under test is often given an impact and the response measured to determine the system's vibration properties. The impact is often created by hitting the test specimen with a hammer containing a device for measuring the impact force. Use of the impulse response for vibration testing is also discussed in Chapter 7. In practice, a force is considered to be an impulse if its duration (Δt) is very short compared with the period, $T = 2\pi/\omega$, associated with the structure's undamped natural frequency. In typical vibration tests, Δt is on the order of 10^{-3} s.

Example 3.1.1

In vibration testing an instrumented hammer is often used to hit a device to excite it and to measure the impact force simultaneously. If the device being tested is a single-degree-of-freedom system, plot the responses given $m = 1$ kg, $c = 0.5$ kg/s, $k = 4$ N/m, and $\hat{F} = 2$ N · s. It is often difficult to provide a single impact with a hammer. Sometimes a "double hit" occurs, so the exciting force may have the form

$$F(t) = 2\delta(t) + \delta(t - \tau)$$

Plot the response of the same system with a double hit and compare it to the response to a single impact.

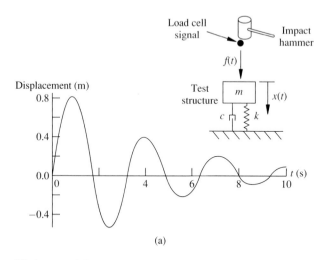

(a)

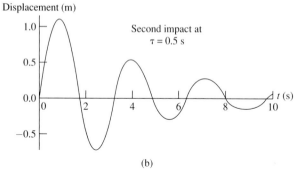

(b)

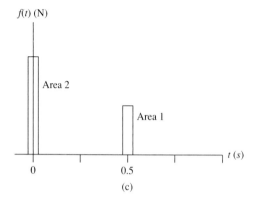

(c)

Figure 3.2 Response of a single-degree-of-freedom system to (a) a single impact and (b) a double impact for $\tau = 0.5$ s. Plot (c) indicates the force versus time curve for the double impact. Note the larger amplitude of (b).

Solution   The solution to the single unit impact and time $t = 0$ is given by equations (3.7) and (3.8) with $\omega = \sqrt{4} = 2$ and $\zeta = c/(2m\omega) = 0.125$. Thus, with $\hat{F} = 2\delta(t)$,

$$x_1(t) = \frac{2}{(1)(2\sqrt{1 - (0.125)^2})} e^{-(0.125)(2)t} \sin 2\sqrt{1 - (0.125)^2}t$$

$$= 1.008 e^{-0.25t} \sin(1.984t)$$

which is plotted in Figure 3.2(a). Similarly, the response to $\delta(t - \tau)$ is calculated from equations (3.7) and (3.9) as

$$x_2(t) = 0.504 e^{-0.25(t-\tau)} \sin 1.984(t - \tau) \qquad t > \tau$$

and $x_2(t) = 0$ for $0 < t < \tau = 0.5$. The force input $f(t)$ is indicated in Figure 3.2(c). It is important to note that no contribution from x_2 occurs until time $t = \tau$. Using the principle of superposition for linear systems, the response to the "double impact" will be the sum of the above to impulse responses:

$$x(t) = x_1(t) + x_2(t)$$

$$= \begin{cases} 1.008 e^{-0.25t} \sin(1.984t) & 0 < t < \tau \\ 1.008 e^{-0.25t} \sin(1.984t) + 0.504 e^{-0.25(t-\tau)} \sin 1.984(t - \tau) & t > \tau \end{cases}$$

This is plotted in Figure 3.2(b) for the value $\tau = 0.5$ s.

Note that the only obvious difference between the two responses is that the "double-hit" response has a "spike" at $\tau = t = 0.5$. The time, τ, represents the time delay between the two hits. The actual value of τ determines exactly how different the two plots of Figure 3.2 look. If $\omega_d \tau$ is near $0, 2\pi, \ldots, n\pi$, the double-hit response of Figure 3.2(b) will look like the sum of two decaying sine waves of the same frequency. Hence Figure 3.2(b) will look like Figure 3.2(a) with a larger amplitude. Similarly, if the quantity $\omega_d \tau$ is near $\pi/2, 3\pi/2, \ldots$, the double-hit response will be the sum of two decaying sine waves that are out of phase and the amplitude of Figure 3.2(b) will be smaller. In the example here, $\omega_d \tau = 0.992$, which is well away from either even or odd multiples of $\pi/2$. $\square$

3.2 RESPONSE TO AN ARBITRARY INPUT

The response of a single-degree-of-freedom system to an arbitrary, general excitation is examined in this section. The response of a single-degree-of-freedom system to an arbitrary force of varying magnitude can be calculated from the concept of the impulse response defined in Section 3.1. The procedure is to divide the exciting force up into infinitesimal impulses, calculate the responses to these individual impulses and add these individual responses to calculate the total response. This is best shown in Figure 3.3, which illustrates an arbitrary applied force $F(t)$ divided into n time intervals of length Δt so that each time increment is defined by $\Delta t = t/n$. At each time interval, t_i, the solution can be calculated by considering the response to be due to an impulse Δt in duration and of force magnitude $F(t)$ [i.e., an impulse of magnitude $F(t_i)\Delta t$].

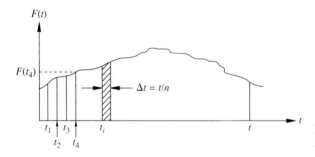

Figure 3.3   Arbitrary excitation force $F(t)$ split up into n impulses.

The response to the impulse following time t_i is then given by equation (3.7) as the increment

$$\Delta x(t_i) = F(t_i)h(t - t_i)\Delta t \tag{3.10}$$

so that the total response after j intervals is the sum

$$x(t_j) = \sum_{i=1}^{j} F(t_i)h(t - t_i)\Delta t \tag{3.11}$$

This again uses the fact that the equation of motion is linear, so that the principle of superposition applies. Forming the sequence of partial sums and finding the limit as $\Delta t \to 0$ ($n \to \infty$) yields

$$x(t) = \int_0^t F(\tau)h(t - \tau)\, d\tau \tag{3.12}$$

from the first fundamental theorem of integral calculus.

For an underdamped single-degree-of-freedom system the impulse response function $h(t - \tau)$ is given by equation (3.9). Substitution of the impulse response function of equation (3.9) into equation (3.12) then yields the result that the response of an underdamped system to an arbitrary input $F(t)$ is given by

$$x(t) = \frac{1}{m\omega_d}e^{-\zeta\omega t} \int_0^t \left[F(\tau)e^{\zeta\omega\tau} \sin \omega_d(t - \tau) \right] d\tau$$

$$= \frac{1}{m\omega_d} \int_0^t F(t - \tau)e^{-\zeta\omega\tau} \sin \omega_d\tau\, d\tau \tag{3.13}$$

as long as the initial conditions are zero. The integral in equation (3.13), known as *Duhamel's integral* or a *convolution integral*, can be used to calculate the response to an arbitrary input as long as it satisfies certain mathematical conditions. The following example illustrates the procedure.

Example 3.2.1

Consider the excitation force of the form given in Figure 3.4. The force is zero until time t_0, when it jumps to a constant level, F_0. This is called the *step function* and when used to excite a single-degree-of-freedom system might model some machine operation or an automobile running over a surface that changes level (such as a curb). Calculate the solution of

$$m\ddot{x} + c\dot{x} + kx = F(t) = \begin{cases} 0 & t_0 > t > 0 \\ F_0 & t \geq t_0 \end{cases} \tag{3.14}$$

with $x_0 = v_0 = 0$ and $F(t)$ as described in Figure 3.4. Here it is assumed that the values of m, c, and k are such that the system is underdamped ($0 < \zeta < 1$).

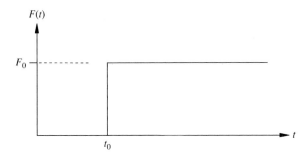

Figure 3.4   Step function of magnitude F_0 applied at time $t = t_0$.

Solution   Applying the convolution integral given by equation (3.13) directly yields

$$x(t) = \frac{1}{m\omega_d} e^{-\zeta\omega t} \left[\int_0^{t_0} (0) e^{\zeta\omega\tau} \sin \omega_d (t - \tau) \, d\tau + \int_{t_0}^t F_0 e^{\zeta\omega\tau} \sin \omega_d (t - \tau) \, d\tau \right]$$

$$= \frac{F_0}{m\omega_d} e^{-\zeta\omega t} \int_{t_0}^t e^{\zeta\omega\tau} \sin \omega_d (t - \tau) \, d\tau$$

Using a table of integrals to evaluate this expression yields

$$x(t) = \frac{F_0}{k} - \frac{F_0}{k\sqrt{1-\zeta^2}} e^{-\zeta\omega(t-t_0)} \cos[\omega_d(t - t_0) - \phi] \qquad t \geq t_0 \qquad (3.15)$$

where

$$\phi = \tan^{-1} \frac{\zeta}{\sqrt{1-\zeta^2}} \qquad (3.16)$$

Note that if $t_0 = 0$, equation (3.15) becomes just

$$x(t) = \frac{F_0}{k} - \frac{F_0}{k\sqrt{1-\zeta^2}} e^{-\zeta\omega t} \cos(\omega_d t - \phi) \qquad (3.17)$$

and if there is no damping ($\zeta = 0$), this expression simplifies further to

$$x(t) = \frac{F_0}{k}(1 - \cos \omega t) \qquad (3.18)$$

Examining the damped response given in equation (3.17), it is obvious that for large time the second term of the response dies out and the steady state is just

$$x_{ss}(t) = \frac{F_0}{k} \qquad (3.19)$$

In fact, the underdamped step response given by equations (3.15) and (3.17) consists of the constant function F_0/k minus a decaying oscillation, as illustrated in Figure 3.5.

Often in the design of vibrating systems subject to a step input, the time it takes for the response to reach the largest value, called the *time to peak* and denoted t_p in Figure 3.5, is used as a measure of the quality of the response. Other quantities used to measure the character of the step response are the *overshoot,* denoted 0.S. in Figure 3.5, which is the largest value of the response "over" the steady-state value, and the *settling time,* denoted by

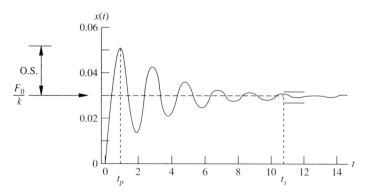

Figure 3.5 Response of an underdamped system to the step excitation of Figure 3.4 for $\zeta = 0.1$ and $\omega = 3.16$ rad/s (with $F_0 = 30$ N, $k = 1000$ N/m, $t_0 = 0$).

t_s, in Figure 3.5. The settling time is the time it takes for the response to get and stay within a certain percentage of the steady-state response. For the case of $t_0 = 0$, t_p and t_s are given by $t_p = \pi/\omega_d$ and $t_s = 3/\zeta\omega$. The peak time is exact (see Problem 3.15) and the settling time is an approximation of when the response stays within 3% of the steady-state value.

□

Example 3.2.2

Another common excitation in vibration is a constant force that is applied for a short period of time and then removed. A rough model of such a force is given in Figure 3.6. Calculate the response of an underdamped system to this excitation.

Solution This pulse-like loading can be written as a combination of step functions calculated in Example 3.2.1, as illustrated in Figure 3.5. The response of a single-degree-of-

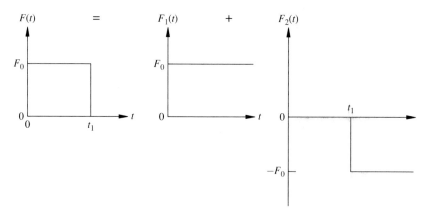

Figure 3.6 Square pulse excitation of magnitude F_0 lasting for t_1 seconds can be written as the sum of a step function starting at zero of magnitude F_0 and a step function starting at t_1 of magnitude $-F_0$ [i.e., $F(t) = F_1(t) + F_2(t)$].

freedom system to $F(t) = F_1(t) + F_2(t)$ is just the sum of the response to $F_1(t)$ plus the response of $F_2(t)$ because the system is linear. First consider the response of an underdamped system to $F_1(t)$. This response is just that calculated in Example 3.2.1 for $t_0 = 0$ and given by (3.17). Next consider the response of the system to $F_2(t)$. This is just the response given by equation (3.15) with F_0 replaced by $-F_0$ and t_0 replaced by t_1. Hence subtracting (3.15) from (3.17) yields the result that the response to the pulse of Figure 3.6 is

$$x(t) = \frac{F_0 e^{-\zeta \omega t}}{k\sqrt{1 - \zeta^2}} \left\{ e^{\zeta \omega t_1} \cos[\omega_d(t - t_1) - \phi] - \cos(\omega_d t - \phi) \right\} \qquad t > t_1$$

where ϕ is as defined in equation (3.16). A plot of this response is given in Figure 3.7 for different pulse widths t_1. Note that the response is much different for $t_1 > \pi/\omega$, and has a maximum magnitude of about five times the maximum magnitude of the time response for $t_1 < \pi/\omega$. Also note that the steady-state response (i.e., the response for large time), is zero in this case.   ☐

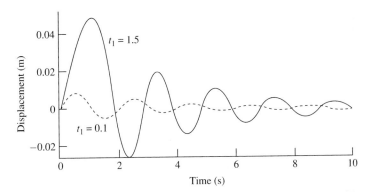

Figure 3.7   Response of an underdamped system to a pulse input of width t_1. The dashed line is for $t_1 = 0.1 < \pi/\omega$ and the solid line is for $t_1 = 1.5 > \pi/\omega$. Both plots are for the case $F_0 = 30$ N, $k = 1000$ N/m, $\zeta = 0.1$, and $\omega = 3.16$ rad/s.

Example 3.2.3

A load of dirt of mass m_d is dropped on the floor of a truck bed. The truck bed is modeled as a spring–mass–damper system (of values k, m, and c respectively). The load is modeled as applying a force $F(t) = m_d g$ to the spring mass damper system as illustrated in Figure 3.8. This allows a crude analysis of the response of the truck's suspension when the truck is being loaded. Calculate the vibration response of the truck bed and compare the maximum deflection to the static load on the truck bed.

Solution   In this case the input force is just a constant [i.e., $F(t) = m_d g$], so that the equation of vibration becomes

$$m\ddot{x} + c\dot{x} + kx = \begin{cases} m_d g & t > 0 \\ 0 & t \leq 0 \end{cases}$$

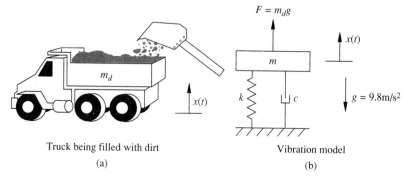

Truck being filled with dirt   Vibration model

(a)   (b)

Figure 3.8   A model of a truck being filled with a load of dirt of weight $m_d g$ and a vibration model that considers the mass of the dirt as an applied constant force.

From equation (3.15) of Example 3.2.1, the response (let $F_0 = m_d g$) is just

$$x(t) = \frac{m_d g}{k} \left[1 - \frac{1}{\sqrt{1 - \zeta^2}} e^{-\zeta \omega t} \cos(\omega_d t - \phi) \right]$$

To obtain a rough idea about the nature of this expression, its undamped value is

$$x(t) = \frac{m_d g}{k} (1 - \cos \omega t)$$

which has a maximum amplitude (when t is such that $\cos \omega t = -1$) of

$$x_{max} = 2 \frac{m_d g}{k}$$

This is twice the static displacement (i.e., twice the deflection the truck would experience if the dirt were placed gently and slowly onto the truck). Thus if the truck were designed with springs based only on the static load, with no margins of safety, the springs in the truck would potentially break, or permanently deform, when subjected to the same mass applied dynamically (i.e., dropped) to the truck. Hence it is important to consider the vibration (dynamic) response in designing structures that could be loaded dynamically. □

It should be noted that the response of a single-degree-of-freedom system to any impact can be calculated numerically, even if the integral in equation (3.12) cannot be evaluated in closed form as done in the preceding examples. Such general numerical procedures, based roughly on equation (3.10), are discussed in Section 9.7.

3.3 RESPONSE TO AN ARBITRARY PERIODIC INPUT

The specific case of periodic inputs is considered in this section. The response to periodic inputs can be calculated by the methods of Section 3.2. However, periodic disturbances occur quite often and merit special consideration. In Chapter 2 the response to a harmonic input is considered. The phrase harmonic input refers to a sinusoidal driving function at a single frequency. Here, the response to any periodic input is considered. A periodic function is any function that repeats itself in time [i.e., any function for which there exists a fixed time T called the period such that $f(t) = f(t + T)$ for all values of t]. A simple example is a forcing function that is the sum of two sinusoids of different frequency. An example of a general periodic forcing function, $F(t)$, of period T is given in Figure 3.9. Note from the figure that the periodic force does not look periodic at all if examined in an interval less than the period T. However, the forcing function does repeat itself every T seconds.

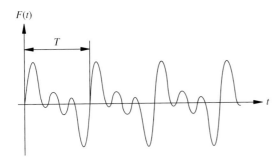

Figure 3.9 Example of a general periodic function of period T.

According to the theory developed by Fourier, any periodic function $F(t)$, with period T, may be represented by an infinite series of the form

$$F(t) = \frac{a_0}{2} + \sum_{n=1}^{\infty} (a_n \cos n\omega_T t + b_n \sin n\omega_T t) \qquad (3.20)$$

where $\omega_T = 2\pi/T$ and where the coefficients a_0, a_n, and b_n for a given periodic function $F(t)$ are calculated by the formulas

$$a_0 = \frac{2}{T} \int_0^T F(t)\, dt \qquad (3.21)$$

$$a_n = \frac{2}{T} \int_0^T F(t) \cos n\omega_T t\, dt \quad n = 1, 2, \ldots \qquad (3.22)$$

$$b_n = \frac{2}{T} \int_0^T F(t) \sin n\omega_T t\, dt \quad n = 1, 2, \ldots \qquad (3.23)$$

Note that the first coefficient a_0 is twice the average of the function $F(t)$ over one cycle. The coefficients a_0, a_n, and b_n are called *Fourier coefficients*. The series of equation (3.20) is the *Fourier series*. A more complete discussion of Fourier series can be found in most introductory differential equation texts (e.g., Boyce and DiPrima, 1986).

The Fourier series is useful and relatively straightforward to work with because of a special property of the trigonometric functions used in the series. This special property, called *orthogonality*, can be stated as follows:

$$\int_0^T \sin n\omega_T t \sin m\omega_T t \, dt = \begin{cases} 0 & m \neq n \\ T/2 & m = n \end{cases} \tag{3.24}$$

$$\int_0^T \cos n\omega_T t \cos m\omega_T t \, dt = \begin{cases} 0 & m \neq n \\ T/2 & m = n \end{cases} \tag{3.25}$$

and

$$\int_0^T \cos n\omega_T t \sin m\omega_T t \, dt = 0 \tag{3.26}$$

The m and n here are integers. The truth of these three orthogonality conditions follows from direct integration, as the reader is invited to try in Problem 3.19. The orthogonality property (i.e., the integral of the product of two functions is zero) is used repeatedly in vibration analysis. In particular, orthogonality is used extensively in Chapters 4, 6, 7, 8, and 9. Orthogonality is also used in statics and dynamics (i.e., the unit vectors are orthogonal).

In Fourier analysis, the orthogonality of the sine and cosine functions on the interval $0 < t < T$ is used to derive the formulas given in equations (3.21), (3.22), and (3.23). These coefficient values are derived as follows. The Fourier coefficients a_n are determined by multiplying equation (3.20) by $\cos m\omega_T t$ and integrating over the period T. Similarly, the coefficients b_n are determined by multiplying by $\sin m\omega_T t$ and integrating. The summation on the right side of equation (3.20) vanishes except for one term because of the orthogonality properties of the trigonometric function. Examining the orthogonality conditions given above illustrates that all terms of the integrated product $\int_0^T F(t) \sin m\omega_T \, dt$, where $F(t)$ is represented by the Fourier series expressed in equation (3.20), will be zero except the term containing b_m. This yields equation (3.23). Likewise all the terms in the series $\int_0^T F(t) \cos m\omega_T \, dt$ are zero except the term containing a_m. This yields equation (3.22). Furthermore, this procedure can be repeated for each of the values of n in the summation in the Fourier series. The procedure for calculating the Fourier coefficient of a simple force is illustrated in the following example.

Example 3.3.1

A triangular wave of period T is illustrated in Figure 3.10 and is described by

$$F(t) = \begin{cases} \dfrac{4}{T} t - 1 & 0 \leq t \leq \dfrac{T}{2} \\ 1 - \dfrac{4}{T}\left(t - \dfrac{T}{2}\right) & \dfrac{T}{2} \leq t \leq T \end{cases}$$

Determine the Fourier coefficients for this function.

Solution   Straightforward integration of equation (3.21) yields that

$$a_0 = \frac{2}{T}\int_0^{T/2}\left(\frac{4}{T}t - 1\right) dt + \frac{2}{T}\int_{T/2}^T \left[1 - \frac{4}{T}\left(t - \frac{T}{2}\right)\right] dt = 0$$

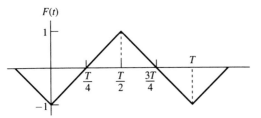

Figure 3.10   Plot of a triangular wave of period T.

which is also the average value of the triangular wave over one period. Similarly, integration of equation (3.23) yields the result that $b_n = 0$ for every n. Equation (3.22) yields

$$a_n = \frac{2}{T} \int_0^{T/2} \left(\frac{4}{T}t - 1 \right) \cos n\omega_T t \, dt + \frac{2}{T} \int_{T/2}^{T} \left[1 - \frac{4}{T}\left(t - \frac{T}{2} \right) \right] \cos n\omega_T t \, dt$$

$$= \begin{cases} 0 & n \text{ even} \\ \dfrac{-8}{\pi^2 n^2} & n \text{ odd} \end{cases}$$

Thus the Fourier representation of this function becomes

$$F(t) = -\frac{8}{\pi^2} \left[\cos \frac{2\pi}{T}t + \frac{1}{9} \cos \frac{6\pi}{T}t + \frac{1}{25} \cos \frac{10\pi}{T}t \cdots \right]$$

which has frequency $2\pi/T$. It is instructive to plot $F(t)$ by adding one term at a time. In this way it is clear how many terms of the infinite series are needed to obtain a reasonable representation of $F(t)$ as plotted in Figure 3.10 (see MATLAB Problem TB3.3). $\square$

Since a general periodic force can be represented as a sum of sines and cosines, and since the system under consideration is linear, the response of a single-degree-of-freedom system is calculated by computing the response to the individual terms of the Fourier series and adding the results. This is similar to the procedure used to solve the base excitation problem of equation (2.33), where the input to a single-degree-of-freedom system consisted of the sum of a single sine and cosine term. This is how superposition and Fourier series are used together to compute the solution for any periodic input. Thus the particular solution $x(t)$ of

$$m\ddot{x} + c\dot{x} + kx = F(t) \tag{3.27}$$

where $F(t)$ is periodic, can be written as

$$x_p(t) = x_1(t) + \sum_{n=1}^{\infty} [x_{cn}(t) + x_{sn}(t)] \tag{3.28}$$

Here the particular solution $x_1(t)$ satisfies the equation

$$m\ddot{x}_1(t) + c\dot{x}_1(t) + kx_1(t) = \frac{a_0}{2} \tag{3.29}$$

the particular solution $x_{cn}(t)$ satisfies the equation

$$m\ddot{x}_{cn} + c\dot{x}_{cn} + kx_{cn} = a_n \cos n\omega_T t \tag{3.30}$$

for all values of n, and the particular solution $x_{sn}(t)$ satisfies the equation

$$m\ddot{x}_{sn} + c\dot{x}_{sn} + kx_{sn} = b_n \sin n\omega_T t \tag{3.31}$$

for all values of n. The solution to equations (3.30) and (3.31) are calculated in Section 2.2 and the solution to equation (3.29) is calculated in Section 3.2. If the system is subject to nonzero initial conditions, this must also be taken into consideration.

The particular solution to equation (3.29) is that of the step response calculated in equation (3.17) with $F_0 = a_0/2m$. This yields

$$x_1(t) = \frac{a_0}{2k} \tag{3.32}$$

The particular solution of equation (3.30) is calculated in equation (2.28) to be

$$x_{cn}(t) = \frac{a_n/m}{\left[[\omega^2 - (n\omega_T)^2]^2 + (2\zeta\omega n\omega_T)^2\right]^{1/2}} \cos(n\omega_T t - \phi_n) \tag{3.33}$$

where

$$\phi_n = \tan^{-1} \frac{2\zeta\omega n\omega_T}{\omega^2 - (n\omega_T)^2}$$

Similarly, the particular solution of equation (3.31) is calculated to be

$$x_{sn}(t) = \frac{b_n/m}{\left[[\omega^2 - (n\omega_T)^2]^2 + (2\zeta\omega n\omega_T)^2\right]^{1/2}} \sin(n\omega_T t - \phi_n) \tag{3.34}$$

The total particular solution of equation (3.27) is then given by the sum of equations (3.32), (3.33), and (3.34) as indicated by equation (3.28). The total solution $x(t)$ is the sum of the particular solution $x_p(t)$ calculated above and the homogeneous solution obtained in Section (1.3). For the underdamped case $(0 < \zeta < 1)$ this becomes

$$x(t) = Ae^{-\zeta\omega t} \sin(\omega_d t - \theta) + \frac{a_0}{2k} + \sum_{n=1}^{\infty} [x_{cn}(t) + x_{sn}(t)] \tag{3.35}$$

where A and θ are determined by the initial conditions. As in the case of a simple harmonic input as described in equation (2.29), the constants A and θ describing the transient response will be different than calculated for the free-response case given in equation (1.38). This is because part of the transient term of equation (3.35) is the result of initial conditions and part is due to the excitation force $F(t)$.

Example 3.3.2

Consider the base excitation problem of Section 2.3 (see Window 3.2) and calculate the total response of the system to initial conditions $x_0 = 0.01$ m and $v_0 = 0$. Assume that $\omega_b = 3$ rad/s, $m = 1$ kg, $c = 10$ kg/s, $k = 1000$ N/m, and $Y = 0.05$ m.

Window 3.2
Review of the Base Excitation Problem of Section 2.3

The base excitation problem is to solve the expression

$$\ddot{x} + 2\zeta\omega\dot{x} + \omega^2 x = 2\zeta\omega\omega_b Y \cos\omega_b t + \omega^2 Y \sin\omega_b t \qquad (2.33)$$

for the motion of a mass, $x(t)$, excited by a harmonic displacement of frequency ω_b and amplitude Y through its spring–damper connections. This has particular solution indicated by the second term on the right-hand side of equation (3.37).

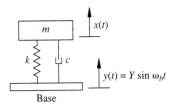

Solution   The equation of motion is given by equation (2.33), which has a periodic input of

$$F(t) = cY\omega_b \cos\omega_b t + kY \sin\omega_b t \qquad (3.36)$$

Comparing coefficients with the Fourier expansion of equation (3.20) yields $a_0 = 0$, $a_n = b_n = 0$ for all $n > 1$ and

$$a_1 = cY\omega_b = (10 \text{ kg/s})(0.05 \text{ m})(3 \text{ rad/s}) = 1.5\text{N}$$

$$b_1 = kY = (1000 \text{ N/m})(0.05 \text{ m}) = 50 \text{ N}$$

The solution for $x_{c1}(t)$ from equation (3.30) is given by equations (2.35) and (2.36), and the solution for $x_{s1}(t)$ from equation (3.31) is given by equations (2.37) and (2.36). The summation of solutions indicated in equation (3.28) is then given by equation (2.38) and the total solution becomes

$$x(t) = Ae^{-\zeta\omega t}\sin(\omega_d t - \theta) + \omega Y \left[\frac{\omega^2 + (2\zeta\omega_b)^2}{\left(\omega^2 - \omega_b^2\right)^2 + (2\zeta\omega\omega_b)^2}\right]^{1/2} \cos(\omega_b t - \phi_1 - \phi_2) \quad (3.37)$$

where A and θ are to be determined by the initial conditions and ϕ_1 and ϕ_2 are as defined by equations (2.36) and (2.39). Since $\zeta = c/(2\sqrt{km}) = 0.158$ and $\omega = \sqrt{k/m} = 31.62$ rad/s, these phase angles become

$$\phi_1 = \tan^{-1}\left(\frac{2(0.158)(31.62)(3)}{(31.62)^2 - (3)^2}\right) = 0.03 \text{ rad}$$

$$\phi_2 = \tan^{-1}\left(\frac{31.62}{(2)(0.158)(3)}\right) = 1.5 \text{ rad}$$

and the magnitude becomes

$$\omega Y \left[\frac{\omega^2 + (2\zeta\omega_b)^2}{(\omega^2 - \omega_b^2)^2 + (2\zeta\omega\omega_b)^2} \right]^{1/2} = (31.62)(0.05) \left\{ \frac{(31.62)^2 + [2(0.158)(3)]^2}{[(31.62)^2 - (3)^2]^2 + [2(0.158)(3)(31.62)]^2} \right\}^{1/2}$$

$$= 0.05 \text{ m}$$

The solution given in equation (3.37) takes the form

$$x(t) = Ae^{-5t} \sin(31.22t - \theta) + 0.05 \cos(3t - 1.53) \tag{3.38}$$

where $\omega_d = \omega\sqrt{1 - \zeta^2} = 31.22$ rad/s. At $t = 0$ this becomes

$$x(0) = A \sin(-\theta) + 0.05 \cos(-1.53)$$

or

$$0.01 \text{ m} = -A \sin\theta + (0.05)(0.04) \tag{3.39}$$

Differentiating $x(t)$ yields

$$\dot{x}(t) = Ae^{-5t} \cos(31.22t - \theta)(31.22) - 5Ae^{-5t} \sin(31.22t - \theta) - 0.15 \sin(3t - 1.53)$$

At $t = 0$ this becomes

$$0 = (31.22)A \cos(-\theta) - 5A \sin(-\theta) - 0.15 \sin(-1.53) \tag{3.40}$$

Equations (3.39) and (3.40) represent two equations in the two unknown constants of integration A and θ. Solving these yields $A = -0.11$ m and $\theta = 0.07$ rad, so that the total solution is

$$x(t) = -0.11e^{-5t} \sin(31.22t - 0.07) + 0.05 \cos(3t - 1.53)$$

This is plotted in Figure 3.11. Note that the transient term is not noticeable after 1 s.

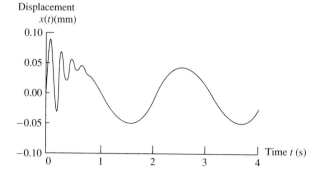

Displacement
$x(t)$(mm)

Figure 3.11   Total time response of a spring–mass–damper system under base excitation as calculated in Example 3.3.2.

3.4 TRANSFORM METHODS

The Laplace transform was introduced briefly in Section 2.5 as an alternative method of solving for the forced harmonic response of a single-degree-of-freedom system. The Laplace transform technique is even more useful for calculating the responses of systems to a variety of force excitations, both periodic and nonperiodic. The usefulness of the Laplace transform technique of solving differential equations and in particular, solving for the forced response lies in the availability of tabulated Laplace transform pairs. Using tabulated Laplace transform pairs reduces the solution of forced vibration problems to algebraic manipulations and table "lookup." In addition, the Laplace transform approach provides certain theoretical advantages and leads to a formulation that is very useful for experimental vibration measurements.

The definition of a Laplace transform of the function $f(t)$ is

$$L[f(t)] = F(s) = \int_0^\infty f(t)e^{-st}\, dt \tag{3.41}$$

for an integrable function $f(t)$ such that $f(t) = 0$ for $t < 0$. The variable s is complex valued. The Laplace transform changes the domain of the function from the positive real number line (t) to the complex number plane (s). The integration in the Laplace transform changes differentiation into multiplication, as the following example illustrates.

Example 3.4.1

Calculate the Laplace transform of the derivative $\dot{f}(t)$.

$$L[\dot{f}(t)] = \int_0^\infty \dot{f}(t)e^{-st}\, dt = \int_0^\infty e^{-st}\frac{d[f(t)]}{dt}\, dt$$

Integration by parts yields

$$L[\dot{f}(t)] = e^{-st}f(t)\Big|_0^\infty + s\int_0^\infty e^{-st}f(t)\, dt$$

Recognizing that the integral in the last term above is the definition of $F(s)$ yields

$$L[\dot{f}(t)] = sF(s) - f(0)$$

where $F(s)$ denotes the Laplace transform of $f(t)$. Repeating this procedure on $\ddot{f}(t)$ yields $L[\ddot{f}(t)] = s^2 F(s) - sf(0) - \dot{f}(0)$.

☐

Example 3.4.2

Calculate the Laplace transform of the unit step function defined by the right-hand side of equation (3.14) and denoted by $\mu(t)$ for the case $t_0 = 0$.

Solution

$$L[\mu(t)] = \int_0^\infty e^{-st}\, dt = -\frac{e^{-st}}{s}\Big|_0^\infty = -\frac{e^{-\infty}}{s} + \frac{e^{-0}}{s} = \frac{1}{s}$$

☐

The procedure for solving for the forced response of a mechanical system is first to take the Laplace transform of the equation of motion. Next, the transformed expression is algebraically solved for $X(s)$, the Laplace transform of the response. The inverse transform of this expression is found by using a table of Laplace transforms to yield the desired time history of the response $x(t)$. This is illustrated in the following example. A sample table of Laplace transform pairs is given in Table 3.1.

TABLE 3.1 COMMON LAPLACE TRANSFORMS FOR ZERO INITIAL CONDITIONS AND[a]

$F(s)$	$f(t)$
1. 1	$\delta(0)$ unit impulse
2. $1/s$	1, unit step
3. $\frac{1}{s+a}$	e^{-at}
4. $\frac{1}{(s+a)(s+b)}$	$\frac{1}{b-a}(e^{-at} - e^{-bt})$
5. $\frac{\omega}{s^2+\omega^2}$	$\sin \omega t$
6. $\frac{s}{s^2+\omega^2}$	$\cos \omega t$
7. $\frac{1}{s(s^2+\omega^2)}$	$\frac{1}{\omega^2}(1 - \cos \omega t)$
8. $\frac{1}{s^2+2\zeta\omega s+\omega^2}$	$\frac{1}{\omega_d}e^{-\zeta\omega t}\sin\omega_d t,\ \zeta < 1, \omega_d = \omega\sqrt{1-\zeta^2}$
9. $\frac{\omega^2}{s(s^2+2\zeta\omega s+\omega^2)}$	$1 - \frac{\omega}{\omega_d}e^{-\zeta\omega t}\sin(\omega_d t + \phi),\ \phi = \cos^{-1}\zeta, \zeta < 1$

[a]A more complete table appears in Appendix B.

Example 3.4.3

Calculate the forced response of an undamped spring–mass system to a unit step function. Assume that both initial conditions are zero.

Solution   The equation of motion is

$$m\ddot{x}(t) + kx(t) = \mu(t)$$

Taking the Laplace transform of this equation yields

$$(ms^2 + k)X(s) = \frac{1}{s}$$

Solving algebraically for $X(s)$ yields

$$X(s) = \frac{1}{s(ms^2 + k)} = \frac{1/m}{s(s^2 + \omega^2)}$$

Examining the definition of the Laplace transform, note that the coefficient $1/m$ passes through the transform. The time function corresponding to the value of $X(s)$ above can be found as entry 7 in Table 3.1. This implies that

$$x(t) = \frac{1/m}{\omega^2}(1 - \cos \omega t) = \frac{1}{k}(1 - \cos \omega t)$$

which, of course, agrees with the solution given by equation (3.18) with $F_0 = 1$.

□

Example 3.4.4

Calculate the response of an underdamped spring–mass system to a unit impulse. Assume zero initial conditions.

Solution   The equation of motion is

$$m\ddot{x} + c\dot{x} + kx = \delta(t)$$

Taking the Laplace transform of both sides of this expression using the results of Example 3.4.1 and entry 1 in Table 3.1 yields

$$(ms^2 + cs + k)X(s) = 1$$

Solving for $X(s)$ yields

$$X(s) = \frac{1/m}{s^2 + 2\zeta\omega s + \omega^2}$$

Assuming that $\zeta < 1$ and consulting entry 8 of Table 3.1 yields

$$x(t) = \frac{1/m}{\omega\sqrt{1 - \zeta^2}} e^{-\zeta\omega t} \sin(\omega\sqrt{1 - \zeta^2}t) = \frac{1}{m\omega_d} e^{-\zeta\omega t} \sin\omega_d t$$

in agreement with equation (3.6).   □

It should be noted that the Laplace transform can be used for problems with untabulated pairs by inverting the integration indicated in equation (3.41). The inversion integral is

$$x(t) = \frac{1}{2\pi j} \int_{-\infty}^{\infty} X(s)e^{st}\, ds \tag{3.42}$$

where $j = \sqrt{-1}$. The inverse Laplace transform is discussed in greater detail in Appendix B.

A related transform is the *Fourier transform*, which arises from considering the Fourier series of a nonperiodic function. The Fourier transform of a function $x(t)$ is denoted by $X(\omega)$ and is defined by

$$X(\omega) = \int_{-\infty}^{\infty} x(t)e^{-j\omega t}\, dt \tag{3.43}$$

which transforms the variable $x(t)$ from a function of time into a function of frequency ω. The inversion of this transform is performed by the integral

$$x(t) = \frac{1}{2\pi} \int_{-\infty}^{\infty} X(\omega)e^{j\omega t}\, d\omega \tag{3.44}$$

The Fourier transform integral defined by equation (3.43) arises from the Fourier series representation of a function described by equation (3.20) by writing the series in complex form and allowing the period to go to infinity.

Note that the definition of the Fourier transform and the Laplace transform are similar. In fact, the form of the Fourier transform pairs given in equations (3.43) and

(3.44) can be obtained by substituting $s = j\omega$ into the Laplace transform pair given by equations (3.41) and (3.42). Although this does not constitute a rigorous definition, it does provide a connection between the two types of transforms.

Fourier transforms are not used as frequently for solving vibration problems as are Laplace transforms. However, the Fourier transform is used extensively in discussing random vibration problems and in the measurement of vibration parameters. Appendix B discusses additional details of transforms. A rigorous description of the use of various transforms, their properties, and their applications can be found in Churchill (1972).

3.5 RESPONSE TO RANDOM INPUTS

So far all the driving forces considered have been deterministic functions of time. That is, given a value of the time t, the value of $F(t)$ is precisely known. Here the response of a system subject to a random force input $F(t)$ is investigated. Disturbances are often characterized as random if the value of $F(t)$ for a given value of t is known only statistically. That is, a random signal has no obvious pattern. For random signals it is not possible to focus on the details of the signal, as it is with a pure deterministic signal. Hence random signals are classified and manipulated in terms of their statistical properties.

Randomness in vibration analysis can be thought of as the result of a series of experiments all performed in an identical fashion under identical circumstances, each of which produces a different response. One record or time history is not enough to describe such a vibration; rather, a statistical description of all possible responses is required. In this case a vibration response $x(t)$ should not be thought of as a single signal, but rather as a collection, or ensemble, of possible time histories resulting from the same conditions (i.e., same system, same controlled environment, same length of time). A single element of such an ensemble is called a *sample* function (or response).

Consider a random signal $x(t)$, or sample, as pictured in Figure 3.12(d). The first distinction to be made about a random time history is whether or not the signal is *stationary*. A random signal is *stationary* if its statistical properties (usually, its average or mean) do not change with time. The *average* of the random, ergodic signal $x(t)$ is defined and denoted by

$$\bar{x} = \lim_{T \to \infty} \frac{1}{T} \int_0^T x(t)\, dt \tag{3.45}$$

as introduced in Section 1.2, equation (1.20), for deterministic signals. Here, it is convenient to consider signals with a zero average or mean [i.e., $\bar{x}(t) = 0$]. This is not too restrictive an assumption, since if $\bar{x}(t) \neq 0$, a new variable $x' = x - \bar{x}$ can be defined. The new variable x' now has zero mean.

The *mean-square value* of the random variable $x(t)$ is denoted by $\bar{x}^2$ and is defined by

$$\bar{x}^2 = \lim_{T \to \infty} \frac{1}{T} \int_0^T x^2(t)\, dt \tag{3.46}$$

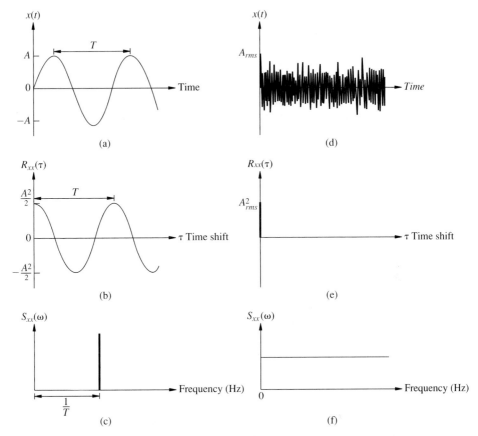

Figure 3.12 (a) Simple sine function; (b) its autocorrelation; (c) its power spectral density. (d) Random signal; (e) its autocorrelation; (f) its power spectral density.

as introduced in Section 1.2, equation (1.21), for deterministic signals. In the case of random signals this is also called the *variance* and provides a measure of the magnitude of the fluctuations in the signal $x(t)$. A related quantity, called the *root-mean-square (rms)* value, is just the square root of the variance:

$$x_{rms} = \sqrt{\bar{x}^2} \tag{3.47}$$

This definition can be applied to the value of a single response over its time history or to an ensemble value at a fixed time.

Another measure of interest in random variables is how fast the value of the variable changes. This addresses the issue of how long it takes to measure enough samples of the variable before a meaningful statistical value can be calculated. As many measured vibration signals are random, such an indication of how quickly a variable may change is very useful. The *autocorrelation function*, denoted by $R_{xx}(\tau)$ and defined by

$$R_{xx}(\tau) = \lim_{T \to \infty} \frac{1}{T} \int_0^T x(t)x(t + \tau)\,dt \tag{3.48}$$

provides a measure of how fast the signal $x(t)$ is changing. The value τ is the time difference between the values at which the signal $x(t)$ is sampled. The prefix *auto* refers to the fact that the term $x(t)x(t+\tau)$ is the product of values of the same sample at two different times. The autocorrelation is a function of the time difference τ only in the special case of stationary random signals. Figure 3.12(e) illustrates the autocorrelation of a random signal, and Figure 3.12(b) illustrates that of a sine function. The Fourier transform of the autocorrelation function defines the *power spectral density* (PSD). Denoting the PSD by $S_{xx}(\omega)$ and repeating the definition of equation (3.43) results in

$$S_{xx}(\omega) = \frac{1}{2\pi} \int_{-\infty}^{\infty} R_{xx}(\tau)e^{-j\omega\tau}\,d\tau \tag{3.49}$$

Note that this integral of $R_{xx}(\tau)$ changes the real number τ into a frequency-domain value ω. Figure 3.12(c) illustrates the PSD of a pure sine signal, and Figure 3.12(f) illustrates the PSD of a random signal. The autocorrelation and power spectral density, defined by equations (3.48) and (3.49), respectively, can be used to examine the response of a spring–mass system to a random excitation.

Recall from Section 3.2 that the response $x(t)$ of a spring–mass–damper system to an arbitrary forcing function $F(t)$ can be represented by using the impulse response function $h(t-\tau)$ given by equation (3.9) for underdamped systems. The Fourier transform of the function $h(t - \tau)$ can be used to relate the PSD of the random input of an underdamped system to the PSD of the system's response. First note from equation (3.8), Example 3.4.4, and entry 8 of Table 3.1 that the Laplace transform of $h(t)$ for a single-degree-of-freedom system is

$$L[(h(t)] = L\left[\frac{1}{m\omega_d}e^{-\zeta\omega t}\sin\omega_d t\right] = \frac{1}{m\omega_d}L[e^{-\zeta\omega t}\sin\omega_d t]$$

$$= \frac{1}{ms^2 + cs + k} = H(s) \tag{3.50}$$

where $H(s)$ is the system transfer function as defined by equation (2.68). In this case the Fourier transform of $h(t)$ can be obtained from the Laplace transform by setting $s = j\omega$ in equation (3.50). This yields simply

$$H(j\omega) = \frac{1}{k - \omega^2 m + c\omega j} \tag{3.51}$$

which upon comparison with equation (2.61) is also the frequency response function for the single-degree-of-freedom oscillator. Let $X(\omega)$ denote the Fourier transform of the impulse response function, $h(t)$; then, from equations (3.43) and (3.51),

$$X(\omega) = \int_{-\infty}^{\infty} h(t)e^{-j\omega t}\,dt = H(\omega) \tag{3.52}$$

where the j is dropped from the argument of H for convenience. Thus the frequency response function of Section 2.5 can be related to the Fourier transform of the impulse response function. This becomes extremely significant in Chapter 7.

Next recall the formulation of the solution of a vibration problem using the impulse response function. From equation (3.12), the response $x(t)$ to a driving force $F(t)$ is simply

$$x(t) = \int_0^t F(\tau)h(t - \tau)\,d\tau \tag{3.53}$$

Note that since $h(t - \tau) = 0$ for $t < \tau$, the lower limit can be extended to minus infinity, so that expression (3.53) can be rewritten as

$$x(t) = \int_{-\infty}^t F(\tau)h(t - \tau)\,d\tau \tag{3.54}$$

Next, the variable of integration τ can be changed to θ by using $\tau = t - \theta$, and hence $d\tau = -d\theta$. Using this change of variables, the previous integral can be written

$$x(t) = -\int_{+\infty}^0 F(t - \theta)h(\theta)\,d\theta = \int_{-\infty}^\infty F(t - \theta)h(\theta)\,d\theta \tag{3.55}$$

which provides an alternative form of the solution of a forced vibration problem in terms of the impulse response function.

Finally, consider the PSD of the response $x(t)$ given by equation (3.49) as

$$S_{xx}(\omega) = \frac{1}{2\pi} \int_{-\infty}^\infty R_{xx}(\tau)e^{-j\omega\tau}\,d\tau \tag{3.56}$$

Upon substitution of the definition of $R_{xx}(\tau)$ from equation (3.48) this becomes

$$S_{xx}(\omega) = \frac{1}{2\pi} \int_{-\infty}^\infty \left[\lim_{T\to\infty} \frac{1}{T} \int_0^T x(\sigma)x(\sigma + \tau)\,d\sigma \right] e^{-j\omega\tau}\,d\tau \tag{3.57}$$

The expressions for $x(t)$ in the integral are evaluated next using equation (3.55), which results in

$$S_{xx}(\omega) =$$

$$\frac{1}{2\pi} \int_{-\infty}^\infty \left[\lim_{T\to\infty} \frac{1}{T} \int_0^T \left[\int_{-\infty}^\infty F(\sigma - \theta)h(\theta)\,d\theta \int_{-\infty}^\infty F(\sigma - \theta + \tau)h(\theta)\,d\theta \right] d\sigma \right] e^{-j\omega\tau}\,d\tau \tag{3.58}$$

$$= \frac{1}{2\pi} \int_{-\infty}^\infty \lim_{T\to\infty} \frac{1}{T} \int_0^T \left[F(\hat{t})F(\hat{t} + \tau) \int_{-\infty}^\infty h(\theta)e^{-j\omega\theta}\,d\theta \int_{-\infty}^\infty h(\theta)e^{j\omega\theta}\,d\theta \right] d\sigma e^{-j\omega\tau}\,d\tau \tag{3.59}$$

where $e^{(\hat{t} - \hat{t})j\omega} = 1$ has been inserted inside the inner integrals and a subsequent change of variables ($\hat{t} = \sigma - \theta$) has been performed on the argument of F, which is subsequently moved outside the integral. The two integrals inside the brackets in equation (3.59) are $H(\omega)$ and its complex conjugate $H(-\omega)$, according to equation (3.52). Recognizing the frequency response functions $H(\omega)$ and $H(-\omega)$ in equation (3.59), this expression can be rewritten as

$$S_{xx}(\omega) = |H(\omega)|^2 \left[\frac{1}{2\pi} \int_{-\infty}^{\infty} R_{ff}(\tau) e^{-j\omega\tau} \, d\tau \right]$$

or simply

$$S_{xx}(\omega) = |H(\omega)|^2 S_{ff}(\omega) \tag{3.60}$$

Here R_{ff} denotes the autocorrelation function for $F(t)$ and S_{ff} denotes the PSD of the forcing function $F(t)$. The notation $|H(\omega)|^2$ indicates the square of the magnitude of the complex frequency response function. A more rigorous derivation of the result can be found in Newland (1984). It is more important to study the result [i.e., equation (3.60)], then the derivations at this level.

Equation (3.60) represents an important connection between the power spectral density of the driving force, the dynamics of the structure, and the power spectral density of the response. In the deterministic case, a solution was obtained relating the harmonic force applied to the system and the resulting response (Chapter 2). In the case where the input is a random excitation, the statement equivalent to a solution is equation (3.60), which indicates how fast the response $x(t)$ changes in relation to how fast the (random) driving force is changing. In the deterministic case of a sinusoidal driving force, the solution allowed the conclusion that the response was also sinusoidal with a new magnitude and phase, but of the same frequency as the driving force. In a way, equation (3.60) makes the equivalent statement for a random excitation (see Window 3.3). It states that when the excitation is a stationary random process, the response will be a stationary random process and the response changes as rapidly as the driving force, but with a modified amplitude. In both the deterministic case and the random case the amplitude of the response is related to the frequency response function of the structure.

Example 3.5.1

Consider the single-degree-of-freedom system of Window 3.1 subject to a random (white noise) force input $F(t)$. Calculate the power spectral density of the response $x(t)$ given that the PSD of the applied force is the constant value S_0.

Solution   The equation of motion is

$$m\ddot{x} + c\dot{x} + kx = F(t)$$

From equation (2.62) or equation (3.51) the frequency response function is

$$H(\omega) = \frac{1}{k - m\omega^2 + c\omega j}$$

Thus

$$|H(\omega)|^2 = \left| \frac{1}{k - m\omega^2 + c\omega j} \right|^2 = \frac{1}{(k - m\omega^2) + c\omega j} \cdot \frac{1}{(k - m\omega^2) - c\omega j}$$

$$= \frac{1}{(k - m\omega^2)^2 + c^2\omega^2}$$

From equation (3.60) the PSD of the response becomes

$$S_{xx} = |H(\omega)|^2 S_{ff} = \frac{S_0}{(k - m\omega^2)^2 + c^2\omega^2}$$

This states that if a single-degree-of-freedom system is excited by a stationary random force (of constant mean and rms value) that has a constant power-spectral density of value S_0, the response of the system will also be random with nonconstant (i.e., frequency dependent) PSD of $S_{xx}(\omega) = S_0/[(k - m\omega^2)^2 + c^2\omega^2]$.

$\square$

Another useful quantity in discussing the response of a system to random vibration is the expected value. The *expected value* (or more appropriately, the *ensemble average*) of $x(t)$ is denoted $E[x]$ and defined by

$$E[x] = \lim_{T \to \infty} \int_0^T \frac{x(t)}{T} \, dt \tag{3.61}$$

which, from equation (3.45), is also the mean value, $\bar{x}$. The expected value is also related to the probability that $x(t)$ lies in a given interval through the *probability density function* $p(x)$. An example of $p(x)$ is the familiar Gaussian distribution function (bell-shaped curve). In terms of the probability density function, the expected value is defined by

$$E[x] = \int_{-\infty}^{\infty} xp(x) \, dx \tag{3.62}$$

The average of the product of the two functions $x(t)$ and $x(t + \tau)$ describes how the function $x(t)$ changes with time, and for a stationary random process is the autocorrelation function:

$$E[x(t)x(t + \tau)] = \lim_{T \to \infty} \frac{1}{T} \int_0^T x(t)x(t + \tau) \, dt = R_{xx}(\tau) \tag{3.63}$$

upon comparison with equation (3.48). From equation (3.46) the mean-square value becomes

$$\bar{x}^2 = R_{xx}(0) = E[x^2] \tag{3.64}$$

The mean-square value can, in turn, be related to the power spectral density function by inverting equation (3.49) using the Fourier transform pair of equations (3.43) and (3.44). This yields

$$R_{xx}(\tau)_{\tau=0} = \int_{-\infty}^{\infty} S_{xx}(\omega) \, d\omega = E[x^2] \tag{3.65}$$

Equation (3.60) relates the PSD of the response $x(t)$ to the PSD of the driving force $F(t)$ and the frequency response function. Combining equations (3.60) and (3.65) yields

$$E[x^2] = \int_{-\infty}^{\infty} |H(\omega)|^2 S_{ff}(\omega) \, d\omega \tag{3.66}$$

This expression relates the mean-square value of the response to the PSD of the (random) driving force and the dynamics of the system. Equations (3.66) and (3.60) form the basis for random vibration analysis for stationary random driving forces. These expressions represent the equivalent of using the impulse response function and frequency response functions to describe deterministic vibration excitations (see Window 3.3).

Window 3.3
***Comparison between Calculations for the Response of a
Spring–Mass–Damper System to Deterministic and Random Excitations***

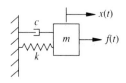

$$transfer\ function = G(s) = \frac{1}{ms^2 + cs + k}$$

Frequency response function: $G(j\omega) = H(\omega) = \frac{1}{k - m\omega^2 + c\omega j}$

Impulse response function: $h(t) = \frac{1}{m\omega_d} e^{-\zeta\omega t} \sin \omega_d t$

Which has the Laplace transform

$$L[h(t)] = \frac{1}{ms^2 + cs + k} = G(s)$$

and the Fourier transform of the impulse response function is just the frequency response function $H(\omega)$. These quantities relate the input and response by

For deterministic $f(t)$:   For random $f(t)$:

$X(s) = G(s)F(s)$ $\leftrightarrow$ $S_{xx}(\omega) = |H(\omega)|^2 S_{ff}(\omega)$

$x(t) = \int_0^t h(t - \tau)f(\tau)\,d\tau$ $\leftrightarrow$ $E[x^2] = \int_{-\infty}^{\infty} |H(\omega)|^2 S_{ff}(\omega)\,d\omega$

To use equation (3.66), the integral involving $|H(\omega)|^2$ must be evaluated. In many useful cases, $S_{ff}(\omega)$ is constant. Hence values of $\int |H(\omega)|^2 d\omega$ have been tabulated (see Newland, 1984). For example,

$$\int_{-\infty}^{\infty} \left| \frac{B_0}{A_0 + j\omega A_1} \right|^2 d\omega = \frac{\pi B_0^2}{A_0 A_1} \tag{3.67}$$

and

$$\int_{-\infty}^{\infty} \left| \frac{B_0 + j\omega B_1}{A_0 + j\omega A_1 - \omega^2 A_2} \right|^2 d\omega = \frac{\pi(A_0 B_1^2 + A_2 B_0^2)}{A_0 A_1 A_2} \tag{3.68}$$

Such integrals, along with equation (3.66), allow computation of the expected value, as the following example illustrates.

Example 3.5.2

Calculate the mean-square value of the response of the system described in Example 3.5.1.

Solution   Since the PSD of the forcing function is the constant S_0, equation (3.66) becomes

$$E[x^2] = S_0 \int_{-\infty}^{\infty} \left| \frac{1}{k - m\omega^2 + jc\omega} \right|^2 d\omega$$

Comparison with (3.68) yields $B_0 = 1$, $B_1 = 0$, $A_0 = k$, $A_1 = c$, and $A_2 = m$. Hence

$$E[x^2] = S_0 \frac{\pi m}{kcm} = \frac{\pi S_0}{kc}$$

Hence, if a spring–mass–damper system is excited by a random force described by constant PSD, S_0, it will have a random response, $x(t)$, with mean-square value $\pi S_0/kc$. $\qquad \square$

Two basic relationships used in analyzing spring–mass–damper systems excited by random inputs are illustrated in this section. The output or response of a randomly excited system is also random, and unlike deterministic systems, cannot be exactly predicted. Hence the response is related to the driving force through the statistical quantities of power spectral density and mean-square values. See Window 3.3 for a comparison between response calculations for deterministic and random inputs.

3.6 SHOCK SPECTRUM

Many disturbances are abrupt or sudden in nature. The impulse is an example of a force applied suddenly. Such a sudden application of a force or other form of disturbance resulting in a transient response is referred to as a *shock*. Because of the common occurrence of shock inputs, a special characterization of the response to a shock has developed as a standard design and analysis tool. This characterization is called the *response spectrum* and consists of a plot of the maximum absolute value of the system's time response versus the natural frequency of the system.

The impulse response discussed in Section 3.1 provides a mechanism for studying the response of a system to a shock input. Recall that the impulse response function, $h(t)$, was derived from considering a force input, $\delta(t)$, of large magnitude and short duration and can be used to calculate the response of a system to any input. The impulse response function forms the basis for calculating the response spectrum introduced here.

Recall from equation (3.12) that the response of a system to an arbitrary input $F(t)$ can be written as

$$x(t) = \int_0^t F(\tau)h(t - \tau)\, d\tau \tag{3.69}$$

where $h(t - \tau)$ is the impulse response function for the system. For an underdamped system $h(t - \tau)$ is given by equation (3.9):

$$h(t - \tau) = \frac{1}{m\omega_d} e^{-\zeta\omega(t-\tau)} \sin \omega_d(t - \tau) \qquad t > \tau \qquad (3.70)$$

which becomes

$$h(t - \tau) = \frac{1}{m\omega} \sin \omega(t - \tau) \qquad (3.71)$$

in the undamped case. The response spectrum is defined to be a plot of the peak or maximum value of the response versus frequency. For an undamped system, equations (3.69) and (3.71) can be combined to yield the maximum value of the displacement response as

$$x(t)_{max} = \frac{1}{m\omega} \left| \int_0^t F(\tau) \sin[\omega(t - \tau)] \, d\tau \right|_{max} \qquad (3.72)$$

Calculating a response spectrum then involves substitution of the appropriate $F(t)$ into equation (3.72) and plotting $x(t)_{max}$ versus the undamped natural frequency. This is usually done on a computer; however, the following example illustrates the procedure by hand calculation.

Example 3.6.1

Calculate the response spectrum for the forcing function given in Figure 3.13. The abruptness of the response is characterized by the time t_1.

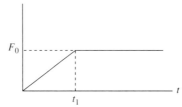

Figure 3.13 Step disturbance with rise time of t_1 seconds.

Solution   As in Example 3.2.2, the response $F(t)$ sketched in Figure 3.13 can be written as the sum of two other simple functions. In this case the input is the sum of

$$F_1(t) = \frac{t}{t_1} F_0$$

and

$$F_2(t) = \begin{cases} 0 & 0 < t < t_1 \\ -\frac{t-t_1}{t_1} F_0 & t \geq t_1 \end{cases}$$

Following the steps taken in Example 3.2.2, the response is calculated by evaluating the response to $F_1(t)$ and separately to $F_2(t)$. Linearity is then used to obtain the total response to $F(t) = F_1(t) + F_2(t)$. The response to $F_1(t)$, denoted $x_1(t)$, calculated using equations (3.69) and (3.71), becomes

$$x_1(t) = \frac{\omega}{k} \int_0^t \frac{F_0\tau}{t_1} \sin \omega(t - \tau) \, d\tau = \frac{F_0}{k} \left(\frac{t}{t_1} - \frac{\sin \omega t}{\omega t_1} \right) \qquad (3.73)$$

Similarly, the response to $F_2(t)$, denoted $x_2(t)$, becomes

$$x_2(t) = \int_0^t F_2(\tau) \frac{1}{m\omega} \sin \omega(t - \tau)\, d\tau = \frac{-F_0}{m\omega} \int_{t_1}^t \frac{\tau - t_1}{t_1} \sin \omega(t - \tau)\, d\tau$$

which becomes

$$x_2(t) = -\frac{F_0}{k} \left[\frac{t - t_1}{t_1} - \frac{\sin \omega(t - t_1)}{\omega t_1} \right] \tag{3.74}$$

so that the total response becomes the sum $x(t) = x_1(t) + x_2(t)$:

$$x(t) = \begin{cases} \dfrac{F_0}{k} \left(\dfrac{t}{t_1} - \dfrac{\sin \omega t}{\omega t_1} \right) & t < t_1 \\[3mm] \dfrac{F_0}{k\omega t_1} [\omega t_1 - \sin \omega t + \sin \omega(t - t_1)] & t \geq t_1 \end{cases} \tag{3.75}$$

Equation (3.75) is the response of an undamped system to the excitation of Figure 3.13. To find the maximum response the derivative of equation (3.75) is set equal to zero and solved for the time t_p at which the maximum occurs. This time t_p is then substituted into the response $x(t_p)$ given by equation (3.75) to yield the maximum response $x(t_p)$. Differentiating equation (3.75) yields $\dot{x}(t_p) = 0$ or

$$-\cos \omega t_p + \cos \omega(t_p - t_1) = 0 \tag{3.76}$$

Using simple trigonometry formulas and solving for ωt_p yields

$$\tan \omega t_p = \frac{1 - \cos \omega t_1}{\sin \omega t_1} \qquad \text{or} \qquad \omega t_p = \tan^{-1} \frac{1 - \cos \omega t_1}{\sin \omega t_1} \tag{3.77}$$

where t_p denotes the time to the first peak [i.e., the time for which the maximum value of equation (3.75) occurs]. Expression (3.77) corresponds to a right triangle of sides $(1 - \cos \omega t_1)$, $\sin \omega t_1$, and hypotenuse

$$\sqrt{\sin^2 \omega t_1 + (1 - \cos \omega t_1)^2} = \sqrt{2(1 - \cos \omega t_1)} \tag{3.78}$$

Hence $\sin \omega t_p$ can be calculated from $(\omega t_p > \pi)$

$$\sin \omega t_p = -\sqrt{\frac{1}{2}(1 - \cos \omega t_1)} \tag{3.79}$$

and

$$\cos \omega t_p = \frac{-\sin \omega t_1}{\sqrt{2(1 - \cos \omega t_1)}} \tag{3.80}$$

Substitution of this expression into solution (3.75) evaluated at t_p yields after some manipulation that [here $x(t_p) = x_{max}$]

$$\frac{x_{max}k}{F_0} = 1 + \frac{1}{\omega t_1} \sqrt{2(1 - \cos \omega t_1)} \tag{3.81}$$

where the left side represents the dimensionless maximum displacement. It is customary to plot the response spectrum (dimensionless) versus the dimensionless frequency

$$\frac{t_1}{T} = \frac{\omega t_1}{2\pi} \tag{3.82}$$

where T is the structure's natural period. This provides a scale related to the characteristic time, t_1, of the input. Figure 3.14 is a plot of the response spectrum for the ramp input force of Figure 3.13. Note that each point on the plot corresponds to a different rise time, t_1, of the excitation. The vertical scale is an indication of the relationship between the structure and the rise time of the excitation.

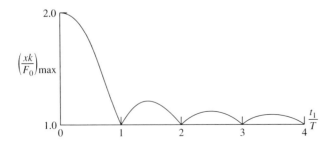

Figure 3.14   Response spectrum for the input force of Figure 3.13. The vertical axis is the dimensionless maximum response and the horizontal axis is dimensionless frequency (or delay time).

3.7 MEASUREMENT VIA TRANSFER FUNCTIONS

The forced response of a vibrating system is very useful in measuring the physical parameters of a system. As indicated in Section 1.6, the measurement of a system's damping coefficient can only be made dynamically. In some systems the damping is large enough that the vibration does not last long enough for a free decay measurement to be taken. This section examines the use of the forced response, transfer functions, and random vibration analysis introduced in previous sections to measure the mass, damping, and stiffness of a system.

The use of transfer functions to measure the properties of structures comes from electrical engineering. In circuit applications a function generator is used to apply a sinusoidal voltage signal to a circuit. The output is measured for a range of input frequencies. The ratio of the Laplace transform of the two signals then yields the transfer function of the test circuit. A similar experiment may be performed on mechanical structures. A signal generator is used to drive a force-generating device (called a *shaker*) that drives the structure sinusoidally through a range of frequencies at a known amplitude and phase. Both the response (either acceleration, velocity, or displacement) and the input force are measured using various transducers. The transform of the input and output signal is calculated and the frequency response function for the system is determined. The physical parameters are then derived from the magnitude and phase of the frequency response function. The details of the measurement procedures are discussed in Chapter 7. The methods of extracting physical parameters from the frequency response function are introduced here.

Several different transfer functions are used in vibration measurement, depending on whether displacement, velocity, or acceleration is measured. The various transfer functions are illustrated in Table 3.2. The table indicates, for example, that the *accelerance transfer function* and corresponding frequency response function are obtained from dividing the transform of the acceleration response by the transform of the driving force.

TABLE 3.2   TRANSFER FUNCTIONS USED
IN VIBRATION MEASUREMENT

Response Measurement	Transfer Function	Inverse Transfer Function
Acceleration	Accelerance	Apparent mass
Velocity	Mobility	Impedance
Displacement	Receptance	Dynamic stiffness

The three transfer functions given in Table 3.2 are related to each other by simple multiplications of the transform variable s, since this corresponds to differentiation. Thus with the *receptance transfer function* denoted by (also called the *compliance* or *admittance*)

$$\frac{X(s)}{F(s)} = H(s) = \frac{1}{ms^2 + cs + k} \tag{3.83}$$

the *mobility transfer function* becomes

$$\frac{sX(s)}{F(s)} = sH(s) = \frac{s}{ms^2 + cs + k} \tag{3.84}$$

because $sX(s)$ is the transform of the velocity. Similarly, $s^2X(s)$ is the transform of the acceleration and the *accelerance transfer function* (*inertance*) becomes

$$\frac{s^2X(s)}{F(s)} = s^2H(s) = \frac{s^2}{ms^2 + cs + k} \tag{3.85}$$

Each of these also defines the corresponding frequency response function by substituting $s = j\omega_{dr}$.

Consider calculating the magnitude of the complex compliance $H(j\omega_{dr})$ from equations (2.62) or (3.83). As expected from equation (2.61), this yields

$$|H(j\omega_{dr})| = \frac{1}{\sqrt{(k - m\omega_{dr}^2)^2 + (c\omega_{dr})^2}} \tag{3.86}$$

Note that the largest value of this magnitude occurs near $k - m\omega_{dr}^2 = 0$, or when the driving frequency is equal to the undamped natural frequency, $\omega_{dr} = \sqrt{k/m}$. Recall from Section 2.2 that this also corresponds to a phase shift of 90°. This argument is used in testing to determine the natural frequency of vibration of a test article from a measured magnitude plot of the system transfer function. This is illustrated in Figure 3.15. The exact value of the peak frequency is derived in Example 2.2.2.

In principle, each of the physical parameters in the transfer function can be determined from the experimental plot of the frequency response function's magnitude. The

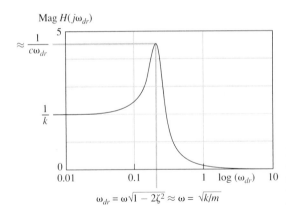

Figure 3.15 Magnitude plot for a spring–mass–damper system for the compliance transfer function indicating the determination of the natural frequency and stiffness.

natural frequency, ω, is determined from the position of the peak. The damping constant c is approximated from the value of the frequency and a measurement of the magnitude $|H(j\omega_{dr})|$ at $\omega_{dr} = \sqrt{k/m}$ since from equation (3.86)

$$\left| H\left(j\sqrt{\tfrac{k}{m}}\right) \right| = \frac{1}{\sqrt{(c\omega)^2}} = \frac{1}{c\omega} \tag{3.87}$$

This formulation for measuring the damping coefficient provides an alternative to the logarithmic decrement technique presented in Section 1.6. Next, the stiffness can be determined from the zero frequency point. For $\omega_{dr} = 0$, equation (3.84) yields

$$|H(0)| = \frac{1}{\sqrt{k^2}} = \frac{1}{k} \tag{3.88}$$

Since $\omega = \sqrt{k/m}$, knowledge of ω and k yields the value for m. In this way m, c, k, ω, and ζ can all be determined from measurements of $|H(j\omega_{dr})|$. More practical methods are discussed in Chapter 7. Of course, m and k can usually be measured by static experiments as well, for comparison.

The analysis above all depends on the experimentally determined function $|H(j\omega_{dr})|$. Most experiments contain several sources of noise, so that a clean plot of $|H(j\omega_{dr})|$ is hard to get. The common approach is to repeat the experiment several times and essentially average the data (i.e., use ensemble averages). In practice, matched sets of input force time histories, $f(t)$, and response time histories, $x(t)$, are averaged to produce $R_{xx}(t)$ and $R_{ff}(t)$ using equation (3.48). The Fourier transform of these averages is then taken using equation (3.49) to get the PSD functions $S_{xx}(\omega)$ and $S_{ff}(\omega)$. Equation (3.60) is then used to calculate $|H(j\omega_{dr})|$ from the PSD values of the measured input and response. This procedure works for averaging noisy data as well as for the case of using a random excitation (zero mean) as the driving force. The transforms and computations required to calculate $|H(j\omega_{dr})|$ are usually made digitally in a dedicated computer used for vibration testing. This is discussed in Chapter 7 in more detail.

3.8 STABILITY

The concept of stability was introduced in Section 1.8 in the context of free vibration. Here the definitions of stability for free vibration are extended to include the forced response case. Recall from equation (1.76) that the free response is stable if it stays within a finite bound for all time (see Window 3.4). This concept of a well-behaved response can also be applied to the forced motion of a vibrating system. In fact, in a sense the inverted pendulum of Example 1.8.1 is an analysis of a forced response if gravity is considered to be the driving force.

Window 3.4
Review of Stability of the Free Response from Section 1.8

A solution $x(t)$ is *stable* if there exists some finite number M such that

$$|x(t)| < M$$

for all $t > 0$. If this bound cannot be satisfied, the response $x(t)$ is said to be *unstable*.

If a response $x(t)$ is stable and $x(t)$ approaches zero as t gets large, the solution $x(t)$ is said to be *asymptotically stable*. An undamped spring–mass system is stable as long as m and k are positive and the value of M is just the amplitude A [i.e., $M = A$, where $x(t) = A\sin(\omega t + \phi)$]. A damped system is asymptotically stable if m, c, and k are all positive [i.e., $x(t) = Ae^{-\zeta\omega t}\sin(\omega_{dt} + \phi)$ goes to zero as t increases].

The stability of the forced response of a system can be defined by considering the nature of the applied force or input. The system

$$m\ddot{x} + c\dot{x} + kx = F(t) \tag{3.89}$$

is defined to be *bounded-input, bounded-output stable* (or simply BIBO stable) if for any bounded input, $F(t)$, the output, or response $x(t)$ is bounded for any arbitrary set of initial conditions. Systems that are BIBO stable are manageable at resonance and do not "blow up."

Note that the undamped version of equation (3.89) is not BIBO stable. To see this, note that if $F(t)$ is chosen to be $F(t) = \sin(k/m)^{1/2}t$ for the case $c = 0$, the response $x(t)$ is clearly not bounded as indicated in Figure 2.4. Also recall that the magnitude of the forced response of an undamped system is $F_0/[\omega^2 - \omega_{dr}^2]$, which approaches infinity as ω_{dr} approaches ω (see Window 3.5). However, the input force $F(t)$ is bounded since

$$|F(t)| = \left|\sin\left(\frac{k}{m}\right)^{1/2}t\right| \le 1 \tag{3.90}$$

Window 3.5

Review of the Response of a Single-Degree-of-Freedom System to Harmonic Excitation

The undamped system:

$$m\ddot{x} + kx = F_0 \cos \omega_{dr} t \quad x(0) = x_0, \qquad \dot{x}(0) = v_0$$

has the solution

$$x(t) = \frac{v_0}{\omega} \sin \omega t + \left(x_0 - \frac{f_0}{\omega^2 - \omega_{dr}^2} \right) \cos \omega t + \frac{f_0}{\omega^2 - \omega_{dr}^2} \cos \omega_{dr} t \qquad (2.11)$$

where $f_0 = F_0/m$, $\omega = \sqrt{k/m}$, and x_0 and v_0 are initial conditions. The underdamped system

$$m\ddot{x} + c\dot{x} + kx = F_0 \cos \omega_{dr} t$$

has steady-state solution

$$x(t) = \frac{f_0}{\sqrt{(\omega^2 - \omega_{dr}^2)^2 + (2\zeta\omega\omega_{dr})^2}} \cos \left(\omega_{dr} t - \tan^{-1} \frac{2\zeta\omega\omega_{dr}}{\omega^2 - \omega_{dr}^2} \right) \qquad (2.28)$$

where the damping ratio ζ satisfies $0 < \zeta < 1$.

---

for all time. Thus there is some bounded force for which the response is not bounded and the definition of BIBO stable is violated. This situation corresponds to resonance. Clearly, an undamped system is poorly behaved at resonance.

Next consider the damped case ($c > 0$). Immediately, the example of resonance above is no longer unbounded. The forced-response magnitude curves given in Figure 2.6 illustrate that the response is always bounded for any bounded periodic driving force. To see that the response of an underdamped system is bounded for any bounded input, recall that the solution for an arbitrary driving force is given in terms of the impulse response in equation (3.12) to be

$$x(t) = \int_0^t f(\tau)h(t - \tau) \, d\tau \qquad (3.91)$$

where $f(t) = F(t)/m$. Taking the absolute value of both sides of this expression yields

$$|x(t)| = \left| \int_0^t f(\tau)h(t - \tau) \, d\tau \right| \leq \int_0^t |f(\tau)h(t - \tau)| \, d\tau \qquad (3.92)$$

where the inequality results from the definition of integrals as a limit of summations. Noting that $|hf| \leq |h||f|$ yields

$$|x(t)| \leq \int_0^t |f(\tau)| \, |h(t - \tau)| \, d\tau \leq M \int_0^t |h(t - \tau)| \, d\tau \qquad (3.93)$$

where $f(t)$ (and hence $F(t)$), is assumed to be bounded by M [i.e., $|f(t)| < M$]. Note that the choice of the constant M is arbitrary and is always chosen as a matter of convenience. Next consider evaluating the integral on the right of inequality (3.93) for the underdamped case. The impulse response for an underdamped system is given by equation (3.9). Substitution of equation (3.9) into (3.93) yields

$$|x(t)| \leq M \int_0^t \frac{1}{m\omega_d} |e^{-\zeta\omega(t-\tau)}| |\sin \omega_d(t-\tau)| \, d\tau$$

$$\leq \frac{M}{m\omega_d} e^{-\zeta\omega t} \int_0^t e^{\zeta\omega\tau} \, d\tau = \frac{M}{m\zeta\omega\omega_d}(1 - e^{-\zeta\omega t}) \leq M \qquad (3.94)$$

since $|\sin \omega_d(t-\tau)| \leq 1$ for all $t > 0$, $m\zeta\omega\omega_d > 1$ and $1 - e^{-\zeta\omega t} < 1$. Thus

$$|x(t)| \leq M$$

Hence as long as the input force is bounded (say, by M) the calculation above illustrates that the response $x(t)$ of an underdamped system is also bounded, and the system is BIBO stable.

The results of Section 2.1 clearly indicate that the response of an undamped system is well behaved, or bounded, as long as the harmonic input is *not* at or near the natural frequency (see Window 3.5). In fact the response given by equation (2.11) illustrates that the maximum magnitude will be less than some constant as long as $\omega_{dr} \neq \omega$. To see this, take the absolute value of equation (2.11), which yields

$$|x(t)| \leq \left|\frac{v_0}{\omega}\right| + \left|x_0 - \frac{f_0}{\omega^2 - \omega_{dr}^2}\right| + \left|\frac{f_0}{\omega^2 - \omega_{dr}^2}\right| < M \qquad (3.95)$$

where M is finite as long as $\omega_{dr} \neq \omega$, since each term is finite. Here v_0 and x_0 are the initial velocity and displacement, respectively. Thus the undamped forced response is sometimes well behaved and sometimes not. Such systems are said to be *Lagrange stable*. Specifically, a system is defined to be Lagrange stable, or bounded, with respect to a *given* input if the response is bounded for any set of initial conditions. Undamped systems are Lagrange stable with respect to many inputs. This definition is useful when $F(t)$ is known completely or known to fall into a specific class of functions. Both the damped and undamped solutions given in Window 3.5 are Lagrange stable for $\omega \neq \omega_{dr}$.

In general, if the homogeneous solution is asymptotically stable, the forced response will be BIBO stable. If the homogeneous response is stable (marginally stable), the forced response will only be Lagrange stable. The forced response of an unstable homogeneous system can still be BIBO stable, as illustrated in the following example.

Example 3.8.1

Consider the inverted pendulum of Example 1.8.1 and discuss its stability properties.

Solution   Considering the small-angle approximation of the inverted pendulum equation results in the equation of motion

$$ml^2\ddot{\theta}(t) + \frac{kl^2}{2}\theta(t) = mgl\theta$$

If $mgl\theta$ is considered to be an applied force, the homogeneous solution is stable since m, l, and k are all positive. The forced response, however, is not bounded unless $kl > 2mg$ and was shown in Example 1.8.1 to be divergent (unbounded) in this case. Hence the forced response of this system is Lagrange stable for $F(t) = mgl\theta$ if $kl > 2mg$, and unbounded (unstable) if $kl < 2mg$.

$\square$

Example 3.8.2

Consider again the inverted pendulum of Example 1.8.1. Design an applied force $F(t)$ such that the response is bounded for $kl < 2mg$.

Solution   The problem is to find $F(t)$ such that θ satisfying

$$ml^2\ddot{\theta}(t) + \left(\frac{kl^2}{2} - mgl\right)\theta(t) = F(t)$$

is bounded. As a starting point assume that $F(t)$ has the form

$$F(t) = -a\theta - b\dot{\theta}$$

where a and b are to be determined by the design for stability. This form is attractive because it changes the inhomogeneous problem into a homogeneous problem. The equation of motion then becomes

$$ml^2\ddot{\theta}(t) + b\dot{\theta} + \left(\frac{kl^2}{2} - mgl + a\right)\theta = 0$$

From Section 1.8 it is known that if each of the coefficients is positive, the response is asymptotically stable, which is certainly bounded. Hence choose $b > 0$ and a such that

$$\frac{kl^2}{2} - mgl + a > 0$$

and the forced response will be bounded.

$\square$

An applied force can also cause a stable (or asymptotically stable) system response to become unstable. To see this, consider the system

$$\ddot{x} + \dot{x} + 4x = f(t) \tag{3.96}$$

where $f(t) = ax + b\dot{x}$. If a is chosen to be 2 and $b = 2$, the equation of motion becomes

$$\ddot{x} - \dot{x} + 2x = 0 \tag{3.97}$$

which has a solution that grows exponentially and illustrates flutter instability. At first glance this seems to violate the statement above that asymptotically stable homogeneous system are BIBO stable. This example, however, does not violate the definition because the input $f(t)$ is not bounded. The applied force is a function of the displacement and velocity, which grow without bound.

PROBLEMS

Section 3.1

3.1. Calculate the solution to

$$\ddot{x} + 2\dot{x} + 2x = \delta(t - \pi)$$

$$x(0) = 1 \qquad \dot{x}(0) = 0$$

and plot the response.

3.2. Calculate the solution to

$$\ddot{x} + 2\dot{x} + 3x = \sin t + \delta(t - \pi)$$

$$x(0) = 0 \qquad \dot{x}(0) = 1$$

and plot the response.

3.3. Calculate the impulse response function for a critically damped system.

3.4. Calculate the impulse response of an overdamped system.

3.5. Derive equation (3.6) from equations (1.36) and (1.38).

3.6. Consider a simple model of an airplane wing given in Figure 3.16. The wing is approximated as vibrating back and forth in its plane, massless compared to the missile carriage system

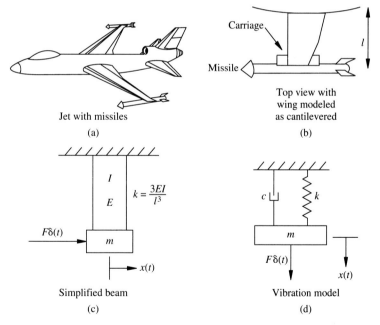

Figure 3.16 Modeling of wing vibration resulting from the release of a missile. (a) system of interest; (b) simplification of the detail of interest; (c) crude model of the wing: a cantilevered beam section (recall Figure 1.23); (d) vibration model used to calculate the response neglecting the mass of the wing.

(of mass m). The modulus and moment of inertia of the wing are approximated by E and I, respectively, and l is the length of the wing. The wing is modeled as a simple cantiliver for the purpose of estimating the vibration resulting from the release of the missile, which is approximated by the impulse function $F\delta(t)$. Calculate the response and plot your results for the case of an aluminum wing 2 m long with $m = 100$ kg, $\zeta = 0.01$, and $I = 0.25\ m^4$. Model F as 1000 N lasting over 10^{-3} s.

3.7. A cam in a large machine can be modeled as applying a 10,000 N-force over an interval of 0.005 s. This can strike a valve that is modeled as having physical parameters: $m = 10$ kg, $c = 18$ N $\cdot$ s/m, and stiffness $k = 9000$ N/m. The cam strikes the valve once every 1 s. Calculate the vibration response, $x(t)$, of the valve once it has been impacted by the cam. The valve is considered to be closed if the distance between its rest position and its actual position is less than 0.0001 m. Is the valve closed the very next time it is hit by the cam?

3.8. The vibration of packages dropped from a height of h meters can be approximated by considering Figure 3.17. Calculate the vibration of the mass m after the system falls and hits the ground. Assume that the system is underdamped.

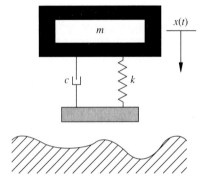

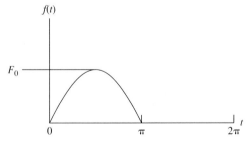

Figure 3.17   Vibration model of a package being dropped onto the ground.

Section 3.2

3.9. Calculate the response of an overdamped single-degree-of-freedom system to an arbitrary non-periodic excitation.

3.10. Calculate the response of an underdamped system to the excitation given in Figure 3.18.

Figure 3.18   Plot of a pulse input of the form $f(t) = F_0 \sin t$.

3.11. Calculate and plot the response of an undamped system to a step function with a finite rise time of t_1 for the case $m = 1$ kg, $k = 1$ N/m, $t_1 = 4$ s and $F_0 = 20$ N. This function is described by

$$F(t) = \begin{cases} \dfrac{F_0 t}{t_1} & 0 \le t \le t_1 \\ F_0 & t > t_1 \end{cases}$$

3.12. A wave consisting of the wake from a passing boat impacts a seawall. It is desired to calculate the resulting vibration. Figure 3.19 illustrates the situation and suggests a model. This force in Figure 3.19 can be expressed as

$$F(t) = \begin{cases} F_0 \left(1 - \dfrac{t}{t_0} \right) & 0 \le t \le t_0 \\ 0 & t > t_0 \end{cases}$$

Calculate the response of the sea wall–dike system to such a load.

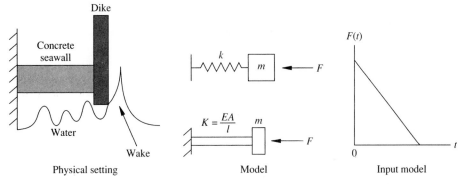

Figure 3.19   Model of a wave hitting a seawall modeled as a nonperiodic force exciting an undamped single-degree-of-freedom system.

3.13. Determine the response of an undamped system to a ramp input of the form $F(t) = F_0 t$, where F_0 is a constant. Plot the response for three periods for the case $m = 1$ kg, $k = 100$ N/m and $F_0 = 50$ N.

3.14. A machine resting on an elastic support can be modeled as a single-degree-of-freedom spring–mass system arranged in the vertical direction. The ground is subject to a motion $y(t)$ of the form illustrated in Figure 3.20. The machine has a mass of 5000 kg and the support has stiffness 1.5×10^3 N/m. Calculate the resulting vibration of the machine.

3.15. Consider the step response described in Figure 3.5. Calculate t_p by noting that it occurs at the first peak, or critical point, of the curve.

3.16. Calculate the value of the overshoot (o.s.), for the system of Figure 3.5.

3.17. It is desired to design a system so that its step response has a settling time of 3 s and a time to peak of 1 s Calculate the appropriate natural frequency and damping ratio to use in the design.

3.18. Plot the response of a spring–mass–damper system for this input of Figure 3.6 for the case that the pulse width is the natural period of the system (i.e., $t_1 = \pi/\omega$).

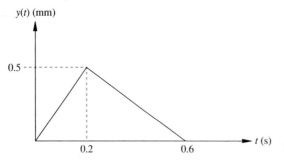

Figure 3.20   Triangular pulse input.

Section 3.3

3.19. Derive equations (3.24), (3.25), and (3.26) and hence verify the equations for the Fourier coefficient given by equations (3.21), (3.22), and (3.23).

3.20. Calculate b_n from Example 3.3.1 and show that $b_n = 0$, $n = 1, 2, \ldots, \infty$ for the triangular force of Figure 3.10. Also verify the expression for a_n by completing the integration indicated. (*Hint*: Change the variable of integration from t to $x = 2\pi nt/T$.)

3.21. Determine the Fourier series for the rectangular wave illustrated in Figure 3.21.

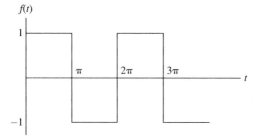

Figure 3.21   Rectangular periodic signal.

3.22. Determine the Fourier series representation of the sawtooth curve illustrated in Figure 3.22.

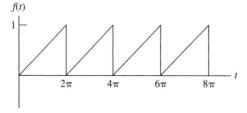

Figure 3.22   Sawtooth periodic signal.

3.23. Calculate and plot the response of the base excitation problem with base motion specified by the velocity

$$\dot{y}(t) = 3e^{-t/2}\mu(t) \quad \text{m/s}$$

where $\mu(t)$ is the unit step function and $m = 10$ kg, $\zeta = 0.01$, and $k = 1000$ N/m. Assume that the initial conditions are both zero.

3.24. Calculate and plot the response of the spring–mass–damper system of Figure 2.1 with $m = 100$ kg, $\zeta = 0.1$ and $k = 1000$ N/m to the signal of Figure 3.10, with maximum force of 1 N. Assume that the initial conditions are zero and let $T = 2\pi s$.

3.25. Calculate the total response of the system of Example 3.3.2 for the case of a base motion driving frequency of $\omega_b = 3.162$ rad/s.

Section 3.4

3.26. Calculate the response of

$$m\ddot{x} + c\dot{x} + kx = F_0\mu(t)$$

where $\mu(t)$ is the unit step function for the case with $x_0 = v_0 = 0$. Use the Laplace transform method and assume that the system is underdamped.

3.27. Using the Laplace transform method, calculate the response of the system of Example 3.4.4 for the overdamped case ($\zeta > 1$). Plot the response for $m = 1$ kg, $k = 100$ N/m and $\zeta = 1.5$.

3.28. Calculate the response of the underdamped system given by

$$m\ddot{x} + c\dot{x} + kx = F_0 e^{-at}$$

using the Laplace transform method. Assume that $a > 0$ and that the initial conditions are all zero.

Section 3.5

3.29. Calculate the mean-square response of a system to an input force of constant PSD, S_0, and frequency response function $H(\omega) = 10/(3 + 2j\omega)$.

3.30. Consider the base excitation problem of Section 2.3 as applied to an automobile model of Example 2.3.1 and illustrated in Figure 2.10. In this problem let the road have a random stationary cross section producing a PSD of S_0. Calculate the PSD of the response and the mean-square value of the response.

3.31. Calculate the PSD of the triangular wave of Figure 3.10.

3.32. Verify that the average of $x - \bar{x}$ is zero by using the definition given in equation (3.45).

Section 3.6

3.33. A power-line pole with a transformer is modeled by

$$m\ddot{x} + kx = -\ddot{y}$$

where x and y are as indicated in Figure 3.23. Calculate the response of the relative displacement $(x - y)$ if the pole is subject to an earthquake base excitation of (assume the initial conditions are zero)

$$\ddot{y}(t) = \begin{cases} A\left(1 - \dfrac{t}{t_0}\right) & 0 \le t \le 2t_0 \\ 0 & t > 2t_0 \end{cases}$$

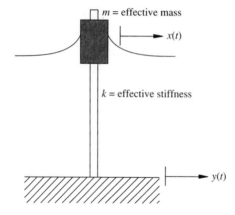

Figure 3.23 Vibration model of a power-line pole with a transformer mounted on it.

3.34. Calculate the response spectrum of an undamped system to the forcing function

$$F(t) = \begin{cases} F_0 \sin \frac{\pi t}{t_1} & 0 \le t \le t_1 \\ 0 & t > t_1 \end{cases}$$

assuming the initial conditions are zero.

Section 3.7

3.35. Using complex algebra, derive equation (3.86) from (3.83) with $s = j\omega_{dr}$.

3.36. Using the values indicated in Figure 3.15, plot the inertance transfer function's magnitude and phase.

3.37. Using the values indicated in Figure 3.15, plot the mobility transfer function's magnitude and phase.

3.38. Calculate the compliance transfer function for a system described by

$$a\ddddot{x} + b\dddot{x} + c\ddot{x} + d\dot{x} + ex = f(t)$$

where $f(t)$ is the input force and $x(t)$ is a displacement.

3.39. Calculate the frequency response function for the compliance of Problem 38.

3.40. Plot the magnitude of the frequency response function for the system of Problem 38 for

$$a = 1, b = 4, c = 11, d = 16, \text{ and } e = 8.$$

3.41. An experimental (compliance) magnitude plot is illustrated in Figure 3.24. determine ω, ζ, c, m and k. Assume that the units correspond to m/N along the vertical axis.

Section 3.8

3.42. Show that a critically damped system is BIBO stable.

3.43. Show that an overdamped system is BIBO stable.

3.44. Is the solution of $2\ddot{x} + 18x = 4\cos 2t + \cos t$ Lagrange stable?

3.45. Calculate the response of equation (3.96) for $x_0 = 0$, $v_0 = 1$ for the case that $a = 4$ and $b = 0$. Is the response bounded?

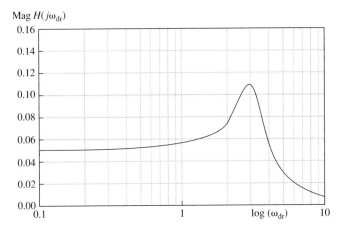

Mag $H(j\omega_{dr})$

Figure 3.24

M ATLAB VIBRATION TOOLBOX

You may use the files contained in the *Vibration Toolbox*, first discussed at the end of Chapter 1 (immediately following the problems), to help solve many of the exercises listed above. If you have not yet used the *Vibration Toolbox*, return to Chapter 1 for an introduction and open the file "INTRO.TXT" on the disk on the inside back cover. The files contained in folder VTB3 may be used to help understand the nature of the general forced response of a single-degree-of-freedom system as discussed in this chapter and the dependence of this response on various parameters. The following problems are suggested to help build some intuition regarding the material on general forced response and to become familiar with the various formulas.

TOOLBOX PROBLEMS

TB3.1. Use file VTB3_1, to solve for the response of a system with a 10-kg mass, damping $c = 2.1$ kg/s, and stiffness $k = 2100$ N/m, subject to an impulse at time $t = 0$ of magnitude 10 N. Next vary the value of c, first increasing it, then decreasing it, and note the effect in the responses.

TB3.2. Use file VTB3_2 to reproduce the plot of Figure 3.5. Then see what happens to the response as the damping coefficient is varied by trying an overdamped and critically damped value of ζ, and examining the resulting response.

TB3.3. If you are confident with MATLAB, try using the plot command to plot (say, for $T = 6$)

$$-\frac{8}{\pi^2}\cos\frac{2\pi}{T}t, \qquad -\frac{8}{\pi^2}\left(\cos\frac{2\pi}{T}t + \frac{1}{9}\cos\frac{6\pi}{T}t\right)$$

then

$$-\frac{8}{\pi^2}\left(\cos\frac{2\pi}{T}t + \frac{1}{9}\cos\frac{6\pi}{T}t + \frac{1}{25}\cos\frac{10\pi}{T}t\right)$$

and so on, until you are satisfied that the Fourier series computed in Example 3.3.1 converges to the function plotted in Figure 3.10. If you are not familiar enough with MATLAB to try this on your own, run VTB3_3, which is a demo that does this for you.

TB3.4. Using file VTB3_4, examine the effect of varying the system's natural frequency on the response spectrum for the force given in Figure 3.13. Pick the frequencies $f = 10$ Hz, 100 Hz, and 1000 Hz and compare the various response spectrum plots.

4 Multiple-Degree-of-Freedom Systems

This chapter presents the vibration analysis of systems with more than one degree of freedom (e.g., more than one moving part). Such models are required to analyze the vibration of complicated devices such as the automobile pictured here. This particular automobile is fitted with a state-of-the-art active suspension system that is capable of changing the various stiffness and damping values of the car's suspension system under computer control to adapt to various road and load conditions. The analysis methods presented in this chapter were used in the design of this active system. Active systems are discussed in Section 5.8. Automobile vibrations are discussed in example 4.8.2 and in problems 4.13, 4.41, and 4.71.

The bottom photo is of the Sears tower in Chicago, which experiences vibrations induced by wind. Vibration of buildings is often modeled by multiple-degree-of-freedom systems as discussed in this chapter: Examples 4.4.3, 4.5.2, and Section 4.6 examine building vibration.

In the preceding chapters, a single coordinate and single second-order differential equation sufficed to describe the vibratory motion of the mechanical device under consideration. However, many mechanical devices and structures cannot be modeled successfully by a single degree of freedom. For example, the base excitation problem of Section 2.3 requires a coordinate for the base as well as the main mass if the base motion is not prescribed, as assumed in Section 2.3. If the base motion is not prescribed then the coordinate, y, will also satisfy a second-order differential equation and the system becomes a two-degree-of-freedom model. Machines with many moving parts have many degrees of freedom.

In this chapter a two-degree-of-freedom example is used to introduce the special phenomena associated with multiple-degree-of-freedom systems. These phenomena are then extended to systems with an arbitrary but finite number of degrees of freedom. To keep a record of each coordinate of the system, vectors are introduced and used along with matrices. This is done both for the ease of notation and to enable vibration theory to take advantage of the disciplines of matrix theory, linear algebra, and computational matrix theory.

4.1 TWO-DEGREE-OF-FREEDOM MODEL (UNDAMPED)

Consider the two-mass system of Figure 4.1(a). This undamped system is similar to the system of Figure 2.7 except that the base motion is not prescribed in this case. Figure 4.1(b) illustrates a single-mass system capable of moving in two directions and hence provides an example of a two-degree-of-freedom system as well. Figure 4.1(c) illustrates a single rigid mass that is capable of moving in translation as well as rotation about its axis. In each of these three cases more than one coordinate is required to describe the vibration of the system. Each of the three parts of Figure 4.1 constitutes a two-degree-of-freedom system. A physical example of each system might be a (a) two-story building, (b) the vibration of a drill press, and (c) the rocking motion of an automobile or aircraft.

A force diagram illustrating the spring forces acting on each mass in Figure 4.1(a) is illustrated in Figure 4.2. The force of gravity is excluded following the reasoning used in Figure 1.14 (i.e., the static deflection balances the gravitational force). Summing

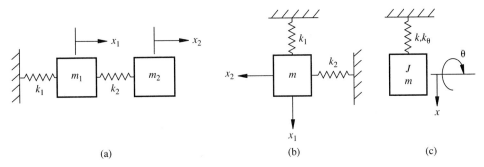

Figure 4.1   (a) Simple two-degree-of-freedom model consisting of two masses connected in series by two springs. (b) Single mass with two degrees of freedom (i.e., the mass moves along both the x_1 and x_2 directions). (c) Single mass with one translational degree of freedom and one rotational degree of freedom.

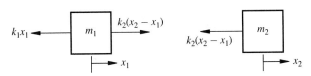

Figure 4.2   Free-body diagrams of each mass in the system of Figure 4.1(a), indicating the restoring force provided by the springs.

forces on each mass in the horizontal direction yields

$$m_1\ddot{x}_1 = -k_1 x_1 + k_2(x_2 - x_1)$$
$$m_2\ddot{x}_2 = -k_2(x_2 - x_1)$$

(4.1)

Rearranging these two equations yields

$$m_1\ddot{x}_1 + (k_1 + k_2)x_1 - k_2 x_2 = 0$$
$$m_2\ddot{x}_2 - k_2 x_1 + k_2 x_2 = 0$$

(4.2)

Equations (4.2) consist of two coupled second-order ordinary differential equations, each of which requires two initial conditions to solve. Hence these two coupled equations are subject to the four initial conditions,

$$x_1(0) = x_{10} \qquad \dot{x}_1(0) = \dot{x}_{10} \qquad x_2(0) = x_{20} \qquad \dot{x}_2(0) = \dot{x}_{20}$$

(4.3)

where the constants $\dot{x}_{10}$, $\dot{x}_{20}$ and x_{10}, x_{20} represent the initial velocities and displacements of each of the two masses. These initial conditions provide the four constants of integration needed to solve the two second-order differential equations for the free response of each mass.

There are several approaches available to solve equations (4.2) and (4.3) for the responses $x_1(t)$ and $x_2(t)$. First note that neither equation can be solved by itself because each equation contains both x_1 and x_2 (i.e., the equations are *coupled*). Physically, this states that the motion of x_1 affects the motion of x_2, and vice versa. A convenient method of solving this system is to use vectors and matrices. The vector approach to solving this simple two-degree-of-freedom problem is also readily extendable to systems with an arbitrary finite number of degrees of freedom. Vectors and matrices are introduced here briefly to solve equation (4.2). A more detailed account can be found in Appendix C.

Define the vector $\mathbf{x}(t)$ to be the column vector consisting of the two responses of interest:

$$\mathbf{x}(t) = \begin{bmatrix} x_1(t) \\ x_2(t) \end{bmatrix} \tag{4.4}$$

This is called a displacement or response vector and is a 2×1 array of functions. Differentiation of a vector is defined here by differentiating each element so that

$$\dot{\mathbf{x}}(t) = \begin{bmatrix} \dot{x}_1(t) \\ \dot{x}_2(t) \end{bmatrix} \quad \text{and} \quad \ddot{\mathbf{x}}(t) = \begin{bmatrix} \ddot{x}_1(t) \\ \ddot{x}_2(t) \end{bmatrix} \tag{4.5}$$

are the velocity and the acceleration vectors, respectively. A square matrix is a square array of numbers, which could be made, for instance, by combining two, 2×1 column vectors to produce a 2×2 matrix. An example of a 2×2 matrix is given by

$$M = \begin{bmatrix} m_1 & 0 \\ 0 & m_2 \end{bmatrix} \tag{4.6}$$

Note here that capital letters are used to denote matrices and bold lowercase letters are used to denote vectors.

Vectors and matrices can be multiplied together in a variety of ways. The method most useful to define the product of a matrix times a vector is to define the result to be a vector with elements consisting of the dot product of the vector with each "row" of the matrix treated as a vector. The *dot product* of a vector is defined by

$$\mathbf{x}^T\mathbf{x} = \begin{bmatrix} x_1 & x_2 \end{bmatrix} \begin{bmatrix} x_1 \\ x_2 \end{bmatrix} = x_1^2 + x_2^2 \tag{4.7}$$

which is a scalar. The symbol $\mathbf{x}^T$ denotes the transpose of the vector and changes a column vector into a row vector. Equation (4.7) is also called the *inner product* or *scalar product* of the vector $\mathbf{x}$ with itself. A scalar, a, times a vector, $\mathbf{x}$, is defined simply as $a\mathbf{x} = \begin{bmatrix} ax_1 & ax_2 \end{bmatrix}^T$ (i.e., a vector of the same dimension with each element multiplied by the scalar). (Recall that a scalar is any real or complex number.) These rules for manipulating vectors should be familiar from introductory mechanics (i.e., statics and dynamics) texts.

Another type of product can be defined between two vectors, namely the *outer product*. The outer product of two vectors $\mathbf{x} = \begin{bmatrix} x_1 & x_2 \end{bmatrix}^T$ and $\mathbf{y} = \begin{bmatrix} y_1 & y_2 \end{bmatrix}^T$ is defined to be the matrix given by

$$\mathbf{x}\mathbf{y}^T = \begin{bmatrix} x_1 \\ x_2 \end{bmatrix} \begin{bmatrix} y_1 & y_2 \end{bmatrix} = \begin{bmatrix} x_1 y_1 & x_1 y_2 \\ x_2 y_1 & x_2 y_2 \end{bmatrix} \tag{4.8}$$

By comparison the inner product of the two vectors $\mathbf{x}$ and $\mathbf{y}$ is the scalar $\mathbf{x}^T\mathbf{y} = x_1 y_1 + x_2 y_2$. The outer product is useful in vibration measurement and is discussed in Chapter 7. Several other useful products can be defined. The following example illustrates the rules for multiplying a matrix times a vector.

Example 4.1.1

Consider the product of the matrix M of equation (4.6) and the acceleration vector $\ddot{\mathbf{x}}$ of equation (4.5). This product becomes

$$M\ddot{\mathbf{x}} = \begin{bmatrix} m_1 & 0 \\ 0 & m_2 \end{bmatrix} \begin{bmatrix} \ddot{x}_1 \\ \ddot{x}_2 \end{bmatrix} = \begin{bmatrix} m_1\ddot{x}_1 + 0\ddot{x}_2 \\ 0\ddot{x}_1 + m_2\ddot{x}_2 \end{bmatrix} = \begin{bmatrix} m_1\ddot{x}_1 \\ m_2\ddot{x}_2 \end{bmatrix}$$

where the first element of the product is defined to be the dot product of the row vector $[m_1 \quad 0]$ with the column vector $\ddot{\mathbf{x}}$ and the second element is the dot product of the row vector $[0 \quad m_2]$ with $\ddot{\mathbf{x}}$. Note that the product of a matrix and a vector is a vector. □

Example 4.1.2

Consider the 2×2 matrix K defined by

$$K = \begin{bmatrix} k_1 + k_2 & -k_2 \\ -k_2 & k_2 \end{bmatrix} \tag{4.9}$$

and calculate the product $K\mathbf{x}$.

Solution   Again the product is formed by considering the first element to be the inner product of the row vector $[k_1 + k_2 \quad -k_2]$ and the column vector $\mathbf{x}$. The second element of the product vector $K\mathbf{x}$ is formed from the inner product of the row vector $[-k_2 \quad k_2]$ and the vector $\mathbf{x}$. This yields

$$K\mathbf{x} = \begin{bmatrix} k_1 + k_2 & -k_2 \\ -k_2 & k_2 \end{bmatrix} \begin{bmatrix} x_1 \\ x_2 \end{bmatrix} = \begin{bmatrix} (k_1 + k_2)x_1 - k_2x_2 \\ -k_2x_1 + k_2x_2 \end{bmatrix} \tag{4.10}$$

□

Two vectors of the same size are said to be equal if and only if each element of one vector is equal to the corresponding element in the other vector. With this in mind, consider the *vector equation*

$$M\ddot{\mathbf{x}} + K\mathbf{x} = \mathbf{0} \tag{4.11}$$

where $\mathbf{0}$ denotes the column vector of zeros:

$$\mathbf{0} = \begin{bmatrix} 0 \\ 0 \end{bmatrix}$$

Substitution of the value for M from equation (4.6) and the value for K from equation (4.9) into equation (4.11) yields

$$\begin{bmatrix} m_1 & 0 \\ 0 & m_2 \end{bmatrix} \begin{bmatrix} \ddot{x}_1 \\ \ddot{x}_2 \end{bmatrix} + \begin{bmatrix} k_1 + k_2 & -k_2 \\ -k_2 & k_2 \end{bmatrix} \begin{bmatrix} x_1 \\ x_2 \end{bmatrix} = \begin{bmatrix} 0 \\ 0 \end{bmatrix}$$

These products can be carried out as indicated in Example 4.1.1 and equation (4.10) to yield

$$\begin{bmatrix} m_1\ddot{x}_1 \\ m_2\ddot{x}_2 \end{bmatrix} + \begin{bmatrix} (k_1 + k_2)x_1 - k_2x_2 \\ -k_2x_1 + k_2x_2 \end{bmatrix} = \begin{bmatrix} 0 \\ 0 \end{bmatrix}$$

Adding the two vectors on the right element by element yields

$$\begin{bmatrix} m_1\ddot{x}_1 + (k_1 + k_2)x_1 - k_2x_2 \\ m_2\ddot{x}_2 - k_2x_1 + k_2x_2 \end{bmatrix} = \begin{bmatrix} 0 \\ 0 \end{bmatrix} \tag{4.12}$$

Equating the corresponding elements of the two vectors yields

$$m_1\ddot{x}_1 + (k_1 + k_2)x_1 - k_2x_2 = 0$$
$$m_2\ddot{x}_2 - k_2x_1 + k_2x_2 = 0 \tag{4.13}$$

which are identical to equations (4.2). Hence the system of equations (4.2) can be written as the vector equation given in (4.11), where the coefficient matrices are defined by the matrices of equations (4.6) and (4.9). The matrix M defined by equation (4.6) is called the *mass matrix*, and the matrix K defined by equation (4.9) is called the *stiffness matrix*. The calculation and comparison above provides an extremely important connection between vibration analysis and matrix analysis. This simple connection allows computers to be used to solve large and complicated vibration problems quickly (discussed in Chapter 9). It also forms the foundation for the rest of this chapter (as well as the rest of the book).

The mass and stiffness matrices, M and K, described above have the special property of being symmetric. A *symmetric matrix* is a matrix that is equal to its transpose. The *transpose* of a matrix, denoted A^T, is formed from interchanging the rows and columns of a matrix. The first row of A^T is the first column of A, and so on. The mass matrix M is also called the *inertia matrix* and the force vector $M\ddot{x}$ corresponds to the inertial forces in the system of Figure 4.1(a). Similarly, the force $K\mathbf{x}$ represents the elastic restoring forces of the system described in Figure 4.1(a).

Example 4.1.3

Consider the matrix A defined by

$$A = \begin{bmatrix} a & b \\ c & d \end{bmatrix}$$

where a, b, c, and d are real numbers. Calculate values of these constants such that the matrix A is symmetric.

Solution   For A to be symmetric $A = A^T$ or

$$A = \begin{bmatrix} a & b \\ c & d \end{bmatrix} = \begin{bmatrix} a & c \\ b & d \end{bmatrix} = A^T$$

Comparing the elements of A and A^T yields that $c = b$ must hold if the matrix A is to be symmetric. Note that the elements in the c and b position of the matrix K given in equation (4.9) are equal so that $K = K^T$.

$\square$

It is useful to note that if $\mathbf{x}$ is a column vector:

$$\mathbf{x} = \begin{bmatrix} x_1 \\ x_2 \end{bmatrix}$$

then $\mathbf{x}^T$ is a row vector (i.e., $\mathbf{x}^T = [x_1 \quad x_2]$). This makes it convenient to write a column vector in one line. For example, the vector $\mathbf{x}$ can also be written as $\mathbf{x} = [x_1 \quad x_2]^T$, a column vector. The act of forming a transpose also undoes itself, so that $(A^T)^T = A$.

The initial conditions can also be written in terms of vectors as

$$\mathbf{x}_0 = \begin{bmatrix} x_1(0) \\ x_2(0) \end{bmatrix} \qquad \dot{\mathbf{x}}_0 = \begin{bmatrix} \dot{x}_1(0) \\ \dot{x}_2(0) \end{bmatrix} \tag{4.14}$$

Here $\mathbf{x}_0$ denotes the initial displacement vector and $\dot{\mathbf{x}}_0$ denotes the initial velocity vector. Equation (4.12) can now be solved by following the procedures used for solving single-degree-of-freedom systems and incorporating a few common results from matrix theory.

Recall from Section 1.2 that the single-degree-of-freedom version of equation (4.11) was solved by assuming a harmonic solution and calculating values for the constants in the assumed form. The same approach is used here. Following the argument used in equations (1.13) to (1.19), a solution is assumed of the form

$$\mathbf{x}(t) = \mathbf{u}e^{j\omega t} \tag{4.15}$$

Here $\mathbf{u}$ is a vector of constants to be determined, ω is a constant to be determined, and $j = \sqrt{-1}$. Recall that the scalar $e^{j\omega t}$ represents harmonic motion since $e^{j\omega t} = \cos \omega t + j \sin \omega t$.

Substitution of this assumed form of the solution into the vector equation of motion yields

$$(-\omega^2 M + K)\mathbf{u}e^{j\omega t} = \mathbf{0} \tag{4.16}$$

where the common factor $\mathbf{u}e^{j\omega t}$ has been factored to the right side. Note that the scalar $e^{j\omega t} \neq 0$ for any value of t and hence equation (4.16) yields the fact that ω and $\mathbf{u}$ must satisfy the vector equation

$$(-M\omega^2 + K)\mathbf{u} = \mathbf{0} \tag{4.17}$$

Note that this represents two algebraic equations in the three unknowns: ω, u_1, and u_2 where $\mathbf{u} = [u_1 \quad u_2]^T$.

For this homogeneous set of algebraic equations to have a nonzero solution for the vector $\mathbf{u}$, the inverse of the coefficient matrix $(-M\omega^2 + K)$ must not exist. To see that this is the case, suppose that the inverse of $(-M\omega^2 + K)$ does exist. Then multiplying both sides of equation (4.17) by $(-M\omega^2 + K)^{-1}$ yields $\mathbf{u} = \mathbf{0}$, a trivial solution, as it implies no motion. Hence the solution of equation (4.11) depends in some way on the *matrix inverse*. Matrix inverses are introduced in the following example.

Example 4.1.4

Consider the 2×2 matrix A defined by

$$A = \begin{bmatrix} a & b \\ c & d \end{bmatrix}$$

and calculate its inverse.

Solution   The inverse of a square matrix A is a matrix of the same dimension, denoted A^{-1}, such that

$$AA^{-1} = A^{-1}A = I$$

where I is the identity matrix. In this case I has the form

$$I = \begin{bmatrix} 1 & 0 \\ 0 & 1 \end{bmatrix}$$

The inverse matrix for a general 2×2 matrix is

$$A^{-1} = \frac{1}{\det A} \begin{bmatrix} d & -b \\ -c & a \end{bmatrix} \tag{4.18}$$

provided that $\det A \neq 0$, where $\det A$ denotes the *determinant* of the matrix A. The determinant of the matrix A has the value

$$\det A = ad - bc$$

To see that equation (4.18) is in fact the inverse, note that

$$A^{-1}A = \frac{1}{ad - bc} \begin{bmatrix} d & -b \\ -c & a \end{bmatrix} \begin{bmatrix} a & b \\ c & d \end{bmatrix}$$

$$= \frac{1}{ad - bc} \begin{bmatrix} ad - bc & bd - bd \\ ac - ac & ad - bc \end{bmatrix} = \begin{bmatrix} 1 & 0 \\ 0 & 1 \end{bmatrix}$$

It is important to realize that the matrix A has an inverse if and only if $\det A \neq 0$. Thus requiring $\det A = 0$ forces A not to have an inverse. Matrices that do not have an inverse are called *singular* matrices.

$\square$

Applying the condition of singularity to the coefficient matrix of equation (4.17) yields the result that for a nonzero solution $\mathbf{u}$ to exist,

$$\det(-\omega^2 M + K) = 0 \tag{4.19}$$

which yields one algebraic equation in one unknown (ω). Substituting the values of the matrices M and K into this expression yields

$$\det \begin{bmatrix} -\omega^2 m_1 + k_1 + k_2 & -k_2 \\ -k_2 & -\omega^2 m_2 + k_2 \end{bmatrix} = 0 \tag{4.20}$$

Using the definition of the determinant yields that the unknown quantity ω^2 must satisfy

$$m_1 m_2 \omega^4 - (m_1 k_2 + m_2 k_1 + m_2 k_2)\omega^2 + k_1 k_2 = 0 \tag{4.21}$$

This expression is called the *characteristic equation* for the system and is used to determine the constants ω in the assumed form of the solution given by equation (4.15) once the values of the physical parameters m_1, m_2, k_1, and k_2 are known.

Example 4.1.5

Calculate the solutions for ω of the characteristic equation given by equation (4.21) for the case that the physical parameters have the values: $m_1 = 9, m_2 = 1, k_1 = 24$, and $k_2 = 3$.

Solution   For these values the characteristic equation becomes

$$\omega^4 - 6\omega^2 + 8 = (\omega^2 - 2)(\omega^2 - 4) = 0$$

so that $\omega_1^2 = 2$ and $\omega_2^2 = 4$. There are two roots and each corresponds to two values of the constant ω in the assumed form of the solution:

$$\omega_1 = \pm\sqrt{2} \qquad \omega_2 = \pm 2$$

$\square$

Once the value of ω in equation (4.15) is established, the value of the constant vector $\mathbf{u}$ can be found by solving equation (4.17) for $\mathbf{u}$ given each value of ω^2. That is, for each value of ω^2 (i.e., ω_1^2 and ω_2^2) there is a vector $\mathbf{u}$ satisfying equation (4.17). For ω_1^2 the vector $\mathbf{u}_1$ satisfies

$$(-\omega_1^2 M + K)\mathbf{u}_1 = \mathbf{0} \tag{4.22}$$

and for ω_2^2 the vector $\mathbf{u}_2$ satisfies

$$(-\omega_2^2 M + K)\mathbf{u}_2 = \mathbf{0} \tag{4.23}$$

These expressions can be solved for the direction of the vectors $\mathbf{u}_1$ and $\mathbf{u}_2$ but not for the magnitude. To see that this is true, note that if $\mathbf{u}_1$ satisfies equation (4.22), so does the vector $a\mathbf{u}_1$, where a is any nonzero number. Hence the vectors satisfying (4.22) and (4.23) are of arbitrary magnitude. The following example illustrates the computation of $\mathbf{u}_1$ and $\mathbf{u}_2$ for the values of Example 4.1.5.

Example 4.1.6

Calculate the vectors $\mathbf{u}_1$ and $\mathbf{u}_2$ of equations (4.22) and (4.23) for the values of ω, K, and M of Example 4.1.5.

Solution   Let $\mathbf{u}_1 = [u_{11} \quad u_{12}]^T$. Then equation (4.22) with $\omega^2 = \omega_1^2 = 2$ becomes

$$\begin{bmatrix} 27 - 9(2) & -3 \\ -3 & 3 - (2) \end{bmatrix} \begin{bmatrix} u_{11} \\ u_{12} \end{bmatrix} = \begin{bmatrix} 0 \\ 0 \end{bmatrix}$$

Performing the indicated product and enforcing the equality yields the two equations

$$9u_{11} - 3u_{12} = 0 \qquad \text{and} \qquad -3u_{11} + u_{12} = 0$$

Note that these two equations are dependent and yield the same solution, that is,

$$\frac{u_{11}}{u_{12}} = \frac{1}{3} \qquad \text{or} \qquad u_{11} = \frac{1}{3}u_{12}$$

Only the ratio of the elements is determined here, i.e., only the direction of the vector and not its magnitude is determined by equation (4.17). As mentioned above this happens because if $\mathbf{u}$ satisfies equation (4.17) then so does $a\mathbf{u}$ where a is any nonzero number.

A numerical value for each element of the vector $\mathbf{u}$ may be obtained by arbitrarily assigning one of the elements. For example, let $u_{12} = 1$; then the value of $\mathbf{u}_1$ is

$$\mathbf{u}_1 = \begin{bmatrix} \frac{1}{3} \\ 1 \end{bmatrix}$$

This procedure is repeated using $\omega_2^2 = 4$ to yield that the elements of $\mathbf{u}_2$ must satisfy

$$-9u_{21} - 3u_{22} = 0 \qquad \text{or} \qquad u_{21} = -\frac{1}{3}u_{22}$$

Choosing $u_{22} = 1$ yields

$$\mathbf{u}_2 = \begin{bmatrix} -\frac{1}{3} \\ 1 \end{bmatrix}$$

which is the vector satisfying equation (4.23). There are several other ways of fixing the magnitude of a vector besides the one illustrated here. Some other methods are presented in Example 4.2.2 and equation (4.46).

$\square$

The solution of equation (4.2) subject to initial conditions $\mathbf{x}_0$ and $\dot{\mathbf{x}}_0$ can be constructed in terms of the numbers $\pm\omega_1$, $\pm\omega_2$ and the vectors $\mathbf{u}_1$ and $\mathbf{u}_2$. This is similar to the construction of the solution of the single-degree-of-freedom case discussed in Section 1.2. Since the equations to be solved are linear, the sum of any solution is also a solution. From the calculation above there are four solutions in the form of equation (4.15):

$$\mathbf{x}(t) = \mathbf{u}_1 e^{-j\omega_1 t}, \qquad \mathbf{u}_1 e^{+j\omega_1 t}, \qquad \mathbf{u}_2 e^{-j\omega_2 t}, \quad \text{and} \quad \mathbf{u}_2 e^{+j\omega_2 t} \qquad (4.24)$$

Thus a general solution is the sum of these:

$$\mathbf{x}(t) = (ae^{j\omega_1 t} + be^{-j\omega_1 t})\mathbf{u}_1 + (ce^{j\omega_2 t} + de^{-j\omega_2 t})\mathbf{u}_2 \qquad (4.25)$$

where a, b, c, and d are arbitrary constants of integration to be determined by the initial conditions.

Applying Euler formulas for the sine function to equation (4.25) yields an alternative form of the solution:

$$\mathbf{x}(t) = A_1 \sin(\omega_1 t + \phi_1)\mathbf{u}_1 + A_2 \sin(\omega_2 t + \phi_2)\mathbf{u}_2 \qquad (4.26)$$

where the constants of integration are now in the form of two amplitudes, A_1 and A_2, and two phase shifts, ϕ_1 and ϕ_2. These constants can be calculated from the initial conditions $\mathbf{x}_0$ and $\dot{\mathbf{x}}_0$. Equation (4.26) is the two-degree-of-freedom analog of equation (1.19) for a single-degree-of-freedom case.

The form of equation (4.26) gives physical meaning to the solution. It states that each mass in general oscillates at two frequencies: ω_1 and ω_2. These are called the *natural frequencies* of the system. Furthermore, suppose that the initial conditions are chosen such that $A_2 = 0$. With such initial conditions each mass oscillates at only one frequency, ω_1, and the relative positions of the masses at any given instant of time is determined by the elements of the vector $\mathbf{u}_1$. Hence $\mathbf{u}_1$ is called the *first mode shape* of the system. Similarly, if the initial conditions are chosen such that A_1 is zero, both coordinates oscillate at frequency ω_2, with relative positions given by the vector $\mathbf{u}_2$, called the *second mode shape*. The mode shapes and natural frequencies are clarified further in the exercises and in the sections following. Mode shapes have become a standard in vibration engineering and are used extensively in the following. The concepts of natural frequencies and mode shapes are extremely important.

Example 4.1.7

Calculate the solution of the system of Example 4.1.5 for the initial conditions $x_{10} = 1$ mm, $x_{20} = 0$, and $\dot{x}_{10} = \dot{x}_{20} = 0$.

Solution   To solve this equation (4.26) is written as

$$\begin{bmatrix} x_1(t) \\ x_2(t) \end{bmatrix} = \begin{bmatrix} \dfrac{1}{3} A_1 \sin(\sqrt{2}t + \phi_1) - \dfrac{1}{3} A_2 \sin(2t + \phi_2) \\ A_1 \sin(\sqrt{2}t + \phi_1) + A_2 \sin(2t + \phi_2) \end{bmatrix} \tag{4.27}$$

At $t = 0$ this yields

$$\begin{bmatrix} 1 \\ 0 \end{bmatrix} = \begin{bmatrix} \dfrac{1}{3} A_1 \sin \phi_1 - \dfrac{1}{3} A_2 \sin \phi_2 \\ A_1 \sin \phi_1 + A_2 \sin \phi_2 \end{bmatrix} \tag{4.28}$$

Differentiating equation (4.27) and evaluating the resulting expression at $t = 0$ yields

$$\begin{bmatrix} \dot{x}_1(0) \\ \dot{x}_2(0) \end{bmatrix} = \begin{bmatrix} 0 \\ 0 \end{bmatrix} = \begin{bmatrix} \dfrac{\sqrt{2}}{3} A_1 \cos \phi_1 - \dfrac{2}{3} A_2 \cos \phi_2 \\ \sqrt{2} A_1 \cos \phi_1 + 2A_2 \cos \phi_2 \end{bmatrix} \tag{4.29}$$

Equations (4.28) and (4.29) represent four equations in the four unknown constants of integrations A_1, A_2, ϕ_1, and ϕ_2. Writing out these four equations yields

$$3 = A_1 \sin \phi_1 - A_2 \sin \phi_2 \tag{4.30}$$

$$0 = A_1 \sin \phi_1 + A_2 \sin \phi_2 \tag{4.31}$$

$$0 = \sqrt{2} A_1 \cos \phi_1 - 2A_2 \cos \phi_2 \tag{4.32}$$

$$0 = \sqrt{2} A_1 \cos \phi_1 + 2A_2 \cos \phi_2 \tag{4.33}$$

Adding equations (4.32) and (4.33) yields that

$$2\sqrt{2} A_1 \cos \phi_1 = 0$$

so that $\phi_1 = \pi/2$. Since $\phi_1 = \pi/2$, equation (4.33) reduces to

$$2A_2 \cos \phi_2 = 0$$

so that $\phi_2 = \pi/2$. Substitution of the values of ϕ_1 and ϕ_2 into equations (4.30) and (4.31) yields

$$3 = A_1 - A_2 \quad \text{and} \quad 0 = A_1 + A_2$$

which has solutions $A_1 = 3/2$, $A_2 = -3/2$. Thus

$$x_1(t) = 0.5 \sin\left(\sqrt{2}t + \frac{\pi}{2}\right) + 0.5 \sin\left(2t + \frac{\pi}{2}\right) = 0.5(\cos\sqrt{2}t + \cos 2t)$$

$$x_2(t) = \frac{3}{2}\sin\left(\sqrt{2}t + \frac{\pi}{2}\right) - \frac{3}{2}\sin\left(2t + \frac{\pi}{2}\right) = 1.5(\cos\sqrt{2}t - \cos 2t)$$

(4.34)

These are plotted in Figure 4.3. More efficient ways to calculate the solutions are presented in later sections. The numerical aspects of calculating a solution are discussed in Chapter 9.

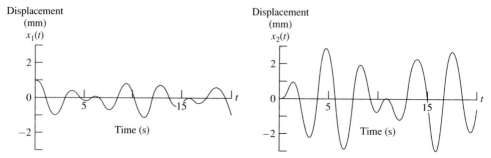

Figure 4.3　Plots of the responses of $x_1(t)$ and $x_2(t)$ for the problem of Example 4.1.7.

Note that in this case, the response of each mass contains both frequencies of the system. That is, the responses for both $x_1(t)$ and $x_2(t)$ are combinations of signals containing the two frequencies ω_1 and ω_2 (i.e., the sum of two harmonic signals). Note from the development of equation (4.34) that the mode shapes determine the relative magnitude of these two harmonic signals.

□

It is interesting and important to note that the two natural frequencies ω_1 and ω_2 of the two-degree-of-freedom system are not equal to either of the natural frequencies of the two single-degree-of-freedom systems constructed from the same components. To see this, note that in the Example 4.1.5, $\sqrt{k_1/m_1} = 1.63$, which is not equal to ω_1 or ω_2 (i.e., $\omega_1 = \sqrt{2}$, $\omega_2 = 2$). Similarly, $\sqrt{k_2/m_2} = 1.732$, which does not coincide with either frequency of the two-degree-of-freedom system composed of the same springs and masses, each attached to ground.

4.2 EIGENVALUES AND NATURAL FREQUENCIES

The method of solution indicated in Section 4.1 can be extended and formalized to take advantage of the algebraic eigenvalue problem. This allows the power of mathematics to be used in solving vibration problems and sets the background needed for analyzing systems with an arbitrary number of degrees of freedom. In addition, the important concepts of mode shapes and natural frequencies can be generalized by connecting the undamped vibration problem to the mathematics of the algebraic eigenvalue problem.

In solving a single-degree-of-freedom system it was useful to divide the equation of motion by the mass. Hence consider resolving the system of two equations described in matrix form by equation (4.11) by making a coordinate transformation that is equivalent to dividing the equations of motion by the mass in the system. To that end consider the *matrix square root* defined to be the matrix $M^{1/2}$ such that $M^{1/2}M^{1/2} = M$, the mass matrix. For the simple example of the mass matrix given in equation (4.6), the mass matrix is diagonal and the matrix square root becomes simply

$$M^{1/2} = \begin{bmatrix} \sqrt{m_1} & 0 \\ 0 & \sqrt{m_2} \end{bmatrix} \tag{4.35}$$

The rule for calculating the square root of a matrix for more complicated matrices is given in Section 9.6. When the mass matrix M is not diagonal, the system is said to be dynamically coupled and the calculation of the square root of the mass matrix is more complicated. This is discussed at the end of Section 4.7.

The inverse of the diagonal matrix $M^{1/2}$, denoted $M^{-1/2}$, becomes simply

$$M^{-1/2} = \begin{bmatrix} \dfrac{1}{\sqrt{m_1}} & 0 \\ 0 & \dfrac{1}{\sqrt{m_2}} \end{bmatrix} \tag{4.36}$$

This matrix allows the vibration problem of equation (4.11) to be transformed into a symmetric eigenvalue problem, allowing results from the mathematics of eigenvalue problems to be applied to vibration analysis.

To accomplish this transformation, let the vector $\mathbf{x}$ in equation (4.11) be replaced with

$$\mathbf{x} = M^{-1/2}\mathbf{q} \tag{4.37}$$

and multiply the resulting equation by $M^{-1/2}$. This yields

$$M^{-1/2}MM^{-1/2}\ddot{\mathbf{q}}(t) + M^{-1/2}KM^{-1/2}\mathbf{q}(t) = \mathbf{0} \tag{4.38}$$

Since $M^{-1/2}MM^{-1/2} = I$, the identity matrix, expression (4.38) reduces to

$$I\ddot{\mathbf{q}}(t) + \tilde{K}\mathbf{q}(t) = \mathbf{0} \tag{4.39}$$

The matrix $\tilde{K} = M^{-1/2}KM^{-1/2}$, like the matrix K, is a symmetric matrix.

Equation (4.39) is solved, as before, by assuming a solution of the form $\mathbf{q}(t) = \mathbf{v}e^{j\omega t}$, where $\mathbf{v}$ is a vector of constants. Substitution of this form into equation (4.39) yields

$$\tilde{K}\mathbf{v} = \omega^2\mathbf{v} \tag{4.40}$$

upon dividing by the nonzero scalar $e^{j\omega t}$. Here it is important to note that the constant vector $\mathbf{v}$ cannot be zero if motion is to result. Next let $\lambda = \omega^2$ in equation (4.40). This yields

$$\tilde{K}\mathbf{v} = \lambda\mathbf{v} \tag{4.41}$$

where $\mathbf{v} \neq \mathbf{0}$. This is precisely the statement of the algebraic eigenvalue problem. The scalar λ satisfying equation (4.41) for nonzero vectors $\mathbf{v}$ is called the *eigenvalue* and $\mathbf{v}$ is called the (corresponding) *eigenvector*. Since the matrix $\tilde{K}$ is symmetric, this is called the *symmetric eigenvalue problem*. The eigenvector $\mathbf{v}$ generalizes the concept of a mode shape $\mathbf{u}$ used in Section 4.1.

If the system being modeled has n lumped masses each free to move with a single displacement labeled $x_i(t)$, the matrices M, K and hence $\tilde{K}$ will be $n \times n$ and the vectors $\mathbf{x}$, $\mathbf{q}$, and $\mathbf{v}$ will be $n \times 1$ in dimension. Each subscript i denotes a single degree of freedom where i ranges from 1 to n, and the vector $\mathbf{x}$ denotes the collection of the n degrees of freedom. It is also convenient to label the frequencies ω and eigenvectors $\mathbf{v}$ with subscripts i, so that ω_i and $\mathbf{v}_i$ denote the ith natural frequency and corresponding ith eigenvector, respectively. In this section, only $n = 2$ is considered, but the notation is useful and valid for any number of degrees of freedom.

Equation (4.41) connects the problem of calculating the free vibration response of a conservative system with the mathematics of symmetric eigenvalue problems. This allows the developments of mathematics to be applied directly to vibration. The theoretical advantage of this relationship is developed here. The computational advantage, which is substantial, is discussed in Chapter 9, in particular in Section 9.5.

Example 4.2.1

Calculate the matrix $\tilde{K}$ using the values of Example 4.1.5.

Solution   Matrix products are defined here for matrices of the same size by extending the idea of a matrix times a vector outlined in Example 4.1.1. The result is a third matrix of the same size. The first column of the matrix product AB is the product of the matrix A with the first column of B considered as a vector, and so on. To illustrate this consider the product $KM^{-1/2}$, where $M^{-1/2}$ is defined by equation (4.36).

$$KM^{-1/2} = \begin{bmatrix} 27 & -3 \\ -3 & 3 \end{bmatrix} \begin{bmatrix} \dfrac{1}{3} & 0 \\ 0 & 1 \end{bmatrix} = \begin{bmatrix} (27)\left(\dfrac{1}{3}\right) + (-3)(0) & (27)(0) + (-3)(1) \\ (-3)\left(\dfrac{1}{3}\right) + 3(0) & (-3)(0) + (3)(1) \end{bmatrix}$$

$$= \begin{bmatrix} 9 & -3 \\ -1 & 3 \end{bmatrix}$$

Multiplying this by $M^{-1/2}$ yields

$$M^{-1/2}KM^{-1/2} = \begin{bmatrix} \dfrac{1}{3} & 0 \\ 0 & 1 \end{bmatrix} \begin{bmatrix} 9 & -3 \\ -1 & 3 \end{bmatrix}$$

$$= \begin{bmatrix} \left(\dfrac{1}{3}\right)(9) + (0)(-1) & \left(\dfrac{1}{3}\right)(-3) + (0)(3) \\ (0)(9) + (1)(-1) & (0)(-3) + (1)(3) \end{bmatrix} = \begin{bmatrix} 3 & -1 \\ -1 & 3 \end{bmatrix}$$

Note that $(M^{-1/2}KM^{-1/2})^T = M^{-1/2}KM^{-1/2}$, so that $\tilde{K}$ is symmetric, but that $(KM^{-1/2})^T \neq (KM^{-1/2})$:

$$(KM^{-1/2})^T = \begin{bmatrix} 9 & -1 \\ -3 & 3 \end{bmatrix} \neq \begin{bmatrix} 9 & -3 \\ -1 & 3 \end{bmatrix} = KM^{-1/2}$$

so that this matrix is not symmetric.

$\square$

The symmetric eigenvalue problem has several advantages. For example, it can readily be shown that the solutions of equation (4.41) are real numbers. Furthermore, it can be shown that the eigenvectors satisfying equation (4.41) are orthogonal just like the unit vectors ($\mathbf{i}$, $\mathbf{j}$, $\mathbf{k}$, $\mathbf{e}_1$, $\mathbf{e}_2$, $\mathbf{e}_3$) used in the vector analysis of forces (regardless of whether or not the eigenvalues are repeated). Two vectors $\mathbf{v}_1$ and $\mathbf{v}_2$ are defined to be *orthogonal* if their dot product is zero, that is, if

$$\mathbf{v}_1^T \mathbf{v}_2 = 0 \tag{4.42}$$

The eigenvectors satisfying (4.41) are of arbitrary length just like the vectors $\mathbf{u}_1$ and $\mathbf{u}_2$ of Section 4.1. Following the analogy of unit vectors from statics (introductory mechanics), the eigenvectors can be normalized so that the length is 1. The norm of a vector is denoted by $||\mathbf{x}||$ and defined by

$$||\mathbf{x}|| = \sqrt{\mathbf{x}^T\mathbf{x}} = \left[\sum_{i=1}^{n} (x_i^2) \right]^{1/2} \tag{4.43}$$

A set of vectors that satisfy both (4.42) and $||\mathbf{x}|| = 1$ are called *orthonormal*. The unit vectors from a Cartesian coordinate system form an *orthonormal* set of vectors (recall that $\mathbf{i} \cdot \mathbf{i} = 1$, $\mathbf{i} \cdot \mathbf{j} = 0$, etc.). A summary of vector inner products is given in Window 4.1. The following example illustrates the eigenvalue problem.

Example 4.2.2

Solve the eigenvalue problem for the two-degree-of-freedom system of Example 4.2.1.

Solution   The eigenvalue problem is to calculate the eigenvalues λ and eigenvectors $\mathbf{v}$ that satisfy equation (4.41). Rewriting equation (4.41) yields

$$(\tilde{K} - \lambda I)\mathbf{v} = \mathbf{0}$$

or

$$\begin{bmatrix} 3 - \lambda & -1 \\ -1 & 3 - \lambda \end{bmatrix} \mathbf{v} = \mathbf{0} \tag{4.44}$$

where $\mathbf{v}$ must be nonzero. Hence the matrix coefficient must be singular and therefore its determinant must be zero.

$$\det \begin{bmatrix} 3 - \lambda & -1 \\ -1 & 3 - \lambda \end{bmatrix} = \lambda^2 - 6\lambda + 8 = 0$$

This last expression is the characteristic equation and has the two roots

$$\lambda_1 = 2 \quad \text{and} \quad \lambda_2 = 4$$

Window 4.1
Summary of Vector Dot or Inner Product

Let $\mathbf{x}$ and $\mathbf{y}$ be two different $n \times 1$ column vectors:

$$\mathbf{x} = \begin{bmatrix} x_1 \\ x_2 \\ . \\ . \\ . \\ x_n \end{bmatrix} \qquad \mathbf{y} = \begin{bmatrix} y_1 \\ y_2 \\ . \\ . \\ . \\ y_n \end{bmatrix}$$

Then $\mathbf{x}^T$ is a row vector (i.e., $\mathbf{x}^T = [x_1 \quad x_2 \cdots x_n]$). The dot product of $\mathbf{x}$ with itself is a scalar given by

$$\mathbf{x}^T\mathbf{x} = x_1^2 + x_2^2 + \cdots x_n^2 = \sum_{i=1}^{n} x_i^2$$

The dot product of $\mathbf{x}$ and $\mathbf{y}$ is also a scalar given by

$$\mathbf{x}^T\mathbf{y} = x_1 y_1 + x_2 y_2 + \cdots + x_n y_n = \sum_{i=1}^{n} x_i y_i = \mathbf{y}^T\mathbf{x}$$

The product of a square $n \times n$ matrix M times a $n \times 1$ vector $\mathbf{x}$ is a $n \times 1$ matrix or the vector $\mathbf{y} = M\mathbf{x}$. Hence the triple product $\mathbf{x}^T M\mathbf{x}$ is really the product of two vectors $\mathbf{x}^T\mathbf{y}$ and thus is a scalar.

---

which are the eigenvalues of the matrix $\tilde{K}$. Note that these are also the squares of the natural frequencies, ω_i^2, as calculated in Example 4.1.5.

The eigenvector associated with λ_1 is calculated from equation (4.44) with $\lambda = \lambda_1 = 2$ and $\mathbf{v}_1 = [v_1 \quad v_2]^T$:

$$(\tilde{K} - \lambda_1 I)\mathbf{v}_1 = \mathbf{0} = \begin{bmatrix} 3-2 & -1 \\ -1 & 3-2 \end{bmatrix} \begin{bmatrix} v_1 \\ v_2 \end{bmatrix} = \begin{bmatrix} 0 \\ 0 \end{bmatrix}$$

This results in the two dependent scalar equations

$$v_1 - v_2 = 0 \qquad \text{and} \qquad -v_1 + v_2 = 0$$

Hence $v_1 = v_2$, which defines the direction of the vector $\mathbf{v}_1$.

To fix a value for the elements of $\mathbf{v}_1$, the normalization condition of (4.43) is used to force $\mathbf{v}_1$ to have a magnitude of 1. This results in (setting $v_1 = v_2$)

$$1 = \|\mathbf{v}_1\| = \sqrt{v_2^2 + v_2^2} = \sqrt{2}v_2$$

Solving for v_2 yields

$$v_2 = \frac{1}{\sqrt{2}}$$

so that the normalized vector $\mathbf{v}_1$ becomes

$$\mathbf{v}_1 = \frac{1}{\sqrt{2}} \begin{bmatrix} 1 \\ 1 \end{bmatrix}$$

Similarly, substitution of $\lambda_2 = 4$ into (4.44) solving for the elements of $\mathbf{v}_2$ and normalizing the result yields

$$\mathbf{v}_2 = \frac{1}{\sqrt{2}}\begin{bmatrix} 1 \\ -1 \end{bmatrix}$$

Now note that the product $\mathbf{v}_1^T\mathbf{v}_2$ yields

$$\mathbf{v}_1^T\mathbf{v}_2 = \frac{1}{\sqrt{2}}\frac{1}{\sqrt{2}}[1 \quad 1]\begin{bmatrix} 1 \\ -1 \end{bmatrix} = \frac{1}{2}(1-1) = 0$$

so that the set of vectors $\mathbf{v}_1$ and $\mathbf{v}_2$ are orthogonal as well as normal. Hence the two vectors $\mathbf{v}_1$ and $\mathbf{v}_2$ form an orthonormal set.

□

Note that the eigenvalues calculated in Example 4.2.2 are the same as the frequencies squared as calculated in Example 4.1.5, but that the eigenvectors look different from the mode shapes of Example 4.1.6 (i.e., $\mathbf{u}_1 \neq \mathbf{v}_1$ and $\mathbf{u}_2 \neq \mathbf{v}_2$). The eigenvectors and mode shapes are related through equation (4.37) by $\mathbf{u}_1 = M^{-1/2}\mathbf{v}_1$ or by $\mathbf{v}_1 = M^{1/2}\mathbf{u}_1$ or

$$\mathbf{v}_1 = \begin{bmatrix} 3 & 0 \\ 0 & 1 \end{bmatrix}\begin{bmatrix} \frac{1}{3} \\ 1 \end{bmatrix} = \begin{bmatrix} 1 \\ 1 \end{bmatrix}$$

This vector $\mathbf{v}_1 = M^{1/2}\mathbf{u}_1$ can also be normalized to illustrate that it is entirely equivalent to the vector $\mathbf{v}_1$ calculated in Example 4.2.2. To normalize the vector $\mathbf{v}_1 = [1 \quad 1]^T$, or any vector for that matter, an unknown scalar α is sought such that the scaled vector $\alpha\mathbf{v}_1$ has unit norm, that is, such that

$$(\alpha\mathbf{v}_1)^T(\alpha\mathbf{v}_1) = 1$$

Solving this expression for $\mathbf{v}_1 = [1 \quad 1]^T$ yields

$$\alpha^2(1+1) = 1$$

or $\alpha = 1/\sqrt{2}$. Thus the new vector $\alpha\mathbf{v}_1 = (1/\sqrt{2})[1 \quad 1]^T$ is the normalized version of the vector $M^{1/2}\mathbf{u}_1$ and is exactly the same as the normalized $\mathbf{v}_1$ calculated in Example 4.2.2 directly from the matrix $\tilde{K}$ and normalized. Remember that the eigenvalue problem determines only the direction of the eigenvector, leaving its magnitude arbitrary. The process of normalization is just a systematic way to scale each eigenvector or mode shape.

In general, any vector $\mathbf{x}$ can be normalized simply by calculating

$$\frac{1}{\sqrt{\mathbf{x}^T\mathbf{x}}}\mathbf{x} \tag{4.45}$$

Equation (4.45) can be used to normalize any nonzero real vector of any length.

An alternative approach to normalizing mode shapes is often used. This method is based on equation (4.17). Each vector $\mathbf{u}_i$ corresponding to each natural frequency ω_i is normalized with respect to the mass matrix M by scaling the mode shape vector, which satisfies equation (4.17) such that the vector $\mathbf{w}_i = \alpha_i \mathbf{u}_i$ satisfies

$$(\alpha_i \mathbf{u}_i)^T M (\alpha_i \mathbf{u}_i) = 1$$

or   (4.46)

$$\alpha_i^2 \mathbf{u}_i^T M \mathbf{u}_i = \mathbf{w}_i^T M \mathbf{w}_i = 1$$

This yields the special choice of $\alpha_i = 1/\sqrt{\mathbf{u}_i^T M \mathbf{u}_i}$.

The vector $\mathbf{w}_i$ is said to be *mass normalized*. Multiplying equation (4.17) by the scalar α_i yields (for $i = 1$ and 2)

$$-\omega_i^2 M \mathbf{w}_i + K \mathbf{w}_i = \mathbf{0}$$ (4.47)

Multiplying this by $\mathbf{w}_i^T$ yields the two scalar relations (for $i = 1$ and 2)

$$\omega_i^2 = \mathbf{w}_i^T K \mathbf{w}_i \qquad i = 1, 2$$ (4.48)

where the mass normalization $\mathbf{w}_i^T M \mathbf{w}_i = 1$ of equation (4.46) was used to evaluate the left side.

Next consider the vector $\mathbf{v}_i = M^{1/2} \mathbf{u}_i$, where $\mathbf{v}_i$ is normalized so that $\mathbf{v}_i^T \mathbf{v}_i = 1$. Then by substitution

$$\mathbf{v}_i^T \mathbf{v}_i = (M^{1/2} \mathbf{u}_i)^T M^{1/2} \mathbf{u}_i$$

$$= \mathbf{u}_i^T M^{1/2} M^{1/2} \mathbf{u}_i$$

$$= \mathbf{u}_i^T M \mathbf{u}_i = 1$$

so that the vector $\mathbf{u}_i$ is mass normalized. In this last argument the property of the transpose is used [i.e., $(M^{1/2}\mathbf{u})^T = \mathbf{u}^T (M^{1/2})^T$] and the fact that $M^{1/2}$ is symmetric, so that $(M^{1/2})^T = M^{1/2}$.

As indicated in Example 4.2.2, the eigenvectors of a symmetric matrix are orthogonal and can always be calculated to be normal. Such vectors are called orthonormal, as summarized in Window 4.2. This fact can be used to decouple the equations of motion of any order undamped system by making a new matrix P out of the normalized eigenvectors, such that each vector forms a column. Thus the matrix P is defined by

$$P = [\mathbf{v}_1 \quad \mathbf{v}_2 \quad \mathbf{v}_3 \cdots \mathbf{v}_n]$$ (4.49)

where n is the number of degrees of freedom in the system ($n = 2$ for the examples of this section). Note the matrix P has the unique property that $P^T P = I$, which follows directly from considering the matrix product definition that the ijth element of $P^T P$ is the product of the ith row of P^T with the jth column of P. Matrices that satisfy the equation $P^T P = I$ are called *orthogonal matrices*. This matrix P composed of the eigenvectors of $\tilde{K}$ is often called the *modal matrix*.

Window 4.2
Summary of Orthonormal Vectors

Two vectors x_1 and x_2 are normal if

$$\mathbf{x}_1^T \mathbf{x}_1 = 1 \qquad \text{and} \qquad \mathbf{x}_2^T \mathbf{x}_2 = 1$$

and orthogonal if $\mathbf{x}_1^T \mathbf{x}_2 = 0$. If x_1 and x_2 are normal and orthogonal, they are said to be *orthonormal*. This is abbreviated

$$\mathbf{x}_i^T \mathbf{x}_j = \delta_{ij} \qquad i = 1, 2, \quad j = 1, 2$$

where δ_{ij} is the *Kronecker delta*, defined by

$$\delta_{ij} = \left\{ \begin{matrix} 0 \text{ if } i \neq j \\ 1 \text{ if } i = j \end{matrix} \right\}$$

If a set of n vectors $\{\mathbf{x}_i\}_{i=1}^n$ is orthonormal, it is denoted by

$$\mathbf{x}_i^T \mathbf{x}_j = \delta_{ij} \qquad i, j = 1, 2, \ldots n$$

Be careful not to confuse δ_{ij}, the Kronecker delta used here, with $\delta = \ln[x(t)/x(t + T)]$, the logarithm decrement of Section 1.6, or with $\delta(t - t_j)$, the Dirac delta or impulse function of Section 3.1, or with the static deflection δ_s of Section 5.2.

Example 4.2.3

Write out the matrix P for the system of Example 4.2.2 and calculate $P^T P$.

Solution   Using the values for the vectors v_1 and v_2 from Example 4.2.2 yields

$$P = \frac{1}{\sqrt{2}} \begin{bmatrix} 1 & 1 \\ 1 & -1 \end{bmatrix}$$

so that $P^T P$ becomes

$$P^T P = \left(\frac{1}{\sqrt{2}} \right) \left(\frac{1}{\sqrt{2}} \right) \begin{bmatrix} 1 & 1 \\ 1 & -1 \end{bmatrix} \begin{bmatrix} 1 & 1 \\ 1 & -1 \end{bmatrix}$$

$$= \frac{1}{2} \begin{bmatrix} 1+1 & 1-1 \\ 1-1 & 1+1 \end{bmatrix} = \begin{bmatrix} 1 & 0 \\ 0 & 1 \end{bmatrix} = I$$

□

Another interesting and useful matrix calculation is to consider the product of the three matrices $P^T \tilde{K} P$. It can be shown (see Appendix C) that this product results in a diagonal matrix. Furthermore, the diagonal entries are the eigenvalues of the matrix $\tilde{K}$. This is denoted by

$$\Lambda = \text{diag}(\lambda_i) = P^T \tilde{K} P \tag{4.50}$$

and is called the *spectral matrix* of $\tilde{K}$. The following example illustrates this calculation.

Example 4.2.4

Calculate the matrix $P^T \tilde{K} P$ for the two-degree-of-freedom system of Example 4.2.1.

$$P^T \tilde{K} P = \frac{1}{\sqrt{2}} \begin{bmatrix} 1 & 1 \\ 1 & -1 \end{bmatrix} \begin{bmatrix} 3 & -1 \\ -1 & 3 \end{bmatrix} \frac{1}{\sqrt{2}} \begin{bmatrix} 1 & 1 \\ 1 & -1 \end{bmatrix}$$

$$= \frac{1}{2} \begin{bmatrix} 1 & 1 \\ 1 & -1 \end{bmatrix} \begin{bmatrix} 2 & 4 \\ 2 & -4 \end{bmatrix}$$

$$= \frac{1}{2} \begin{bmatrix} 4 & 0 \\ 0 & 8 \end{bmatrix} = \begin{bmatrix} 2 & 0 \\ 0 & 4 \end{bmatrix} = \Lambda$$

Note that the diagonal elements of the spectral matrix Λ are the natural frequencies ω_1 and ω_2 squared. That is, from Example 4.2.2, $\omega_1^2 = 2$ and $\omega_2^2 = 4$, so that $\Lambda = \text{diag}(\omega_1^2 \ \ \omega_2^2) = \text{diag}(2 \ \ 4)$.

$\square$

Examining the solution of Example 4.2.4 and comparing it to the natural frequencies of Example 4.1.5, suggests that in general

$$\Lambda = \text{diag}(\lambda_i) = \text{diag}(\omega_i^2) = \text{diag}(2, 4) \tag{4.51}$$

This expression connects the eigenvalues with the natural frequencies (i.e., $\lambda_i = \omega_i^2$). The following example illustrates the matrix methods for vibration analysis presented in this section and provides a summary.

Example 4.2.5

Consider the system of Figure 4.4. Write the dynamic equations in matrix form, calculate $\tilde{K}$, its eigenvalues and eigenvectors, and hence determine the natural frequencies of the system (use $m_1 = 1$ kg, $m_2 = 4$ kg, $k_1 = k_3 = 10$ N/m, and $k_2 = 2$ N/m). Also calculate the matrices P and Λ.

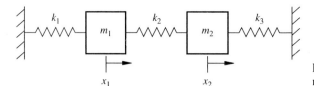

Figure 4.4   Two-degree-of-freedom model of a structure fixed at both ends.

Solution   Using free-body diagrams of each of the two masses yields the following equations of motion:

$$m_1 \ddot{x}_1 + (k_1 + k_2)x_1 - k_2 x_2 = 0$$
$$m_2 \ddot{x}_2 - k_2 x_1 + (k_2 + k_3)x_2 = 0 \tag{4.52}$$

In matrix form this becomes

$$\begin{bmatrix} m_1 & 0 \\ 0 & m_2 \end{bmatrix} \ddot{\mathbf{x}} + \begin{bmatrix} k_1 + k_2 & -k_2 \\ -k_2 & k_2 + k_3 \end{bmatrix} \mathbf{x} = \mathbf{0} \tag{4.53}$$

Using the numerical values for the physical parameters m_i and k_i yields that

$$M = \begin{bmatrix} 1 & 0 \\ 0 & 4 \end{bmatrix} \qquad K = \begin{bmatrix} 12 & -2 \\ -2 & 12 \end{bmatrix}$$

The matrix $M^{-1/2}$ becomes

$$M^{-1/2} = \begin{bmatrix} 1 & 0 \\ 0 & \dfrac{1}{2} \end{bmatrix}$$

so that

$$\tilde{K} = M^{-1/2}(KM^{-1/2}) = \begin{bmatrix} 1 & 0 \\ 0 & \dfrac{1}{2} \end{bmatrix} \begin{bmatrix} 12 & -1 \\ -2 & 6 \end{bmatrix} = \begin{bmatrix} 12 & -1 \\ -1 & 3 \end{bmatrix}$$

Note that $\tilde{K}$ is symmetric (i.e., $\tilde{K}^T = \tilde{K}$), as expected. The eigenvalues of $\tilde{K}$ are calculated from

$$\det(\tilde{K} - \lambda I) = \det \begin{bmatrix} 12 - \lambda & -1 \\ -1 & 3 - \lambda \end{bmatrix} = \lambda^2 - 15\lambda + 35 = 0$$

This quadratic equation has the solution

$$\lambda = \frac{15}{2} \pm \frac{1}{2}\sqrt{85}$$

so that (using a hand calculator and rounding after four decimal places)

$$\lambda_1 = 2.8902 \qquad \lambda_2 = 12.1098$$

The eigenvectors are calculated from (for λ_1)

$$\begin{bmatrix} 12 - 2.8902 & -1 \\ -1 & 3 - 2.8902 \end{bmatrix} \begin{bmatrix} v_1 \\ v_2 \end{bmatrix} = \begin{bmatrix} 0 \\ 0 \end{bmatrix}$$

so that the vector $\mathbf{v}_1 = [v_1 \quad v_2]^T$ satisfies

$$9.1098 v_1 = v_2$$

Normalizing the vector $\mathbf{v}_1$ yields

$$1 = \|\mathbf{v}_1\| = \sqrt{v_1^2 + v_2^2} = \sqrt{v_1^2 + (9.1098)^2 v_1^2}$$

so that

$$v_1 = \frac{1}{\sqrt{1 + (9.1098)^2}} = 0.1091$$

and

$$v_2 = 9.1098 v_1 = 0.9939$$

Thus the normalized vector $\mathbf{v}_1 = [0.1091 \quad 0.9939]^T$.

Similarly, the vector $\mathbf{v}_2$ corresponding to the eigenvalue λ_2 in normalized form becomes $\mathbf{v}_2 = [0.9940 \quad -0.1091]^T$. Note that $\mathbf{v}_1^T \mathbf{v}_2 = -0.00001$, which is effectively zero for the accuracy of an inexpensive calculation. Similarly, $\sqrt{\mathbf{v}_1^T \mathbf{v}_1} = 0.99997$, which rounds off to 1 for four decimal places. The matrix of eigenvectors P becomes

$$P = [\mathbf{v}_1 \quad \mathbf{v}_2] = \begin{bmatrix} 0.1091 & 0.9940 \\ 0.9939 & -0.1091 \end{bmatrix}$$

Thus the matrix Λ becomes

$$\Lambda = P^T \tilde{K} P = \begin{bmatrix} 0.1091 & 0.9939 \\ 0.9940 & -0.1091 \end{bmatrix} \begin{bmatrix} 12 & -1 \\ -1 & 3 \end{bmatrix} \begin{bmatrix} 0.1091 & 0.9940 \\ 0.9939 & -0.1091 \end{bmatrix}$$

$$= \begin{bmatrix} 0.1091 & 0.9939 \\ 0.9940 & -0.1091 \end{bmatrix} \begin{bmatrix} 0.3153 & 12.0371 \\ 2.8726 & -1.3213 \end{bmatrix} = \begin{bmatrix} 2.8895 & 0.0000 \\ 0.0000 & 12.1090 \end{bmatrix}$$

where all the products and sums have been rounded to four decimal places. Note that $\Lambda =$ diag (2.8895, 12.1090) is not quite equal to the eigenvalues calculated directly for the characteristic equation, because of round-off used in the computations. This is discussed in more detail in Chapter 9, along with efficient ways of calculating eigenvalues and eigenvectors (see Section 9.5). Note that in Example 4.2.4, based on the values given in Example 4.2.1, the values for Λ came out exactly. The exact values result because the values of m_i and k_i were chosen to yield only integer arithmetic with no round-off required. The round-off problem is also evident in calculating $P^T P$, which should be the identity matrix. Instead,

$$P^T P = \begin{bmatrix} 0.9997 & 0.0000 \\ 0.0000 & 1.0000 \end{bmatrix}$$

☐

Computation of ω_i, P, and so on, is systematic and can easily be performed on a computer as is done on the disk included on the inside back cover. Such numerical treatments are introduced in the *Vibration Toolbox* at the end of the problem section of this chapter. The matrix analysis of a two-degree-of-freedom system is illustrated in this section. The use of these matrix computations in computing and analyzing the solution for the vibration response in an organized fashion is presented in the following section.

4.3 MODAL ANALYSIS

The matrix of eigenvectors P calculated in Section 4.2 can be used to decouple the equations of vibration into two separate equations. The two separate equations are then second-order-single-degree-of-freedom equations that can be solved and analyzed using the methods of Chapter 1. The matrices P and $M^{-1/2}$ can be used again to transform the solution back in the original coordinate system. The matrices P and $M^{-1/2}$ can also be called *transformations*, which is appropriate in this case because they are used to transform the vibration problem between different coordinate systems. This procedure is called *modal analysis*, because the transformation P, often called the modal matrix, is related to the mode shapes of the vibrating system.

Consider the matrix form of the equation of vibration

$$M\ddot{\mathbf{x}} + K\mathbf{x} = 0 \tag{4.54}$$

subject to the initial conditions

$$\mathbf{x}(0) = \mathbf{x}_0 \qquad \dot{\mathbf{x}}(0) = \dot{\mathbf{x}}_0$$

where $\mathbf{x}_0 = [x_1(0) \quad x_2(0)]^T$ is the vector of initial displacements, and $\dot{\mathbf{x}}_0 = [\dot{x}_1(0) \quad \dot{x}_2(0)]^T$ is the vector of initial velocities. As outlined in Section 4.2, substitution of $\mathbf{x} = M^{-1/2}\mathbf{q}(t)$ into equation (4.54) and multiplying from the left by $M^{-1/2}$ yields

$$I\ddot{\mathbf{q}}(t) + \tilde{K}\mathbf{q}(t) = \mathbf{0} \tag{4.55}$$

where $\ddot{\mathbf{x}} = M^{-1/2}\ddot{\mathbf{q}}$, since the matrix M is constant and where $\tilde{K} = M^{-1/2}KM^{-1/2}$, as before. The transformation $M^{-1/2}$ simply transforms the problem from the coordinate system defined by $\mathbf{x} = [x_1(t) \quad x_2(t)]^T$ to a new coordinate system defined by $\mathbf{q} = [q_1(t) \quad q_2(t)]^T$ as defined in equation (4.57).

Note that the initial conditions in the new coordinate system are found by solving

$$\mathbf{x}(t) = M^{-1/2}\mathbf{q}(t) \tag{4.56}$$

for the vector $\mathbf{q}$ and setting $t = 0$. To recover $\mathbf{q}(t)$, equation (4.56) is multiplied by $M^{1/2}$, which yields

$$\mathbf{q}(t) = M^{1/2}\mathbf{x}(t) \tag{4.57}$$

Differentiating this with respect to time yields $\dot{\mathbf{q}} = M^{1/2}\dot{\mathbf{x}}$, since $M^{1/2}$ is a constant matrix. Thus the initial conditions in the new coordinate system become

$$\mathbf{q}(0) = M^{1/2}\mathbf{x}_0 \qquad \text{and} \qquad \dot{\mathbf{q}}(0) = M^{1/2}\dot{\mathbf{x}}_0 \tag{4.58}$$

Next define a second coordinate system $\mathbf{r}(t) = [r_1(t) \quad r_2(t)]^T$ by

$$\mathbf{q}(t) = P\mathbf{r}(t) \tag{4.59}$$

where P is the matrix composed of the orthonormal eigenvectors of $\tilde{K}$ as defined in equation (4.49). Substitution of the vector $\mathbf{q} = P\mathbf{r}(t)$ into equation (4.55) and multiplying from the left by the matrix P^T yields

$$P^TP\ddot{\mathbf{r}}(t) + P^T\tilde{K}P\mathbf{r}(t) = \mathbf{0} \tag{4.60}$$

Using the result $P^TP = I$ and equation (4.50), this can be reduced to

$$I\ddot{\mathbf{r}}(t) + \Lambda\mathbf{r}(t) = \mathbf{0} \tag{4.61}$$

Following the argument for equations (4.58), the transformed initial conditions become (recall $\mathbf{q} = P\mathbf{r}$, so that $\mathbf{r} = P^T\mathbf{q}$ since $P^TP = I$)

$$\mathbf{r}(0) = P^T\mathbf{q}(0) \qquad \text{and} \qquad \dot{\mathbf{r}}(0) = P^T\dot{\mathbf{q}}(0) \tag{4.62}$$

Equation (4.61) can be written out by performing the indicated matrix calculations as

$$\begin{bmatrix} 1 & 0 \\ 0 & 1 \end{bmatrix}\begin{bmatrix} \ddot{r}_1(t) \\ \ddot{r}_2(t) \end{bmatrix} + \begin{bmatrix} \omega_1^2 & 0 \\ 0 & \omega_2^2 \end{bmatrix}\begin{bmatrix} r_1(t) \\ r_2(t) \end{bmatrix} = \begin{bmatrix} 0 \\ 0 \end{bmatrix} \tag{4.63}$$

$$\begin{bmatrix} \ddot{r}_1(t) + \omega_1^2 r_1(t) \\ \ddot{r}_2(t) + \omega_2^2 r_2(t) \end{bmatrix} = \begin{bmatrix} 0 \\ 0 \end{bmatrix} \tag{4.64}$$

The equality of the two vectors in this last expression implies the two decoupled equations

$$\ddot{r}_1(t) + \omega_1^2 r_1(t) = 0 \tag{4.65}$$

and

$$\ddot{r}_2(t) + \omega_2^2 r_2(t) = 0 \tag{4.66}$$

These two equations are subject to the initial conditions given in equation (4.62). Equations (4.65) and (4.66) are called the *modal equations* and the coordinate system $\mathbf{r}(t) = [r_1(t) \quad r_2(t)]^T$ is called the *modal coordinate system*. Equations (4.65) and (4.66) are said to be decoupled because each depends only on a single coordinate. Hence each equation can be solved independently by using the method of Sections 1.1 and 1.2 (see Window 4.3). Denoting the initial conditions individually by $r_{10}, \dot{r}_{10}, r_{20}$, and $\dot{r}_{20}$ and using equation (1.10), the solution of each of the modal equations (4.65) and (4.66) is simply

$$r_1(t) = \frac{\sqrt{\omega_1^2 r_{10}^2 + \dot{r}_{10}^2}}{\omega_1} \sin\left(\omega_1 t + \tan^{-1} \frac{\omega_1 r_{10}}{\dot{r}_{10}}\right) \tag{4.67}$$

$$r_2(t) = \frac{\sqrt{\omega_2^2 r_{20}^2 + \dot{r}_{20}^2}}{\omega_2} \sin\left(\omega_2 t + \tan^{-1} \frac{\omega_2 r_{20}}{\dot{r}_{20}}\right) \tag{4.68}$$

Window 4.3
Review of the Solution to a Single-Degree-of-Freedom
Undamped System from Section 1.1 and Window 1.2

The solution to $m\ddot{x} + kx = 0$ or $\ddot{x} + \omega^2 x = 0$ subject to $x(0) = x_0$ and $\dot{x}(0) = v_0$ is

$$x(t) = \sqrt{x_0^2 + \frac{v_0^2}{\omega^2}} \sin\left(\omega t + \tan^{-1} \frac{\omega x_0}{v_0}\right) \tag{1.10}$$

where $\omega = \sqrt{k/m}$.

Once the modal solutions (4.67) and (4.68) are known, the transformations $M^{1/2}$ and P can be used on the vector $\mathbf{r}(t) = [r_1(t) \quad r_2(t)]^T$ to recover the solution $\mathbf{x}(t)$ in the physical coordinates $x_1(t)$ and $x_2(t)$. To obtain the vector $\mathbf{x}$ from the vector $\mathbf{r}$, substitute equations (4.59) into (4.56) to get

$$\mathbf{x} = M^{-1/2}\mathbf{q} = M^{-1/2}P\mathbf{r}(t) \tag{4.69}$$

The matrix product $M^{-1/2}P$ is again a matrix, which is denoted by S:

$$S = M^{-1/2}P \tag{4.70}$$

and is the same size as M and $P(2 \times 2)$. The matrix S can also be used to reduce the number of steps required to perform the procedure indicated above. This procedure, referred to as *modal analysis*, provides a means of calculating the solution to a two-degree-of-freedom vibration problem by performing a number of matrix calculations. The usefulness of this approach is that these matrix computations can be easily programmed on a computer (even on some calculators). In addition, the modal analysis procedure is easily extended to systems with an arbitrary number of degrees of freedom. This is developed in the next section. Figure 4.5 illustrates the coordinate transformation used in modal analysis. The following summarizes the procedure of modal analysis using the matrix transformation S. This is followed by an example that re-solves Example 4.1.6 using modal methods.

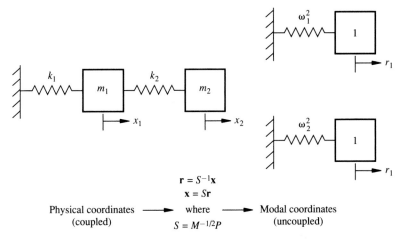

Figure 4.5   Schematic illustration of decoupling equations of motion using modal analysis.

Taking the transpose of equation (4.70) yields

$$S^T = (M^{-1/2}P)^T = P^T M^{-1/2} \tag{4.71}$$

since $(AB)^T = B^T A^T$ (see Appendix C). In addition, the inverse of a matrix product is given by $(AB)^{-1} = B^{-1}A^{-1}$ (see Appendix C), so that

$$S^{-1} = (M^{-1/2}P)^{-1} = P^{-1}M^{1/2} \tag{4.72}$$

However, the matrix P has as its inverse P^T since $P^T P = I$. Thus equation (4.72) becomes

$$S^{-1} = P^T M^{1/2} \tag{4.73}$$

These matrix results can be used to aid in the solution of equation (4.54) by modal analysis.

The modal analysis of equation (4.54) starts with the substitution of $\mathbf{x} = S\mathbf{r}(t)$ into equation (4.54). Multiplying the result by S^T yields

$$S^T M S \ddot{\mathbf{r}}(t) + S^T K S \mathbf{r}(t) = \mathbf{0} \tag{4.74}$$

Expanding the matrix S in equation (4.74) into its factors as given by equation (4.70) yields

$$P^T M^{-1/2} M M^{-1/2} P \ddot{\mathbf{r}}(t) + P^T M^{-1/2} K M^{-1/2} P \mathbf{r}(t) = \mathbf{0} \tag{4.75}$$

or

$$P^T P \ddot{\mathbf{r}}(t) + P^T \tilde{K} P \mathbf{r}(t) = \mathbf{0} \tag{4.76}$$

Using the properties of the matrix P, this becomes

$$\ddot{\mathbf{r}}(t) + \Lambda \mathbf{r}(t) = \mathbf{0} \tag{4.77}$$

which represents the two decoupled equations (4.65) and (4.66). Recall that these are called the *modal equations*, $\mathbf{r}(t)$ is called the *modal coordinate system*, and the diagonal matrix Λ contains the squares of the natural frequencies.

The initial conditions for $\mathbf{r}(t)$ are calculated by solving for $\mathbf{r}(t)$ in equation (4.69). Multiplying equation (4.69) by S^{-1} yields $\mathbf{r}(t) = S^{-1}\mathbf{x}(t)$, which becomes, after using equation (4.73),

$$\mathbf{r}(t) = P^T M^{1/2} \mathbf{x}(t) \tag{4.78}$$

The initial conditions on $\mathbf{r}(t)$ are thus

$$\mathbf{r}(0) = P^T M^{1/2} \mathbf{x}_0 \quad \text{and} \quad \dot{\mathbf{r}}(0) = P^T M^{1/2} \dot{\mathbf{x}}_0 \tag{4.79}$$

With the appropriate initial conditions given by equations (4.79), the modal equations (4.67) and (4.68) yield the solution vector $\mathbf{r}(t)$ in modal coordinates. To calculate the solution in physical coordinates $\mathbf{x}(t)$, the transformation S is used as indicated by equation (4.69). These steps are summarized in Window 4.4 and illustrated in the following examples.

Example 4.3.1

Calculate the solution of the two-degree-of-freedom system given by

$$M = \begin{bmatrix} 9 & 0 \\ 0 & 1 \end{bmatrix} \quad K = \begin{bmatrix} 27 & -3 \\ -3 & 3 \end{bmatrix} \quad \mathbf{x}(0) = \begin{bmatrix} 1 \\ 0 \end{bmatrix} \quad \dot{\mathbf{x}}(0) = \mathbf{0}$$

Compare the result to that obtained in Example 4.1.6 for the same system and initial conditions.

Window 4.4
Steps in Solving Equation (4.54) by Modal Analysis

1. Calculate $M^{-1/2}$.
2. Calculate $\tilde{K} = M^{-1/2}KM^{-1/2}$.
3. Calculate the symmetric eigenvalue problem for $\tilde{K}$ to get ω_i^2 and $\mathbf{v}_i$.
4. Calculate $S = M^{-1/2}P$ and $S^{-1} = P^T M^{1/2}$.
5. Calculate the modal initial conditions: $\mathbf{r}(0) = S^{-1}\mathbf{x}_0$, $\dot{\mathbf{r}}(0) = S^{-1}\dot{\mathbf{x}}_0$.
6. Substitute the components of $\mathbf{r}(0)$ and $\dot{\mathbf{r}}(0)$ into equations (4.67) and (4.68) to get the solution in modal coordinate $\mathbf{r}(t)$.
7. Multiply $\mathbf{r}(t)$ by S to get the solution $\mathbf{x}(t) = S\mathbf{r}(t)$.

Solution   From Examples 4.1.5, 4.2.1, 4.2.3, and 4.2.4 the following have been calculated

$$M^{-1/2} = \begin{bmatrix} \frac{1}{3} & 0 \\ 0 & 1 \end{bmatrix} \qquad \tilde{K} = \begin{bmatrix} 3 & -1 \\ -1 & 3 \end{bmatrix}$$

$$P = \frac{1}{\sqrt{2}} \begin{bmatrix} 1 & 1 \\ 1 & -1 \end{bmatrix} \qquad \Lambda = \text{diag}\,(2,4)$$

which provides the information required in the first three steps of Window 4.4. The next step is to calculate the matrix S and its inverse.

$$S = M^{-1/2}P = \frac{1}{\sqrt{2}} \begin{bmatrix} \frac{1}{3} & 0 \\ 0 & 1 \end{bmatrix} \begin{bmatrix} 1 & 1 \\ 1 & -1 \end{bmatrix} = \frac{1}{\sqrt{2}} \begin{bmatrix} \frac{1}{3} & \frac{1}{3} \\ 1 & -1 \end{bmatrix}$$

$$S^{-1} = P^T M^{1/2} = \frac{1}{\sqrt{2}} \begin{bmatrix} 1 & 1 \\ 1 & -1 \end{bmatrix} \begin{bmatrix} 3 & 0 \\ 0 & 1 \end{bmatrix} = \frac{1}{\sqrt{2}} \begin{bmatrix} 3 & 1 \\ 3 & -1 \end{bmatrix}$$

The reader should verify that $SS^{-1} = I$, as a check. In addition, note that $S^T MS = I$. The modal initial conditions are calculated from

$$\mathbf{r}(0) = S^{-1}\mathbf{x}_0 = \frac{1}{\sqrt{2}} \begin{bmatrix} 3 & 1 \\ 3 & -1 \end{bmatrix} \begin{bmatrix} 1 \\ 0 \end{bmatrix} = \begin{bmatrix} \dfrac{3}{\sqrt{2}} \\ \dfrac{3}{\sqrt{2}} \end{bmatrix}$$

$$\dot{\mathbf{r}}(0) = S^{-1}\dot{\mathbf{x}}_0 = S^{-1}\mathbf{0} = \mathbf{0}$$

so that $r_{10} = r_{20} = 3/\sqrt{2}$ and $\dot{r}_{10} = \dot{r}_{20} = 0$. Equations (4.67) and (4.68) yield that the modal solutions are

$$r_1(t) = \frac{3}{\sqrt{2}} \sin\left(\sqrt{2}t + \frac{\pi}{2}\right) = \frac{3}{\sqrt{2}} \cos\sqrt{2}t$$

$$r_2(t) = \frac{3}{\sqrt{2}} \sin\left(2t + \frac{\pi}{2}\right) = \frac{3}{\sqrt{2}} \cos 2t$$

The solution in the physical coordinate system $\mathbf{x}(t)$ is calculated from

$$\mathbf{x}(t) = S\mathbf{r}(t) = \frac{1}{\sqrt{2}}\begin{bmatrix} \frac{1}{3} & \frac{1}{3} \\ 1 & -1 \end{bmatrix}\begin{bmatrix} \frac{3}{\sqrt{2}}\cos\sqrt{2}t \\ \frac{3}{\sqrt{2}}\cos 2t \end{bmatrix} = \begin{bmatrix} (0.5)(\cos\sqrt{2}t + \cos 2t \\ (1.5)(\cos\sqrt{2}t - \cos 2t \end{bmatrix}$$

This is, of course, identical to the solution obtained in Example 4.1.6 and plotted in Figure 4.3.

□

Example 4.3.2

Calculate the response of the system of Example 4.2.5 illustrated in Figure 4.4 to the initial displacement $\mathbf{x}(0) = [1 \quad 1]^T$ [with $\dot{\mathbf{x}}(0) = \mathbf{0}$] using modal analysis.

Solution   Again following the steps illustrated in Window 4.4, the matrices $M^{-1/2}$ and $\tilde{K}$ become

$$M^{-1/2} = \begin{bmatrix} 1 & 0 \\ 0 & \frac{1}{2} \end{bmatrix} \qquad \tilde{K} = \begin{bmatrix} 12 & -1 \\ -1 & 3 \end{bmatrix}$$

Solving the symmetric eigenvalue problem for $\tilde{K}$ (this time using a computer and commercial code) yields

$$P = \begin{bmatrix} -0.1091 & -0.9940 \\ -0.9940 & 0.1091 \end{bmatrix} \qquad \Lambda = \text{diag } (2.8902, 12.1098)$$

Here the arithmetic is held to eight decimal places but only four are shown. The matrices S and S^{-1} become

$$S = \begin{bmatrix} -0.1091 & -0.9940 \\ -0.4970 & 0.0546 \end{bmatrix} \qquad S^{-1} = \begin{bmatrix} -0.1091 & -1.9881 \\ -0.9940 & 0.2182 \end{bmatrix}$$

As a check, note that

$$P^T\tilde{K}P = \begin{bmatrix} 2.8902 & 0 \\ 0 & 12.1098 \end{bmatrix} \qquad P^TP = I$$

The modal initial conditions become

$$\mathbf{r}_0(t) = S^{-1}\mathbf{x}_0 = \begin{bmatrix} -0.1091 & -0.9881 \\ -0.9940 & 0.2182 \end{bmatrix}\begin{bmatrix} 1 \\ 1 \end{bmatrix} = \begin{bmatrix} -2.0972 \\ -0.7758 \end{bmatrix}$$

$$\dot{\mathbf{r}}_0(t) = S^{-1}\mathbf{x}_0 = \mathbf{0}$$

Using these values of $r_{10}, r_{20}, \dot{r}_{20}$, and $\dot{r}_{10}$ in equations (4.67) and (4.68) yields the modal solutions

$$r_1(t) = -2.0972\cos(1.7001t)$$

$$r_2(t) = -0.7758\cos(3.4799t)$$

Using the transformation $\mathbf{x} = S\mathbf{r}(t)$ yields that the solution in physical coordinates is

$$\mathbf{x}(t) = \begin{bmatrix} 0.2288\cos(1.7001t) + 0.7712\cos(3.4799t) \\ 1.0424\cos(1.7001t) - 0.0424\cos(3.4799t) \end{bmatrix}$$

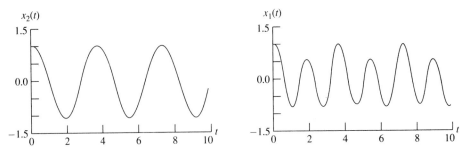

Figure 4.6   Plot of the solutions given in Example 4.3.2.

Note that this satisfies the initial conditions as it should. A plot of the responses is given in Figure 4.6.

The plot of $x_2(t)$ in the figure illustrates that the mass m_2 is not much affected by the second frequency. This is because the particular initial condition does not cause the first mass to be excited very much at $\omega_2 = 3.4799$ rad/s [note the coefficient of $\cos(3.4799)$ in the equation for $x_2(t)$]. However the plot of $x_1(t)$ clearly indicates the presence of both frequencies, because the initial condition strongly excites both modes (i.e., both frequencies) in this coordinate. The effects of changing the initial condition on the response can be examined by using the program VTB4_2 in the disk enclosed with the book to solve Problem TB4.3 at the end of the chapter.   □

Note that the calculations for the frequencies and modal matrix in Example 4.3.2 work well in the sense that $P^T \tilde{K} P$ is diagonal and $P^T P$ is the identity matrix. When these two conditions were applied in the calculations for the same problem in Example 4.2.5, these two checks did not quite agree. In fact, the frequencies calculated from Λ become different. This is because the calculations in Example 4.2.5 were made by rounding after four decimals, while those made in Example 4.3.2 are rounded after eight decimal places. This illustrates the importance of numerical methods in computing vibration properties. This is discussed in more detail in Chapter 9.

4.4 MORE THAN TWO DEGREES OF FREEDOM

Many structures, machines, and mechanical devices require numerous coordinates to describe their vibrational motion. For instance, an automobile suspension was modeled in earlier chapters as a single degree of freedom. However, a car has four wheels; hence a more accurate model is to use four degrees of freedom or coordinates. Since an automobile can roll, pitch, and yaw, it may be appropriate to use even more coordinates to describe the motion. Systems with any finite number of degrees of freedom can be analyzed by using the modal analysis procedure outlined in Window 4.4.

For each mass in the system, there corresponds a coordinate $x_i(t)$ describing its motion in one dimension, this gives rise to an $n \times 1$ vector $\mathbf{x}(t)$, with $n \times n$ mass matrix M and stiffness matrix K satisfying

$$M\ddot{\mathbf{x}} + K\mathbf{x} = 0 \tag{4.80}$$

The form of equation (4.80) also holds if each mass is allowed to rotate or move in the y, z, or pitch and yaw directions. In this situation, the vector $\mathbf{x}$ could reflect up to six coordinates for each mass and the mass and stiffness matrices would be modified to reflect the additional inertia and stiffness quantities. Figure 4.7 illustrates the possibilities of coordinates for a simple element. However, for the sake of simplicity of explanation, the initial discussion is confined to mass elements that are free to move in only one direction.

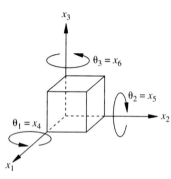

Figure 4.7 Single mass element illustrating all the possible degrees of freedom. The six degrees of freedom of the rigid body consist of three rotational and three translational motions. If the predominant forward motion of the body is in the x_2 direction, such as in an airplane, then θ_2 is called *roll*, θ_3 is called *yaw*, and θ_1 is called *pitch*.

As a generic example consider the n masses connected by n springs in Figure 4.8. Summing the forces on each of the n masses yields n equations of the form

$$m_i\ddot{x}_i + k_i(x_i - x_{i-1}) - k_{i+1}(x_{i+1} - x_i) = 0 \qquad i = 1, 2\ldots, n \tag{4.81}$$

where m_i denotes the ith mass and k_i the ith spring coefficient. In matrix form these equations take the form of equation (4.80) where

$$M = \text{diag}\,(m_1, m_2, \ldots, m_n) \tag{4.82}$$

and

$$K = \begin{bmatrix} k_1 + k_2 & -k_2 & 0 & 0 & \cdots & 0 \\ -k_2 & k_2 + k_3 & -k_3 & & & \vdots \\ 0 & -k_3 & k_3 + k_4 & & & \\ \vdots & & & & k_{n-1} + k_n & -k_n \\ 0 & & \cdots & & -k_n & k_n \end{bmatrix} \tag{4.83}$$

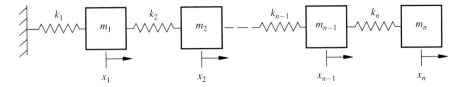

Figure 4.8   Example of an n-degree-of-freedom system.

The $n \times 1$ vector $\mathbf{x}(t)$ becomes

$$\mathbf{x}(t) = \begin{bmatrix} x_1(t) \\ x_2(t) \\ \cdot \\ \cdot \\ \cdot \\ x_n(t) \end{bmatrix} \tag{4.84}$$

The notation of Window 4.4 can be used to solve n-degree-of-freedom problems. Each of the steps is exactly the same, however, the matrix computations are all $n \times n$ and the resulting modal equation becomes the n decoupled equations:

$$\ddot{r}_1(t) + \omega_1^2 r_1(t) = 0$$
$$\ddot{r}_2(t) + \omega_2^2 r_2(t) = 0$$
$$\vdots \tag{4.85}$$
$$\ddot{r}_n(t) + \omega_n^2 r_n(t) = 0$$

There are now n natural frequencies, ω_i, which correspond to the eigenvalues of the $n \times n$ matrix $M^{-1/2} K M^{-1/2}$.

   The n eigenvalues are determined from the characteristic equation given by

$$\det (\lambda I - \tilde{K}) = 0 \tag{4.86}$$

which gives rise to an nth-order polynomial in λ. The determinant of an $n \times n$ matrix A is given by

$$\det A = \sum_{s=1}^{n} a_{ps} |A_{ps}| \tag{4.87}$$

for any fixed value of p between 1 and n. Here a_{ps} is the element of the matrix A at the intersection of the pth row and sth column, and $|A_{ps}|$ is the determinant of the submatrix formed from A by striking out the pth row and sth column, multiplied by $(-1)^{p+s}$. The following example illustrates the use of equation (4.87).

Example 4.4.1

Expand equation (4.87) for $p = 1$ to calculate the following determinant:

$$\det A = \det \begin{bmatrix} 1 & 3 & -2 \\ 0 & 1 & 1 \\ 2 & 5 & 3 \end{bmatrix} = 1[(1)(3) - (1)(5)] - 3[(0)(3) - (2)(1)] - 2[(0)(5) - (1)(2)] = 8$$

$\square$

Once the ω_i^2 are determined from equation (4.86), the normalized eigenvectors are obtained following the methods suggested in Example 4.2.2. The key differences between using the modal approach described by Window 4.4 for multiple-degree-of-freedom systems and for a two-degree-of-freedom system is in the computation of the characteristic equation, its solution to get ω_i^2, and solving for the normalized eigenvectors. With the exception of calculating the matrix P and Λ, the rest of the procedure is simple matrix multiplication. The following example illustrates the procedure for a three-degree-of-freedom system.

Example 4.4.2

Calculate the solution of the n-degree-of-freedom system of Figure 4.8 for $n = 3$ by modal analysis. Use the values $m_1 = m_2 = m_3 = 4$ kg and $k_1 = k_2 = k_3 = 4$ N/m, and the initial condition $x_1(0) = 1$ m with all other initial displacements and velocities zero.

Solution   The mass and stiffness matrices for $n = 3$ for the values given become

$$M = 4I \qquad K = \begin{bmatrix} 8 & -4 & 0 \\ -4 & 8 & -4 \\ 0 & -4 & 4 \end{bmatrix}$$

Following the steps suggested in Window 4.4 yields

1. $M^{-1/2} = \dfrac{1}{2}I$

2. $\tilde{K} = M^{-1/2}KM^{-1/2} = \dfrac{1}{4}\begin{bmatrix} 8 & -4 & 0 \\ -4 & 8 & -4 \\ 0 & -4 & 4 \end{bmatrix} = \begin{bmatrix} 2 & -1 & 0 \\ -1 & 2 & -1 \\ 0 & -1 & 1 \end{bmatrix}$

3.

$$\det(\lambda I - \tilde{K}) = \det\left(\begin{bmatrix} \lambda - 2 & 1 & 0 \\ 1 & \lambda - 2 & 1 \\ 0 & 1 & \lambda - 1 \end{bmatrix}\right)$$

$$= (\lambda - 2)\det\left(\begin{bmatrix} \lambda - 2 & 1 \\ 1 & \lambda - 1 \end{bmatrix}\right) - (1)\det\left(\begin{bmatrix} 1 & 1 \\ 0 & \lambda - 1 \end{bmatrix}\right) + (0)\det\left(\begin{bmatrix} 1 & \lambda - 2 \\ 0 & 1 \end{bmatrix}\right)$$

$$= (\lambda - 2)[(\lambda - 2)(\lambda - 1) - 1] - 1[(\lambda - 1) - 0]$$

$$= (\lambda - 1)(\lambda - 2)^2 - (\lambda - 2) - \lambda + 1 = \lambda^3 - 5\lambda^2 + 6\lambda - 1 = 0$$

The roots of this cubic equation are

$$\lambda_1 = 0.1981 \qquad \lambda_2 = 1.5550 \qquad \lambda_3 = 3.2470$$

Thus the system's natural frequencies are

$$\omega_1 = 0.4450 \qquad \omega_2 = 1.2470 \qquad \omega_3 = 1.8019$$

To calculate the first eigenvector, substitute $\lambda_1 = 0.1981$ into $(\tilde{K} - \lambda I)\mathbf{v}_1 = \mathbf{0}$ and solve for the vector $\mathbf{v}_1 = [v_1 \ v_2 \ v_3]^T$. This yields

$$\begin{bmatrix} 2 - 0.1981 & -1 & 0 \\ -1 & 2 - 0.1981 & -1 \\ 0 & -1 & 1 - 0.1981 \end{bmatrix} \begin{bmatrix} v_1 \\ v_2 \\ v_3 \end{bmatrix} = \begin{bmatrix} 0 \\ 0 \\ 0 \end{bmatrix}$$

Multiplying out this last expression yields three equations, only two of which are independent,

$$(1.8019)v_1 - v_2 = 0$$

$$-v_1 + (1.8019)v_2 - v_3 = 0$$

$$-v_2 + (0.8019)v_3 = 0$$

Solving the first and third equations yields

$$v_1 = 0.4450v_3 \quad \text{and} \quad v_2 = 0.8019v_3$$

The second equation is dependent and does not yield any new information. Substituting these values into the vector $\mathbf{v}_1$ yields

$$\mathbf{v}_1 = v_3 \begin{bmatrix} 0.4450 \\ 0.8019 \\ 1 \end{bmatrix}$$

Normalizing the vector yields

$$\mathbf{v}_1^T \mathbf{v}_1 = v_3^2 \left[(0.4450)^2 + (0.8019)^2 + 1^2 \right] = 1$$

Solving for v_3 and substituting back into the expression for $\mathbf{v}_1$ yields the normalized version of the eigenvector $\mathbf{v}_1$ as

$$\mathbf{v}_1 = \begin{bmatrix} 0.3280 \\ 0.5910 \\ 0.7370 \end{bmatrix}$$

Similarly, $\mathbf{v}_2$ and $\mathbf{v}_3$ can be calculated and normalized to be

$$\mathbf{v}_2 = \begin{bmatrix} -0.7370 \\ -0.3280 \\ 0.5910 \end{bmatrix} \quad \mathbf{v}_3 = \begin{bmatrix} -0.5910 \\ 0.7370 \\ -0.3280 \end{bmatrix}$$

The matrix P is then

$$P = \begin{bmatrix} 0.3280 & -0.7370 & -0.5910 \\ 0.5910 & -0.3280 & 0.7370 \\ 0.7370 & 0.5910 & -0.3280 \end{bmatrix}$$

(The reader should verify that $P^T P = I$ and $P^T \tilde{K} P = \Lambda$)

4. The matrix $S = M^{-1/2}P = \dfrac{1}{2}IP$ or

$$S = \begin{bmatrix} 0.1640 & -0.3685 & -0.2955 \\ 0.2955 & -0.1640 & 0.3685 \\ 0.3685 & 0.2955 & -0.1640 \end{bmatrix}$$

and

$$S^{-1} = P^T M^{1/2} = P^T 12 = \begin{bmatrix} 0.6560 & 1.1820 & 1.4740 \\ -1.4740 & -0.6560 & 1.1820 \\ -1.1820 & 1.4740 & -0.6560 \end{bmatrix}$$

(Again the reader should verify that $S^{-1}S = I$.)

5. The initial conditions in modal coordinates become

$$\dot{\mathbf{r}}(0) = S^{-1}\dot{\mathbf{x}}_0 = S^{-1}\mathbf{0} = \mathbf{0}$$

and

$$\mathbf{r}(0) = S^{-1}\mathbf{x}_0 = \begin{bmatrix} 0.6560 & 1.1820 & 1.4740 \\ -1.4740 & -0.6560 & 1.1820 \\ -1.1820 & 1.4740 & -0.6560 \end{bmatrix} \begin{bmatrix} 1 \\ 0 \\ 0 \end{bmatrix} = \begin{bmatrix} 0.6560 \\ -1.4740 \\ -1.1820 \end{bmatrix}$$

6. The modal solutions of equation (4.85) are each of the form given by equation (4.67) and can now be determined as

$$r_1(t) = (0.6560)\sin\left(0.4450t + \frac{\pi}{2}\right) = 0.6560\cos(0.4450t)$$

$$r_2(t) = (-1.4740)\sin\left(1.247t + \frac{\pi}{2}\right) = -1.4740\cos(1.2470t)$$

$$r_3(t) = (-1.1820)\sin\left(1.8019t + \frac{\pi}{2}\right) = -1.1820\cos(1.8019t)$$

7. The solution in physical coordinates is next calculated from

$$\mathbf{x} = S\mathbf{r}(t) = \begin{bmatrix} 0.1640 & -0.3685 & -0.2955 \\ 0.2955 & -0.1640 & 0.3685 \\ 0.3685 & 0.2955 & -0.1640 \end{bmatrix} \begin{bmatrix} 0.6560\cos(0.4450t) \\ -1.4740\cos(1.2470t) \\ -1.1820\cos(1.8019t) \end{bmatrix}$$

$$\begin{bmatrix} x_1(t) \\ x_2(t) \\ x_3(t) \end{bmatrix} = \begin{bmatrix} 0.1075\cos(0.4450t) + 0.5443\cos(1.2470t) + 0.3492\cos(1.8019t) \\ 0.1938\cos(0.4450t) + 0.2417\cos(1.2470t) - 0.4355\cos(1.8019t) \\ 0.2417\cos(0.4450t) - 0.4355\cos(1.2470t) + 0.1935\cos(1.8019t) \end{bmatrix}$$

These calculations can be programmed; however, solving the characteristic equation is not the best way to calculate eigenvalues. The methods discussed in Section 9.5 should be consulted for larger-order systems. The computer files mentioned in the *Vibration Toolbox* at the end of this chapter can be used to check these calculations and to plot the responses. $\square$

Another approach to modal analysis is to use the *mode summation* or *expansion method*. This procedure is based on a fact from linear algebra—that the eigenvectors of a real symmetric matrix form a complete set (i.e., that any n-dimensional vector can be represented as a linear combination of the eigenvectors of an $n \times n$ symmetric matrix). Recall the symmetric statement of the vibration problem:

$$I\ddot{\mathbf{q}}(t) + \tilde{K}\mathbf{q}(t) = \mathbf{0} \tag{4.88}$$

Let $\mathbf{v}_i$ denote the n eigenvectors of the matrix $\tilde{K}$ and let λ_i denote the corresponding eigenvalues. According to the argument preceding equation (4.41) a solution of (4.88) is

$$\mathbf{q}_i(t) = \mathbf{v}_i e^{\pm\sqrt{\lambda_i}jt} \tag{4.89}$$

since $\lambda_i = \omega_i^2$. This represents two solutions that can be added together following the argument used for equation (1.18) to yield

$$\mathbf{q}_i(t) = (a_i\, e^{-\sqrt{\lambda_i}\, jt} + b_i e^{\sqrt{\lambda_i}\, jt})\mathbf{v}_i \tag{4.90}$$

or using Euler's formula,

$$\mathbf{q}_i(t) = d_i \sin(\omega_i t + \phi_i)\mathbf{v}_i \tag{4.91}$$

where d_i and ϕ_i are constants to be determined by initial conditions. Since the set of vectors $\mathbf{v}_i, i = 1, 2, \ldots, n$ are eigenvectors of a symmetric matrix, a linear combination can be used to represent any $n \times 1$ vector, and in particular, the solution vector $\mathbf{q}(t)$. Hence

$$\mathbf{q}(t) = \sum_{i=1}^{n} d_i \sin(\omega_i t + \phi_i)\mathbf{v}_i \tag{4.92}$$

The constant d_i and ϕ_i can be evaluated from the initial conditions

$$\mathbf{q}(0) = \sum_{i=1}^{n} d_i \sin \phi_i \mathbf{v}_i \tag{4.93}$$

and

$$\dot{\mathbf{q}}(0) = \sum_{i=1}^{n} \omega_i d_i \cos \phi_i \mathbf{v}_i \tag{4.94}$$

Multiplying equation (4.93) by $\mathbf{v}_j^T$ and using the orthogonality (see Window 4.2) of the vector $\mathbf{v}_i$ (i.e., $\mathbf{v}_j^T \mathbf{v}_i = 0$ for all values of the summation index $i = 1, 2, \ldots, n$ except for $i = j$) yields

$$\mathbf{v}_j^T \mathbf{q}(0) = d_j \sin \phi_j \tag{4.95}$$

for each value of $j = 1, \ldots, n$. Similarly, multiplying equation (4.94) by $\mathbf{v}_j^T$ yields

$$\mathbf{v}_j^T \dot{\mathbf{q}}(0) = \omega_j d_j \cos \phi_j \tag{4.96}$$

for each $j = 1, 2, \ldots, n$. Combining equations (4.95) and (4.96) and renaming the index yields

$$\phi_i = \tan^{-1} \frac{\omega_i \mathbf{v}_i^T \mathbf{q}(0)}{\mathbf{v}_i^T \dot{\mathbf{q}}(0)} \qquad i = 1, 2, \ldots, n \tag{4.97}$$

and

$$d_i = \frac{\mathbf{v}_i^T \mathbf{q}(0)}{\sin \phi_i} \qquad i = 1, 2 \ldots, n \tag{4.98}$$

Equations (4.92), (4.97) and (4.98) represent the solution in modal summation form. Equation (4.92) is sometimes called the *expansion theorem* and is equivalent to writing a function as a Fourier series. The constants d_i are sometimes referred to as expansion coefficients.

Note as an immediate consequence of equation (4.97) that if the system has zero initial velocity, $\dot{\mathbf{q}} = \mathbf{0}$, each coordinate has a phase shift of 90°. The initial conditions in the coordinate system $\mathbf{q}(t)$ can also be chosen such that $d_i = 0$ for all $i = 2, \ldots, n$. In this case the summation in equation (4.92) reduces to the single term:

$$\mathbf{q}(t) = d_1 \sin(\omega_1 t + \phi_1)\mathbf{v}_1 \tag{4.99}$$

This states that each coordinate $q_i(t)$ oscillates with the same frequency and phase. To obtain the solution in physical coordinates, recall that $\mathbf{x} = M^{-1/2}\mathbf{q}$ so that

$$\mathbf{x}(t) = d_1 \sin(\omega_1 t + \phi_1)M^{-1/2}\mathbf{v}_1 \tag{4.100}$$

The product of a matrix and a vector is another vector. Defining $\mathbf{u}_1 = M^{-1/2}\mathbf{v}_1$, equation (4.100) becomes

$$\mathbf{x}(t) = d_1 \sin(\omega_1 t + \phi_1)\mathbf{u}_1 \tag{4.101}$$

This states that if the system is given a set of initial conditions such that $d_2 = d_3 = \cdots = d_n = 0$, then (4.101) is that total solution and each mass oscillates at the first natural frequency (ω_1). Furthermore, the vector $\mathbf{u}_1$ specifies the relative magnitudes of oscillation of each mass with respect to the rest position. Hence $\mathbf{u}_1$ is called the *first mode shape*. Note that the first mode shape is related to the first eigenvector of $\tilde{K}$ by

$$\mathbf{u}_1 = M^{-1/2}\mathbf{v}_1 \tag{4.102}$$

This argument can be repeated for each of the indices i so that $\mathbf{u}_2 = M^{-1/2}\mathbf{v}_2, \mathbf{u}_3 = M^{-1/2}\mathbf{v}_3$, and so on, are the second, third, and so on, mode shapes and the series solution

$$\mathbf{x}(t) = \sum_{i=1}^{n} d_i \sin(\omega_i t + \phi_i)\mathbf{u}_i \tag{4.103}$$

illustrates how each mode shape contributes to forming the total response of the system.

The initial condition required to excite a system into a single mode can be determined from equation (4.98). Because of the mutual orthoganility of the eigenvectors, if $\mathbf{q}(0)$ is chosen to be one of the eigenvectors, $\mathbf{v}_j$, then each d_i is zero except for the index $i = j$. Hence to excite the structure in, say, the second mode, choose $\mathbf{q}(0) = \mathbf{v}_2$ and $\dot{\mathbf{q}}(0) = \mathbf{0}$. Then the solution (4.92) becomes

$$\mathbf{q}(t) = d_2 \sin\left(\omega_2 t + \frac{\pi}{2}\right)\mathbf{v}_2 \tag{4.104}$$

and each coordinate of $\mathbf{q}$ oscillates with frequency ω_2. To transform $\mathbf{q}(0) = \mathbf{v}_2$ into physical coordinates, note that $\mathbf{x} = M^{-1/2}\mathbf{q}$, so that $\mathbf{x}(0) = M^{-1/2}\mathbf{q}(0) = M^{-1/2}\mathbf{v}_2 = \mathbf{u}_2$. Hence exciting the system by imposing an initial displacement equal to the second mode shape results in each mass oscillating at the second natural frequency. The modal summation method is illustrated in the next example.

Example 4.4.3

Consider a simple model of the horizontal vibration of a four-story building as illustrated in Figure 4.9, subject to a wind that gives the building an initial displacement of $\mathbf{x}(0) = [0.025 \quad 0.020 \quad 0.01 \quad 0.001]^T$.

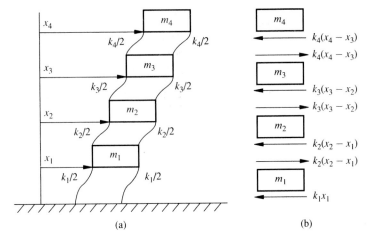

Figure 4.9   (a) Simple model of the horizontal vibration of a four-story building. Here each floor is modeled as a lumped mass and the walls are modeled as providing horizontal stiffness. (b) Restoring forces acting on each mass (floor).

In modeling buildings it is known that most of the mass is in the floor of each section and that the walls can be treated as massless columns providing lateral stiffness. From Figure 4.9 the equations of motion of each floor are

$$m_1\ddot{x}_1 + (k_1 + k_2)x_1 - k_2x_2 = 0$$

$$m_2\ddot{x}_2 - k_2x_1 + (k_2 + k_3)x_2 - k_3x_3 = 0$$

$$m_3\ddot{x}_3 - k_3x_2 + (k_3 + k_4)x_3 - k_4x_4 = 0$$

$$m_4\ddot{x}_4 - k_4x_3 + k_4x_4 = 0$$

In matrix form, these four equations can be written as

$$\begin{bmatrix} m_1 & 0 & 0 & 0 \\ 0 & m_2 & 0 & 0 \\ 0 & 0 & m_3 & 0 \\ 0 & 0 & 0 & m_4 \end{bmatrix} \ddot{x} + \begin{bmatrix} k_1+k_2 & -k_2 & 0 & 0 \\ -k_2 & k_2+k_3 & -k_3 & 0 \\ 0 & -k_3 & k_3+k_4 & -k_4 \\ 0 & 0 & -k_4 & k_4 \end{bmatrix} x = 0$$

Some reasonable values for a building are $m_1 = m_2 = m_3 = m_4 = 4000$ kg and $k_1 = k_2 = k_3 = k_4 = 5000$ N/m. In this case the numerical values of M and K become

$$M = 4000I \qquad K = \begin{bmatrix} 10,000 & -5000 & 0 & 0 \\ -5000 & 10,000 & -5000 & 0 \\ 0 & -5000 & 10,000 & -5000 \\ 0 & 0 & -5000 & 5000 \end{bmatrix}$$

To simplify the calculations, each matrix is divided by 1000. Since the equation of motion is homogeneous, this corresponds to dividing both sides of the matrix equation by 1000 so that the equality is preserved. The initial conditions are

$$\mathbf{x}(0) = \begin{bmatrix} 0.025 \\ 0.020 \\ 0.010 \\ 0.001 \end{bmatrix} \qquad \dot{\mathbf{x}} = \begin{bmatrix} 0 \\ 0 \\ 0 \\ 0 \end{bmatrix}$$

The matrices $M^{-1/2}$ and $\tilde{K}$ become

$$M^{-1/2} = \frac{1}{2}I \qquad \tilde{K} = \begin{bmatrix} 2.5 & -1.25 & 0 & 0 \\ -1.25 & 2.5 & -1.25 & 0 \\ 0 & -1.25 & 2.5 & -1.25 \\ 0 & 0 & -1.25 & 1.25 \end{bmatrix}$$

The matrix $M^{1/2}$ and the initial condition on $\mathbf{q}(t)$ become

$$M^{1/2} = 2I \qquad \mathbf{q}(0) = M^{1/2}\mathbf{x}(0) = \begin{bmatrix} 0.050 \\ 0.040 \\ 0.020 \\ 0.002 \end{bmatrix} \qquad \dot{\mathbf{q}}(0) = M^{1/2}\mathbf{0} = \mathbf{0}$$

Using an eigenvalue solver (see Chapter 9 for details and/or use the files discussed at the end of this chapter and contained in VTB.4 on the disk) the eigenvalue problem for $\tilde{K}$ yields

$$\lambda_1 = 0.1508 \qquad \lambda_2 = 1.2500 \qquad \lambda_3 = 2.9341 \qquad \lambda_4 = 4.4151$$

$$\mathbf{v}_1 = \begin{bmatrix} 0.2280 \\ 0.4285 \\ 0.5774 \\ 0.6565 \end{bmatrix} \qquad \mathbf{v}_2 = \begin{bmatrix} 0.5774 \\ 0.5774 \\ 0.0 \\ -0.5774 \end{bmatrix} \qquad \mathbf{v}_3 = \begin{bmatrix} 0.6565 \\ -0.2280 \\ -0.5774 \\ 0.4285 \end{bmatrix} \qquad \mathbf{v}_4 = \begin{bmatrix} -0.4285 \\ 0.6565 \\ -0.5774 \\ 0.2280 \end{bmatrix}$$

Converting this into natural frequencies and mode shapes ($\omega_i = \sqrt{\lambda_i}$ and $\mathbf{u}_i = M^{-1/2}\mathbf{v}_i$) yields $\omega_1 = 0.3883$, $\omega_2 = 1.1180$, $\omega_3 = 1.7129$, $\omega_4 = 2.1012$, and

$$\mathbf{u}_1 = \begin{bmatrix} 0.1140 \\ 0.2143 \\ 0.2887 \\ 0.3283 \end{bmatrix} \qquad \mathbf{u}_2 = \begin{bmatrix} 0.2887 \\ 0.2887 \\ 0.0 \\ -0.2887 \end{bmatrix} \qquad \mathbf{u}_3 = \begin{bmatrix} 0.3283 \\ -0.1140 \\ -0.2887 \\ 0.2143 \end{bmatrix} \qquad \mathbf{u}_4 = \begin{bmatrix} -0.2143 \\ 0.3283 \\ -0.2887 \\ 0.1140 \end{bmatrix}$$

Since $\dot{\mathbf{q}}(0) = \mathbf{0}$, equation (4.97) yields that each of the phase shifts is $\phi_i = \pi/2$ and equation (4.98) becomes

$$d_i = \frac{\mathbf{v}_i^T \mathbf{q}(0)}{\sin(\pi/2)} = \mathbf{v}_i^T \mathbf{q}(0)$$

Substituting $\mathbf{v}_i^T$ and $\mathbf{q}(0)$ into the expansion above yields the following values for the expansion coefficients:

$$d_1 = 0.0414 \qquad d_2 = 0.0508 \qquad d_3 = 0.0130 \qquad d_4 = -0.0063$$

The solution given by equation (4.103) then becomes (in meters)

$$\mathbf{x}(t) = \begin{bmatrix} 0.0047 \\ 0.0089 \\ 0.0120 \\ 0.0136 \end{bmatrix} \cos(0.3883t) + \begin{bmatrix} 0.0147 \\ 0.0147 \\ 0 \\ -0.0147 \end{bmatrix} \cos(1.1180t)$$

$$+ \begin{bmatrix} 0.0043 \\ -0.0015 \\ -0.0038 \\ 0.0028 \end{bmatrix} \cos(1.7129t) + \begin{bmatrix} 0.0013 \\ -0.0021 \\ 0.0018 \\ -0.0007 \end{bmatrix} \cos(2.1012t)$$

The mode shapes $\mathbf{u}_1$, $\mathbf{u}_2$, $\mathbf{u}_3$, and $\mathbf{u}_4$ are plotted in Figure 4.10. Figure 4.11 plots the response of each floor of the building in centimeters.

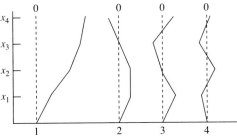

Figure 4.10   Plot of the four mode shapes associated with the solution for the system of Figure 4.9.

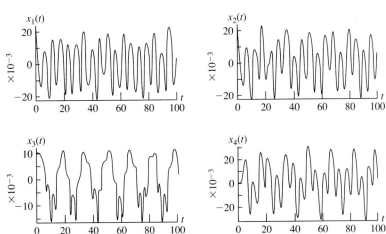

Figure 4.11   Time response of each of the four floors for the building model given in Figure 4.9.

Note that in all the previous examples, the stiffness matrix K is *banded* (i.e., the matrix has nonzero elements on the diagonal and one element above and below the diagonal, the other elements being zero). This is typical of structural models but is not necessarily the case for machine parts or other mechanical devices.

The concept of mode shapes presented in this section and illustrated in Example 4.4.3 is extremely important. The language of modes, mode shapes, and natural frequencies forms the basis for discussing vibration phenomena of complex systems. The word *mode* generally refers to both the natural frequency and its corresponding mode shape. A mode shape is a mathematical description of a deflection. It forms a pattern that describes the shape of vibration if the system were to vibrate only at the corresponding natural frequency. It is neither tangible nor simple to observe; however, it provides a simple way to discuss and understand the vibration of complex objects. Its physical significance lies in the fact that every vibrational response of a system consists of some combinations of mode shapes. An entire industry has been formed around the concept of modes.

4.5 SYSTEMS WITH VISCOUS DAMPING

Viscous energy dissipation can be introduced to the modal analysis solution suggested above in two ways. Again, as in modeling single-degree-of-freedom systems, viscous damping is introduced more as a mathematical convenience rather than a physical truth. However, viscous damping does provide an excellent model in many physical situations and represents a significant improvement over the undamped model. The simplest method of modeling damping is to use *modal damping*. Modal damping places an energy dissipation term of the form

$$2\zeta_i\omega_i\dot{r}_i(t) \tag{4.105}$$

in equations (4.85). This form is chosen largely for its mathematical convenience. Here $\dot{r}_i(t)$ denotes the velocity of the ith modal coordinate, ω_i is the ith natural frequency, and ζ_i is the ith *modal damping ratio*. The modal damping ratios, ζ_i, are assigned by "experience" to be some number between 0 and 1, or by making measurements of the response and estimating ζ_i. Usually, ζ_i is small unless the structure contains viscoelastic material or a hydraulic damper is present. Common values are $0 \leq \zeta < 0.05$. An automobile shock absorber, which uses fluid, may yield values as high as $\zeta = 0.5$.

Once the modal damping ratios are assigned, equations (4.85) become

$$\ddot{r}_i(t) + 2\zeta_i\omega_i\dot{r}_i(t) + \omega_i^2 r_i(t) = 0 \qquad i = 1, 2, \ldots, n \tag{4.106}$$

which have solutions of the form $(0 < \zeta_i < 1)$

$$r_i(t) = A_i e^{-\zeta_i\omega_i t} \sin(\omega_{di} t + \phi_i) \qquad i = 1, 2, \ldots, n \tag{4.107}$$

where A_i and ϕ_i are constants to be determined by the initial conditions and $\omega_{di} = \omega_i \sqrt{1 - \zeta_i^2}$ as given in Window 4.5. Once this solution is established, the modal analysis method of solution suggested in Window 4.4 is used.

Window 4.5
Review of a Damped Single-Degree-of-Freedom System

The solution of $m\ddot{x} + c\dot{x} + kx = 0$, $x(0) = x_0, \dot{x}(0) = \dot{x}_0$, or $\ddot{x} + 2\zeta\omega\dot{x} + \omega^2 x = 0$ is (for the underdamped case $0 < \zeta < 1$)

$$x(t) = Ae^{-\zeta\omega t} \sin(\omega_d t + \theta)$$

where $\omega = \sqrt{k/m}, \zeta = c/(2m\omega), \omega_d = \omega\sqrt{1 - \zeta^2}$, and

$$A = \left[\frac{(\dot{x}_0 + \zeta\omega x_0)^2 + (x_0\omega_d)^2}{\omega_d^2} \right]^{1/2} \qquad \theta = \tan^{-1} \frac{x_0\omega_d}{\dot{x}_0 + \zeta\omega x_0}$$

from equations (1.36), (1.37), and (1.38).

The only difference is that equation (4.107) replaces step 6 where

$$A_i = \left[\frac{(\dot{r}_{i0} + \zeta_i\omega_i r_{i0})^2 + (r_{i0}\omega_{di})^2}{\omega_{di}^2} \right]^{1/2} \tag{4.108}$$

$$\phi_i = \tan^{-1} \frac{r_{i0}\omega_{di}}{\dot{r}_{i0} + \zeta_i\omega_i r_{i0}} \tag{4.109}$$

Here r_{i0} and $\dot{r}_{i0}$ are the ith elements of $\mathbf{r}(0)$ and $\dot{\mathbf{r}}(0)$, respectively. Equations (4.108) and (4.109) are directly from equation (1.38) derived for a single-degree-of-freedom system of the same form as equation (4.107). These equations are correct only if each modal damping ratio is underdamped. The following example illustrates the solution technique for a system with assumed modal damping.

Example 4.5.1

Consider again the system of Example 4.3.1, and calculate the solution of the same system if modal damping of the form $\zeta_1 = 0.05$ and $\zeta_2 = 0.1$ is assumed.

Solution   From Example 4.3.1, $\omega_1 = \sqrt{2}$ and $\omega_2 = 2$, so that $\omega_{d1} = \sqrt{2}[1 - (0.05)^2]^{1/2} = 1.4124$ and $\omega_{d2} = 2[1 - (0.1)^2]^{1/2} = 1.9899$. As calculated in Example 4.3.1, the modal initial conditions are $r_{10} = r_{20} = 3/\sqrt{2}$ and $\dot{r}_{10} = \dot{r}_{20} = 0$. Substitution of these values into equations (4.108) and (4.109) yields

$$A_1 = 2.1240 \qquad \phi_1 = 1.52 \text{ rad} \qquad (87.13°)$$

$$A_2 = 2.1320 \qquad \phi_2 = 1.47 \text{ rad} \qquad (84.26°)$$

Note that compared to the constant A_i and ϕ_i in the magnitude and phase of the undamped system, only a small change occurs in the amplitude. The phase, however, changes $3°$ and $6°$, respectively, because of the damping. The solution is then of the form

$$\mathbf{x}(t) = \mathbf{S}\mathbf{r}(t) = \frac{1}{\sqrt{2}}\begin{bmatrix} \frac{1}{3} & \frac{1}{3} \\ 1 & -1 \end{bmatrix}\begin{bmatrix} 2.1240e^{-0.0706t}\sin(1.4124t + 1.52) \\ 2.1320e^{-0.989t}\sin(1.9899t + 1.47) \end{bmatrix}$$

or

$$\mathbf{x}(t) = \begin{bmatrix} 1.0013e^{-0.0706t}\sin(1.4124t + 1.52) + 1.0051e^{-0.1989t}\sin(1.9899t + 1.47) \\ 3.0038e^{-0.0706t}\sin(1.4124t + 1.52) - 3.0151e^{-0.1989t}\sin(1.9899t + 1.47) \end{bmatrix}$$

A plot of $x_1(t)$ versus t and $x_2(t)$ versus t is given in Figure 4.12.

□

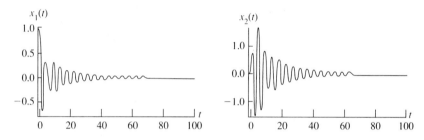

Figure 4.12   Plot of the damped response of the system of Example 4.5.1.

The modal damping ratio approach is also easily applicable to the mode summation method of Section 4.4. In this case equation (4.91) is replaced by the damped version

$$\mathbf{q}_i(t) = d_i e^{-\zeta_i \omega_i t}\sin(\omega_{di} t + \phi_i)\mathbf{v}_i \tag{4.110}$$

The initial displacement condition calculation becomes

$$\mathbf{q}_i(0) = d_i \sin(\phi_i)\mathbf{v}_i \tag{4.111}$$

so that equation (4.95) still holds. However, the velocity becomes

$$\dot{\mathbf{q}}_i(t) = d_i\left[\omega_{di}e^{-\zeta_i\omega_i t}\cos(\omega_{di}t + \phi_i) - \zeta_i\omega_i e^{-\zeta_i\omega_i t}\sin(\omega_i t + \phi_i)\right]\mathbf{v}_i \tag{4.112}$$

or at $t = 0$,

$$\dot{\mathbf{q}}_i(0) = d_i(\omega_{di}\cos\phi_i - \zeta_i\omega_i \sin\phi_i)\mathbf{v}_i \tag{4.113}$$

Multiplying equations (4.111) and (4.113) by $\mathbf{v}_i^T$ from the left and solving for the constants ϕ_i and d_i yields

$$\phi_i = \tan^{-1}\frac{\omega_{di}\mathbf{v}_i^T \mathbf{q}(0)}{\mathbf{v}_i^T \dot{\mathbf{q}}(0) + \zeta_i\omega_i \mathbf{v}_i^T \mathbf{q}(0)}$$

$$d_i = \frac{\mathbf{v}_i^T \mathbf{q}(0)}{\sin\phi_i} \tag{4.114}$$

Again, the values of ζ_i are assigned based on experience or on measurement, then the calculations of Section 4.4 are used with the initial conditions given by equation (4.114). The solution given by equation (4.103) is replaced by

$$\mathbf{x}(t) = \sum_{i=1}^{n} d_i e^{-\zeta_i \omega_i t} \sin(\omega_{di} t + \phi_i) \mathbf{u}_i \tag{4.115}$$

which yields the damped response.

Example 4.5.2

Repeat Example 4.4.3 assuming that the damping in the building is measured to be about $\zeta = 0.01$ in each mode.

Solution   Each of the steps of the solution to Example 4.4.3 is the same until the initial conditions are calculated. From equation (4.114) with $\dot{\mathbf{q}}(0) = M^{1/2}\mathbf{x}(0) = \mathbf{0}$, for each i,

$$\phi_i = \tan^{-1} \frac{\omega_{di}}{\zeta_i \omega_i} = \tan^{-1} \frac{\sqrt{1 - \zeta_i^2}}{\zeta_i}$$

For $\zeta_i = 0.01$, this becomes

$$\phi_i = 89.42° \qquad i = 1, 2, 3, 4 \qquad \text{(or 1.56 rad)}$$

Since $1/\sin(89.42°) = 1.00005$, the expansion coefficients d_i are taken to be $\mathbf{v}_i^T \mathbf{q}(0)$, as calculated in Example 4.4.3. The mode shapes are unchanged, so the solution becomes

$$\mathbf{x}(t) = \begin{bmatrix} 0.0047 \\ 0.0089 \\ 0.0120 \\ 0.0136 \end{bmatrix} e^{-0.0038t} \sin(0.3883t + 1.56) + \begin{bmatrix} 0.0147 \\ 0.0147 \\ 0 \\ -0.0147 \end{bmatrix} e^{-0.0117t} \sin(1.1180t + 1.56)$$

$$+ \begin{bmatrix} 0.0043 \\ -0.0015 \\ -0.0038 \\ 0.0028 \end{bmatrix} e^{-0.0171t} \sin(1.7129t + 1.56) + \begin{bmatrix} 0.0013 \\ -0.0021 \\ 0.0018 \\ -0.0007 \end{bmatrix} e^{-0.0210t} \sin(2.1012t + 1.56)$$

Each coordinate of $\mathbf{x}(t)$ is plotted in Figure 4.13. Note that each plot shows the effects of multiple frequencies.   ☐

Damping can also be modeled directly. For instance, consider the system of Figure 4.14. The equation of motion of this system can be found from summing the forces on each mass, as before. This yields the following equations of motion in matrix form:

$$\begin{bmatrix} m_1 & 0 \\ 0 & m_2 \end{bmatrix} \ddot{\mathbf{x}} + \begin{bmatrix} c_1 + c_2 & -c_2 \\ -c_2 & c_2 \end{bmatrix} \dot{\mathbf{x}} + \begin{bmatrix} k_1 + k_2 & -k_2 \\ -k_2 & k_2 \end{bmatrix} \mathbf{x} = \mathbf{0} \tag{4.116}$$

where $\mathbf{x} = [x_1(t) \quad x_2(t)]^T$. Equation (4.116) yields an example of a damping matrix C, defined by

$$C = \begin{bmatrix} c_1 + c_2 & -c_2 \\ -c_2 & c_2 \end{bmatrix} \tag{4.117}$$

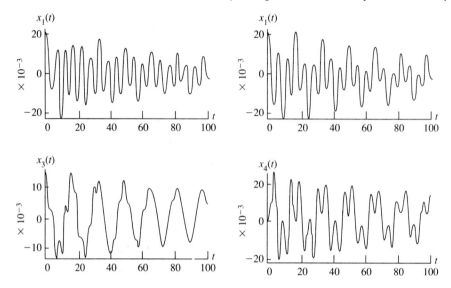

Figure 4.13 Plot of the damped response of the system of Example 4.5.2, which is the building of Example 4.4.3 with damping of $\zeta = 0.01$ in each mode.

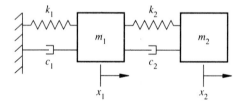

Figure 4.14 Two-degree-of-freedom system with viscous damping.

Here c_1 and c_2 refer to the damping coefficients indicated in Figure 4.14. The damping matrix C is symmetric and in a general n-degree-of-freedom system will be an $n \times n$ matrix. Thus a damped n-degree-of-freedom system is modeled by equations of the form

$$M\ddot{\mathbf{x}} + C\dot{\mathbf{x}} + K\mathbf{x} = \mathbf{0} \tag{4.118}$$

The difficulty with modeling damping in this fashion is that modal analysis cannot in general be used to solve equation (4.118). This is true because the damping provides additional coupling between the equations of motion which cannot always be decoupled by the modal transformation S (see Caughey and O'Kelly, 1965). Other methods can be used to solve equation (4.118) as discussed in Sections 9.5 and 9.6.

Modal analysis can be used directly to solve equation (4.118) if the damping matrix C can be written as a linear combination of the mass and stiffness matrix, that is, if

$$C = \alpha M + \beta K \tag{4.119}$$

where α and β are constants. This form of damping is called *proportional damping*. Substitution of equation (4.119) into equation (4.118) yields

$$M\ddot{\mathbf{x}}(t) + (\alpha M + \beta K)\dot{\mathbf{x}}(t) + K\mathbf{x}(t) = \mathbf{0} \tag{4.120}$$

Substitution of $\mathbf{x} = M^{-1/2}\mathbf{q}$ and multiplying by $M^{-1/2}$ yields

$$\ddot{\mathbf{q}}(t) + (\alpha I + \beta \tilde{K})\dot{\mathbf{q}}(t) + \tilde{K}\mathbf{q}(t) = \mathbf{0} \tag{4.121}$$

Continuing to follow the steps of Window 4.4, substituting of $\mathbf{q}(t) = P\mathbf{r}(t)$ and premultiplying by P^T where P is the matrix of eigenvectors of $\tilde{K}$ yields

$$\ddot{\mathbf{r}}(t) + (\alpha I + \beta \Lambda)\dot{\mathbf{r}}(t) + \Lambda\mathbf{r}(t) = \mathbf{0} \tag{4.122}$$

This corresponds to the n decoupled modal equations

$$\ddot{r}_i(t) + 2\zeta_i\omega_i\dot{r}_i(t) + \omega_i^2 r_i(t) = 0 \tag{4.123}$$

where $2\zeta_i\omega_i = \alpha + \beta\omega_i^2$ or

$$\zeta_i = \frac{\alpha}{2\omega_i} + \frac{\beta\omega_i}{2} \qquad i = 1, 2, \ldots, n \tag{4.124}$$

Here α and β can be chosen to produce some measured (or desired, in the design case) values of the modal damping ratio ζ_i. On the other hand, if α and β are known, equation (4.124) determines the value of the modal damping ratios ζ_i. The solution of equation (4.123) for the underdamped case ($0 < \zeta_i < 1$) is

$$r_i(t) = A_i e^{-\zeta_i\omega_i t} \sin(\omega_{di} t + \phi_i) \tag{4.125}$$

where A_i and ϕ_i are determined by applying the initial conditions on $\mathbf{r}(t)$. The solution in physical coordinates is then calculated from $\mathbf{x}(t) = S\mathbf{r}(t)$, where $S = M^{-1/2}P$ as before.

4.6 MODAL ANALYSIS OF THE FORCED RESPONSE

The forced response of a multiple-degree-of-freedom system can also be calculated by use of modal analysis. For example, consider the building system of Figure 4.9 with a force $F_4(t)$ applied to the fourth floor. This force could be the result of an out-of-balance rotating machine on the fourth floor, for example. The equation of motion takes the form

$$M\ddot{\mathbf{x}} + C\dot{\mathbf{x}} + K\mathbf{x} = \mathbf{F}(t) \tag{4.126}$$

where $\mathbf{F}(t) = [0 \ \ 0 \ \ 0 \ \ F_4(t)]^T$. On the other hand, if the wind is time varying, it will be applied in some way to each floor of the building and the applied force $\mathbf{F}(t)$ will take the form

$$\mathbf{F}(t) = \begin{bmatrix} F_1(t) \\ F_2(t) \\ F_3(t) \\ F_4(t) \end{bmatrix} \tag{4.127}$$

The approach of modal analysis again follows Window 4.4 and uses transformations to reduce equation (4.126) to a set of n decoupled modal equations, which in this case will be inhomogenuous. Then the methods of Chapter 3 can be applied to solve for the individual forced response in the modal coordinate system. The modal solution is then transformed back into the physical coordinate system.

To this end, assume that the damping matrix C is proportional of the form given by equation (4.119). Following the procedure in Window 4.4, let $\mathbf{x}(t) = M^{-1/2}\mathbf{q}(t)$ in equation (4.126) and multiply by $M^{-1/2}$. This yields

$$I\ddot{\mathbf{q}}(t) + \tilde{C}\dot{\mathbf{q}}(t) + \tilde{K}\mathbf{q}(t) = M^{-1/2}\mathbf{F}(t) \tag{4.128}$$

where $\tilde{C} = M^{-1/2}CM^{-1/2}$. Next calculate the eigenvalue problem for $\tilde{K}$. Let $\mathbf{q}(t) = P\mathbf{r}(t)$, where P is the matrix of eigenvectors of $\tilde{K}$ and multiply by P^T. This yields

$$\ddot{\mathbf{r}}(t) + \ \text{diag}[2\zeta_i\omega_i]\dot{\mathbf{r}}(t) + \Lambda\mathbf{r}(t) = P^T M^{-1/2}\mathbf{F}(t) \tag{4.129}$$

where the matrix $\text{diag}[2\zeta_i\omega_i]$ follows from equation (4.123). The vector $P^T M^{-1/2}\mathbf{F}(t)$ has elements $f_i(t)$ which will be linear combinations of the forces F_i applied to each mass. Hence the decoupled modal equations take the form

$$\ddot{r}_i(t) + 2\zeta_i\omega_i\dot{r}_i(t) + \omega_i^2 r_i(t) = f_i(t) \tag{4.130}$$

Referring to Section 3.2, this has the solution (reviewed in Window 4.6 for the underdamped case)

$$r_i(t) = d_i e^{-\zeta_i\omega_i t} \sin(\omega_{di} t + \phi_i) + \frac{1}{\omega_{di}} e^{-\zeta_i\omega_i t} \int_0^t f_i(\tau) e^{\zeta_i\omega_i\tau} \sin\omega_{di}(t - \tau) \, d\tau \tag{4.131}$$

where d_i and ϕ_i must be determined by the modal initial conditions and $\omega_{di} = \omega_i\sqrt{1 - \zeta_i^2}$ as before. Note that f_i may represent a sum of forces if more than one force is applied to the system. In addition, if a force is applied to only one mass of the system, this force becomes applied to each of the modal equations (4.131) by the transformation S, as illustrated in the following example.

Window 4.6
Forced Response of an Underdamped System from Section 3.2

The response of an underdamped system

$$m\ddot{x} + c\dot{x} + kx = F(t)$$

(with zero initial conditions) is given by (for $0 < \zeta < 1$)

$$x(t) = \frac{1}{m\omega_d}e^{-\zeta\omega t}\int_0^t F(\tau)e^{\zeta\omega\tau}\sin\omega_d(t-\tau)d\tau$$

where $\omega = \sqrt{k/m}$, $\zeta = c/(2m\omega)$, and $\omega_d = \omega\sqrt{1-\zeta^2}$. With nonzero initial conditions this becomes

$$x(t) = Ae^{-\zeta\omega t}\sin(\omega_d t + \theta) + \frac{1}{\omega_d}e^{-\zeta\omega t}\int_0^t f(\tau)e^{\zeta\omega\tau}\sin\omega_d(t-\tau)d\tau$$

where $f = F/m$ and A and θ are constants determined by the initial conditions.

Example 4.6.1

Consider the simple two-degree-of-freedom system with a harmonic force applied to one mass as indicated in Figure 4.15.

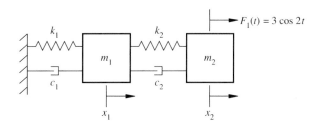

Figure 4.15 Damped two-degree-of-freedom system for Example 4.6.1.

For this example, let $m_1 = 9$ kg, $m_2 = 1$ kg, $k_1 = 24$ N/m, and $k_2 = 3$ N/m. Also assume that the damping is proportional with $\alpha = 0$ and $\beta = 0.1$, so that $c_1 = 2.4$ N·s/m and $c_2 = 0.3$ N·s/m. Calculate the steady-state response.

Solution   The equations of motion in matrix form become

$$\begin{bmatrix} 9 & 0 \\ 0 & 1 \end{bmatrix}\ddot{\mathbf{x}} + \begin{bmatrix} 2.7 & -0.3 \\ -0.3 & 0.3 \end{bmatrix}\dot{\mathbf{x}} + \begin{bmatrix} 27 & -3 \\ -3 & 3 \end{bmatrix}\mathbf{x} = \begin{bmatrix} 0 \\ F_1(t) \end{bmatrix}$$

The matrices $M^{1/2}$ and $M^{-1/2}$ become

$$M^{1/2} = \begin{bmatrix} 3 & 0 \\ 0 & 1 \end{bmatrix} \qquad M^{-1/2} = \begin{bmatrix} \frac{1}{3} & 0 \\ 0 & 1 \end{bmatrix}$$

so that

$$\tilde{C} = M^{-1/2}CM^{-1/2} = \begin{bmatrix} 0.3 & -0.1 \\ -0.1 & 0.3 \end{bmatrix} \quad \text{and} \quad \tilde{K} = \begin{bmatrix} 3 & -1 \\ -1 & 3 \end{bmatrix}$$

The eigenvalue problem for $\tilde{K}$ yields

$$\lambda_1 = 2 \qquad \lambda_2 = 4 \qquad P = 0.7071 \begin{bmatrix} 1 & -1 \\ 1 & 1 \end{bmatrix}$$

Hence the natural frequencies of the system are $\omega_1 = \sqrt{2}$ and $\omega_2 = 2$, the matrices $P^T\tilde{C}P$ and $P^T\tilde{K}P$ become

$$P^T\tilde{C}P = \begin{bmatrix} 0.2 & 0 \\ 0 & 0.4 \end{bmatrix} \qquad \text{and} \qquad P^T\tilde{K}P = \begin{bmatrix} 2 & 0 \\ 0 & 4 \end{bmatrix}$$

The vector $\mathbf{f}(t) = P^T M^{-1/2}\mathbf{F}(t)$ becomes

$$\mathbf{f}(t) = \begin{bmatrix} 0.2357 & 0.7071 \\ -0.2357 & 0.7071 \end{bmatrix} \begin{bmatrix} 0 \\ F_1(t) \end{bmatrix} = 0.7071 \begin{bmatrix} F_1(t) \\ F_1(t) \end{bmatrix}$$

Hence the decoupled modal equations become

$$\ddot{r}_1 + 0.2\dot{r}_1 + 2r_1 = 0.7071(3)\cos 2t = 2.1213\cos 2t$$

$$\ddot{r}_2 + 0.4\dot{r}_2 + 4r_2 = 0.7071(3)\cos 2t = 2.1213\cos 2t$$

Comparing the coefficient of $\dot{r}_i$ in each case to $2\zeta_i\omega_i$ yields

$$\zeta_1 = \frac{0.2}{2\sqrt{2}} = 0.0707$$

$$\zeta_2 = \frac{0.4}{2(2)} = 0.1000$$

Thus the damped natural frequencies become

$$\omega_{d1} = \omega_1\sqrt{1 - \zeta_1^2} = 1.4106 \approx 1.41$$

$$\omega_{d2} = \omega_2\sqrt{1 - \zeta_2^2} = 1.9899 \approx 1.99$$

Note that while the force F_1 is applied only to mass m_2, it becomes applied to both co-ordinates when transformed to modal coordinates. The modal equations for r_1 and r_2 can be solved by equation (4.131), or since in this case of a simple harmonic excitation, the particular solution is given directly by equation (2.28) as

$$r_{1p}(t) = \frac{2.1213}{\sqrt{(2-4)^2 + [2(0.0707)\sqrt{2}(2)]^2}} \cos\left(2t - \tan^{-1}\frac{2(0.0707)\sqrt{2}(2)}{\sqrt{2}^2 - 2^2}\right)$$

$$= (1.040)\cos(2t + 0.1974) = 1.040\cos(2t + 0.1974)$$

and

$$r_{2p}(t) = \frac{2.1213}{\sqrt{(4-4)^2 + [2(0.1)(2)(2)]^2}} \cos\left(2t - \tan^{-1}\frac{2(0.1)(2)(2)}{2^2 - 2^2}\right)$$

$$= 2.6516\cos\left(2t - \frac{\pi}{2}\right) = 2.6516\sin 2t$$

Here r_{ip} is used to denote the particular solution of the ith modal equation. Note that $r_2(t)$ is excited at its resonance frequency but has high damping, so that the larger but finite amplitude for $r_{2p}(t)$ is not unexpected. If the transient response is ignored [it dies out per equation (2.30)], the solution above yields the steady-state response. The solution in the physical coordinate system is

$$\mathbf{x}_{ss}(t) = M^{-1/2}P\mathbf{r}(t) = \begin{bmatrix} 0.2357 & -0.2357 \\ 0.7071 & 0.7071 \end{bmatrix} \begin{bmatrix} 1.040\cos(2t + 0.1974) \\ 2.6516\sin 2t \end{bmatrix}$$

so that in the steady state

$$x_1(t) = 0.2451\cos(2t + 0.1974) - 0.6249\sin 2t$$

$$x_2(t) = 0.7354\cos(2t + 0.1974) + 1.8749\sin 2t$$

Note that even though there is a fair amount of damping in the resonant mode, the coordinates each have a large component vibrating near the resonant frequency.

$\square$

4.7 LAGRANGE'S EQUATIONS

So far the multiple-degree-of-freedom systems considered have been simple to model using Newton's law (force balance) directly. More complicated machine parts and structures are often better considered by using the methods of analytical mechanics. These methods are based on calculating the work and kinetic energy of the system as a whole and seeking a minimum much like the energy method of Section 1.4. A brief "working" description is given here. A more precise account can be found in D'Souza and Garg (1984), for instance.

The procedure begins by assigning a generalized coordinate to each moving part. The standard rectangular coordinate system is an example of a generalized coordinate, but any length, angle, or other coordinate that uniquely defines the position of the part at any time forms a generalized coordinate. It is usually desirable to choose coordinates that are independent. It is customary to designate each coordinate by the letter q with a subscript so that a set of n generalized coordinates is written as $q_1, q_2, \dots, q_n$.

An example of generalized coordinates is illustrated in Figure 4.16. In the figure the location of the two masses can be described by the set of four coordinates x_1, y_1, x_2, and y_2 or the two coordinates θ_1 and θ_2. The coordinates θ_1 and θ_2 are taken to be generalized coordinates because they are independent. The Cartesian coordinates (x_1, x_2, y_1, y_2) are not independent and hence would not make a desirable choice of generalized coordinates. Note that

$$x_1^2 + y_1^2 = l_1^2 \qquad \text{and} \qquad (x_2 - x_1)^2 + (y_2 - y_1)^2 = l_2^2 \qquad (4.132)$$

express the dependence of the Cartesian coordinates on each other. The relationships in equation (4.132) are called equations of constraint.

A new configuration of the double pendulum of Figure 4.16 can be obtained by changing the generalized coordinates $q_1 = \theta_1$ and $q_2 = \theta_2$ by an amount δq_1 and δq_2, respectively. Here δq_i are referred to as *virtual displacements*, which are defined to be infinitesimal displacements that do not violate constraints and such that there is no

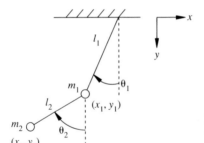

Figure 4.16 Example of generalized coordinates for a double pendulum illustrating an example of constraints.

significant change in the system's geometry. The *virtual work*, denoted δW, is the work done in causing the virtual displacement. The principal of virtual work states that if a system at rest (or at equilibrium) under the action of a set of forces is given a virtual displacement, the virtual work done by the forces is zero. The generalized force (or moment) at the ith coordinate, denoted Q_i, is related to the work done in changing q_i by the amount δq_i and is defined to be

$$Q_i = \frac{\delta W}{\delta q_i} \qquad (4.133)$$

The quantity Q_i will be a moment if q_i is a rotational coordinate and a force if it is a translational coordinate.

The Lagrange formulation states that the equations of motion of a vibrating system can be derived from

$$\frac{d}{dt}\left(\frac{\partial T}{\partial \dot{q}_i}\right) - \frac{\partial T}{\partial q_i} + \frac{\partial U}{\partial q_i} = Q_i \qquad i = 1, 2, \ldots, n \qquad (4.134)$$

where $\dot{q}_i = \partial q_i / \partial t$ is the generalized velocity, T is the kinetic energy of the system, U is the potential energy of the system, and Q_i represents all the nonconservative forces corresponding to q_i. Here $\partial / \partial q_i$ denotes the partial derivative with respect to the coordinate q_i. For conservative systems, $Q_i = 0$ and equation (4.134) becomes

$$\frac{d}{dt}\left(\frac{\partial T}{\partial \dot{q}_i}\right) - \frac{\partial T}{\partial q_i} + \frac{\partial U}{\partial q_i} = 0 \qquad i = 1, 2, \ldots, n \qquad (4.135)$$

Equations (4.134) and (4.135) represent one equation for each generalized coordinate. These expressions allow the equation of motion of complicated systems to be derived without using free-body diagrams and summing forces and moments. The Lagrange equation can be rewritten in a slightly simplified form by defining the Lagrangian, L, to be $L = (T - U)$, the difference between the kinetic and potential energies. Then if $\partial U / \dot{q}_i = 0$, the Lagrange equation becomes

$$\frac{d}{dt}\left(\frac{\partial L}{\partial \dot{q}_i}\right) - \frac{\partial L}{\partial q_i} = 0 \qquad i = 1, 2 \ldots, n \qquad (4.136)$$

The following examples illustrate the procedure.

Example 4.7.1

Derive the equations of motion of the system of Figure 4.17 using the Lagrange equation.

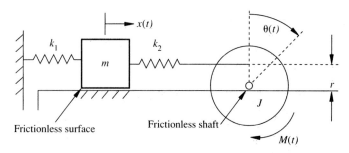

Figure 4.17   Vibration model of a simple machine part. The quantity $M(t)$ denotes an applied moment. The disk rotates without translation.

Solution   The motion of this system can be described by the two coordinates x and θ, so a good choice of generalized coordinates is $q_1(t) = x(t)$ and $q_2(t) = \theta(t)$. The kinetic energy becomes

$$T = \frac{1}{2}m\dot{q}_1^2 + \frac{1}{2}J\dot{q}_2^2$$

The potential energy becomes

$$U = \frac{1}{2}k_1 q_1^2 + \frac{1}{2}k_2(rq_2 - q_1)^2$$

Here $Q_1 = 0$ and $Q_2 = M(t)$. Using equation (4.135) yields for $i = 1$,

$$\frac{d}{dt}(m\dot{q}_1 + 0) - 0 + k_1 q_1 + k_2(rq_2 - q_1)(-1) = 0$$

or

$$m\ddot{q}_1 + (k_1 + k_2)q_1 - k_2 rq_2 = 0 \qquad (4.137)$$

Similarly, for $i = 2$, equation (4.134) yields

$$J\ddot{q}_2 + k_2 r^2 q_2 - k_2 rq_1 = M(t) \qquad (4.138)$$

Combining equations (4.137) and (4.138) into matrix form yields

$$\begin{bmatrix} m & 0 \\ 0 & J \end{bmatrix}\ddot{\mathbf{q}}(t) + \begin{bmatrix} k_1 + k_2 & -rk_2 \\ -rk_2 & r^2 k_2 \end{bmatrix}\mathbf{q}(t) = \begin{bmatrix} 0 \\ M(t) \end{bmatrix} \qquad (4.139)$$

$\square$

Example 4.7.2

A machine part consists of three levers connected by lightweight linkages. A vibration model of this part is given in Figure 4.18. Use the Lagrange method to obtain the equation of vibration. Take the angles to be the generalized coordinates. Linearize the result and put it in matrix form.

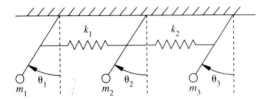

Figure 4.18 Vibration model of three coupled levers. The length of the lever is l and the spring is attached at α units from the pivot point.

Solution   The kinetic energy is

$$T = \frac{1}{2}m_1 l^2 \dot{\theta}_1^2 + \frac{1}{2}m_2 l^2 \dot{\theta}_2^2 + \frac{1}{2}m_3 l^2 \dot{\theta}_3^2$$

The potential energy becomes

$$U = m_1 gl(1 - \cos\theta_1) + m_2 gl(1 - \cos\theta_2) + m_3 gl(1 - \cos\theta_3)$$

$$+ \frac{1}{2}k_1(\alpha\theta_2 - \alpha\theta_1)^2 + \frac{1}{2}k_2(\alpha\theta_3 - \alpha\theta_2)^2$$

Applying the Lagrange equation for $i = 1$ yields

$$m_1 l^2 \ddot{\theta}_1 + m_1 gl \sin\theta_1 - \alpha k_1(\alpha\theta_2 - \alpha\theta_1) = 0$$

For $i = 2$, the Lagrange equation yields

$$m_2 l^2 \ddot{\theta}_2 + m_2 gl \sin\theta_2 + \alpha k_1(\alpha\theta_2 - \alpha\theta_1) - \alpha k_2(\alpha\theta_3 - \alpha\theta_2) = 0$$

and for $i = 3$, the Lagrange equation becomes

$$m_3 l^2 \ddot{\theta}_3 + m_3 gl \sin\theta_3 + \alpha k_2(\alpha\theta_3 - \alpha\theta_2) = 0$$

These three equations can be linearized by assuming θ is small so that $\sin\theta \sim \theta$. Note that this linearization occurs after the equations have been derived. In matrix form this becomes

$$\begin{bmatrix} m_1 l^2 & 0 & 0 \\ 0 & m_2 l^2 & 0 \\ 0 & 0 & m_3 l^2 \end{bmatrix} \ddot{\mathbf{q}}(t)$$

$$+ \begin{bmatrix} m_1 gl + \alpha^2 k_1 & -\alpha^2 k_1 & 0 \\ -\alpha^2 k_1 & m_2 gl + \alpha^2(k_1 + k_2) & -\alpha^2 k_2 \\ 0 & -\alpha^2 k_2 & m_3 gl + \alpha^2 k_2 \end{bmatrix} \mathbf{q}(t) = \mathbf{0}$$

where $\mathbf{q}(t) = [q_1 \quad q_2 \quad q_3]^T = [\theta_1 \quad \theta_2 \quad \theta_3]^T$ is the generalized set of coordinates. $\qquad\square$

Example 4.7.3

Consider the wing vibration model of Figure 4.19. Using the vertical motion of point of attachment of the springs, $x(t)$, and the rotation of this point, $\theta(t)$, determine the equations of motion using Lagrange's method. Use the small-angle approximation (recall the pendulum of Example 1.4.6) and write the equations in matrix form. Note that G denotes the center of mass and e denotes the distance between the point of rotation and the center of mass. Ignore the gravitational force.

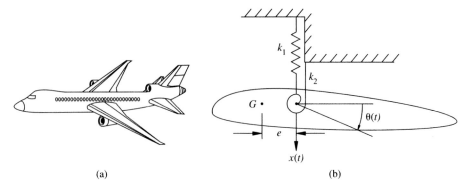

Figure 4.19   An airplane in flight (a) presents a number of different vibration models, one of which is given in part (b). In (b) a vibration model of a wing in flight is sketched which accounts for bending and torsional motion by modeling the wing as attached to ground (the aircraft body in this case) through a linear spring k_1 and a torsional spring k_2.

Solution   Let m denote the mass of the wing section and J denote the rotational inertia about point G. The kinetic energy is

$$T = \frac{1}{2}m\dot{x}_G^2 + \frac{1}{2}J\dot{\theta}^2$$

where x_G is the displacement of the point G. This displacement is related to the coordinate $x(t)$ of the point of attachment of the springs by

$$x_G(t) = x(t) - e\sin\theta(t)$$

which is obtained by examining the geometry of Figure 4.19. Thus $\dot{x}_G(t)$ becomes

$$\dot{x}_G(t) = \dot{x}(t) - e\cos\theta(t)\frac{d\theta}{dt} = \dot{x}(t) - e\dot{\theta}\cos\theta$$

The expression for kinetic energy in terms of the generalized coordinates $q_1 = x$ and $q_2 = \theta$ then becomes

$$T = \frac{1}{2}m[\dot{x} - e\dot{\theta}\cos\theta]^2 + \frac{1}{2}J\dot{\theta}^2$$

The expression for potential energy is

$$U = \frac{1}{2}k_1x^2 + \frac{1}{2}k_2\theta^2$$

which is already in terms of the generalized coordinates. The Lagrangian, L, becomes

$$L = T - U = \frac{1}{2}m[\dot{x} - e\dot{\theta}\cos\theta]^2 + \frac{1}{2}J\dot{\theta}^2 - \frac{1}{2}k_1x^2 - \frac{1}{2}k_2\theta^2$$

Calculating the derivatives required by equation (4.136) for $i = 1$ yields

$$\frac{\partial L}{\partial \dot{q}_1} = \frac{\partial L}{\partial \dot{x}} = m[\dot{x} - e\dot{\theta}\cos\theta]$$

$$\frac{d}{dt}\left(\frac{\partial L}{\partial \dot{x}}\right) = m\ddot{x} - me\ddot{\theta}\cos\theta + me\dot{\theta}^2\sin\theta$$

$$\frac{\partial L}{\partial q_1} = \frac{\partial L}{\partial x} = -k_1 x$$

so that equation (4.136) becomes

$$m\ddot{x} - me\ddot{\theta}\cos\theta + em\dot{\theta}^2\sin\theta + k_1 x = 0$$

Assuming small motions so that the approximation $\cos\theta \to 1$, $\sin\theta \to \theta$ holds and assuming that the term $\dot{\theta}^2\theta$ is small enough to ignore (more about this in Chapter 10) results in a linear equation in $x(t)$ given by

$$m\ddot{x} - me\ddot{\theta} + k_1 x = 0$$

Calculating the derivatives of the Lagrangian required by equation (4.136) for $i = 2$ yields

$$\frac{\partial L}{\partial \dot{q}_2} = \frac{\partial L}{\partial \dot{\theta}} = m[\dot{x} - e\dot{\theta}\cos\theta](-e\cos\theta) + J\dot{\theta} = -me\cos\theta\dot{x} + me^2\dot{\theta}\cos^2\theta + J\dot{\theta}$$

$$\frac{d}{dt}\left(\frac{\partial L}{\partial \dot{q}_2}\right) = \frac{d}{dt}\left(\frac{\partial L}{\partial \dot{\theta}}\right)$$

$$= -me\cos\theta\ddot{x} + me\dot{x}\sin\theta\dot{\theta} + me^2\ddot{\theta}\cos^2\theta - 2me^2\dot{\theta}^2\sin\theta\cos\theta + J\ddot{\theta}$$

$$\frac{\partial L}{\partial q_2} = \frac{\partial L}{\partial \theta} = m[\dot{x} - e\dot{\theta}\cos\theta](e\dot{\theta}\sin\theta) - k_2\theta$$

$$= me\dot{x}\dot{\theta}\sin\theta - me^2\dot{\theta}^2\sin\theta\cos\theta - k_2\theta$$

so that equation (4.136) becomes

$$J\ddot{\theta} + me\cos\theta\ddot{x} + me^2\cos^2\theta\ddot{\theta} - me^2\dot{\theta}^2\sin\theta\cos\theta + k_2\theta = 0$$

Again using the small-angle, small motion approximation (i.e., $\sin\theta \to \theta$, $\cos\theta \to 1$, $\dot{\theta}^2\theta \to 0$) results in a linear equation in $\theta(t)$ given by

$$(J + me^2)\ddot{\theta} - me\ddot{x} + k_2\theta = 0$$

Combining the expression for $i = 1$ and $i = 2$, into one vector equation in the generalized vector $\mathbf{q} = [q_1(t) \quad q_2(t)]^T = [x(t) \quad \theta(t)]^T$ yields

$$\begin{bmatrix} m & -me \\ -me & me^2 + J \end{bmatrix}\begin{bmatrix} \ddot{x}(t) \\ \ddot{\theta}(t) \end{bmatrix} + \begin{bmatrix} k_1 & 0 \\ 0 & k_2 \end{bmatrix}\begin{bmatrix} x(t) \\ \theta(t) \end{bmatrix} = \begin{bmatrix} 0 \\ 0 \end{bmatrix}$$

Note here that the two equations of motion are coupled, not through stiffness terms, as in Example 4.7.2, but rather, through the inertia terms. Such systems are called *dynamically coupled*, meaning that the terms that couple the equation in $\theta(t)$ to the equation in $x(t)$ are in the mass matrix (i.e., meaning that the mass matrix is not diagonal). In all previous examples, the mass matrix is diagonal and the stiffness matrix is not diagonal. Such systems are called *statically coupled*. Dynamically coupled systems have nondiagonal mass matrices, and hence require a more complicated method of calculating the matrix $M^{-1/2}$. This method is discussed in Section 9.6. The MATLAB programs at the end of this chapter can be applied to compute $M^{-1/2}$ for dynamically coupled systems, even though the theory is not developed until Chapter 9.

$\square$

Example 4.7.3 not only illustrates a dynamically coupled system but also presents a system that is easier to approach using Lagrange's method than by using a force balance to obtain the equation of motion. In fact, several vibration texts have reported an incorrect set of equations of motion for the problem of the preceding example by using the sum of forces and moments rather than taking a Lagrangian approach.

4.8 EXAMPLES

Several examples of multiple-degree-of-freedom systems, their schematics, and equations of motion are presented in this section. The "art" in vibration analysis and design is often related to choosing an appropriate mathematical model to describe a given structure or machine. The following examples are intended to provide additional "practice" in modeling and analysis.

Example 4.8.1

A drive shaft for a belt-driven machine such as a lathe is illustrated in Figure 4.20(a). The vibration model of this system is indicated in Figure 4.20(b), along with a free-body diagram of the machine. Write the equations of motion in matrix form and solve for the case $J_1 = J_2 = J_3 = 10$ kg m^2/rad, $k_1 = k_2 = k_3 = 10^3$N $\cdot$ m/rad, $c = 2$ N $\cdot$ m $\cdot$s/rad for zero initial conditions and where the applied moment $M(t)$ is a unit impulse function.

Solution   In Figure 4.20(a) the bearings and shaft lubricant are modeled as lumped viscous damping, and the shafts are modeled as torsional springs. The pulley and machine disks are modeled as rotational inertias. The motor is modeled simply as supplying a moment to the pulley. Figure 4.20(b) illustrates a free-body diagram for each of the three disks, where the damping is assumed to act in proportion to the relative motion of the masses and of the same value at each coordinate (other damping models may be more appropriate, but this choice yields an easy form to solve).

Examining the free-body diagram of Figure 4.20(b) and summing the moments on each of the disks yields

$$J_1\ddot{\theta}_1 = k_1(\theta_2 - \theta_1) + c(\dot{\theta}_2 - \dot{\theta}_1)$$

$$J_2\ddot{\theta}_2 = k_2(\theta_3 - \theta_2) + c(\dot{\theta}_3 - \dot{\theta}_2) - k_1(\theta_2 - \theta_1) - c(\dot{\theta}_2 - \dot{\theta}_1)$$

$$J_3\ddot{\theta}_3 = -k_2(\theta_3 - \theta_2) - c(\dot{\theta}_3 - \dot{\theta}_2) + M(t)$$

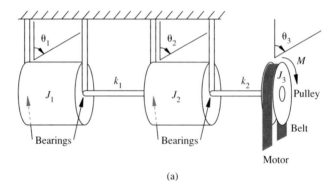

(a)

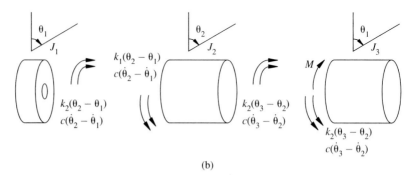

(b)

Figure 4.20 (a) Schematic of the moving parts of lathe. The bearings that support the rotating shaft are modeled as providing viscous damping while the shafts provide stiffness and the belt drive provides an applied torque. (b) Free-body diagrams of the three inertias in the rotating system of part (a). The shafts are modeled as providing stiffness, or as rotational springs, and the bearings are modeled as rotational dampers.

where θ_1, θ_2 and θ_3 are the rotational coordinates as indicated in Figure 4.20. The unit for θ is radians. Rearranging these equations yields

$$J_1\ddot{\theta}_1 + c\dot{\theta}_1 + k_1\theta_1 - c\dot{\theta}_2 - k_1\theta_2 = 0$$

$$J_2\ddot{\theta}_2 + 2c\dot{\theta}_2 - c\dot{\theta}_1 - c\dot{\theta}_3 + (k_1 + k_2)\theta_2 - k_1\theta_1 - k_2\theta_3 = 0$$

$$J_3\ddot{\theta}_3 + c\dot{\theta}_3 - c\dot{\theta}_2 - k_2\theta_2 + k_2\theta_3 = M(t)$$

In matrix form this becomes

$$\begin{bmatrix} J_1 & 0 & 0 \\ 0 & J_2 & 0 \\ 0 & 0 & J_3 \end{bmatrix}\ddot{\boldsymbol{\theta}} + \begin{bmatrix} c & -c & 0 \\ -c & 2c & -c \\ 0 & -c & c \end{bmatrix}\dot{\boldsymbol{\theta}} + \begin{bmatrix} k_1 & -k_1 & 0 \\ -k_1 & k_1 + k_2 & -k_2 \\ 0 & -k_2 & k_2 \end{bmatrix}\boldsymbol{\theta} = \begin{bmatrix} 0 \\ 0 \\ M(t) \end{bmatrix}$$

where $\boldsymbol{\theta}(t) = [\theta_1(t)\ \ \theta_2(t)\ \ \theta_3(t)]^T$. Using the values for the coefficients given above this becomes

$$10I\ddot{\boldsymbol{\theta}} + 2\begin{bmatrix} 1 & -1 & 0 \\ -1 & 2 & -1 \\ 0 & -1 & 1 \end{bmatrix}\dot{\boldsymbol{\theta}} + 10^3\begin{bmatrix} 1 & -1 & 0 \\ -1 & 2 & -1 \\ 0 & -1 & 1 \end{bmatrix}\boldsymbol{\theta} = \begin{bmatrix} 0 \\ 0 \\ \delta(t) \end{bmatrix}$$

Note that the damping matrix is proportional to the stiffness matrix, so that modal analysis can be used to calculate the solution. Also note that

$$\tilde{C} = 0.2 \begin{bmatrix} 1 & -1 & 0 \\ -1 & 2 & -1 \\ 0 & -1 & 1 \end{bmatrix} \qquad \tilde{K} = 10^2 \begin{bmatrix} 1 & -1 & 0 \\ -1 & 2 & -1 \\ 0 & -1 & 1 \end{bmatrix} \qquad M^{-1/2}\mathbf{F} = \frac{1}{\sqrt{10}} \begin{bmatrix} 0 \\ 0 \\ \delta(t) \end{bmatrix}$$

Following the steps of Example 4.6.1, the eigenvalue problem for $\tilde{K}$ yields

$$\lambda_1 = 0 \qquad \lambda_2 = 100 \qquad \lambda_3 = 300$$

Note that one of the eigenvalues is zero. Thus the matrix $\tilde{K}$ is singular. The physical meaning of this is interpreted in this example. The normalized eigenvectors of $\tilde{K}$ yield

$$P = \begin{bmatrix} 0.5774 & 0.7071 & 0.4082 \\ 0.5774 & 0 & -0.8165 \\ 0.5774 & -0.7071 & 0.4082 \end{bmatrix} \qquad P^T = \begin{bmatrix} 0.5774 & 0.5774 & 0.5774 \\ 0.7071 & 0 & -0.7071 \\ 0.4082 & -0.8165 & 0.4082 \end{bmatrix}$$

Further computation yields

$$P^T\tilde{C}P = \text{diag } [0 \quad 0.2 \quad 0.6]$$

$$P^T\tilde{K}P = \text{diag } [0 \quad 100 \quad 300]$$

$$P^T M^{-1/2}\mathbf{F}(t) = \begin{bmatrix} 0.1826 \\ -0.2236 \\ 0.1291 \end{bmatrix} \delta(t)$$

The decoupled modal equations are

$$\ddot{r}_1(t) = 0.1826\delta(t)$$

$$\ddot{r}_2(t) + 0.2\dot{r}_2(t) + 100r_2(t) = -0.2236\delta(t)$$

$$\ddot{r}_3(t) + 0.6\dot{r}_3(t) + 300r_3(t) = 0.1291\delta(t)$$

Obviously, $\omega_1 = 0$, $\omega_2 = 10$ rad/s and $\omega_3 = 17.3205$ rad/s.

Comparing coefficients of $\dot{r}_i$ with $2\zeta_i\omega_i$ yields the three modal damping ratios

$$\zeta_1 = 0$$

$$\zeta_2 = \frac{0.2}{2(10)} = 0.01$$

$$\zeta_3 = \frac{0.6}{2(17.3205)} = 0.01732$$

so that the second two modes are underdamped. Hence the two damped natural frequencies become

$$\omega_{d2} = \omega_2\sqrt{1 - \zeta_2^2} = 9.9995 \text{ rad/s}$$

$$\omega_{d3} = \omega_3\sqrt{1 - \zeta_3^2} = 17.3179 \text{ rad/s}$$

As was the case in Example 4.6.1, while the moment is applied to only one physical location, it is applied to each of the three modal coordinates. The modal equation for r_2 and r_3 have solutions given by equation (3.6). The solution corresponding to the zero eigenvalue

($\omega_1 = 0$) can be calculated by direct integration or by using the Laplace transform method. Taking the Laplace transform yields

$$s^2 r(s) = 0.1826 \quad \text{or} \quad r(s) = \frac{0.1826}{s^2}$$

The inverse Laplace transform of this last expression yields (see Appendix B)

$$r_1(t) = (0.1826)t$$

Physically, this is interpreted as the unconstrained motion of the shaft (i.e., the shaft rotates or spins continuously through 360°). This is also called the *rigid-body mode* or *zero mode* and results from $\tilde{K}$ being singular (i.e., from the zero eigenvalue). Such systems are also called *semidefinite*, as explained in Appendix C.

Following equation (3.6), the solution for $r_2(t)$ and $r_3(t)$ becomes

$$r_2(t) = \frac{f_2}{\omega_{d2}} e^{-\zeta_2 \omega_2 t} \sin \omega_{d2} t = -0.0224 e^{-0.1t} \sin 9.9995t$$

$$r_3(t) = \frac{f_3}{\omega_{d3}} e^{-\zeta_3 \omega_3 t} \sin \omega_{d3} t = 0.0075 e^{-0.2999t} \sin 17.3179t$$

The total solution in physical coordinates is then calculated from $\boldsymbol{\theta}(t) = M^{-1/2} P \mathbf{r}(t) = 1/\sqrt{10} P \mathbf{r}(t)$ or

$$\boldsymbol{\theta}(t) = \begin{bmatrix} 0.0333t - 0.0050e^{-0.1t} \sin(9.9995t) + 0.0010e^{-0.2999t} \sin(17.3179t) \\ 0.0333t - 0.0019e^{-0.2999t} \sin(17.3179t) \\ 0.0333t + 0.0053e^{-0.1t} \sin(9.9995t) + 0.0010e^{-0.2999t} \sin(17.3179t) \end{bmatrix}$$

The three solutions $\theta_1(t)$, $\theta_2(t)$, and $\theta_3(t)$ are plotted in Figure 4.21. Figure 4.22 plots the three solutions without the rigid-body term. This represents the vibrations experienced by each disk as they rotate.

□

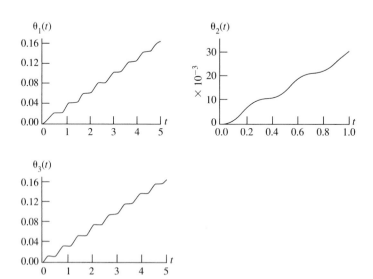

Figure 4.21 Response of each of the disks of Figure 4.20 to an impulse at θ_3, illustrating the effects of a rigid-body rotation.

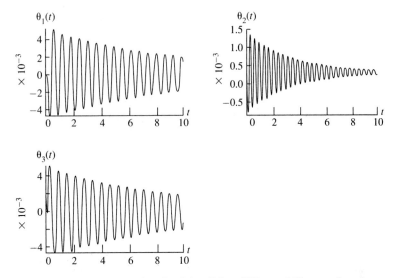

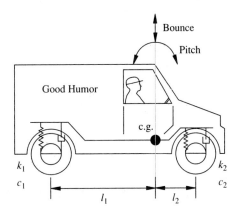

Figure 4.22 Response of each of the disks of Figure 4.20 to an impulse at θ_3 without the rigid-body mode, illustrating the vibration that occurs in each disk.

Example 4.8.2

In Figure 2.10 a vehicle is modeled as a single-degree-of-freedom system. In this example a two-degree-of-freedom model is used for a vehicle which allows for bounce and pitch motion. This model can be determined from the schematic of Figure 4.23. Determine the equations of motion, solve them by modal analysis, and determine the response to the engine being shut off, which is modeled as an impulse moment applied to $\theta(t)$.

Figure 4.23 Sketch of the side section of a vehicle used to suggest a vibration model for examining its angular (pitch) and up-and-down (bounce) motion. The center of gravity is denoted c.g.

Solution The sketch of the vehicle of Figure 4.23 can be simplified by modeling the entire mass of the system as concentrated at the center of gravity (c.g.). The tire and wheel assembly is approximated as a simple spring–dashpot arrangement as illustrated in Figure 4.24. The rotation of the vehicle in the x–y plane is described by the angle $\theta(t)$

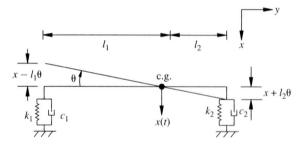

Figure 4.24   Vehicle of Figure 4.23 modeled as having all of its mass at its c.g. and two degrees of freedom, consisting of the pitch, $\theta(t)$, about the c.g. and a translation $x(t)$ of the c.g.

and the up-and-down motion is modeled by $x(t)$. The angle $\theta(t)$ is taken to be positive in the clockwise direction and the vertical displacement is taken as positive in the downward direction. Rigid translation in the y direction is ignored for the sake of concentrating on the vibration characteristics of the vehicle (e.g., Example 4.8.1 illustrates the concept of ignoring rigid-body motion).

Summing the forces in the x direction yields

$$m\ddot{x} = -c_1(\dot{x} - l_1\dot{\theta}) - c_2(\dot{x} + l_2\dot{\theta}) - k_1(x - l_1\theta) - k_2(x + l_2\theta) \tag{4.140}$$

since the spring k_1 experiences a displacement $x - l_1\theta$ and k_2 experiences a displacement $x + l_2\theta$. Similarly, the velocity experienced by the damper c_1 is $\dot{x} - l_1\dot{\theta}$ and that of c_2 is $\dot{x} + l_2\dot{\theta}$. Taking moments about the center of gravity yields

$$J\ddot{\theta} = c_1l_1(\dot{x} - l_1\dot{\theta}) - c_2l_2(\dot{x} + l_2\dot{\theta}) + k_1l_1(x - l_1\theta) - k_2l_2(x + l_2\theta) \tag{4.141}$$

where $J = mr^2$. Here r is the radius of gyration of the vehicle (recall Example 1.4.6). Equations (4.140) and (4.141) can be rewritten as

$$m\ddot{x} + (c_1 + c_2)\dot{x} + (l_2c_2 - l_1c_1)\dot{\theta} + (k_1 + k_2)x + (l_2k_2 - l_1k_1)\theta = 0$$
$$mr^2\ddot{\theta} + (c_2l_2 - c_1l_1)\dot{x} + (l_2^2c_2 + l_1^2c_1)\dot{\theta} + (k_2l_2 - k_1l_1)x + (l_1^2k_1 + l_2^2k_2)\theta = 0 \tag{4.142}$$

In matrix form, these two coupled equations become

$$m\begin{bmatrix} 1 & 0 \\ 0 & r^2 \end{bmatrix}\ddot{\mathbf{x}} + \begin{bmatrix} c_1 + c_2 & l_2c_2 - l_1c_1 \\ l_2c_2 - l_1c_1 & l_2^2c_2 + l_1^2c_1 \end{bmatrix}\dot{\mathbf{x}} + \begin{bmatrix} k_1 + k_2 & k_2l_2 - k_1l_1 \\ k_2l_2 - k_1l_1 & l_1^2k_1 + l_2^2k_2 \end{bmatrix}\mathbf{x} = \mathbf{0} \tag{4.143}$$

where the vector $\mathbf{x}$ is defined by

$$\mathbf{x} = \begin{bmatrix} x(t) \\ \theta(t) \end{bmatrix}$$

Reasonable values for a truck are

$$r^2 = 0.64 \text{ m}^2 \qquad m = 4000 \text{ kg} \qquad c_1 = c_2 = 2000 \text{ N} \cdot \text{s/m}$$

$$k_1 = k_2 = 20{,}000 \text{ N/m} \qquad l_1 = 0.9 \text{ m} \qquad l_2 = 1.4 \text{ m}$$

With these values, equation (4.143) becomes

$$\begin{bmatrix} 4000 & 0 \\ 0 & 2560 \end{bmatrix}\ddot{\mathbf{x}} + \begin{bmatrix} 4000 & 1000 \\ 1000 & 5540 \end{bmatrix}\dot{\mathbf{x}} + \begin{bmatrix} 40{,}000 & 10{,}000 \\ 10{,}000 & 55{,}400 \end{bmatrix}\mathbf{x} = \begin{bmatrix} 0 \\ 0 \end{bmatrix} \tag{4.144}$$

Note that $C = (0.1)K$, so that the damping is proportional. If a moment $M(t)$ is applied to the angular coordinate $\theta(t)$ the equations of motion become

$$\begin{bmatrix} 4000 & 0 \\ 0 & 2560 \end{bmatrix} \ddot{\mathbf{x}} + \begin{bmatrix} 4000 & 1000 \\ 1000 & 5540 \end{bmatrix} \dot{\mathbf{x}} + \begin{bmatrix} 40{,}000 & 10{,}000 \\ 10{,}000 & 55{,}400 \end{bmatrix} \mathbf{x} = \begin{bmatrix} 0 \\ \delta(t) \end{bmatrix}$$

Following the usual procedures of modal analysis, calculation of $M^{-1/2}$ yields

$$M^{-1/2} = \begin{bmatrix} 0.0158 & 0 \\ 0 & 0.0198 \end{bmatrix}$$

Thus

$$\tilde{C} = \begin{bmatrix} 1.0000 & 0.3125 \\ 0.3125 & 2.1641 \end{bmatrix} \quad \text{and} \quad \tilde{K} = \begin{bmatrix} 10.000 & 3.1250 \\ 3.1250 & 21.6406 \end{bmatrix}$$

Solving the eigenvalue problem for $\tilde{K}$ yields

$$P = \begin{bmatrix} 0.9698 & 0.2439 \\ -0.2439 & 0.9698 \end{bmatrix} \quad \text{and} \quad P^T = \begin{bmatrix} 0.9698 & -0.2439 \\ 0.2439 & 0.9698 \end{bmatrix}$$

with eigenvalues $\lambda_1 = 9.2141$ and $\lambda_2 = 22.4265$, so that the natural frequencies are

$$\omega_1 = 3.0355 \text{ rad/s} \quad \text{and} \quad \omega_2 = 4.7357 \text{ rad/s}$$

Thus

$$P^T \tilde{K} P = \text{diag } [9.2141 \quad 22.4265] \quad \text{and} \quad P^T \tilde{C} P = \text{diag } [0.9214 \quad 2.2426]$$

Comparing the elements of $P^T \tilde{C} P$ to ω_1 and ω_2 yields the modal damping ratios

$$\zeta_1 = \frac{0.9214}{2(3.0355)} = 0.1518 \quad \text{and} \quad \zeta_2 = \frac{2.2436}{2(4.7357)} = 0.2369$$

Using the formula from Window 4.6 for damped natural frequencies yields $\omega_{d1} = 3.0003$ rad/s and $\omega_{d2} = 4.6009$ rad/s . The modal forces are calculated from

$$P^T M^{-1/2} \begin{bmatrix} 0 \\ \delta(t) \end{bmatrix} = \begin{bmatrix} 0.0153 & -0.0048 \\ 0.0039 & -0.0192 \end{bmatrix} \begin{bmatrix} 0 \\ \delta(t) \end{bmatrix} = \begin{bmatrix} -0.0048\delta(t) \\ 0.0192\delta(t) \end{bmatrix}$$

The decoupled modal equations become

$$\ddot{r}_1(t) + (0.9214)\dot{r}_1(t) + (9.2141)r_1(t) = -0.0048\delta(t)$$

$$\ddot{r}_2(t) + (2.2436)\dot{r}_2(t) + (22.4265)r_2(t) = 0.0192\delta(t)$$

From equations (3.7) and (3.8) these have solutions

$$r_1(t) = \frac{1}{m\omega_{d1}} e^{-\zeta_1\omega_1 t} \sin \omega_{d1} t = \frac{1}{(1)(3.0003)} e^{-(0.1518)(9.2141)t} \sin(3.0003t)$$

$$= 0.3333 e^{-1.3987t} \sin(3.0003t)$$

$$r_2(t) = \frac{1}{4.6009} e^{-(0.2369)(4.7357)t} \sin(4.6009t)$$

$$= 0.2173 e^{-1.1219t} \sin(4.6009t)$$

The solution in physical coordinates is obtained from

$$\begin{bmatrix} x(t) \\ \theta(t) \end{bmatrix} = M^{-1/2} P \mathbf{r}(t)$$

which yields

$$x(t) = 0.0050e^{-1.3987t} \sin(3.0003t) + 0.0008e^{-1.1219t} \sin(4.6009t)$$

$$\theta(t) = 0.0016e^{-1.3987t} \sin(3.0003t) + 0.0042e^{-1.1219t} \sin(4.6009t)$$

These coordinates are plotted in Figure 4.25.

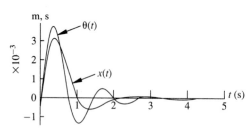

Figure 4.25 Plot of the bounce and pitch vibration of a vehicle of Figure 4.23 as the result of the engine being shut off (in meters versus seconds).

Example 4.8.3

The punch press of Figure 4.26 can be modeled for vibration analysis in the x direction as indicated by the three-degree-of-freedom system of Figure 4.27. Discuss the solution for the response due to an impact at m_1 using modal analysis.

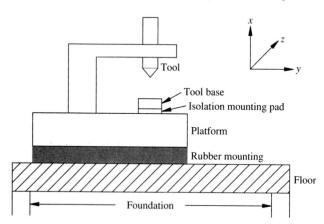

Figure 4.26 Schematic of a punch press machine.

Solution The mass and stiffness of the various components can be easily approximated using the static methods suggested in Chapter 1. However, it is very difficult to estimate values for the damping coefficients. Hence an educated guess is made for the modal damping ratios. Such guesses are often made based on experience or from measurements such as the logarithmic decrement. In this case the values of various masses and stiffness coefficients are [in mks units and $f(t) = 1000\delta(t)$]

$$m_1 = 400 \text{ kg} \qquad m_2 = 2000 \text{ kg} \qquad m_3 = 8000 \text{ kg}$$

$$k_1 = 300{,}000 \text{ N/m} \qquad k_2 = 80{,}000 \text{ N/m} \qquad k_3 = 800{,}000 \text{ N/m}$$

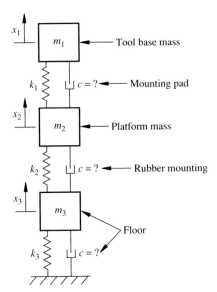

Figure 4.27   Vibration model of the punch press of Figure 4.26.

From free-body diagrams of each mass, the summing of forces in the x direction yields the three coupled equations

$$m_1\ddot{x}_1 = -k_1(x_1 - x_2) + f(t)$$

$$m_2\ddot{x}_2 = k_1(x_1 - x_2) - k_2(x_2 - x_3)$$

$$m_3\ddot{x}_3 = -k_3x_3 + k_2(x_2 - x_3)$$

Rewriting this set of coupled equations in matrix form yields

$$\begin{bmatrix} m_1 & 0 & 0 \\ 0 & m_2 & 0 \\ 0 & 0 & m_3 \end{bmatrix} \ddot{\mathbf{x}} + \begin{bmatrix} k_1 & -k_1 & 0 \\ -k_1 & k_1 + k_2 & -k_2 \\ 0 & -k_2 & k_2 + k_3 \end{bmatrix} \mathbf{x} = \begin{bmatrix} f(t) \\ 0 \\ 0 \end{bmatrix}$$

where $\mathbf{x} = [x_1(t) \quad x_2(t) \quad x_3(t)]^T$. Substituting the numerical values for m_i and k_i yields

$$(10^3)\begin{bmatrix} 0.4 & 0 & 0 \\ 0 & 2 & 0 \\ 0 & 0 & 8 \end{bmatrix} \ddot{\mathbf{x}} + (10^4)\begin{bmatrix} 30 & -30 & 0 \\ -30 & 38 & -8 \\ 0 & -8 & 88 \end{bmatrix} \mathbf{x} = \begin{bmatrix} 1000\delta(t) \\ 0 \\ 0 \end{bmatrix}$$

Following the modal analysis procedure for an undamped system yields

$$M^{1/2} = \begin{bmatrix} 20 & 0 & 0 \\ 0 & 44.7214 & 0 \\ 0 & 0 & 89.4427 \end{bmatrix} \quad M^{-1/2} = \begin{bmatrix} 0.0500 & 0 & 0 \\ 0 & 0.0224 & 0 \\ 0 & 0 & 0.0112 \end{bmatrix}$$

and

$$\tilde{K} = \begin{bmatrix} 750 & -335.4102 & 0 \\ -335.4102 & 190 & -20 \\ 0 & -20 & 110 \end{bmatrix}$$

Solving the eigenvalue problem for $\tilde{K}$ yields

$$P = \begin{bmatrix} -0.4116 & -0.1021 & 0.9056 \\ -0.8848 & -0.1935 & -0.4239 \\ -0.2185 & 0.9758 & 0.0106 \end{bmatrix} \qquad P^T = \begin{bmatrix} -0.4116 & -0.8848 & -0.2185 \\ -0.1021 & -0.1935 & 0.9758 \\ 0.9056 & -0.4239 & 0.0106 \end{bmatrix}$$

and

$$\lambda_1 = 29.0223 \qquad \omega_1 = 5.3872$$

$$\lambda_2 = 113.9665 \qquad \omega_2 = 10.6755$$

$$\lambda_3 = 907.0112 \qquad \omega_3 = 30.1166$$

The modal force vector becomes

$$P^T M^{-1/2} \begin{bmatrix} 1000\delta(t) \\ 0 \\ 0 \end{bmatrix} = \begin{bmatrix} -20.5805 \\ -5.1026 \\ 45.2814 \end{bmatrix} \delta(t)$$

Hence, the undamped modal equations are

$$\ddot{r}_1(t) + 29.0223 r_1(t) = -20.5805\delta(t)$$

$$\ddot{r}_2(t) + 113.9665 r_2(t) = -5.1026\delta(t)$$

$$\ddot{r}_3(t) + 907.0112 r_3(t) = 45.2814\delta(t)$$

To model the damping, note that each mode shape is dominated by one element. From examining the first column of the matrix P, the second element is larger than the other two elements. Hence if the system were vibrating only in the first mode, the motion of $x_2(t)$ would dominate. This element corresponds to the platform mass, which receives high damping from the rubber support. Hence it is given a large damping ratio of $\zeta_1 = 0.1$ (rubber provides a lot of damping). Similarly, the second mode is dominated by its third element corresponding to the motion of $x_3(t)$. This is a predominantly metal part, so it is given a low damping ratio of $\zeta_2 = 0.01$. The third mode shape is dominated by the first element, which corresponds to the mounting pad. Hence it is given a medium damping ratio of $\zeta_3 = 0.05$. Recalling that the velocity coefficient in modal coordinates has the form $2\zeta_i\omega_i$, the damped modal coordinates become $2\zeta_1\omega_1 = 2(0.1)(5.3872)$, $2\zeta_2\omega_2 = 2(0.01)(10.6755)$, and $2\zeta_3\omega_3 = 2(0.05)(30.1166)$. Therefore, the damped modal equations become

$$\ddot{r}_1(t) + 1.0774\dot{r}_1(t) + 29.0223 r_1(t) = -20.5805\delta(t)$$

$$\ddot{r}_2(t) + 0.2135\dot{r}_2(t) + 113.9665 r_2(t) = -5.1026\delta(t)$$

$$\ddot{r}_3(t) + 3.0117\dot{r}_3(t) + 907.0112 r_3(t) = 45.2814\delta(t)$$

These have solutions given by equation (3.6) as

$$r_1(t) = -3.8395e^{-0.5387t}\sin(5.3602t)$$

$$r_2(t) = -0.4780e^{-0.1068t}\sin(10.6750t)$$

$$r_3(t) = 1.5054e^{-1.5058t}\sin(30.0789t)$$

Using the transformation $\mathbf{x}(t) = M^{-1/2}P\mathbf{r}(t)$ yields

$$x_1(t) = 0.0790e^{-0.5387t}\sin(5.3602t) + 0.0024e^{-0.1068t}\sin(10.6750t) + 0.0682e^{-1.5058t}\sin(30.0789t)$$

$$x_2(t) = 0.0760e^{-0.5387t}\sin(5.3602t) + 0.0021e^{-0.1068t}\sin(10.6750t) - 0.0143e^{-1.5058t}\sin(30.0789t)$$

$$x_3(t) = 0.0094e^{-0.5387t}\sin(5.3602t) - 0.0052e^{-0.1068t}\sin(10.6750t) + 0.0002e^{-1.5058t}\sin(30.0789t)$$

These solutions are plotted in Figure 4.28.

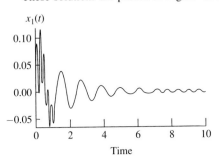
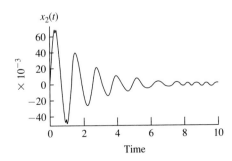

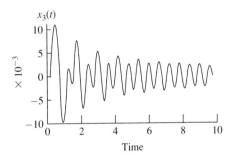

Figure 4.28 Numerical simulation of the vibration of the punch press of Figures 4.26 and 4.27 as the result of the machine tool impacting the tool base.

This example illustrates a method of assigning modal damping to an analytical model. This is a somewhat arbitrary procedure that falls in the category of an educated guess. A more sophisticated method is to measure the modal damping. This is discussed in Chapter 7. Note that the floor, $x_3(t)$, vibrates much longer than the machine parts do. This is something to consider in designing how and where the machine is mounted to the floor of a building. $\square$

PROBLEMS

Section 4.1

4.1. Consider the system of Figure 4.29. For $c_1 = c_2 = c_3 = 0$. Derive the equation of motion and calculate the mass and stiffness matrices. Note that setting $k_3 = 0$ in your solution should result in the stiffness matrix given by equation (4.9).

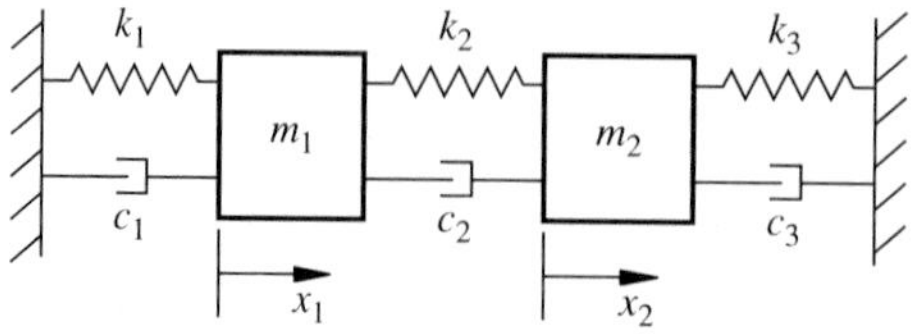

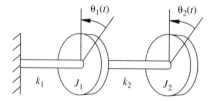

Figure 4.29

4.2. Calculate the characteristic equation from Problem 4.1 for the case

$$m_1 = 9 \text{ kg} \qquad m_2 = 1 \text{ kg} \qquad k_1 = 24 \text{ N/m} \qquad k_2 = 3 \text{ N/m} \qquad k_3 = 3 \text{ N/m}$$

and solve for the system's natural frequencies.

4.3. Calculate the vectors $\mathbf{u}_1$ and $\mathbf{u}_2$ for problem 4.2.

4.4. For initial conditions $\mathbf{x}(0) = [1 \quad 0]^T$ and $\dot{\mathbf{x}}(0) = [0 \quad 0]^T$ calculate the free response of the system of Problem 4.2. Plot the response x_1 and x_2.

4.5. Calculate the response of system of Example 4.1.7 to the initial condition $\mathbf{x}(0) = \mathbf{0}$, $\dot{\mathbf{x}}(0) = [1 \quad 0]^T$, plot the response and compare the result to Figure 4.3.

4.6. Repeat Problem 4.1 from the case that $k_1 = k_3 = 0$.

4.7. Calculate and solve the characteristic equation for Problem 4.6 with $m_1 = 9, m_2 = 1, k_2 = 10$.

4.8. Calculate the solution to the problem of Example 4.1.7, to the initial conditions

$$\mathbf{x}(0) = \begin{bmatrix} 1 \\ 3 \\ 1 \end{bmatrix} \qquad \dot{\mathbf{x}}(0) = \mathbf{0}$$

Plot the response and compare it to that of Figure 4.3.

4.9. Calculate the solution to Example 4.1.7 for the initial condition

$$\mathbf{x}(0) = \begin{bmatrix} -\dfrac{1}{3} \\ 1 \end{bmatrix} \qquad \dot{\mathbf{x}}(0) = \mathbf{0}$$

Plot the response and compare it to that of Figure 4.3 and Problem 4.8.

4.10. Determine the equation of motion in matrix form, then calculate the natural frequencies and mode shapes of the torsional system of Figure 4.30. Assume that the torsional stiffness values provided by the shaft are equal ($k_1 = k_2$) and that disk 1 has three times the inertia as that of disk 2($J_1 = 3J_2$).

Figure 4.30   Torsional system with two disks and hence two degrees of freedom.

4.11. Two subway cars of Figure 4.31 have 2000 kg mass of each and are connected by a coupler. The coupler can be modeled as a spring of stiffness of $k = 280,000$ N/m. Write the equation of motion and calculate the natural frequencies and mode shapes.

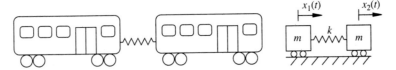

Figure 4.31   Vibration model of two subway cars connected by a coupling device modeled as a massless spring.

4.12. Suppose that the subway cars of Problem 4.11 are given the initial position of $x_{10} = 0$, $x_{20} = 0.1$ m and initial velocities of $v_{10} = v_{20} = 0$. Calculate the response of the cars.

4.13. A more sophisticated model of an vehicle suspension system is given in Figure 4.32. Write the equations of motion in matrix form. Calculate the natural frequencies for $k_1 = 10^3$ N/m, $k_2 = 10^4$ N/m, $m_2 = 50$ kg, and $m_1 = 2000$ kg.

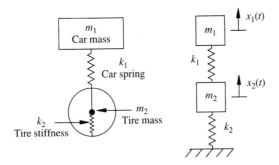

Figure 4.32   Two-degree-of-freedom model of a vehicle suspension system.

4.14. Examine the effect of the initial condition of the system of Figure 4.1(a) on the responses x_1 and x_2 by repeating the solution of Example 4.1.7, first for $x_{10} = 0$, $x_{20} = 1$ with $\dot{x}_{10} = \dot{x}_{20} = 0$ and then for $x_{10} = x_{20} = \dot{x}_{10} = 0$ and $\dot{x}_{20} = 1$. Plot the time response in each case and compare your results against Figure 4.3.

4.15. Refer to the system of Figure 4.1(a). Using the initial conditions of Example 4.1.7, resolve and plot $x_1(t)$ and $x_2(t)$ for the cases that k_2 takes on the values 0.3, 30, and 300. In each case compare the plots of x_1 and x_2 to those obtained in Figure 4.3. What can you conclude?

4.16. Consider the system of Figure 4.1(a) described in matrix form by equations (4.11), (4.9), and (4.6). Determine the natural frequencies in terms of the parameters m_1, m_2, k_1, and k_2. How do these compare to the two single-degree-of-freedom frequencies $\omega = \sqrt{k_1/m_1}$ and $\omega = \sqrt{k_2/m_2}$?

Section 4.2

4.17. Calculate the square root of the matrix

$$M = \begin{bmatrix} 13 & -10 \\ -10 & 8 \end{bmatrix}$$

[Hint: Let $M^{1/2} = \begin{bmatrix} a & -b \\ -b & b \end{bmatrix}$; calculate $(M^{1/2})^2$ and compare to M.]

4.18. Normalize the vectors

$$\begin{bmatrix} 1 \\ -2 \end{bmatrix}, \begin{bmatrix} 0 \\ 5 \end{bmatrix}, \begin{bmatrix} -0.1 \\ 0.1 \end{bmatrix}$$

first with respect to unity (i.e., $\mathbf{x}^T\mathbf{x} = 1$) and then again with respect to the matrix M (i.e., $\mathbf{x}^T M\mathbf{x} = 1$), where

$$M = \begin{bmatrix} 3 & -0.1 \\ -0.1 & 2 \end{bmatrix}$$

4.19. For the example illustrated in Figure 4.29 with $c_1 = c_2 = c_3 = 0$, calculate the matrix $\tilde{K}$ and illustrate that it is symmetric.

4.20. Repeat Example 4.2.5 using eight decimal places. Does $P^T P = I$, and does $P^T \tilde{K} P = \Lambda =$ diag $[\omega_1^2 \quad \omega_2^2]$ exactly?

4.21. Discuss the relationship or difference between a mode shape of equation (4.54) and an eigenvector of $\tilde{K}$.

4.22. Calculate the units of the elements of the matrix $\tilde{K}$.

4.23. Calculate the spectral matrix Λ and the modal matrix P for the vehicle model of Problem 4.13, Figure 4.32.

4.24. Calculate the spectral matrix Λ and the modal matrix P for the subway car system of Problem 4.11, Figure 4.31.

4.25. Calculate $\tilde{K}$ for the torsional vibration example of Problem 4.10. What are the units of $\tilde{K}$?

4.26. Consider the following system:

$$\begin{bmatrix} 1 & 0 \\ 0 & 4 \end{bmatrix} \ddot{\mathbf{x}} + \begin{bmatrix} 3 & -1 \\ -1 & 1 \end{bmatrix} \mathbf{x} = 0$$

where m is in kg and k is in N/m. **(a)** Calculate the eigenvalues of the system. **(b)** Calculate the eigenvectors and normalize them.

4.27. The torsional vibration of the wing of an airplane is modeled in Figure 4.33. Write the equation of motion in matrix form and calculate the natural frequencies in terms of the rotational inertia and stiffness of the wing (see Figure 1.21).

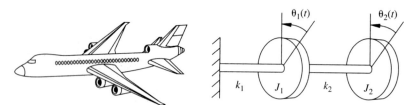

airplane wing with engines

Wing modeled as two shafts and two disks for torsional vibration

Figure 4.33   Crude model of the torsional vibration of an wing consisting of a two-shaft two-disk system of Problem 4.10, to allow an estimate of the natural frequencies of the wing in torsion. The disk inertias approximate the engine inertias.

4.28. Calculate the value of the scalar a such that $\mathbf{x}_1 = [a \quad -1 \quad 1]^T$ and $\mathbf{x}_2 = [1 \quad 0 \quad 1]^T$ are orthogonal.

4.29. Normalize the vectors of Problem 4.28. Are they still orthogonal?

4.30. Which of the following vectors are normal? Orthogonal?

$$\mathbf{x}_1 = \begin{bmatrix} \dfrac{1}{\sqrt{2}} \\ 0 \\ \dfrac{1}{\sqrt{2}} \end{bmatrix} \qquad \mathbf{x}_2 = \begin{bmatrix} 0.1 \\ 0.2 \\ 0.3 \end{bmatrix} \qquad \mathbf{x}_3 = \begin{bmatrix} 0.3 \\ 0.4 \\ 0.3 \end{bmatrix}$$

Section 4.3

4.31. Solve Problem 4.10 by modal analysis for the case where the rods have equal stiffness (i.e., $k_1 = k_2$), $J_1 = 3J_2$, and the initial conditions are $\mathbf{x}(0) = [0 \quad 1]^T$ and $\dot{\mathbf{x}}(0) = \mathbf{0}$.

4.32. Consider the system of Example 4.3.1. Calculate a value of $\mathbf{x}(0)$ and $\dot{\mathbf{x}}(0)$ such that both masses of the system oscillate with a single frequency of 2 rad/s .

4.33. Consider the system of Figure 4.34 consisting of two pendulums coupled by a spring. Determine the natural frequency and mode shapes. Plot the mode shapes as well as the solution to an initial condition consisting of the first mode shape for $k = 20$ N/m, $l = 0.5$ m and $m = 10$ kg, $a = 0.1$ m along the pendulum.

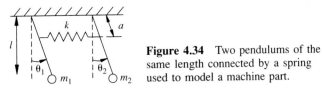

Figure 4.34   Two pendulums of the same length connected by a spring used to model a machine part.

4.34. Resolve Example 4.3.2 with m_2 changed to 10 kg. Plot the response and compare the plots to those of Figure 4.6.

4.35. Use modal analysis to calculate the solution of Problem 4.26 for the initial conditions.

$$\mathbf{x}(0) = [0 \quad 1]^T \,(\text{mm}) \quad \text{and} \quad \dot{\mathbf{x}}(0) = [0 \quad 0]^T \,(\text{mm/s})$$

4.36. For the matrices

$$M^{-1/2} = \begin{bmatrix} \dfrac{1}{\sqrt{2}} & 0 \\ 0 & 4 \end{bmatrix} \qquad \text{and} \qquad P = \dfrac{1}{\sqrt{2}}\begin{bmatrix} 1 & 1 \\ -1 & 1 \end{bmatrix}$$

calculate $M^{-1/2}P$, $(M^{-1/2}P)^T$, and $P^T M^{-1/2}$ and hence verify that the computations in equation (4.71) make sense.

4.37. Consider Example 4.3.1. Calculate the response of the system to the initial condition

$$\mathbf{x}_0 = \dfrac{1}{\sqrt{2}}\begin{bmatrix} 1 \\ 1 \end{bmatrix} \qquad \dot{\mathbf{x}}_0 = \mathbf{0}$$

What is unique about your solution?

4.38. Repeat Problem 4.37 with

$$\mathbf{x}_0 = \dfrac{1}{\sqrt{2}}\begin{bmatrix} 1 \\ -1 \end{bmatrix} \qquad \dot{\mathbf{x}}_0 = \mathbf{0}$$

Examining the solution to this problem and Problem 4.37 can you draw any conclusions?

4.39. Consider the system of Problem 4.1. Let $k_1 = 10,000$ N/m, $k_2 = 15,000$ N/m, and $k_3 = 10,000$ N/m. Assume that both masses are 100 kg. Solve for the free response of this system using modal analysis and the initial conditions

$$\mathbf{x}(0) = [1 \quad 0]^T \qquad \dot{\mathbf{x}}(0) = \mathbf{0}$$

4.40. Consider the model of an vehicle given in Problem 4.13 and illustrated in Figure 4.32. Suppose that the tire hits a bump which corresponds to an initial condition of

$$\mathbf{x}(0) = \begin{bmatrix} 0 \\ 0.01 \end{bmatrix} \qquad \dot{\mathbf{x}}(0) = \mathbf{0}$$

Use modal analysis to calculate the response of the car $x_1(t)$. Plot the response for three cycles.

Section 4.4

4.41. A vibration model of the drive train of a vehicle is illustrated as the three-degree-of-freedom system of Figure 4.35. Calculate the undamped free response [i.e., $M(t) = F(t) = 0, c_1 = c_2 = 0$] for the initial condition $\mathbf{q}(0) = \mathbf{0}$, $\dot{\mathbf{q}}(0) = [0 \quad 0 \quad 1]^T$. Assume that the hub stiffness is 10,000 N/m and that the axle/suspension stiffness is 20,000 N/m. Assume the rotational element J is modeled as a translational mass of 75 kg.

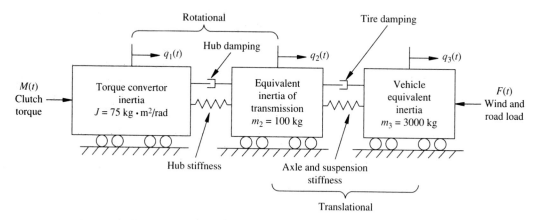

Figure 4.35 Simplified model of an automobile for vibration analysis of the drive train. The parameter values given are representative and should not be considered as exact.

4.42. Calculate the natural frequencies and mode shapes of

$$\begin{bmatrix} 4 & 0 & 0 \\ 0 & 2 & 0 \\ 0 & 0 & 1 \end{bmatrix} \ddot{\mathbf{x}} + \begin{bmatrix} 4 & -1 & 0 \\ -1 & 2 & -1 \\ 0 & -1 & 1 \end{bmatrix} \mathbf{x} = \mathbf{0}$$

4.43. The vibration is the vertical direction of an airplane and its wings can be modeled as a three-degree-of-freedom system with one mass corresponding to the right wing, one mass for the left wing, and one mass for the fuselage. The stiffness connecting the three masses corresponds to that of the wing and is a function of the modulus E of the wing. The equation of motion is

$$m \begin{bmatrix} 1 & 0 & 0 \\ 0 & 4 & 0 \\ 0 & 0 & 1 \end{bmatrix} \begin{bmatrix} \ddot{x}_1 \\ \ddot{x}_2 \\ \ddot{x}_3 \end{bmatrix} + \frac{EI}{l^3} \begin{bmatrix} 3 & -3 & 0 \\ -3 & 6 & -3 \\ 0 & -3 & 3 \end{bmatrix} \begin{bmatrix} x_1 \\ x_2 \\ x_3 \end{bmatrix} = \begin{bmatrix} 0 \\ 0 \\ 0 \end{bmatrix}$$

The model is given in Figure 4.36. Calculate the natural frequencies and mode shapes. Plot the mode shapes and interpret them according to the airplane's deflection.

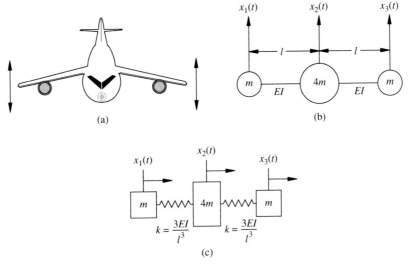

Figure 4.36 Model of the wing vibration of an airplane: (a) vertical wing vibration; (b) lumped mass/beam deflection model; (c) spring-mass model.

4.44. Solve for the free response of the system of Problem 4.43. Where $E = 6.9 \times 10^9$ N/m², $l = 2$ m, $m = 3000$ kg, and $I = 5.2 \times 10^{-6}$ m⁴. Let the initial displacement correspond to a gust of wind that causes an initial condition of $\dot{\mathbf{q}}(0) = \mathbf{0}$, $\mathbf{q}(0) = [0.2 \quad 0 \quad 0]^T$ m. Discuss your solution.

4.45. Consider the two-mass system of Figure 4.37. This system is free to move in the x_1–x_2 plane. Hence each mass has two degrees of freedom. Derive the equations of motion, write them in matrix form, and calculate the eigenvalues and eigenvectors for $m = 10$ kg and $k = 100$ N/m.

4.46. Consider again the system discussed in Problem 4.45. Use modal analysis to calculate the solution if the mass on the left is raised along the x_2 direction exactly 0.01 m and let go.

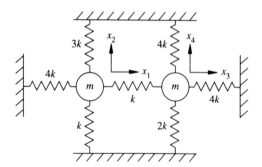

Figure 4.37 Two-mass system free to move in two directions.

4.47. The vibration of a floor in a building containing heavy machine parts is modeled in Figure 4.38. Each mass is assumed to be evenly spaced and significantly larger than the mass of the floor. The equation of motion then becomes ($m_1 = m_2 = m_3 = m$)

$$
m\ddot{\mathbf{q}} + \frac{EI}{l^3}
\begin{bmatrix}
\dfrac{9}{64} & \dfrac{1}{6} & \dfrac{13}{192} \\[2mm]
\dfrac{1}{6} & \dfrac{1}{3} & \dfrac{1}{6} \\[2mm]
\dfrac{13}{192} & \dfrac{1}{6} & \dfrac{9}{64}
\end{bmatrix}
\begin{bmatrix} q_1 \\ q_2 \\ q_3 \end{bmatrix} = \mathbf{0}
$$

Calculate the natural frequencies and mode shapes. Assume that in placing box m_2 on the floor (slowly) the resulting vibration is calculated by assuming that the initial displacement at m_2 is 0.05 m. If $l = 2$ m, $m = 200$ kg, $E = 0.6 \times 10^9$ N/m^2, $I = 4.17 \times 10^{-5}$ m^4. Calculate the response and plot your results.

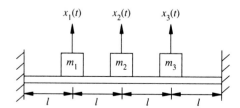

Figure 4.38 Lumped-mass model of boxes loaded on a floor of a building.

4.48. Recalculate the solution to Problem 4.47 for the case that m_2 is increased in mass to 2000 kg. Compare your results to those of Problem 4.47. Do you think it makes a difference where the heavy mass is placed?

4.49. Repeat Problem 4.43 for the case that the airplane body is $10m$ instead of $4m$ as indicated in the figure. What effect does this have on the response, and which design ($4m$ or $10m$) do you think is better as to vibration?

4.50. Often in the design of car, certain parts cannot be reduced in mass. For example, consider the drive train model illustrated in Figure 4.35 of Problem 4.41. The mass of the torque converter and transmission are relatively the same from car to car. However, the mass of the car could change as much as 1000 kg (e.g., a two-seater sports car versus a family sedan). With this in mind, resolve Problem 4.41 for the case that the vehicle inertia is reduced to 2000 kg. Which case has the smallest amplitude of vibration?

Section 4.5

4.51. Consider the example of the automobile drive train system discussed in Problem 4.41. Add 10% modal damping to each coordinate, calculate, and plot the system response.

4.52. Consider the model of an airplane discussed in Problem 4.44, Figure 4.36. (**a**) Resolve the problem assuming that the damping provided by the wing rotation is $\zeta_i = 0.01$ in each mode and recalculate the response. (**b**) If the aircraft is in flight, the damping forces may increase dramatically to $\zeta_i = 0.1$. Recalculate the response and compare it to the more lightly damped case of part (a).

4.53. Repeat the floor vibration problem of Problem 4.47 using modal damping ratios of

$$\zeta_1 = 0.01 \qquad \zeta_2 = 0.1 \qquad \zeta_3 = 0.2$$

4.54. Repeat Problem 4.53 with constant modal damping of $\zeta_1 = \zeta_2 = \zeta_3 = 0.1$ and compare this with the solution of Problem 4.53.

4.55. Consider the damped system of Figure 4.29. Determine the damping matrix and use the formula of equation (4.119) to determine values of the damping coefficient c_i for which this system would be proportionally damped.

4.56. Let $k_3 = 0$ in Problem 4.55. Also let $m_1 = 1, m_2 = 4, k_1 = 2, k_2 = 1$ and calculate c_1, c_2, and c_3 such that $\zeta_1 = 0.01$ and $\zeta_2 = 0.1$.

4.57. Calculate the constants α and β for the two-degree-of-freedom system of Problem 4.26 such that the system has modal damping of $\zeta_1 = \zeta_2 = 0.3$.

4.58. Equation (4.124) represents n equations in only two unknowns and hence cannot be used to specify all the modal damping ratios for a system with $n > 2$. If the floor vibration system of Problem 4.48 has measured damping of $\zeta_1 = 0.01$ and $\zeta_2 = 0.05$, determine ζ_3.

4.59. Does the following system decouple? If so, calculate the mode shapes and write the equation in decoupled form.

$$\begin{bmatrix} 1 & 0 \\ 0 & 1 \end{bmatrix} \ddot{\mathbf{x}} + \begin{bmatrix} 5 & -3 \\ -3 & 3 \end{bmatrix} \dot{\mathbf{x}} + \begin{bmatrix} 5 & -1 \\ -1 & 1 \end{bmatrix} \mathbf{x} = \mathbf{0}$$

4.60. Calculate the damping matrix for the system of Problem 4.58. What are the units of the elements of the damping matrix?

Section 4.6

4.61. Calculate the response of the system of Figure 4.15 discussed in Example 4.6.1 if $F_1(t) = \delta(t)$ and the initial conditions are set to zero. This might correspond to a two-degree-of-freedom model of a car hitting a bump.

4.62. For an undamped two-degree-of-freedom system, show that resonance occurs at one or both of the systems natural frequencies.

4.63. Use modal analysis to calculate the response of the drive train system of Problem 4.41 to a unit impulse on the car body (i.e., and location q_3). Use the modal damping of Problem 4.51. Calculate the solution in terms of physical coordinates, and after subtracting the rigid-body modes, compare the responses of each part.

4.64. Consider the machine tool of Figure 4.27. Resolve Example 4.8.3 if the floor, mass $m = 1000$ kg, is subject to a force of $10 \sin t$ (in Newtons). Calculate the response. How much does this floor vibration affect the machine's toolhead?

4.65. Consider the airplane of Figure 4.36 with damping as described in Problem 4.52 with $\zeta_i = 0.1$. Suppose that the airplane hits a gust of wind which applies an impulse of $3\delta(t)$ at the end of the left wing and $\delta(t)$ and the end of the right wing. Calculate the resulting vibration of the cabin $[x_2(t)]$.

4.66. Consider again the airplane of Figure 4.36 with the modal damping model of Problem 4.52 ($\zeta_i = 0.1$). Suppose that this is a propeller-driven airplane with an internal combustion engine mounted in the nose. At a cruising speed the engine mounts transmit an applied force to the cabin mass ($4m$ at x_2) which is harmonic of the form $50 \sin 10t$. Calculate the effect of this harmonic disturbance at the nose and on the wing tips after subtracting out the translational or rigid motion.

4.67. Consider the automobile model of Problem 4.13 illustrated in Figure 4.32. Add modal damping to this model of $\zeta_1 = 0.01$ and $\zeta_2 = 0.2$ and calculate the response of the body $[x_1(t)]$ to a harmonic input at the second mass of $10 \sin 3t$ N.

Section 4.7

4.68. Use Lagrange's equation to derive the equations of motion of the lathe of Figure 4.20 for the undamped case.

4.69. Use Lagrange's equations to rederive the equations of motion for the automobile of Example 4.8.2 illustrated in Figure 4.24 for the case $c_1 = c_2 = 0$.

4.70. Use Lagrange's equations to rederive the equations of motion for the building model presented in Figure 4.9 of Example 4.4.3 for the undamped case.

4.71. Consider again the model of the vibration of an automobile of Figure 4.24. In this case include the tire dynamics as indicated in Figure 4.39. Derive the equations of motion using Lagrange formulation for the undamped case.

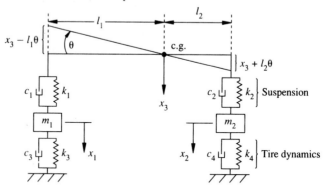

Figure 4.39   Model of vibration for an automobile, including tire dynamics.

MATLAB VIBRATION TOOLBOX

If you have not yet used the *Vibration Toolbox* program, return to the end of Chapter 1 for a brief introduction to using MATLAB files or open the folder/directory INTRO.TXT on the disk on the inside back cover. The MATLAB program language introduced at the end of Chapter 1 is matrix based and hence is ideally suited to vibration analysis of multiple-degree-of-freedom systems. Entering a matrix into MATLAB is as easy as typing the matrix. For example, the stiffness matrix of Example 4.1.5 is

$$K = \begin{bmatrix} 27 & -3 \\ -3 & 3 \end{bmatrix}$$

This is entered into MATLAB simply by typing

```
K = [ 27    -3;    -3    3 ]
```

where the semicolon separates the rows of the matrix (no dimensioning is required!). Section 9.6 discusses this in detail along with some rules for manipulating matrices (Table 9.1). However, all that is needed to use the *Vibration Toolbox* for solving multiple-degree-of-freedom vibration problems is the format above for entering matrices and vectors. A column vector, say of initial conditions, is entered by typing

```
X0 = [  1    0  ]'
```

where the apostrophe denotes the transpose, so this last line enters the vector of conditions

$$\mathbf{x}_0 = \begin{bmatrix} 1 \\ 0 \end{bmatrix}$$

The following exercises are intended to familiarize the reader with the tool-box associated with Chapter 4 which can be found in folder VTB4. The files in this toolbox folder are useful for solving the problems above as well as for building a familiarity with the concepts of this chapter and the nature of the response of multiple-degree-of-freedom systems. This would also be a good place to read Section 9.6 if you have time, and are interested in computing on your own.

TOOLBOX PROBLEMS

TB4.1. Calculate the natural frequencies and mode shapes of the system of Example 4.1.5 using file VTB4_1.

TB4.2. Recalculate Example 4.2.5 using file VTB4_1 and compare your answer with that of the example obtained with a calculator. Verify that $P^T \tilde{K} P = \Lambda$ and $P^T P = I$.

TB4.3. Consider Example 4.3.1 and investigate the effect of the initial condition on the response by using file VTB4_2 and plotting the responses to the following initial displacements (initial velocities all zero):

$$\mathbf{x}_0 = \begin{bmatrix} 1 \\ 1 \end{bmatrix} \quad \mathbf{x}_0 = \begin{bmatrix} 0 \\ 1 \end{bmatrix} \quad \mathbf{x}_0 = \begin{bmatrix} 1 \\ -1 \end{bmatrix} \quad \mathbf{x}_0 = \begin{bmatrix} 0.1 \\ 0.9 \end{bmatrix} \quad \mathbf{x}_0 = \begin{bmatrix} 0.2 \\ 0.8 \end{bmatrix} \quad \text{etc.}$$

Discuss the results by formulating a short paragraph summarizing what you observed.

TB4.4. Check the calculation of Example 4.4.2.

TB4.5. Using file VTB4_2, examine the effect of increasing the mass m_4 in the building vibration problem of Example 4.4.3. Do this by doubling m_4 and recalculating the solution. Notice what happens to the various responses. Try doubling m_4 until the response does not change, or the program fails. Discuss your observations.

TB4.6. Consider the system of Example 4.7.3. Choose the values $m = 10$, $J = 5$, $e = 1$, and $k_1 = 1000$ and calculate the eigenvalues as k_2 varies from 10 to 10,000, in increments of 100. What can you conclude?

5 Design for Vibration Suppression

This chapter presents methods of designing systems so that they suppress vibration. The articles pictured on the left are isolation materials used in such applications as motor mount for an automobile to isolate engine vibration from the occupant of an automobile, as discussed in Sections 5.2, 5.8, and 5.9. The hot air balloon pictured here can hit the ground, upon landing with a rather harsh impact. The basket which holds the occupants is made of wicker because it is lightweight and also because the wicker absorbs the shock of landing. Shock isolation is discussed in Section 5.2 and in Section 5.8.

In this chapter it is assumed that vibration is undesirable and is to be suppressed. The topics of the previous chapters present a number of techniques and methods for analyzing the vibration response of various systems subject to various inputs. Here the focus is to use the skills developed in the preceding chapters to determine ways of adjusting the physical parameters of a system or device in such a way that the vibration response meets some specified shape or performance criteria. This is called *design*; introduced in Section 1.7, it is the focus of this chapter.

Vibration can often lead to a number of undesirable circumstances. For example, vibration of an automobile or truck can lead to driver discomfort and eventually, fatigue. Structural or mechanical failure can often result from sustained vibration (e.g., cracks in airplane wings). Electronic components used in airplanes, automobiles, machines, and so on, may also fail because of vibration, shock, and/or sustained vibration input.

The "fragility level" of devices, how much vibration a given device can withstand, is addressed by the International Organization of Standards (ISO) as well as by some national agencies. Almost every device manufactured for use by the military must meet certain military specifications ("mil specs") regarding the amount of vibration it can withstand. In addition to government and international agencies, manufacturers also set desired vibration performance standards for some products. If a given device does not meet these regulations, it must be redesigned so that it does. This chapter presents several formulations that are useful for designing and redesigning various devices and structures to meet desired vibration standards.

Design is a difficult subject that does not always lend itself to simple formulations. Design problems typically do not have a unique solution. Many different designs may all give acceptable results. Sometimes a design may simply consist of putting together a number of existing (off-the-shelf) devices to create a new device with desired properties. Here we focus on design as it refers to adjusting a system's physical parameters to cause its vibration response to behave in a desired fashion.

5.1 ACCEPTABLE LEVELS OF VIBRATION

To design a device in terms of its vibration response, the desired response must be clearly stated. Many different methods of measuring and describing acceptable levels of vibration have been proposed. Whether or not the criteria should be established in terms of displacement, velocity, or acceleration, and exactly how these should be measured needs to be clarified before a design can begin. These choices often depend on the specific application. For instance, in practice it is generally accepted that the best indication of potential structural damage is the amplitude of the structure's velocity, while acceleration amplitude is the most perceptible by humans. Some common ranges of vibration frequency and displacement are given in Table 5.1.

TABLE 5.1 RANGES OF FREQUENCY AND DISPLACEMENT OF VIBRATION

	Frequency (Hz)	Displacement amplitude (mm)
Atomic vibration	1012	10^{-7}
Threshold of human perception	1–8	10^{-2}
Machinery and building vibration	10–100	10^{-2}–1
Swaying of tall buildings	1–5	10–1000

ISO 2372 (International Organization for Standardization, 1974) is a published standard of acceptable levels of vibration which has the intent of providing a mechanism to facilitate communications between manufacturers and consumers. The standards are tested in terms of rms values of displacement, velocity, and acceleration. Recall that the rms (root mean square) value (defined in Section 1.2) is the square root of the time average of the square of a quantity. For the displacement $x(t)$, the rms value is given in equation (1.21) as

$$x_{\mathrm{rms}} = \left[\lim_{T \to \infty} \frac{1}{T} \int_0^T x^2(t)\, dt \right]^{1/2}$$

A convenient way to express the acceptable values of vibration allowed under ISO 2372 is to plot them on a nomograph as illustrated in Figure 1.7. Such a plot is repeated in Figure 5.1 with several standards indicated on the plot.

The nomograph of Figure 5.1 is a graphical representation of the relationship between displacement, velocity, acceleration, and frequency for an undamped single-degree-of-freedom system. The solution for the displacement is given by equation (1.19) as

$$x(t) = A \sin \omega t$$

(for zero phase), which has amplitude A. Differentiating the displacement solution yields the velocity

$$v(t) = \dot{x}(t) = A\omega \cos \omega t$$

which has amplitude ωA. Differentiating again yields the acceleration

$$a(t) = \ddot{x}(t) = -A\omega^2 \sin \omega t$$

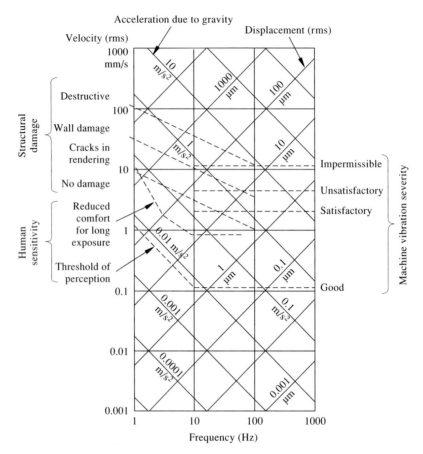

Figure 5.1 Example of an acceptable threshold of sinusoidal vibration for structural damage, machinery vibration, and human perception. Lines of constant velocity are indicated by horizontal lines (–), lines of constant acceleration are indicated by slanted bars (\), lines of constant displacement are indicated by slanted bars (/), and lines of constant frequency are indicated by vertical lines (|).

which has amplitude $A\omega^2$. These three expressions for the magnitude, along with the definition of rms value, allow the nomograph of Figure 5.1 to be constructed.

Example 5.1.1

An automobile has a mass of 1000 kg and a stiffness of 400,000 N/m. If it hits a bump that is 0.1 mm high, it will sustain a sinusoidal vibration. Assuming that the height of the bump corresponds to the rms value of displacement, what velocity and acceleration correspond to this vibration? Is this vibration perceived by the passenger? If the vibration is perceptible and hence not desirable, suggest a means of redesigning the system so that the vibration does not disturb the passengers.

Solution Since the frequency of oscillation of a simple spring–mass system is $\omega = \sqrt{k/m} = 20$ rad/s, or in hertz,

$$f = \frac{\omega}{2\pi} = 3.18 \text{ Hz}$$

the vertical line at $f = 3.18$ Hz is examined. It crosses lines of displacement amplitude of 0.1 mm rms, just above the line of perceptible vibration for humans. Hence the passengers will feel the vibration resulting from hitting the bump. The point corresponding to a displacement amplitude of 0.1 mm and a frequency of 3.18 Hz also corresponds to a horizontal line at about $v = 1$ mm/s, which specifies the car's vibration velocity (in the up/down direction) and an rms value for acceleration of approximately 0.01 m/s^2.

   If the point of intersection between the frequency and displacement line were moved below the line of perceptible vibration, the passengers would not feel the bump. This could be accomplished by lowering the frequency of the car, either by increasing the mass or decreasing the stiffness. According to the line of constant displacement, if the frequency is reduced even just a little, the response would not be perceptible. Reducing the frequency to 3 Hz would provide a redesign capable of protecting the passengers from feeling a 0.1-mm bump. This can be accomplished by adjusting the spring rate by solving

$$f = \frac{1}{2\pi}\sqrt{k/m} = \frac{1}{2\pi}\sqrt{k/1000}$$

for the stiffness. Solving for the stiffness, k, yields

$$k = (6\pi)^2 (1000) \text{ N/m} = 355,305 \text{ N/m}$$

Hence the stiffness must be reduced by 11%. The reduction in stiffness may be accomplished by changing the tire pressure and/or by using different springs.

$\square$

   The design procedure suggested above is oversimplified but helps introduce the ad hoc nature of many design problems. Analysis is used as a tool. Here the desired vibration criteria are provided by an ISO standard (No. 2631 in this case) represented in a nomograph. This plot, together with the formula $\omega = \sqrt{k/m}$, provides the analysis tools.

   In using any thought process to perform a design it is important to think through any potential flaws in the procedure. In the example above there are several possible points of error. One important issue is how well the simple single-degree-of-freedom spring–mass model captures the dynamics of the car. It might be that a more sophisticated multiple-degree-of-freedom model is required (recall Example 4.8.2). Another possible problem with the proposed design change is that the stiffness of the car may not be allowed to be lower than a certain value because of load requirements (static deflection) or other design constraints. In this case the mass might be raised, but that too, may have other constraints on it. In some cases it just may not be possible to design a system to have this desired vibration response. That is, not all design problems have a solution.

   The range of vibrations with which an engineer is concerned, is usually from about 10^{-4} mm at between 0.1 and 1 Hz for objects such as optical benches or sophisticated medical equipment, to a meter displacement for tall buildings in the range 0.1 to 5 Hz. Machine vibrations can range between 10 and 1000 Hz, with deflections between fractions of a millimeter and several centimeters. As technology advances, limitations and acceptable levels of vibration change. Thus these numbers should be considered as rough indications of common values.

Example 5.1.2

Calculate and compare the natural frequency, damping ratio, and damped natural frequency of the single-degree-of-freedom model of a stereo turntable and of the automobile given in Figure 5.2. Also plot and compare their frequency response functions and their impulse response functions. Discuss the similarities and differences of these two devices.

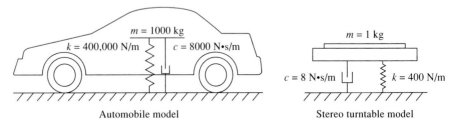

Automobile model   Stereo turntable model

Figure 5.2   Single-degree-of-freedom model of an automobile and a stereo turntable, each with the same frequency.

Solution   To calculate the undamped natural frequency, damping ratio, and damped natural frequency of the car is simple. From the definitions:

$$\omega = \sqrt{\frac{k}{m}} = \sqrt{\frac{400,000}{1000}} = 20 \text{ rad/s}$$

$$\zeta = \frac{c}{2m\omega} = \frac{8000}{2(1000)(20)} = 0.2$$

$$\omega_d = \omega\sqrt{1 - \zeta^2} = 20\sqrt{1 - (0.2)^2} = 19.5959 \text{ rad/s}$$

The same calculations for the stereo turntable are

$$\omega = \sqrt{\frac{k}{m}} = \sqrt{\frac{400}{1}} = 20 \text{ rad/s}$$

$$\zeta = \frac{c}{2m\omega} = \frac{8}{2(1)(20)} = 0.2$$

$$\omega_d = \omega\sqrt{1 - \zeta^2} = 20\sqrt{1 - (0.2)^2} = 19.5959 \text{ rad/s}$$

This illustrates that two objects of very different size can have the same natural frequencies and damping ratios.

In Figure 5.3 the transfer function of each device is plotted as well as the impulse response function for each. Note that these plots do indicate a difference in devices. The phase plots of the transfer function of both the car and the stereo are identical, while the magnitude plots differ by a constant. The impulse response function of the car has a smaller amplitude, although the responses both die out at the same time since the decay rate (ζ) and hence log decrement are the same for each device. The acceptable levels of vibration

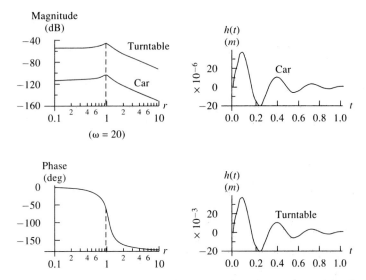

Figure 5.3 Frequency response function and the impulse response of that for the car and stereo turntable of Figure 5.2.

for the car will be much larger than those of the stereo. For instance, a displacement amplitude of 100 mm for the stereo would cause its needle to skip out of the groove in a record and hence not perform properly (a reason why phonograph records never made it in automobiles—thank goodness for CDs and tapes). On the other hand, a similar amplitude of vibration for the car is below the perception level on the nomograph of Figure 5.1.

An important consideration in specifying vibration response is to specify the nature of the input or driving force that causes the response. Disturbances, or inputs, are normally classified as either *shock* or as vibration, depending on how long the input lasts. An input is considered to be a shock if the disturbance is a sharp, aperiodic one lasting a relatively short time. In contrast, an input is considered to be a vibration if it lasts for a long time and has some oscillatory features.

The distinction between shock and vibration is not always clear as the sources of shock and vibration disturbances are numerous and very difficult to place into categories. In Chapter 2 only vibration inputs (e.g., harmonic inputs) are discussed. In Chapter 3 the response of a single-degree-of-freedom system to a variety of inputs, including an impulse, which is a shock, and general periodic inputs (vibration) are discussed. These input signals may result from bumps in the road (for cars), turbulence (for airplanes), rotating machinery, or simply from dropping something.

Often, inputs are a combination of the types discussed above and those in Chapter 3. In addition, inputs are often not known precisely but rather, are known to be of less than a certain magnitude and lasting less than a certain time. For instance, a given shock input to an automobile due to its hitting a bump may take the form of a single-valued curve falling somewhere inside the shaded region of Figure 5.4.

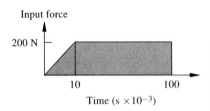

Input force

200 N

10   100

Time (s $\times 10^{-3}$)

Figure 5.4 Sample of a shock input to an automobile, illustrating that the form is not entirely known—that only the bounds of the force's magnitude and time history are known.

Vibrations exist in many devices that are not harmful. For instance, automobiles continually experience vibration without being damaged or causing harm to passengers. However, some vibrations are extremely damaging, such as severe vibration from an earthquake or a badly out of balance tire on a car. The difficult issue for design engineers is deciding between acceptable levels of vibration and those that will cause damage or become so annoying that consumers will not use the device. Once acceptable levels are established, several techniques can be used to limit and alter the shock and vibration response of mechanical systems and structures. These are discussed in the following section.

5.2 VIBRATION ISOLATION

The most effective way to reduce unwanted vibration is to stop or modify the source of the vibration. If this cannot be done, it is sometimes possible to design a *vibration isolation system* to isolate the source of vibration from the system of interest or to isolate the device from the source of vibration. This can be done by using highly damped materials such as rubber to change the stiffness and damping between the source of vibration and the device that is to be protected from the vibrations. The problem of isolating a device from a source of vibration is analyzed in terms of reducing vibration displacement transmitted through base motion as discussed in Section 2.3 and summarized in Window 5.1. The problem of isolating a source of vibration from its surroundings is analyzed in terms of reducing the force transmitted by the source through its mounting points as discussed in Section 2.4 and summarized in Window 5.1. Both force transmissibility and displacement transmissibility are called *isolation problems*.

The analysis tool used to design isolators is the concept of force and/or displacement transmissibility introduced in Section 2.3. By way of review, consider the problem of calculating the *transmissibility ratio*, denoted T.R., defined as the ratio of the magnitude of the force transmitted through the spring and dashpot to the fixed base to the sinusoidal force applied (see Window 5.1) by the machine (modeled as a mass). Symbolically, T.R. $= F_T/F_0$. To calculate the value of T.R., first consider the force transmitted. The force transmitted to the fixed base in Window 5.1 is denoted $F_T(t)$ and is the force applied to the base acting through the spring and dashpot, that is,

$$F_T(t) = kx(t) + c\dot{x}(t) \qquad (5.1)$$

Window 5.1
Summary of Vibration Isolation Transmissibility
Formulas for Both Force and Displacement

The moving-base model on the left is used in designing isolation to protect the device from motion of its point of attachment (base). The model on the right is used to protect the point of attachment (ground) from vibration of the mass.

Displacement Transmissibility

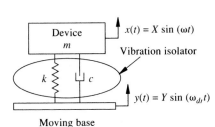

Force Transmissibility

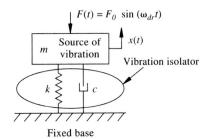

Moving base
Vibration source modeled as base motion

Fixed base
Vibration source mounted on isolator

Here $y(t) = Y \sin \omega_{dr} t$ is the disturbance and from equation (2.41)

$$\frac{X}{Y} = \left[\frac{1 + (2\zeta r)^2}{(1 - r^2)^2 + (2\zeta r)^2} \right]^{1/2}$$

defines the displacement transmissibility and is plotted in Figure 2.8 From equation (2.47),

$$\frac{F_T}{kY} = r^2 \left[\frac{1 + (2\zeta r)^2}{(1 - r^2)^2 + (2\zeta r)^2} \right]^{1/2}$$

defines the related force transmissibility and is plotted in Figure 2.9.

Here $F(t) = F_0 \sin \omega_{dr} t$ is the disturbance and

$$\frac{F_T}{F_0} = \left[\frac{1 + (2\zeta r)^2}{(1 - r^2)^2 + (2\zeta r)^2} \right]^{1/2}$$

defines the force transmissibility for isolating the source of vibration as derived in Section 5.2.

The solution for the case that the driving force is harmonic of the form $F_0 \cos \omega_{dr} t$ is given in Section 2.2, equation (2.29), to be of the form

$$x(t) = Ae^{-\zeta \omega t} \sin(\omega_d t + \theta) + A_0 \cos(\omega_{dr} t - \phi)$$

In steady state (i.e., after some time has elapsed) the first term decays to zero and the response is modeled by

$$x(t) = A_0 \cos(\omega_{dr} t - \phi) \tag{5.2}$$

Differentiating equation (5.2) the velocity in steady state becomes

$$\dot{x}(t) = -\omega_{dr} A_0 \sin(\omega_{dr} t - \phi) \tag{5.3}$$

Substitution of equations (5.2) and (5.3) into equation (5.1) for the force transmitted at steady state yields

$$
\begin{aligned}
F_T(t) &= k A_0 \cos(\omega_{dr} t - \phi) - c \omega_{dr} A_0 \sin(\omega_{dr} t - \phi) \\
&= k A_0 \cos(\omega_{dr} t - \phi) + c \omega_{dr} A_0 \cos(\omega_{dr} t - \phi + \pi/2)
\end{aligned} \tag{5.4}
$$

The magnitude of $F_T(t)$, denoted by, F_T, can be calculated from equation (5.4) by noting that the two terms are $90°\,(\pi/2)$ out of phase with each other and hence can be thought of as two perpendicular vectors (recall Figure 2.13 of Section 2.5). Hence the magnitude of $F_T(t)$ is calculated by taking the vector sum of the two terms on the right of equation (5.4). This yields that the magnitude of the transmitted force is

$$F_T = \sqrt{(k A_0)^2 + (c \omega_{dr} A_0)^2} = A_0 \sqrt{k^2 + c^2 \omega_{dr}^2} \tag{5.5}$$

The value of A_0, the amplitude of vibration at steady state, is given in Window 5.2 to be

$$A_0 = \frac{f_0}{[(\omega^2 - \omega_{dr}^2)^2 + (2\zeta\omega\omega_{dr})^2]^{1/2}} = \frac{F_0/k}{\left[(1 - r^2)^2 + (2\zeta r)^2\right]^{1/2}}$$

Window 5.2
Review of the Steady-State Response of an Underdamped System
Subject to Harmonic Excitation as Discussed in Section 2.2

The steady-state response of

$$\ddot{x} + 2\zeta\omega\dot{x} + \omega^2 x = f_0 \cos \omega_{dr} t$$

where $\omega = \sqrt{k/m}$, $\zeta = c/(2m\omega)$, and $f_0 = F_0/m$, is $x(t) = A_0 \cos(\omega_{dr} t - \phi)$.
Here

$$A_0 = \frac{f_0}{\sqrt{(\omega^2 - \omega_{dr}^2)^2 + (2\zeta\omega\omega_{dr})^2}}, \qquad \phi = \tan^{-1} \frac{2\zeta\omega\omega_{dr}}{\omega^2 - \omega_{dr}^2}$$

---

where $r = \omega_{dr}/\omega$, as before. Substituting this value of A_0 into equation (5.5) yields

$$
\begin{aligned}
F_T &= \frac{F_0/k}{\left[(1 - r^2)^2 + (2\zeta r)^2\right]^{1/2}} \sqrt{k^2 + c^2 \omega_{dr}^2} \\
&= F_0 \frac{\sqrt{1 + c^2 \omega_{dr}^2 / k^2}}{\left[(1 - r^2)^2 + (2\zeta r)^2\right]^{1/2}} = F_0 \sqrt{\frac{1 + (2\zeta r)^2}{(1 - r^2)^2 + (2\zeta r)^2}}
\end{aligned} \tag{5.6}
$$

where $c^2\omega_{dr}^2/k^2 = (2m\omega\zeta)^2\omega_{dr}^2/k^2 = (2\zeta r)^2$. The *transmissibility ratio*, or *transmission ratio*, denoted T.R., is defined as the ratio of the magnitude of the transmitted force to the magnitude of the applied force. By a simple manipulation of equation (5.6) this becomes

$$\text{T.R.} = \frac{F_T}{F_0} = \sqrt{\frac{1 + (2\zeta r)^2}{(1 - r^2)^2 + (2\zeta r)^2}} \tag{5.7}$$

A comparison of this force transmissibility expression with the displacement transmissibility given in Window 5.1 indicates that they are identical. It is important to note, however, that even though they have the same value, they come from different isolation problems and hence describe different phenomena.

The displacement transmissibility ratio given in the left column of Window 5.1 describes how a steady-state displacement (Y) of the base of a device mounted on an isolator is transmitted into motion of the device (X). Figure 5.5 is a plot of the T.R. for various values of the damping ratio ζ and frequency ratio r. The larger the value of T.R., the larger the amplitude of motion of the mass. These curves are useful for designing

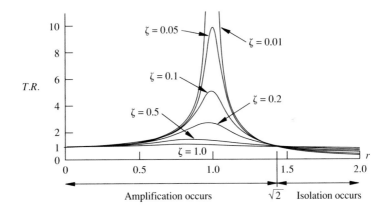

Magnification of the isolation area

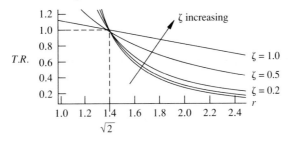

Figure 5.5   Plot of the transmissibility ratio, T.R., indicating the value of T.R. for a variety of choices of the damping ratio ζ and the frequency ratio r. This is a repeat of Figure 2.8 and is a plot of equation (5.7).

the isolators. In particular, the design process consists of choosing ζ and r, within the available isolator's material, such that T.R. is small.

    Note from the figure that if the frequency ratio r is greater than the $\sqrt{2}$, the magnitude of vibration of the device is smaller than the disturbance magnitude Y and vibration isolation occurs. For r less than $\sqrt{2}$, the amplitude X increases (i.e., X is larger than Y). The value of the damping ratio (each curve in Figure 5.5 corresponds to a different ζ) determines how much smaller the amplitude of vibration is for a given frequency ratio. Near resonance, the T.R. is determined completely by the value of ζ (i.e., by the damping in the isolator). In the isolation region the smaller the value of ζ, the smaller the value of T.R. and the better the isolation. Also note that as r is increased for a fixed ω_{dr}, the value of T.R. decreases. This corresponds to increasing the mass or decreasing the stiffness of the isolator.

Example 5.2.1

    An electronic control system for an automobile engine is to be mounted on top of the fender inside the engine compartment of the automobile as illustrated in Figure 5.6. The control module electronically computes and controls the engine timing, fuel/air mixture, and so on, and completely controls the engine. To protect it from fatigue and breakage, it is desirable to isolate the module from the vibration induced in the car body by road and engine vibration. Hence the module is mounted on an isolator. Design the isolator (i.e., pick c and k) if the mass of the module is 3 kg and the dominant vibration of the fender is approximated by $y(t) = (0.01)(\sin 3t)$ m. Here it is desired to keep the displacement of the module less than 0.005 m at all times. Once the design values for isolators are chosen, calculate the magnitude of the force transmitted to the module through the isolator.

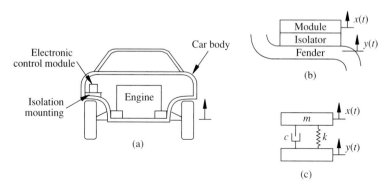

Figure 5.6 (a) Cutaway sketch of the engine compartment of an automobile illustrating the location of the car's electronic control module. (b) Close-up of the control module mounted on the inside fender on an isolator. (c) Vibration model of the module isolator system.

Solution   Since it is desired to keep the vibration of the module, $x(t)$, less than 0.005 m, the response amplitude becomes $X = 0.005$ m. The amplitude of $y(t)$ is $Y = 0.01$ m; hence the desired displacement transmissibility ratio becomes

$$\text{T.R.} = \frac{X}{Y} = \frac{0.005}{0.01} = 0.5$$

Examining the transmissibility curves of Figure 5.5 yields several possible solutions for ζ and ω. A straight horizontal line through T.R. = 0.5, illustrated in Figure 5.7, crosses at several values of ζ and r. For instance, the $\zeta = 0.02$ curve intersects the T.R. = 0.5 line at $r = 1.73$. Thus $r = 1.73$, $\zeta = 0.02$ provides one possible design solution.

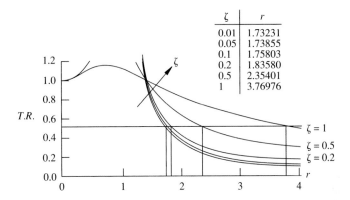

ζ	r
0.01	1.73231
0.05	1.73855
0.1	1.75803
0.2	1.83580
0.5	2.35401
1	3.76976

Figure 5.7   Transmissibility curve of Figure 5.6, repeated here, indicating possible design solution for Example 5.2.1. Each point of intersection with one of the curves of constant ζ yields the desired T.R.

Recalling that $r = \omega_{dr}/\omega = 1.73$ and $\omega_{dr} = 3$, the isolation system's natural frequency is about $\omega = 3/1.73 = 1.73$. Since the mass of the module is $m = 3$ kg, the stiffness is (recall $\omega = \sqrt{k/m}$)

$$k = m\omega^2 = (3 \text{ kg})(1.73)^2 = 9 \text{ N/m}$$

Thus the isolation mount must be made of a material with this stiffness (or add a stiffner). The damping ratio ζ is related to the damping constant c by equation (1.30):

$$c = 2\zeta m\omega = 2(0.02)(3 \text{ kg})(1.73 \text{ rad/s})$$

$$= 0.2076 \text{ kg/s}$$

These values for c and k, together with the geometric size of the module and fender shape, can now be used to choose the isolation mount material. At this point the designer would look through vendor catalogs to search for existing isolation mounts and materials that have approximately these values. If none exactly meet these values, the curves in Figure 5.7 are consulted to see if one of the other solutions corresponds more closely to an existing isolation material. Of course, equation (2.41) or (5.7) can be used to calculate solutions lying in between those illustrated in Figure 5.7.

If many solutions are still available after a search of existing products is made, the choice of a mount can be "optimized" by considering other functions, such as cost, ease of assembly, temperature range, reliability of vendor, availability of product, and quality required. Eventually, a good design must consider all of these aspects.

The electronic module may also have a limit on the amount of force it can withstand. One way to estimate the amount of force is to use the theory developed in Section 2.3, in particular in equation (2.47), which is reproduced in Window 5.1. This expression relates the force transmitted to the module by the motion of the fender through the isolator. Using the values calculated above in equation (2.47) yields

$$F_T = kYr^2 \left[\frac{1 + (2\zeta r)^2}{(1 - r^2)^2 + (2\zeta r)^2} \right]^{1/2} = kYr^2 (\text{T.R.})$$

$$= (9 \text{ N/m})(0.01 \text{ m})(1.73)^2(0.5) = 0.1354 \text{ N}$$

If this force happens to be too large, the design must be redone. With the maximum force transmitted as an additional design consideration, the curves of Figure 2.9 must also be consulted when choosing the values of r and ζ to meet the required design specifications.

$\square$

Example 5.2.1 may have seemed very reasonable. However, many assumptions were made in reaching the final design, and all of these must be given careful thought. For example, the assumption that the motion of the fender is harmonic of the form $y(t) = (0.01) \sin 3t$ is very restrictive. In reality, it is probably random, or at least, the value of ω_{dr} varies through a wide range of frequencies. This is not to say that the solution presented in Example 5.2.1 is useless, just that the designer should keep in mind its limitations. Even though the real input to the system is random, the chosen amplitude of $Y = 0.01$ m and frequency of $\omega_{dr} = 3$ rad/s, might represent a deterministic bound on all possible inputs to the system (i.e., all other disturbance amplitudes may be smaller than $Y = 0.01$ m, and all other driving frequencies might be larger than $\omega_{dr} = 3$ rad/s). Hence in many practical cases the designer is faced with choosing an isolator that will protect the part from, say, $5g$ between 20 and 200 Hz, or the designer will be given a plot of PSD versus ω_r (recall Section 3.5) and try to design the isolator to service these types of inputs. Section 5.9 examines the isolation problem from the practical aspect of working with manufacturers of isolation products.

Another difficulty in base isolation occurs when the disturbance $y(t)$ is a shock rather than a vibration. Isolation for shock disturbances must occur over a wide range of frequencies, whereas the isolation against harmonic inputs, discussed above, occurs for $r > \sqrt{2}$. In this region the damping ratio ζ is chosen to be small for better isolation, whereas shock isolation requires large ζ. Hence a *good vibration isolator* will often be a *poor shock isolator*, and vice versa. To see that this is the case, a shock isolation problem for base excitation is considered next.

The design of shock isolation systems is performed by examining the shock spectrum as introduced in Section 3.6. To make the comparison to vibration isolation clear, the shock spectrum is reconsidered here as a plot of the ratio of the maximum motion of the response acceleration amplitude (i.e., $\omega^2 X$) to the disturbance acceleration amplitude [i.e., the amplitude of $\ddot{y}(t)$] versus the product of the natural frequency and the time duration of the pulse, t_1. Here the disturbance $y(t)$ is modeled as a half sinusoid of the form

$$y(t) = \begin{cases} Y \sin \omega_p t & 0 \le t \le t_1 = \dfrac{\pi}{\omega_p} \\ 0 & t > t_1 = \dfrac{\pi}{\omega_p} \end{cases} \qquad (5.8)$$

as indicated in Figure 5.8. This type of disturbance is often called a *shock pulse*. The frequency ω_p and the corresponding time $t_1 = \pi/\omega_p$ determine how long the shock pulse lasts. The product ωt_1 is used for plotting shock transmissibility rather than plotting the frequency ratio used to design vibration isolators.

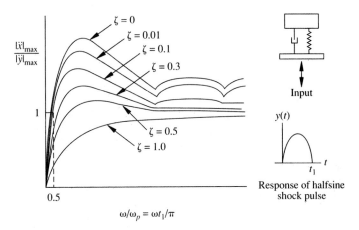

Figure 5.8   Plot of the ratio of output acceleration magnitude to input acceleration magnitude versus a frequency ratio (ω/ω_p) for a single-degree-of-freedom system and a base excitation consisting of a shock pulse for different values of damping ratios. Note that a large ω_p value corresponds to a short pulse width.

A plot of the acceleration amplitude ratio versus the product ωt_1 is given in Figure 5.8. This figure is determined by calculating the acceleration amplitude of the response and comparing it to the magnitude of the acceleration of the input disturbance. Note that as the abscissa increases, corresponding to a longer pulse width, the acceleration experienced by the module is larger than the input acceleration. To reduce the acceleration experienced by the module, the product ωt_1 must be small. For the damping ratio of Example 5.2.1 ($\zeta = 0.02$), the product ωt_1 ratio must be less than 0.5, indicating that for a pulse of width $\frac{1}{3}$ s,

$$\omega t_1 = \frac{\omega}{3} < 0.5 \qquad \text{or} \qquad \omega < 1.5$$

In the example, $m = 3$ kg, so that to minimize the acceleration amplitude of the responses,

$$\sqrt{k/3} < 1.5 \qquad \text{or} \qquad k < 6.75 \text{ N/m}$$

This is in direct contradiction of the result of Example 5.2.1, which concludes that to minimize the steady-state response of the amplitude of vibration due to a harmonic input at a frequency of 3 rad/s, the stiffness must be greater than 9 N/m! Hence the isolator design of Example 5.2.1 provides great protection from harmonic inputs but does not provide isolation against shock. In designing isolation systems, the nature of the driving disturbance is very important. If a variety of pulses and harmonic inputs are present, the design problem becomes much more difficult.

Next consider the problem of isolating a source of harmonic vibration from its surroundings. This is the fixed-base isolation problem illustrated in Window 5.1, where the right side is concerned with reducing the force transmitted through the isolator due to harmonic excitation at the mass. The common example is a rotating machine generating a harmonic force at a constant frequency (recall Figure 2.11). Examples of such machines are electric motors, steam turbines, internal combustion engines, generators, washing machines, and disk drive motors.

Examination of the transmissibility curve of Figure 5.5 indicates that vibration isolation begins for isolation stiffness values such that $\omega_{dr}/\omega > \sqrt{2}$. This gives transmissibility ratios of less than 1, so that the force transmitted to ground is less than the force generated by the rotating machine. Since the mass is usually fixed by the nature of the machine, the isolation mounts are generally chosen based on their stiffness, so that $r > \sqrt{2}$ is satisfied. If this does not give an acceptable solution (for low-frequency excitation), mass can sometimes be added to the machine ($\omega = \sqrt{k/m}$). Since $r = \omega_{dr}/\sqrt{k/m}$, lower stiffness values correspond to larger values of r, which yields better isolation (lower T.R. values).

As damping is increased for a fixed r, the value of T.R. increases, so that low damping is often used. However, some damping is desirable, since when the machine starts up and causes a harmonic disturbance through a range of frequencies, it generally passes through resonance ($r = 1$) and the presence of damping is required to reduce the transmissibility at resonance. Examination of the transmissibility curve indicates that for a large enough frequency ratio (about $r > 3$) and small enough damping ($\zeta < 0.2$) the T.R. value is not affected by damping. Since most springs have very small internal damping (e.g., less than 0.01), the term $(2\zeta r)^2$ is very small [e.g., for $r = 3$, $(2\zeta r)^2 = 0.0036$]. Hence it is common to design a vibration isolation system by neglecting the damping in equation (5.7). In this case T.R. becomes (taking the negative square root for positive values of T.R.)

$$\text{T.R.} = \frac{1}{r^2 - 1} \qquad (r > 3) \tag{5.9}$$

Equation (5.9) can be used to construct design charts for use in choosing vibration isolation pads for mounting rotating machinery.

The driving frequency of a machine is usually specified in terms of its speed of rotation, or revolutions per minute (rpm). If n is the motor speed in rpm,

$$\omega_{dr} = \frac{2\pi n}{60} \tag{5.10}$$

In addition, springs are often classified in terms of their static deflection defined by $\delta_s = W/k = mg/k$, where m is the mass of the machine and g the acceleration due to gravity. It has become very common to design isolators in terms of the machine's rotating speed n and the static deflection δ_s. A third quantity, R, defined as the *reduction* in transmissibility:

$$R = 1 - \text{T.R.} \tag{5.11}$$

is commonly used to quantify the success of the vibration isolator.

Substitution of the (undamped) value of T.R. into equation (5.11) and solving for r yields

$$r = \frac{\omega_{dr}}{\sqrt{k/m}} = \sqrt{\frac{2 - R}{1 - R}} \tag{5.12}$$

Substitution of (5.10) for ω_{dr} and replacing $k = mg/\delta_s$ yields

$$n = \frac{30}{\pi}\sqrt{\frac{g(2 - R)}{\delta_s(1 - R)}} = 29.9093\sqrt{\frac{2 - R}{\delta_s(1 - R)}} \tag{5.13}$$

which relates the motor speed to the reduction factor and the static deflection of the spring. Equation (5.13) can be used to generate design curves, by taking the log of the expression. This yields

$$\log n = -\frac{1}{2}\log \delta_s + \log\left(29.9093\sqrt{\frac{2 - R}{1 - R}}\right) \tag{5.14}$$

which is a straight line on a log-log plot for each value of R. This expression is then used to provide the design chart of Figure 5.9, consisting of plots of motor speed versus static deflection.

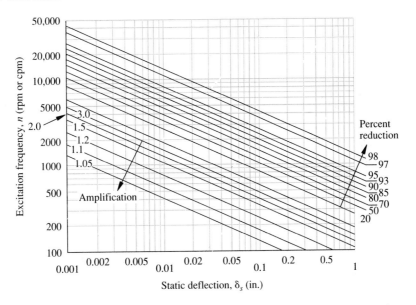

Figure 5.9   Design curves consisting of plots of running speed versus static deflection (or stiffness) for various values of percent reduction in transmitted force.

Example 5.2.2

Consider the computer disk drive of Figure 5.10. The disk drive motor is mounted to the computer chassis through an isolation pad (spring). The motor has a mass of 3 kg and

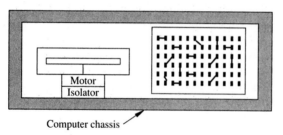

Figure 5.10 Schematic model of a personal computer illustrating the motor running the disk drive system. A small amount of out of balance in the motor can transmit harmonic forces to the chassis and onto circuit boards and other components if not properly isolated.

operates at 5000 rpm. Calculate the value of the stiffness of the isolator needed to provide a 95% reduction in force transmitted to the chassis (considered as ground). How much clearance is needed between the motor and the chassis?

Solution From the chart of Figure 5.9, the line corresponding to a speed of $n = 5000$ rpm hits the curve corresponding to 95% reduction at a static deflection of 0.03 in. or 0.0762 cm. This corresponds to a spring stiffness of

$$k = \frac{mg}{\delta_s} = \frac{(3 \text{ kg})(9.8 \text{ m/s}^2)}{0.000762 \text{ m}} = 38,582 \text{ N/m}$$

The choice of clearance (i.e., distance needed between the motor and the chassis) should be more than twice the static deflection so that the spring has room to extend and compress, providing isolation.

$\square$

The issue of vibration isolation against harmonic forces for large (heavy) machines quickly becomes one of static deflection. For machines requiring extreme isolation, coiled springs often must be used to provide the large static deflection required at low frequency. This can be seen by examining the 98% curves in Figure 5.9 for low values of n. In some cases the static deflection required may be too large to be physically obtainable even in small devices. An example of a similar design constraint is the miniaturization of computers. Manufacturers of laptop computers believe sales are tied to how compact and, in particular, how thin the chassis can be. One constraint could be the isolation system required for the disk drive motor or other components.

5.3 VIBRATION ABSORBERS

Another approach to protecting a device from steady-state harmonic disturbance at a constant frequency is a *vibration absorber*. Unlike the isolator of the previous sections, an absorber consists of a second mass–spring combination added to the primary device to protect it from vibrating. The major effect of adding the second mass–spring system is to change from a single-degree-of-freedom system to a two-degree-of-freedom system. The new system has two natural frequencies (recall Section 4.1). The added spring–mass system is called the absorber. The values of the absorber mass and stiffness are chosen such that the motion of the original mass is a minimum. This is accompanied by substantial motion of the added absorber system as illustrated in the following.

Absorbers are often used on machines that run at constant speed, such as sanders, compactors, reciprocating tools, and electric razors. Probably the most visible vibration absorbers can be seen on transmission lines and telephone lines. A dumbbell-shaped vibration absorber is often used on such wires to provide vibration isolation against wind blowing, which can cause the wire to oscillate at its natural frequency. The presence of the absorber prevents the wire from vibrating so much at resonance that it breaks (or fatigues). Figure 5.11 illustrates a simple vibration absorber attached to a spring–mass system. The equations of motion [summing forces in the vertical direction, (refer to Chapter 4)] are

$$\begin{bmatrix} m & 0 \\ 0 & m_a \end{bmatrix} \begin{bmatrix} \ddot{x} \\ \ddot{x}_a \end{bmatrix} + \begin{bmatrix} k + k_a & -k_a \\ -k_a & k_a \end{bmatrix} \begin{bmatrix} x \\ x_a \end{bmatrix} = \begin{bmatrix} F_0 \sin \omega_{dr} t \\ 0 \end{bmatrix} \qquad (5.15)$$

where $x = x(t)$ is the displacement of the table modeled as having mass m and stiffness k, x_a is the displacement of the absorber mass (of mass m_a and stiffness k_a), and the harmonic force $F_0 \sin \omega_{dr} t$ is the disturbance applied to the table mass. It is desired to design the absorber (i.e., choose m_a and k_a) such that the displacement of the primary system is as small as possible in steady state. Here it is desired to reduce the vibration of the table, which is the primary mass.

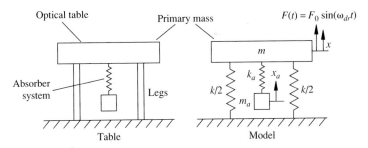

Figure 5.11 An optical table protected by an added vibration absorber. The table and its supporting legs are modeled as a single-degree-of-freedom system with mass m and stiffness k.

In contrast to the solution technique of modal analysis used in Chapter 4, here it is desired to obtain a solution in terms of parameters (m, k, m_a, and k_a) which can then be solved for as part of a design process. To this end, let the steady-state solution of $x(t)$ and $x_a(t)$ be of the form

$$\begin{aligned} x(t) &= X \sin \omega_{dr} t \\ x_a(t) &= X_a \sin \omega_{dr} t \end{aligned} \qquad (5.16)$$

Substitution of these steady-state forms into equation (5.15) yields (after some manipulation)

$$
\begin{bmatrix} k + k_a - m\omega_{dr}^2 & -k_a \\ -k_a & k_a - m_a\omega_{dr}^2 \end{bmatrix} \begin{bmatrix} X \\ X_a \end{bmatrix} \sin \omega_{dr} t = \begin{bmatrix} F_0 \\ 0 \end{bmatrix} \sin \omega_{dr} t \qquad (5.17)
$$

which is an equation in the vector $[X \quad X_a]^T$. Dividing by $\sin \omega_{dr} t$, taking the inverse of the matrix coefficient of $[X \quad X_a]^T$ (see Window 5.3), and multiplying from the left yields

$$
\begin{bmatrix} X \\ X_a \end{bmatrix} = \frac{1}{(k + k_a - m\omega_{dr}^2)(k_a - m_a\omega_{dr}^2) - k_a^2} \begin{bmatrix} k_a - m_a\omega_{dr}^2 & k_a \\ k_a & k + k_a - m\omega_{dr}^2 \end{bmatrix} \begin{bmatrix} F_0 \\ 0 \end{bmatrix}
$$

$$
= \frac{1}{(k + k_a - m\omega_{dr}^2)(k_a - m_a\omega_{dr}^2) - k_a^2} \begin{bmatrix} (k_a - m_a\omega_{dr}^2)F_0 \\ k_a F_0 \end{bmatrix}
$$

$$(5.18)$$

Equating elements of the vector equality given by equation (5.18) yields the result that the magnitude of the steady-state vibration of the device (table) becomes

$$
X = \frac{(k_a - m_a\omega_{dr}^2)F_0}{(k + k_a - m\omega_{dr}^2)(k_a - m_a\omega_{dr}^2) - k_a^2} \qquad (5.19)
$$

while the magnitude of vibration of the absorber mass becomes

$$
X_a = \frac{k_a F_0}{(k + k_a - m\omega_{dr}^2)(k_a - m_a\omega_{dr}^2) - k_a^2} \qquad (5.20)
$$

Note from equation (5.19) that the absorber parameters k_a and m_a can be chosen such that the magnitude of steady-state vibration, X, is exactly zero. This is accomplished by equating the coefficient of F_0 in equation (5.19) to zero:

$$
\omega_{dr}^2 = \frac{k_a}{m_a} \qquad (5.21)
$$

Hence if the absorber parameters are chosen to satisfy the tuning condition of equation (5.21), the steady-state motion of the primary mass is zero (i.e., $X = 0$). In this event the steady-state motion of the absorber mass is calculated from equations (5.20) and (5.16) with $k_a = m_a\omega_{dr}^2$ to be

$$
x_a(t) = -\frac{F_0}{k_a} \sin \omega_{dr} t \qquad (5.22)
$$

Thus the absorber mass oscillates at the driving frequency with amplitude $X_a = F_0/k_a$.

Note that the magnitude of the force acting on the absorber mass is just $k_a x_a = k_a(-F_0/k_a) = -F_0$. Hence when the absorber system is tuned to the driving frequency and has reached steady state, the force provided by the absorber mass is equal in magnitude and opposite in direction to the disturbance force. With zero net force acting on the primary mass, it does not move and the motion is "absorbed" by motion of the absorber mass.

Window 5.3

Recall the inverse of a 2×2 matrix A given by

$$A = \begin{bmatrix} a & b \\ c & d \end{bmatrix}$$

is defined to be

$$A^{-1} = \frac{1}{\det A} \begin{bmatrix} d & -b \\ -c & a \end{bmatrix}$$

where

$$\det A = ad - bc$$

The success of the vibration absorber discussed above depends on several factors. First the harmonic excitation must be well known and not deviate much from its constant value. If the driving frequency drifts much, the tuning condition will not be satisfied, and the primary mass will experience some oscillation. There is also some danger that the driving frequency could shift to one of the combined systems' resonant frequencies, in which case one or the other of the system's coordinates would be driven to resonance and potentially fail. The analysis used to design the system assumes that it can be constructed without introducing any appreciable damping. If damping is introduced, the equations cannot necessarily be decoupled and the magnitude of the displacement of the primary mass will not be zero. In fact, damping defeats the purpose of a tuned vibration absorber and is desirable only if the frequency range of the driving force is too wide for effective operation of the absorber system. This is discussed in the next section. Another key factor in absorber design is that the absorber spring stiffness k_a must be capable of withstanding the full force of the excitation and hence must be capable of the corresponding deflections. The issue of spring size and deflection as well as the value of the absorber mass places a geometric limitation on the design of a vibration absorber system.

The issue of avoiding resonance in absorber design in case the driving frequency shifts can be quantified by examining the mass ratio μ, defined as the ratio of the absorber mass to the primary mass:

$$\mu = \frac{m_a}{m}$$

In addition, it is convenient to define the frequencies

$$\omega_p = \sqrt{\frac{k}{m}}$$ original natural frequency of the primary system without the absorber attached

$$\omega_a = \sqrt{\frac{k_a}{m_a}}$$ the natural frequency of the absorber system before it is attached to the primary system

With these definitions, also note that

$$\frac{k_a}{k} = \mu \frac{\omega_a^2}{\omega_p^2} \tag{5.23}$$

Substitution of the values for μ, ω_p, and ω_a into equation (5.19) for the amplitude of vibration of the primary mass yields (after some manipulation)

$$\frac{Xk}{F_0} = \frac{1 - \omega_{dr}^2/\omega_a^2}{\left[1 + \mu\left(\omega_a/\omega_p\right)^2 - \left(\omega_{dr}/\omega_p\right)^2\right]\left[1 - \left(\omega_{dr}/\omega_a\right)^2\right] - \mu\left(\omega_a/\omega_p\right)^2} \tag{5.24}$$

The absolute value of this expression is plotted in Figure 5.12 for the case $\mu = 0.25$. Such plots can be used to illustrate how much drift in driving frequency can be tolerated by the absorber design. Note that if ω_{dr} should drift to either $0.782\omega_a$ or $1.28\omega_a$, the combined system would experience resonance and fail, since these are the natural frequencies of the combined system. In fact, if the driving frequency shifts such that $|Xk/F_0| > 1$, the force transmitted to the primary system is amplified and the absorber system is not an improvement over the original design of the primary system. The shaded area of Figure 5.12 illustrates the values of ω_{dr}/ω_a such that $|Xk/F_0| \le 1$. This illustrates the useful operating range of the absorber design (i.e., $0.908\omega_a < \omega_{dr} < 1.118\omega_a$). Hence if the driving frequency drifts within this range, the absorber design still offers some protection to the primary system by reducing its steady-state vibration magnitude.

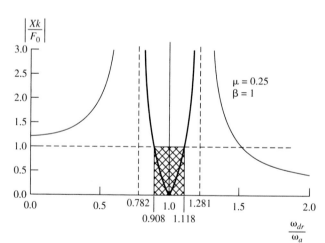

Figure 5.12 Plot of normalized magnitude of the primary mass versus the normalized driving frequency for the case $\mu = 0.25$. The two natural frequencies of the system occur at 0.782 and 1.281.

The design of an absorber can be further illuminated by examining the mass ratio μ, and the frequency ratio β defined as $\beta = \omega_a/\omega_p$. These two dimensionless quantities indirectly specify both the mass and stiffness of the absorber system. The frequency equation (characteristic equation) for the two-mass system is obtained by setting the determinant of the matrix coefficient in equation (5.17) [i.e., the denominator of equation (5.18)] to zero and interpreting ω_{dr} as the system natural frequency ω. Substitution for the value of β and rearranging yields

$$\beta^2 \left(\frac{\omega^2}{\omega_a^2}\right)^2 - [1 + \beta^2(1 + \mu)]\frac{\omega^2}{\omega_a^2} + 1 = 0 \qquad (5.25)$$

which is a quadratic equation in (ω^2/ω_a^2). Solving this yields

$$\left(\frac{\omega}{\omega_a}\right)^2 = \frac{1 + \beta^2(1 + \mu)}{2\beta^2} \pm \frac{1}{2\beta^2}\sqrt{\beta^4(1 + \mu)^2 - 2\beta^2(1 - \mu) + 1} \qquad (5.26)$$

which illustrates how the system's natural frequencies vary with the mass ratio μ and the frequency ratio β. This is plotted for $\beta = 1$ in Figure 5.13. Note that as μ is increased, the natural frequencies split farther apart, and farther from the operating point $\omega_{dr} = \omega_a$ of the absorber. Therefore if μ is too small, the combined system will not tolerate much fluctuation in the driving frequency before it fails. As a rule of thumb, μ is usually taken to be between 0.05 and 0.25 (i.e., $0.05 \leq \mu \leq 0.25$), as larger values of μ tend to indicate a poor design. Vibration absorbers can also fail because of fatigue if $x_a(t)$ and the stresses associated with this motion of the absorber are large. Hence limits are often placed on the maximum value of X_a by the designer. The following example illustrates an absorber design.

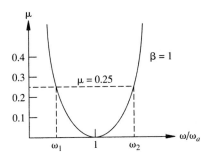

Figure 5.13   Plot of mass ratio versus system natural frequency (normalized to the frequency of the absorber system), illustrating that increasing the mass ratio increases the useful frequency range of a vibration absorber. Here ω_1 and ω_2 indicates the normalized value of the system's natural frequencies.

Example 5.3.1

A radial saw base has a mass of 73.16 kg and is driven harmonically by a motor that turns the saw's blade as illustrated in Figure 5.14. The motor runs at constant speed and

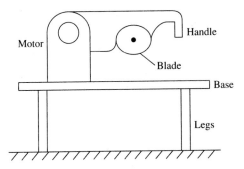

Figure 5.14   Schematic of a radial saw system in need of a vibration absorber.

produces a 13-N force at 180 cycles/min because of a small unbalance in the motor. The resulting forced vibration was not detected until after the saw had been manufactured. The manufacturer wants a vibration absorber designed to drive the table oscillation to zero simply by retrofitting an absorber onto the base. Design the absorber assuming that the effective stiffness provided by the table legs is 2600 N/m. In addition, the absorber must fit inside the table base and hence has a maximum deflection of 0.2 cm.

Solution   To meet the deflection requirement, the absorber stiffness is chosen first. This is calculated by assuming that $X = 0$, so that $|X_a k_a| = |F_0|$ (i.e., so that the mass m_a absorbs all of the applied force). Hence

$$k_a = \frac{F_0}{X_a} = \frac{13 \text{ N}}{0.2 \text{ cm}} = \frac{13 \text{ N}}{0.002 \text{ m}} = 6500 \text{ N/m}$$

Since the absorber is designed such that $\omega_{dr} = \omega_a$,

$$m_a = \frac{k_a}{\omega_{dr}^2} = \frac{6500 \text{ N/m}}{[(180/60)2\pi]^2} = 18.29 \text{ kg}$$

Note in this case that $\mu = 18.29/73.16 = 0.25$.

$\square$

Example 5.3.2

Calculate the bandwidth of operation of the absorber design of Example 5.3.1. Assume that the useful range of an absorber is defined such that $|Xk/F_0| < 1$. For values of $|Xk/F_0| > 1$, the machine could easily drift into resonance and the amplitude of vibration actually becomes an amplification of the effective driving force amplitude.

Solution   From equation (5.24) with $Xk/F_0 = 1$,

$$1 - \left(\frac{\omega_{dr}}{\omega_a}\right)^2 = \left[1 + \mu\left(\frac{\omega_a}{\omega_p}\right)^2 - \left(\frac{\omega_{dr}}{\omega_p}\right)^2\right]\left[1 - \left(\frac{\omega_{dr}}{\omega_a}\right)^2\right] - \mu\left(\frac{\omega_a}{\omega_p}\right)^2$$

Solving this for ω_{dr}/ω_a yields the two solutions

$$\frac{\omega_{dr}}{\omega_a} = \pm\sqrt{1 + \mu}$$

For the system of Example 5.3.1, $\mu = 0.25$, so that the second solution becomes

$$\frac{\omega_{dr}}{\omega_a} = 1.1180$$

The condition that $|Xk/F_0| = 1$ is also satisfied for $Xk/F_0 = -1$. Substitution of this into equation (5.24) followed by some manipulation yields

$$\left(\frac{\omega_a}{\omega_p}\right)^2 \left(\frac{\omega_{dr}}{\omega_a}\right)^4 - \left[2 + (\mu + 1)\left(\frac{\omega_a}{\omega_p}\right)^2\right]\left(\frac{\omega_{dr}}{\omega_a}\right)^2 + 2 = 0$$

which is quadratic in $(\omega_{dr}/\omega_a)^2$. Using the values of $\omega_a^2 = 6500/18.29$, $\omega_p^2 = 2600/73.16$, and $\mu = 0.25$, this simplifies to

$$10\left(\frac{\omega_{dr}}{\omega_a}\right)^4 - 14.5\left(\frac{\omega_{dr}}{\omega_a}\right)^2 + 2 = 0$$

Solving for ω_{dr}/ω_a yields

$$\left(\frac{\omega_{dr}}{\omega_a}\right)^2 = 0.1544, 1.2956 \qquad \text{or} \qquad \frac{\omega_{dr}}{\omega_a} = 0.3929, 1.1382$$

Hence the four roots satisfying $|Xk/F_0| = 1$ are $0, 0.3929, 1.1180$, and 1.1382. Following the example of Figure 5.12 indicates that the driving frequency may vary between $0.3929\omega_a$ and $1.1180\omega_a$, or since $\omega_a = 18.857$,

$$7.4089 < \omega_{dr} < 21.0821 \ \text{(rad/s)}$$

before the response of the primary mass is amplified or the system is in danger of experiencing resonance.

$\square$

The discussion and examples above illustrate the concept of *performance robustness*; that is, the examples illustrate how the design holds up as the parameter values $(k, k_a, \text{etc.})$ drift from the values used in the original design. Example 5.3.2 illustrates that the mass ratio greatly affects the robustness of absorber designs. This is stated in the caption of Figure 5.13; up to a certain point, increasing μ increases the robustness of the absorber design. The effects of damping on absorber design are examined in the next section.

5.4 DAMPING IN VIBRATION ABSORPTION

As mentioned in Section 5.3, damping is often present in devices and has the potential for destroying the ability of a vibration absorber to protect the primary system fully by driving X to zero. In addition, damping is sometimes added to vibration absorbers to prevent resonance or to improve the effective bandwidth of operation of a vibration absorber. Also, a damper by itself is often used as a vibration absorber by dissipating the energy supplied by an applied force. Such devices are called *vibration dampers* rather than absorbers.

First consider the effect of modeling damping in the standard vibration absorber problem. A vibration absorber with damping in both the primary and absorber system is illustrated in Figure 5.15. This system is dynamically equal to the system of Figure 4.14 of Section 4.5. The equations of motion are given in matrix form by equation (4.126) as

$$\begin{bmatrix} m & 0 \\ 0 & m_a \end{bmatrix} \begin{bmatrix} \ddot{x}(t) \\ \ddot{x}_a(t) \end{bmatrix} + \begin{bmatrix} c + c_a & -c_a \\ -c_a & c_a \end{bmatrix} \begin{bmatrix} \dot{x}(t) \\ \dot{x}_a(t) \end{bmatrix}$$
$$+ \begin{bmatrix} k + k_a & -k_a \\ -k_a & k_a \end{bmatrix} \begin{bmatrix} x(t) \\ x_a(t) \end{bmatrix} = \begin{bmatrix} F_0 \\ 0 \end{bmatrix} \sin \omega_{dr} t \qquad (5.27)$$

Note, as was mentioned in Section 4.5, that these equations cannot necessarily be solved by using the modal analysis technique of Chapter 4 because the equations do not decouple $(C \neq \alpha M + \beta K)$. The steady-state solution can be calculated, however, by using a

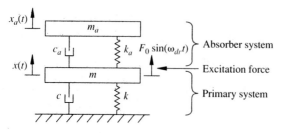

Figure 5.15 Schematic of a vibration absorber with damping in both the primary and absorber system.

combination of the exponential approach discussed in Section 2.5 and the matrix inverse used in previous sections for the undamped case.

To this end, let $F_0 \sin \omega_{dr} t$ be represented in exponential form by $F_0 e^{j\omega_{dr} t}$ in equation (5.27) and assume that the steady-state solution is of the form

$$\mathbf{x}(t) = \mathbf{X} e^{j\omega_{dr} t} = \begin{bmatrix} X \\ X_a \end{bmatrix} e^{j\omega_{dr} t} \tag{5.28}$$

where X is the amplitude of vibration of the primary mass and X_a is the amplitude of vibration of the absorber mass. Substitution into equation (5.27) yields

$$\begin{bmatrix} (k + k_a - m\omega_{dr}^2) + (c + c_a)\omega_{dr} j & -k_a - c_a\omega_{dr} j \\ -k_a - c_a\omega_{dr} j & (k_a - m_a\omega_{dr}^2) + c_a\omega_{dr} j \end{bmatrix} \begin{bmatrix} X \\ X_a \end{bmatrix} e^{j\omega_{dr} t}$$

$$= \begin{bmatrix} F_0 \\ 0 \end{bmatrix} e^{j\omega_{dr} t} \tag{5.29}$$

Note that the coefficient matrix of the vector $\mathbf{X}$ has complex elements. Dividing equation (5.29) by the nonzero scalar $e^{j\omega_{dr} t}$ yields a complex matrix equation in the amplitudes X and X_a. Calculating the matrix inverse using the formula of Example 4.1.4, reviewed in Window 5.3, and multiplying equation (5.29) by the inverse from the right yields

$$\begin{bmatrix} X \\ X_a \end{bmatrix} = \frac{\begin{bmatrix} (k_a - m_a\omega_{dr}^2) + c_a\omega_{dr} j & k_a + c_a\omega_{dr} j \\ k_a + c_a\omega_{dr} j & k + k_a - m\omega_{dr}^2 + (c + c_a)\omega_{dr} j \end{bmatrix} \begin{bmatrix} F_0 \\ 0 \end{bmatrix}}{\det(K - \omega_{dr}^2 M + \omega_{dr} jC)} \tag{5.30}$$

Here the determinant in the denominator is given by (recall Example 4.18)

$$\det(K - \omega_{dr}^2 M + \omega_{dr} jC) = mm_a\omega_{dr}^4 - (c_a c + m_a(k + k_a) + k_a m)\omega_{dr}^2 + k_a k$$
$$+ (kc_a + ck_a)\omega_{dr} - (c_a(m + m_a) + cm_a)\omega_{dr}^3]j \tag{5.31}$$

and the system coefficient matrices M, C, and K are given by

$$M = \begin{bmatrix} m & 0 \\ 0 & m_a \end{bmatrix} \qquad C = \begin{bmatrix} c + c_a & -c_a \\ -c_a & c_a \end{bmatrix} \qquad K = \begin{bmatrix} k + k_a & -k_a \\ -k_a & k_a \end{bmatrix}$$

Simplifying the matrix vector product yields

$$X = \frac{[(k_a - m_a\omega_{dr}^2) + c_a\omega_{dr}j]F_0}{\det(K - \omega_{dr}^2 M + \omega_{dr}jC)} \tag{5.32}$$

$$X_a = \frac{(k_a + c_a\omega_{dr}j)F_0}{\det(K - \omega_{dr}^2 M + \omega_{dr}jC)} \tag{5.33}$$

which expresses the magnitude of the response of the primary mass and absorber mass, respectively. Note that these values are now complex numbers and are multiplied by the complex value $e^{j\omega_{dr}t}$ to get the time responses.

Equations (5.32) and (5.33) are the two-degree-of-freedom version of the frequency response function given for a single-degree-of-freedom system in equation (2.60). The complex nature of these values reflect a magnitude and phase. The magnitude is calculated following the rules of complex numbers and is best done with a symbolic computer code, or after substitution of numerical values for the various physical constants. It is important to note from equation (5.32) that unlike the tuned undamped absorber, the response of the primary system cannot be exactly zero even if the tuning condition is satisfied. Hence the presence of damping ruins the ability of the absorber system to exactly cancel the motion of the primary system.

Equations (5.32) and (5.33) can be analyzed for several specific cases. First, consider the case for which the internal damping of the primary system is neglected ($c = 0$). If the primary system is made of metal, the internal damping is likely to be very low and it is reasonable to neglect it in many circumstances. In this case the determinant of equation (5.31) reduces to the complex number

$$\det(K - \omega_{dr}^2 M - \omega_{dr}Cj)$$

$$= [(-m\omega_{dr}^2 + k)(-m_a\omega_{dr}^2 + k_a) - m_a k_a \omega_{dr}^2] + [(k - (m + m_a)\omega_{dr}^2)c_a\omega_{dr}] j \tag{5.34}$$

The maximum deflection of the primary mass is given by equation (5.32) with the determinant in the denominator evaluated as given in equation (5.34). This is the ratio of two complex numbers and hence is a complex number representing the phase and the amplitude of the response of the primary mass. Using complex arithmetic (see Window 5.4) the amplitude of the motion of the primary mass can be written as the real number

$$\frac{X^2}{F_0^2} = \frac{(k_a - m_a\omega_{dr}^2)^2 + \omega_{dr}^2 c_a^2}{\left[(k - m\omega_{dr}^2)(k_a - m_a\omega_{dr}^2) - m_a k_a \omega_{dr}^2\right]^2 + \left[k - (m + m_a)\omega_{dr}^2\right]^2 c_a^2 \omega_{dr}^2} \tag{5.35}$$

It is instructive to examine this amplitude in terms of the dimensionless ratios introduced in Section 5.3 for the undamped vibration absorber. The amplitude X is written in terms of the static deflection $\delta_{st} = F_0/k$ of the primary system. In addition, consider the mixed "damping ratio" defined by

$$\zeta = \frac{c_a}{2m_a\omega_p} \tag{5.36}$$

Window 5.4
Reminder of Complex Arithmetic

The response magnitude given by equation (5.32) can be written as the ratio of two complex numbers:

$$\frac{X}{F_0} = \frac{A_1 + B_1 j}{A_2 + B_2 j}$$

where A_1, A_2, B_1, and B_2 are real numbers and $j = \sqrt{-1}$. Multiplying this by the conjugate of the denominator divided by itself yields

$$\frac{X}{F_0} = \frac{(A_1 + B_1 j)(A_2 - B_2 j)}{(A_2 + B_2 j)(A_2 - B_2 j)} = \frac{(A_1 A_2 + B_1 B_2)}{A_2^2 + B_2^2} + \frac{B_1 A_2 - A_1 B_2}{A_2^2 + B_2^2} j$$

which indicates how X/F_0 is written as a single complex number of the form $X/F_0 = a + bj$. This is interpreted, as indicated, that the response magnitude has two components: one in phase with the applied force and one out of phase. The magnitude of X/F_0 is the length of the complex number above (i.e., $|X/F_0| = \sqrt{a^2 + b^2}$. This yields

$$\left| \frac{X}{F_0} \right| = \sqrt{\frac{A_1^2 + B_1^2}{A_2^2 + B_2^2}}$$

which corresponds to the expression given in equation (5.35). Also see Appendix A.

---

where $\omega_p = \sqrt{k/m}$ is the original natural frequency of the primary system without the absorber attached. Using the standard frequency ratio $r = \omega_{dr}/\omega_p$, the ratio of natural frequencies $\beta = \omega_a/\omega_p$, (where $\omega_a = \sqrt{k_a/m_a}$), and the mass ratio $\mu = m_a/m$, equation (5.35) can be rewritten as

$$\frac{X}{\delta_{st}} = \frac{Xk}{F_0} = \sqrt{\frac{(2\zeta r)^2 + (r^2 - \beta^2)^2}{(2\zeta r)^2 (r^2 - 1 + \mu r^2)^2 + [\mu r^2 \beta^2 - (r^2 - 1)(r^2 - \beta^2)]^2}} \qquad (5.37)$$

which expresses the dimensionless amplitude of the primary system. Note from examining equation (5.37) that the amplitude of the primary system response is determined by four physical parameter values:

     μ the ratio of the absorber mass to the primary mass

     β the ratio of the decoupled natural frequencies

     r the ratio of the driving frequency to the primary natural frequency

     ζ the ratio of the absorber damping and $2m_a \omega_p$

These four numbers can be considered as design variables and are chosen to give the smallest possible value of the primary mass's response, X, for a given application.

Figure 5.16 illustrates how the damping value, as reflected in ζ, affects the response for a fixed value of μ, β, and r.

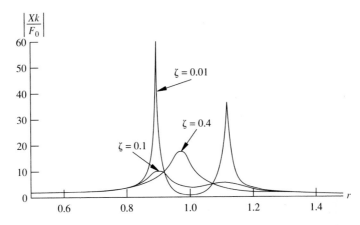

Figure 5.16   Normalized amplitude of vibration of the primary mass as a function of the frequency ratio for several values of the damping in the absorber system for the case of negligible damping in the primary system [i.e., a plot of equation (5.37)].

As mentioned at the beginning of this section, damping is often added to the absorber to improve the bandwidth of operation. This effect is illustrated in Figure 5.16. Recall that if there is no damping in the absorber ($\zeta = 0$), the magnitude of the response of the primary mass as a function of the frequency ratio r is as illustrated in Figure 5.12 (i.e., zero at $r = 1$ but infinite at $r = 0.782$ and $r = 1.281$). Thus the completely undamped absorber has poor bandwidth (i.e., if r changes by a small amount, the amplitude grows). In fact, as noted in Section 5.3, the bandwidth, or useful range of operation of that undamped absorber, is $0.897 \leq r \leq 1.103$. For these values of r, $|Xk/F_0| \leq 1$. However, if damping is added to the absorber ($\zeta \neq 0$), Figure 5.16 results, and the bandwidth, or useful range of operation, is extended. The price for this increased operating region is that $|Xk/F|$ is never zero in the damped case (see Figure 5.16).

Examination of Figure 5.16, shows that as ζ is varied, the amplification of $|Xk/F_0|$ over the range of r can be reduced. The design question now becomes: For what values of the mass ratio μ, the absorber damping ratio ζ, and the frequency ratio β is the magnitude $|Xk/F_0|$ smallest over the region $0 \leq r \leq 2$? Just increasing the damping with μ and β fixed does not necessarily yield the lowest amplitude. Note from Figure 5.16 that $\zeta = 0.1$ produces a smaller amplification over a larger region of r that does the higher ratio, $\zeta = 0.4$. Figures 5.17 and 5.18 yield some hint of how the various parameters affect the magnitude by providing plots of $|Xk/F_0|$ for various combinations of ζ, μ, and β.

A solution of the best choice of μ and ζ is discussed again in Section 5.5. Note from Figure 5.18 that $\mu = 0.25$, $\beta = 0.8$, and $\zeta = 0.27$ yield a minimum value of $|Xk/F_0|$ over a large range of values of r. However, amplification of the response X still

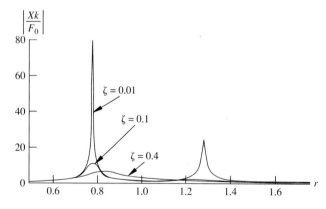

Figure 5.17 Repeat of the plot of Figure 5.16 with $\mu = 0.25$ and $\beta = 1$ for several values of ζ. Note that in this case, $\zeta = 0.4$ yields a lower magnitude than does $\zeta = 0.1$.

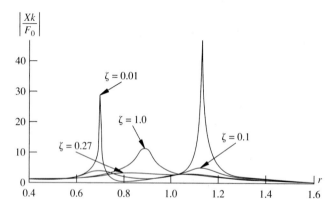

Figure 5.18 Repeat of the plots of Figure 5.16 with $\mu = 0.25$, $\beta = 0.8$ for several values of ζ. In this case $\zeta = 0.27$ yields the lowest amplification over the largest bandwidth.

occurs (i.e., $|Xk/F_0| > 1$ for values of $r < \sqrt{2}$), but no order-of-magnitude increases in $|X|$ occurs as in the case of the undamped absorber.

Next consider the case of an appended absorber mass connected to an undamped primary mass only by a dashpot, an arrangement illustrated in Figure 5.19. Systems of this form arise in the design of vibration reduction devices for rotating systems such as engines, where the operating speed (and hence the driving frequency) varies over a wide range. In such cases a viscous damper is added to the end of the crankshaft (or other rotating device) as indicated in Figure 5.20. The shaft spins through an angle θ_1 with

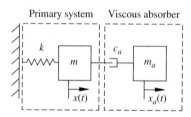

Figure 5.19 Damper–mass system added to a primary mass (with no damping) to form a viscous vibration absorber.

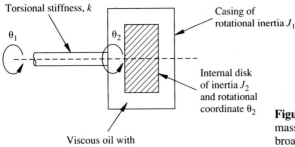

Torsional stiffness, k

θ_1

θ_2

Casing of rotational inertia J_1

Internal disk of inertia J_2 and rotational coordinate θ_2

Viscous oil with damping coefficient c_a

Figure 5.20   Viscous damper and mass added to a rotating shaft for broadband vibration absorption. Often called a *Houdaille damper*.

torsional stiffness k and inertia J_1. The damping inertia J_2 spins through an angle θ_2 in a viscous film providing a damping force $c_a(\dot{\theta}_1 - \dot{\theta}_2)$. If an external harmonic torque is applied of the form $M_0 e^{-\omega_{dr} t j}$, the equation of motion of this system becomes

$$\begin{bmatrix} J_1 & 0 \\ 0 & J_2 \end{bmatrix}\begin{bmatrix} \ddot{\theta}_1 \\ \ddot{\theta}_2 \end{bmatrix} + \begin{bmatrix} c_a & -c_a \\ -c_a & c_a \end{bmatrix}\begin{bmatrix} \dot{\theta}_1 \\ \dot{\theta}_2 \end{bmatrix} + \begin{bmatrix} k & 0 \\ 0 & 0 \end{bmatrix}\begin{bmatrix} \theta_1 \\ \theta_2 \end{bmatrix} = \begin{bmatrix} M_0 \\ 0 \end{bmatrix} e^{-\omega_{dr} t j} \qquad (5.38)$$

This is a rotational equivalent to the translational model given in Figure 5.19. It is easy to calculate the undamped natural frequencies of this two-degree-of-freedom system. They are

$$\omega_p = \sqrt{\frac{k}{J_1}} \text{ and } \omega_a = 0$$

The solution of this set of equations is given by equations (5.32) and (5.33) with m and m_a replaced by J_1 and J_2, respectively, $c = 0, k_a = 0$, and F_0 replaced by M_0. Equation (5.32) is given in nondimensional form as equation (5.37). Hence letting $\beta = \omega_a/\omega_p = 0$ in equation (5.37) yields that amplitude of vibration of the primary inertia J_1 [i.e., the amplitude of $\theta_1(t)$] is described by

$$\frac{Xk}{M_0} = \sqrt{\frac{4\zeta^2 + r^2}{4\zeta^2(r^2 + \mu r^2 - 1)^2 + (r^2 - 1)^2 r^2}} \qquad (5.39)$$

where $\zeta = c/(2J_2\omega_p)$, $r = \omega_{dr}/\omega_p$, and $\mu = J_2/J_1$. Figure 5.21 illustrates several plots of Xk/M_0 for various values of ζ for a fixed μ and r. Note again that the highest damping does not correspond to the largest-amplitude reduction.

The various absorber designs discussed above, excluding the undamped case, result in a number of possible "good" choices for the various design parameters. When faced with a number of good choices, it is natural to ask which is the best choice. Looking for the best possible choice among a number of acceptable or good choices can be made systematic by using methods of optimization introduced in the next section.

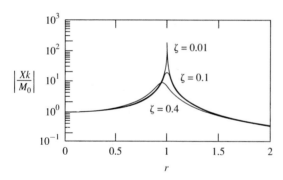

Figure 5.21 Amplitude curves for a system with a viscous absorber, a plot of equation (5.39), for the case $\mu = 0.25$ and for three different values of ζ.

5.5 OPTIMIZATION

In the design of vibration systems the best selection of system parameters is often sought. In the case of the undamped vibration absorber of Section 5.3 the best selection for values of mass and stiffness of the absorber system is obvious from examining the expression for the amplitude of vibration of the primary system. In this case the amplitude could be driven to zero by tuning the absorber mass and stiffness to the driving frequency. In the other cases, especially when damping is included, the choice of parameters to produce the best response is not obvious. In such cases optimization methods can often be used to help select the best performance. Optimization techniques often produce results that are not obvious. An example is in the case of the undamped primary system or the damped absorber system discussed in the preceding section. In this case Figures 5.16 to 5.18 indicate that the best selection of parameters does not correspond to the highest value of the damping in the system as intuition might dictate. These figures essentially represent an optimization by trial and error. In this section a more systematic approach to optimization is suggested by taking advantage of calculus.

Recall from elementary calculus that minimums and maximums of particular functions can be obtained by examining certain derivatives. Namely, if the first derivative vanishes and the second derivative of the function is positive, the function has obtained a minimum value. This section presents a few examples where optimization procedures are used to obtain the best possible vibration reduction for various isolator and absorber systems. A major task of optimization is first deciding what quantity should be minimized to best describe the problem under study. The next question of interest is to decide which variables to allow to vary during the optimization. Optimization methods have developed over the years which allow the parameters during the optimization to satisfy constraints, for example. This approach is often used in design for vibration suppression.

Recall from calculus that a function $f(x)$ experiences a maximum (or minimum) at value of $x = x_m$ given by the solution of

$$f'(x_m) = \frac{d}{dx}[f(x_m)] = 0 \tag{5.40}$$

If this value of x causes the second derivative, $f''(x_m)$, to be less than zero, the value of $f(x)$ at $x = x_m$ is the maximum value that $f(x)$ takes on in the region near $x = x_m$. Similarly, if $f''(x_m)$ is greater than zero, the value of $f(x_m)$ is the smallest or minimum value that $f(x)$ obtains in the interval near x_m. Note that if $f''(x) = 0$, at $x = x_m$, the value $f(x_m)$ is neither a minimum or maximum for $f(x)$. The points where $f'(x)$ vanish are called *critical points*.

These simple rules were used in Section 2.2, Example 2.2.2, for computing the value (r_{peak}) where the maximum value of normalized magnitude of the steady-state response of a harmonically driven single-degree-of-freedom system occurs. The second derivative test was not checked because several plots of the function clearly indicated that the curve contains a global maximum value rather than a minimum. In both absorber and isolator design, plots of the magnitude of the response can be used to avoid having to calculate the second derivative (second derivatives are often unpleasant to calculate).

If the function f to be minimized (or maximized) is a function of two variables [i.e., $f = f(x, y)$], the derivative tests above become slightly more complicated and involve examining the various partial derivatives of the function $f(x, y)$. In this case the critical points are determined from the equations

$$f_x(x, y) = \frac{\partial f(x, y)}{\partial x} = 0$$

$$f_y(x, y) = \frac{\partial f(x, y)}{\partial y} = 0$$

(5.41)

Whether or not these critical points (x, y) are a maximum of the value $f(x, y)$ or a minimum depend on the following:

1. If $f_{xx}(x, y) > 0$ and $f_{xx}(x, y)f_{yy}(x, y) > f_{xy}^2(x, y)$, then $f(x, y)$ has a relative minimum value at x, y.
2. If $f_{xx}(x, y) < 0$ and $f_{xx}(x, y)f_{yy}(x, y) > f_{xy}^2(x, y)$, then $f(x, y)$ has a relative maximum value at x, y.
3. If $f_{xy}^2(x, y) > f_{xx}(x, y)f_{yy}(x, y)$, then $f(x, y)$ is neither a maximum nor a minimum value, the point x, y is a *saddle* point.
4. If $f_{xy}^2(x, y) = f_{xx}(x, y)f_{yy}(x, y)$, the test fails and the point x, y could be any or none of the above.

Plots of $f(x, y)$ can also be used to determine whether or not a given critical point is a maximum, minimum, saddle point or neither. These rules can be used to help solve vibration design problems in some circumstances. As an example of using these optimization formulations for designing a vibration suppression system, recall the damped absorber system of Section 5.4. In this case the magnitude of the primary mass normalized with respect to the input force (moment) magnitude is given in equation (5.39) to be

$$\frac{Xk}{M_0} = \sqrt{\frac{4\zeta^2 + r^2}{4\zeta^2(r^2 + \mu r^2 - 1)^2 + (r^2 - 1)^2 r^2}} = f(r, \zeta)$$

(5.42)

which is considered to be a function of the mixed damping ratio ζ and the frequency ratio r for a fixed mass ratio μ.

In Section 5.4, values of $f(r)$ are plotted versus r for several values of ζ in an attempt to find the value of ζ that yields the smallest maximum value of $f(r, \zeta)$. This is illustrated in Figure 5.21. Figure 5.22 illustrates the magnitude as a function of both ζ and r. From the figure it can be concluded that the derivative $\partial f / \partial r = 0$ yields the maximum value of the magnitude for each fixed ζ.

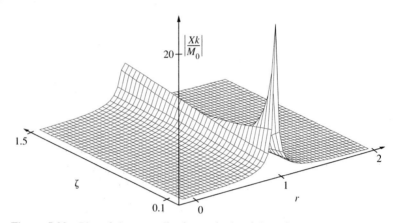

Figure 5.22 Plot of the normalized magnitude of the primary system versus both ζ and r [i.e., a two-dimensional plot of equation (5.42) for $\mu = 0.25$]. This illustrates that the most desirable response is obtained at the *saddle point.*

Looking along the ζ axis, the partial derivative $\partial f / \partial \zeta = 0$ yields the minimum value of $f(r, \zeta)$ for each fixed value of r. The best design, corresponding to the smallest of the largest amplitudes, is thus illustrated in Figure 5.22. This point corresponds to a saddle point and can be calculated by evaluating the appropriate first partial derivatives.

First consider $\partial (Xk/M_0)/\partial \zeta$. From equation (5.42), the function to be differentiated is of the form

$$f = \frac{A^{1/2}}{B^{1/2}} \tag{5.43}$$

where $A = 4\zeta^2 + r^2$ and $B = 4\zeta^2(r^2 + \mu r^2 - 1)^2 + (r^2 - 1)^2 r^2$. Differentiating and equating the resulting derivatives to zero yields

$$\frac{\partial f}{\partial \zeta} = \frac{1}{2} \frac{A^{-1/2} dA}{B^{1/2}} - \frac{1}{2} A^{1/2} \frac{dB}{B^{3/2}} = 0 \tag{5.44}$$

Solving this yields the form $[B \; dA - A \; dB]/2B^{3/2} = 0$ or

$$B \; dA = A \; dB \tag{5.45}$$

where A and B are as defined above and

$$dA = 8\zeta \quad \text{and} \quad dB = 8\zeta(r^2 + \mu r^2 - 1)^2 \tag{5.46}$$

Substitution of these values of A, dA, B, and dB into equation (5.45) yields

$$(1 - r^2)^2 = (1 - r^2 - \mu r^2)^2 \tag{5.47}$$

For $\mu \neq 0, r > 0$, this has the solution

$$r = \sqrt{\frac{2}{2 + \mu}} \tag{5.48}$$

Similarly, differentiating equation (5.42) with respect to r and substituting the value for r obtained above yields

$$\zeta_{op} = \frac{1}{\sqrt{2(\mu + 1)(\mu + 2)}} \tag{5.49}$$

Equation (5.49) reveals the value of ζ that yields the smallest amplitude at the point of largest amplitude (resonance) for the response of the primary mass. The maximum value of the displacement for the optimal damping is given by

$$\left(\frac{Xk}{M_0}\right)_{max} = 1 + \frac{2}{\mu} \tag{5.50}$$

which is obtained by substitution of (5.48) and (5.49) into equation (5.42). This last expression suggests that μ should be as large as possible. However, the practical consideration that the absorber mass should be smaller than the primary mass requires $\mu \leq 1$. The value $\mu = 0.25$ is fairly common.

The second derivative conditions for the function f to have a saddle point (condition 3 above) are too cumbersome to calculate. However, the plot of Figure 5.22 clearly illustrates that these conditions are satisfied. Furthermore, the plot indicates that f as a function of ζ is convex and f as a function of r is concave so that the saddle point condition is also the solution of minimizing the maximum value $f(r, \zeta)$, called the *min-max problem* in applied mathematics and optimization.

Example 5.5.1

A viscous damper–mass absorber is added to the shaft of an engine. The mass moment of inertia of the shaft system is 1.5 kg·m^2/rad and has a torsional stiffness of 6×10^3 N·m/rad. The nominal running speed of the engine is 2000 rpm. Calculate the values of the added damper and mass such that the primary system has a magnification (Xk/M_0) of less than 5 for all speeds and is as small as possible at the running speed.

Solution   Since $\omega_p = \sqrt{k/J}$, the natural frequency of the engine system is

$$\omega_p = \sqrt{\frac{6.0 \times 10^3 \text{N} \cdot \text{m/rad}}{1.5 \text{ kg} \cdot \text{m}^2/\text{rad}}} = 63.24 \text{ rad/s}$$

The running speed of the engine is 2000 rpm or 209.4 rad/s, which is assumed to be the driving frequency (actually, it is a function of the number of cylinders). Hence the frequency ratio is

$$r = \frac{\omega_{dr}}{\omega_p} = \frac{209.4}{63.24} = 3.31$$

so that the running speed is well away from the maximum amplification as illustrated in Figures 5.21 and 5.22 and the absorber is not needed to protect the shaft at its running speed. However, the engine spends some time getting to the running speed and often runs at lower speeds. The peak response occurs at

$$r_{peak} = \frac{\omega_{dr}}{\omega_p} = \sqrt{\frac{2}{2+\mu}}$$

as given by equation (5.48) and has a value of

$$\left(\frac{Xk}{M_0}\right)_{max} = 1 + \frac{2}{\mu}$$

as given by equation (5.50). The magnification is restricted to be 5, so that

$$1 + \frac{2}{\mu} \leq 5, \quad \text{or} \quad \mu \geq 0.5$$

Thus $\mu = 0.5$ is chosen for the design. Since the mass of the primary system is $J_1 = 1.5$ kg $\cdot$ m^2/rad and $\mu = J_2/J_1$, the mass of the absorber is

$$J_2 = \mu J_1 = \frac{1}{2}(1.5) \text{ kg} \cdot \text{m}^2/\text{rad} = 0.75 \text{ kg} \cdot \text{m}^2 \cdot \text{rad}$$

The damping value required for equation (5.50) to hold is given by equation (5.49) or

$$\zeta_{op} = \frac{1}{\sqrt{2(\mu+1)(\mu+2)}} = \frac{1}{\sqrt{2(1.5)(2.5)}} = 0.3651$$

Recall from Section 5.4 [just following equation (5.39)] that $\zeta = c/(2J_2\omega_p)$, so that the optimal damping constant becomes

$$c_{op} = 2\zeta_{op}J_2\omega_p = 2(0.3651)(0.75)(63.24) = 34.638 \text{ N} \cdot \text{m} \cdot \text{s/rad}$$

The two values of J_2 and c given here form an optimal solution to the problem of designing a viscous damper–mass absorber system so that the maximum deflection of the primary shaft is satisfied $|Xk/M_0| < 5$. This solution is optimal in terms of a choice of ζ which corresponds to the saddle point of Figure 5.22 and yields a minimum value of all maximum amplifications.

□

Optimization methods can also be useful in the design of certain types of vibration isolation systems. For example, consider the model of a machine mounted on a elastic damper and spring system as illustrated in Figure 5.23. The equations of motion of the system of Figure 5.23 are

$$m\ddot{x}_1 + c(\dot{x}_1 - \dot{x}_2) + k_1x_1 = F_0 \cos \omega_{dr}t$$
$$c(\dot{x}_1 - \dot{x}_2) = k_2x_2 \tag{5.51}$$

Because no mass term appears in the second equation, the system given by equation (5.51) is of third order. Equation (5.51) can be solved by assuming periodic motions of the form

$$x_1(t) = X_1e^{j\omega_{dr}t}, \quad \text{and} \quad x_2(t) = X_2e^{j\omega_{dr}t} \tag{5.52}$$

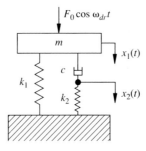

Figure 5.23 Model of a machine mounted on an elastic foundation through an elastic damper to provide vibration isolation.

and considering the exponential representation of the harmonic driving force. Substitution of equation (5.52) into (5.51) yields

$$(k_1 - m\omega_{dr}^2 + jc\omega_{dr})X_1 - jc\omega_{dr}X_2 = F_0$$
$$jc\omega_{dr}X_1 - (k_2 + jc\omega_{dr})X_2 = 0 \tag{5.53}$$

Solving for the amplitudes X_1 and X_2 yields

$$X_1 = \frac{F_0(k_2 + jc\omega_{dr})}{k_2(k_1 - m\omega_{dr}^2) + jc\omega_{dr}(k_1 + k_2 - m\omega_{dr}^2)} \tag{5.54}$$

and

$$X_2 = \frac{c\omega_{dr}F_0 j}{k_2(k_1 - m\omega_{dr}^2) + c\omega_{dr}(k_1 + k_2 - m\omega_{dr}^2)j} \tag{5.55}$$

These two amplitude expressions can be simplified further by substituting the nondimensional quantities $r = \omega_{dr}/\sqrt{k_1/m}, \gamma = k_2/k_1$, and $\zeta = c/(2\sqrt{k_1 m})$. The force transmitted to the base is the vector sum of the two forces $k_1 x_1$ and $k_2 x_2$. Using complex arithmetic and a vector sum (recall Section 2.5) the force transmitted can be written as

$$\text{T.R.} = \frac{F_T}{F_0} = \frac{\sqrt{1 + 4(1+\gamma)^2 \zeta^2 r^2}}{\sqrt{(1-r^2)^2 + 4\zeta^2 r^2(1+\gamma - r^2\gamma)^2}} \tag{5.56}$$

which describes the transmissibility ratio for the system of Figure 5.23.

The force transmissibility ratio can be optimized by viewing the ratio F_T/F_0 as a function of r and ζ. Figure 5.24 yields a plot of F_T/F_0 versus r for $\gamma = 0.333$ and for several values of ζ. This illustrates that the value of the damping ratio greatly affects the transmissibility at resonance. A three-dimensional plot of F_T/F_0 versus r and ζ is given in Figure 5.25, which illustrates that the saddle point value of ζ and r yields the best design for the minimum transmissibility of the maximum force transmitted.

The saddle point illustrated in Figure 5.25 can be found from the derivative of T.R. as given in equation (5.56). These partial derivatives are

$$\frac{\partial (\text{T.R.})}{\partial \zeta} = 0 \quad \text{yields} \quad r_{\max} = \frac{\sqrt{2(1+\gamma)}}{\sqrt{1+2\gamma}} \tag{5.57}$$

and

$$\frac{\partial (\text{T.R.})}{\partial r} = 0 \quad \text{yields} \quad \zeta_{op} = \frac{\sqrt{2(1+2\gamma)/\gamma}}{4(1+\gamma)} \tag{5.58}$$

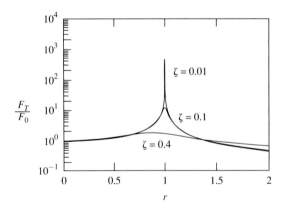

Figure 5.24 Plot of equation (5.56) illustrating the effect of damping on the magnification of force transmitted to ground.

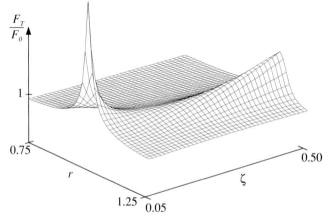

Figure 5.25 Plot of equation (5.56) illustrating F_T/F_0 versus ζ versus r. The plot shows the point where damping minimizes the maximum transmissibility.

These values of r correspond to an optimal design of this type of isolation device. At the saddle point, the value of T.R. becomes

$$(\text{T.R.})_{\max} = 1 + 2\gamma \tag{5.59}$$

which results from substitution of equations (5.57) and (5.58) into equation (5.56). This illustrates that as long as $\gamma < 1$, T.R. < 3 and the isolation system will not cause much difficulty at resonance.

Example 5.5.2

An isolation system is to be designed for a machine modeled by the system of Figure 5.23 (i.e., an elasticity coupled viscous damper). The mass of the machine is $m = 100$ kg and the stiffness $k_1 = 400$ N/m. The driving frequency is 10 rad/s at nominal operating conditions. Design this system (i.e., choose k_2 and c) such that the maximum transmissibility ratio at any speed is 2 (i.e., design the system for "start up" or "run through"). What is the T.R. at the normal operating condition of a driving frequency of 10 rad/s?

Solution   For $m = 100$ kg and $k_1 = 400$ N/m, $\omega = \sqrt{400/100} = 2$ rad/s, so that the normal operating condition is well away from resonance (i.e., $r = \omega_{dr}/\omega = 10/2 = 5$ at running conditions). Equation (5.59) yields that the maximum value for T.R. is

$$(\text{T.R.})_{max} = 1 + 2\gamma \le 2$$

so that $\gamma = 0.5$ and $k_2 = (0.5)(k_1) = (0.5)(400 \text{ N/m}) = 200$ N/m. With $\gamma = 0.5$, the optimal choice of damping ratio is given by equation (5.58) to be

$$\zeta_{op} = \frac{\sqrt{2(1 + 2\gamma)/\gamma}}{4(1 + \gamma)} = 0.4714$$

Hence the optimal choice of damping coefficient is

$$c_{op} = 2\zeta_{op}\omega m = 2(0.4714)(2)(100) = 188.56 \text{ kg/s}$$

The T.R. value at nominal operating frequency of $\omega_{dr} = 10$ rad/s is given by equation (5.56) to be ($r = 10/2 = 5$)

$$\text{T.R.} = \frac{\sqrt{1 + 4(1 + 0.5)^2(0.4714)^2(5)^2}}{\sqrt{(1 - 5^2)^2 + 4(0.4714)^2(5)^2[1 + 0.5 - 5^2(0.5)]^2}} = 0.12$$

Hence the design $k_2 = 200$ N/m and $c = 188.56$ kg/s will protect the surroundings by a T.R. of 0.12 (i.e., only 12% of the applied force is transmitted to ground) and limits the force transmitted near resonance to a factor of 2.

$\square$

5.6 VISCOELASTIC DAMPING TREATMENTS

A common and very effective way to reduce transient and steady-state vibration is to increase the amount of damping in the system so there is greater energy dissipation. This is especially useful in aerospace structures applications, where the added mass of an absorber system may not be practical. While a rigorous derivation of the equations of vibration for structures with damping treatments is beyond the scope of this book, formulas are presented that provide a sample of design calculations for using damping treatments.

A damping treatment consists of adding a layer of viscoelastic material, such as rubber, to an existing structure. The combined system often has a higher damping level and thus reduces unwanted vibration. This procedure is described by using the *complex stiffness* notation. The concept of complex stiffness results from considering the harmonic response of a damped system of the form

$$m\ddot{x} + c\dot{x} + kx = F_0 e^{j\omega_{dr}t} \tag{5.60}$$

Recall from Section 2.5 that the solution to equation (5.60) can be calculated by assuming the form of the solution to be $x(t) = Xe^{j\omega_{dr}t}$, where X is a constant and $j = \sqrt{-1}$. Substitution of the assumed form into equation (5.60) and dividing by the nonzero function $e^{j\omega_{dr}t}$ yields

$$[-m\omega_{dr}^2 + (k + j\omega_{dr}c)]X = F_0 \tag{5.61}$$

This can be written as

$$\left[-m\omega_{dr}^2 + k\left(1 + \frac{\omega_{dr}c}{k}j\right)\right]X = F_0 \tag{5.62}$$

or

$$[-m\omega_{dr}^2 + k^*]X = F_0 \tag{5.63}$$

where $k^* = k(1 + \bar{\eta}j)$. Here $\bar{\eta} = \omega_{dr}c/k$ is called the *loss factor* and k^* is called the *complex stiffness*. This illustrates that in steady state, the viscous damping in a system can be represented as an "undamped" system with a complex-valued stiffness. The imaginary part of the stiffness, $\bar{\eta}$, corresponds to the energy dissipation in the system. Since the loss factor has the form

$$\bar{\eta} = \frac{c}{k}\omega_{dr} \tag{5.64}$$

the loss factor depends on the driving frequency and hence is said to be frequency dependent. Hence the value of the energy dissipation term depends on the value of the driving frequency of the external force exciting the structure.

The concept of complex stiffness developed above is called the *Kelvin–Voigt model* of a material. This corresponds to the standard spring–dashpot configuration as sketched in Figure 5.26 and used extensively in the first four chapters. The difference between the Kelvin–Voigt model used here and the viscous damping model of the previous chapters is that the Kelvin–Voigt model is valid only in steady-state harmonic motion. The complex stiffness and the corresponding frequency-dependent loss factor, $\bar{\eta} = \omega_{dr}c/k$, model the energy dissipation at steady state during harmonic excitation of frequency ω_{dr} only. The viscous dashpot representation introduced in Section 1.3 models energy dissipation in free decay as well as in other transient and forced response excitation. However, the Kelvin–Voigt representation is a more accurate, though limited model of the internal damping in materials.

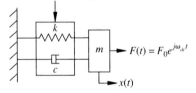

Material with viscoelastic properties

Figure 5.26 Kelvin–Voigt damping model gives rise to the complex stiffness concept of representing damping in steady-state vibration.

The complex stiffness formulation can also be derived from the stress–strain relationship for a linear viscoelastic material. Such materials are called *viscoelastic* because they exhibit both *elastic* behavior and *viscous* behavior, as captured in the Kelvin–Voigt model described in Figure 5.26. Other viscoelastic models exist in addition to this one, but such models are beyond the scope of this book [see Snowden (1968)]. An alternative viscoelastic model is given in Figure 5.23, for instance.

The stress–strain relationship for viscoelastic material can be summarized by extending the modulus of a material, denoted by E, to a complex modulus, denoted E^*, by the relation

$$E^* = E(1 + \eta j) \tag{5.65}$$

where $j = \sqrt{-1}$ as before and η is the loss factor of the viscoelastic material. The complex modulus of a material, as defined in equation (5.65), can be measured and in general is both frequency and temperature dependent over a broad range of values. Some sample values for frequency dependence are given in Figure 5.27 for fixed temperatures.

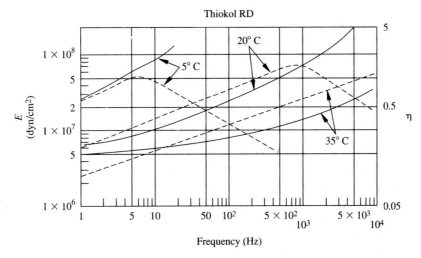

Figure 5.27   Sample plot of elastic modulus (solid lines) and loss factor (dashed lined) versus frequency for several fixed temperatures.

Materials that exhibit viscoelastic behavior are rubber and rubberlike substances (e.g., butyl rubber, neoprene, polyurethane) as well as plexiglass, vinyl, and nylon. A common use of these viscoelastic materials in design is as an additive damping treatment to increase the combined structure's damping or as an isolator. Layers of viscoelastic material are often added to structures composed of lightly damped material such as aluminum or steel to form a new structure that has sufficient stiffness for static loading and sufficient damping for controlling vibration. Table 5.2 lists some values of E and η for a viscoelastic material at two different temperatures and several frequencies.

The loss factor η defined in terms of the complex modulus as given in equation (5.65) is related to the loss factor $\bar{\eta}$ defined by examining the notion of complex stiffness as defined in equation (5.64) in the same way that the stiffness and modulus of a material are related in Table 1.1 and Section 1.5. For example, if the specimen of

TABLE 5.2 SOME COMPLEX MODULUS DATA (I.E., E AND η) FOR PARACRIL-BJ WITH 50 PHRC[a]

E (psi)	η	T (°F)	ω (Hz)	ω (rad/s)	E (N/m²)
3×10^3	0.21	75	10	62.8	2.068×10^7
4×10^3	0.28	75	100	628.3	2.758×10^7
7×10^3	0.55	75	1000	6283.2	4.826×10^7
4×10^3	0.25	50	10	62.8	2.758×10^7
6×10^3	0.5	50	100	628.3	4.137×10^7
13×10^3	1	50	1000	6283.2	8.963×10^7

[a] Nitrite rubber elastomeric material made by U.S. Rubber Company.
Source: Nashif, Jones, and Henderson, 1985, Data Sheet 27.

interest is a cantilevered beam, the stiffness associated with the deflection of the tip in the transverse direction is related to the elastic modulus by

$$k = \frac{3EI}{l^3} \tag{5.66}$$

where I is the area moment of inertia and l is the length of the beam. Hence if the beam is made of viscoelastic material,

$$k^* = \frac{3E^*I}{l^3} = \frac{3I}{l^3}E(1 + \eta\,j) = k(1 + \eta\,j)$$

so that $\eta = \bar{\eta}$ and the two notions of loss factor are identical.

The notion of loss factor η is related to the definition of a damping ratio ζ only at resonance (i.e., $\omega_{dr} = \omega = \sqrt{k/m}$). When the driving frequency is the same as the system's natural frequency, $\eta = 2\zeta$. This simple relationship is often used to describe the free decay of a viscoelastic material (an approximation). The design of structures for reduced vibration magnitude often consists of adding a viscoelastic damping treatment to an existing structure. Many structures are made of metals and alloys that have relatively little internal damping. A viscoelastic damping material (such as rubber) is often added as a layer to the outside surface of a structure (called *free-layer damping* treatment or *unconstrained-layer damping*). A much more effective approach is to cover the free layer with another layer of metal to form a *constrained-layer damping* treatment. In the constrained-layer damping treatment, the damping layer is covered with a (usually thin) layer of metal (stiff) to produce shear deformation in the viscoelastic layer. The constrained-layer approach produces higher loss factors and generally costs more. These damping treatments are manufactured as sheets, tapes, and adhesives for ease of application.

A free-layer damping treatment for a pinned–pinned beam (see Table 6.4) in transverse or bending vibration is illustrated in Figure 5.28. Material 1, the bottom layer, is usually a metal providing the appropriate stiffness. The second layer, denoted as having modulus E_2 and thickness H_2, is the damping treatment. Using the notation of

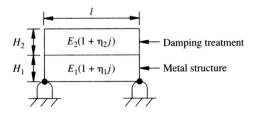

Figure 5.28 Simple supported beam with an unconstrained damping treatment illustrating the geometry and physical parameters.

Figure 5.28, the combined stiffness EI is related to the original stiffness $E_1 I_1$ by

$$\frac{EI}{E_1 I_1} = 1 + e_2 h_2^3 + 3(1 + h_2)^2 \frac{e_2 h_2}{1 + e_2 h_2} \tag{5.67}$$

where $e_2 = E_2/E_1$ and $h_2 = H_2/H_1$ are dimensionless. Note that since all the quantities on the left side of equation (5.67) are positive, the damping treatment increases the stiffness of the system a small amount ($h_2 < 1$). In addition, the combined system's loss factor, η, is given by [assuming that $(e_2 h_2)^2 << e_2 h_2$]

$$\eta = \frac{e_2 h_2 (3 + 6h_2 + 4h_2^2 + 2e_2 h_2^3 + e_2^2 h_2^4)}{(1 + e_2 h_2)(1 + 4e_2 h_2 + 6e_2 h_2^2 + 4e_2 h_2^3 + e_2^2 h_2^4)} \eta_2 \tag{5.68}$$

Equation (5.68) yields a formula that can be used in the design of add-on damping treatments as illustrated in the following example.

Example 5.6.1

An electric motor that drives a cooling fan is mounted on an aluminum shelf (1 cm thick) in a cabinet holding electronic parts (perhaps a mainframe computer) as illustrated in Figure 5.29. The vibration of the motor causes the mounting platform, and hence the surrounding cabinet, to shake. The motor rotates at an effective frequency of 100 Hz. The temperature in the cabinet remains at 75°F. A damping treatment is added to reduce the vibration of the shelf.

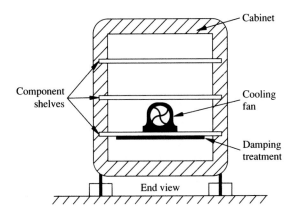

Figure 5.29 Electronic cabinet with cooling fan illustrating the use of a damping treatment.

Solution   The shelf is modeled as a simply supported beam so that equation (5.68) can be used to design the damping treatment. If nitrile rubber is used as the damping treatment, calculate the loss factor of the combined system at 75°F if $H_2 = 1$ cm. Referring to

Table 5.2, the modulus of the rubber at 100 Hz and 75°F is

$$E_2 = 2.758 \times 10^7 \text{Pa}$$

The modulus of aluminum is $E_1 = 7.1 \times 10^{10}$ Pa, so that

$$e_2 = \frac{2.758 \times 10^7}{7.1 \times 10^{10}} = 0.00039$$

The thickness of both the shelf and the damping treatment are taken to be the same, so that $h_2 = 1$. From equation (5.68) the combined loss factor becomes

$$\eta = \frac{(0.00039)[3 + 6 + 4 + 2(0.00039) + (0.00039)^2]}{(1.00039)[1 + 4(0.00039) + 6(0.00039) + 4(0.0039) + (0.00039)^2]}\eta_2$$

$$= 0.0054\eta_2$$

From Table 5.2, $\eta_2 = 0.28$ at 100 Hz and 75°C, so that

$$\eta = 0.00151$$

which is about 50% higher than the loss factor given by pure aluminum.

□

The formula given in equation (5.68) is a bit cumbersome for design work. Often it is approximated by

$$\eta = 14(e_2 h_2^2)\eta_2 \tag{5.69}$$

which is reasonable for many situations. The values of e_2 and η_2 are fixed by the choice of materials and the operating temperature. Once these parameters are fixed, the parameter $h_2 = H_2/H_1$ is the only remaining design choice. Since H_1 is usually determined by stiffness considerations, the remaining design choice is the thickness of the damping layer, H_2.

Example 5.6.2

An aluminum shelf is to be given a damping treatment to raise the system loss factor to $\eta = 0.03$. A rubber material is used with modulus at room temperature of 1% of that of aluminum (i.e., $e_2 = 0.01$). What should the thickness of the damping material be if its loss factor is $\eta_2 = 0.261$ and the aluminum shelf is 1 cm thick?

Solution   From the approximation given by equation (5.69),

$$\eta = 14\eta_2 e_2 h_2^2$$

Using the values given, this becomes

$$0.03 = 14(0.261)(0.01)\frac{H_2^2}{(1 \text{ cm})^2}$$

Solving this for H_2^2 yields

$$H_2^2 = \frac{0.03}{14(0.01)(0.261)}(\text{cm})^2 = 0.82 \text{ cm}^2$$

so that $H_2 = 0.91$ cm will provide the desired loss factor.

□

5.7 CRITICAL SPEEDS OF ROTATING DISKS

Of primary concern in the design of rotating machinery is the vibration phenomenon of *critical speeds*. This phenomenon occurs when a rotating shaft with a disk, such as a jet engine turbine blade rotating about its shaft mounted between two bearings, rotates at a speed that is equivalent to a natural bending frequency of the shaft–disk system. This defines a resonance condition that causes large deflection of the shaft, which in turn causes the system to violently fail (i.e., the engine blows apart). The nature of the resonance and the factors that control the resonance values need to be known and calculated by designers so that they can ensure that a given design is safe for production. The analytical formulation of the critical speed problem also provides some insight into how to avoid such resonance, or critical speeds.

If the rotating mass modeled by the disk is not quite homogeneous or symmetric due to some imperfection, its geometric center and center of gravity will be some distances apart (say, a). This is illustrated in Figure 5.30, which presents a simplified model of a large electric motor's shaft and rotor system (or a bladed turbine engine, etc.). The shaft is constrained from moving in the radial direction by two bearings. As the shaft rotates about its long axis with angular velocity ω, the offset center of gravity pulls the shaft away from the centerline, causing it to bow as it rotates. This is called *whirling*.

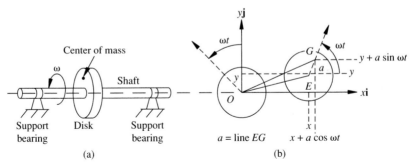

Figure 5.30 Schematic of a model of a disk rotating on a shaft and the corresponding geometry of the center of mass (G) of the disk relative to the neutral axis of shaft (O) and the center of the rotating shaft (E): (a) side view; (b) end view. This diagram is useful in modeling the whirling of rotating machines (engine, turbine compressors, etc.) which are not perfectly balanced (i.e., $a \neq 0$).

The forces acting on the center of mass are the inertial force, any damping force (internal or external), and the elastic force of the shaft. In vector form the force balance yields

$$m\ddot{\mathbf{r}} = -kx\mathbf{i} - ky\mathbf{j} - c\dot{x}\mathbf{i} - c\dot{y}\mathbf{j} \qquad (5.70)$$

where $\mathbf{i}$ and $\mathbf{j}$ are unit vectors, $\mathbf{r}$ the position vector defined by the line OG, m the mass of the disk, c the damping coefficient of the shaft system, and k the stiffness coefficient provided by the shaft system. From examining the end view of Figure 5.30, the vector **r** can also be written in terms of the unit vectors **i** and **j** as

$$\mathbf{r} = (x + a \cos \omega t)\mathbf{i} + (y + a \sin \omega t)\mathbf{j} \tag{5.71}$$

Taking two derivatives yields that the acceleration vector of the center of mass is

$$\ddot{\mathbf{r}} = (\ddot{x} - a\omega^2 \cos \omega t)\mathbf{i} + (\ddot{y} - a\omega^2 \sin \omega t)\mathbf{j} \tag{5.72}$$

Substituting equation (5.72) into equation (5.70) yields

$$(m\ddot{x} - ma\omega^2 \cos \omega t + c\dot{x} + kx)\mathbf{i} + (m\ddot{y} - ma\omega^2 \sin \omega t + c\dot{y} + ky)\mathbf{j} = \mathbf{0} \tag{5.73}$$

Since this is a vector equation, it is equivalent to the two scalar equations

$$m\ddot{x} + c\dot{x} + kx = ma\omega^2 \cos \omega t \tag{5.74}$$

$$m\ddot{y} + c\dot{y} + ky = ma\omega^2 \sin \omega t \tag{5.75}$$

These two equations are exactly the form of equation (2.49) for the response of a spring–mass system to a rotating unbalance discussed in Section 2.4. In this case the x and y motion corresponds to the bending vibration of the shaft instead of the translational motion of a machine in the vertical direction discussed in Section 2.4. Referring to Window 5.5, equation (5.75) has steady-state response magnitude given by equation (2.51); that is, equation (5.75) has the steady-state solution (since $m = m_0$ and $e = a$ in this case)

$$y(t) = \frac{ar^2}{\sqrt{(1 - r^2)^2 + (2\zeta r)^2}} \sin\left(\omega t - \tan^{-1}\frac{2\zeta r}{1 - r^2}\right) \tag{5.76}$$

Window 5.5
Solution of the Rotating Unbalance Equation from Section 2.4

The steady-state solution to

$$m\ddot{x} + c\dot{x} + kx = m_0 e \omega_{dr}^2 \sin \omega_{dr} t$$

where ω_{dr} is the driving frequency of the unbalanced mass, m_0 the mass of the unbalance, and e the distance from m_0 to the center of rotation, is $X \sin(\omega_{dr} t - \phi)$. Here

$$X = \frac{m_0 e}{m} \frac{r^2}{\sqrt{(1 - r^2)^2 + (2\zeta r)^2}} \tag{2.51}$$

and

$$\phi = \tan^{-1}\frac{2\zeta r}{1 - r^2} \tag{2.52}$$

where $r = \omega_{dr}/\sqrt{k/m}$ and $\zeta = c/(2m\omega)$.

Similarly, equation (5.74) has steady-state solution of the form

$$x(t) = \frac{ar^2}{\sqrt{(1-r^2)^2 + (2\zeta r)^2}} \cos\left(\omega t - \tan^{-1}\frac{2\zeta r}{1-r^2}\right) \tag{5.77}$$

since the solution given by equations (2.51) and (2.52) is 90° out of the phase and the phase angle ϕ does not depend on the phase of the exciting force. The angle ϕ given by equation (2.52) becomes the angle between the lines OE and EG. From Figure 5.30 the angle θ made between the x axis and the line $0E$ is

$$\tan\theta = \frac{y}{x} = \frac{\sin(\omega t - \phi)}{\cos(\omega t - \phi)} = \tan(\omega t - \phi) \tag{5.78}$$

or

$$\theta = \omega t - \phi \tag{5.79}$$

Differentiating equation (5.79) with respect to t yields $\dot\theta = \omega$.

The velocity $\dot\theta$ is the velocity of whirling. Whirling is the angular motion of the deflected shaft rotating about the neutral axis of the shaft. The calculation leading to equation (5.79) and its derivative shows that the whirling velocity is the same as the speed with which the disk rotates about the shaft (i.e., $\dot\theta = \omega$). This is called *synchronous whirl*.

The amplitude of motion of the center of the shaft about its neutral axis is the line $\mathbf{r} = OE$ in the end view of Figure 5.30. Note that vector $OE = \mathbf{r} = x\mathbf{i} + y\mathbf{j}$. The magnitude of this vector is just

$$|\mathbf{r}(t)| = \sqrt{x^2 + y^2} = X\sqrt{\sin^2(\omega t - \phi) + \cos^2(\omega t - \phi)} = X \tag{5.80}$$

where X is the magnitude of $x(t)$. Note that $X = Y$, where Y is the magnitude of $y(t)$ as given in equation (5.76). This calculation indicates that the distance between the shaft and its neutral axis is constant and has magnitude

$$X = \frac{ar^2}{\sqrt{(1-r^2)^2 + (2\zeta r)^2}} \tag{5.81}$$

This, of course, is exactly the same form as the magnitude plot of equation (2.51) given in Figure 2.12 for a spring–mass–damper system driven by a rotating out-of-balance mass. This plot is repeated for the rotational amplitude case of interest in Figure 5.31. Note that a resonance phenomenon occurs near $r = 1$, as expected. For lightly damped shafts this corresponds to unacceptably high amplitudes of rotation. The special case of $r = 1$ (i.e., $\omega_r = \sqrt{k/m}$) is called the rotor system's *critical speed*. If a rotor system runs at its critical speed, the large deflection will cause a large force to be transmitted to the bearings and eventually lead to failure. From the design point of view, the running speed, mass, and stiffness are examined for a given rotor and redesigned until $r > 3$, so that the deflections are limited to the size of the distance to the center of the mass of the disk. However, when the rotor system is started up, it must pass through the region near $r = 1$. If this startup procedure occurs too slowly, the resonance phenomena could damage the rotor bearings. Hence some damping in the system is desirable to

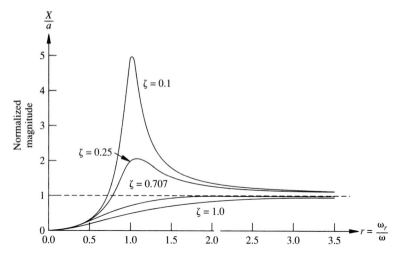

Figure 5.31 Plot of the ratio of radius of deflection (OE) to the distance to the center of mass of the disk (a) versus the frequency ratio for four different values of the damping ratio for the disk and shaft system of Figure 5.30.

avoid excessive amplitude at resonance. Note from Figure 5.31 that as ζ increases, X/a at resonance becomes substantially smaller.

Example 5.7.1

Consider a 55-kg compressor rotor with a shaft stiffness of 1.4×10^7 N/m, with an operation speed of 6000 rpm and a measured internal damping providing a damping ratio of $\zeta = 0.05$. The rotor is assumed to have a worst-case eccentricity of 1000 μm ($a = 0.001$ m). Calculate (a) the rotor's critical speed, (b) the radial amplitude at operating speed, and (c) the whirl amplitude at the system's critical speed.

Solution

(a) The critical speed of the rotor is just the rotor's natural frequency, so that

$$\omega_c = \sqrt{k/m} = \sqrt{\frac{1.4 \times 10^7 \text{ N/m}}{55 \text{ kg}}} = 504.5 \text{ rad/s}$$

which corresponds to a rotor speed of

$$504.5\frac{\text{rad}}{\text{s}} \times \frac{60 \text{ s}}{\text{m}} \times \frac{\text{cycle}}{2\pi \text{ rad}} = 4817.6 \text{ rpm}$$

(b) The value of r at running speed is just

$$r = \frac{\omega}{\sqrt{k/m}} = \frac{(2\pi/60)\omega}{(2\pi/60)\sqrt{k/m}} = \frac{6000}{4817.6} = 1.2454$$

or about 1.25. The value of the radial amplitude of whirl at the operating speed is then given by equation (5.81) with this value of r:

$$X = |\mathbf{r}(t)| = \frac{ar^2}{\sqrt{(1 - r^2)^2 + (2\zeta r)^2}} = \frac{(0.001)(1.25)^2}{\sqrt{[1 - (1.25)^2]^2 + [2(0.05)(1.25)]^2}}$$

$$= 0.0027116 \text{ m}$$

or about 2.7 mm. Here $r = 1.25$, $\zeta = 0.05$, and $a = 0.001$. Note that if $r = 1.2454$ is used, $X = 0.0027455$ m results. This gives some feeling for the sensitivity of the value of X to knowing exact values of r. Minor speed variation of, say, 10% in the running speed would cause r to vary between 1.12 and 1.37.

(c) At critical speed, $r = 1$ and X becomes

$$X = \frac{a}{2\zeta} = \frac{0.001}{2(0.05)} = 0.01 \text{ m}$$

or 1 cm, an order of magnitude larger than the whirl amplitude at running speed.

□

Example 5.7.2

In designing a rotor system, there are many factors besides the deflection calculation indicated above that determine the damping, stiffness, mass and operating speed of the rotor system. Hence the designer concerned about dynamic deflections and critical speeds is often only allowed to change the design a little. Otherwise, an entire redesign must be performed which may become very costly. With this in mind, again consider the rotor of Example 5.7.1. The clearance specification for the rotor inside the compressor housing limits the whirl amplitude at resonance to be 2 mm. Since the whirl amplitude at operating speed is greater than the allowable clearance, what percent of change in mass is required to redesign this system? What percent change in stiffness would result in the same design? Discuss the feasibility of such a change.

Solution   The required mass for a 2-mm deflection can be calculated from equation (5.81) by first determining a value of r corresponding to 2 mm. This yields

$$X = 0.002 \text{ m} = \frac{(0.001)r^2}{\sqrt{(1 - r)^2 + [2(0.05)r^2]^2}}$$

or r must satisfy $r^4 - 2.653r^2 + 1.3332 = 0$. This is a quadratic equation in r^2, which has solutions $r^2 = 0.6737, 1.979$ or $r = 0.8207, 1.406$, since the values of frequency ratio must be positive and real. Examination of the plot in Figure 5.31 of the magnitude yields that the value of r of interest is $r = 1.406$. At running speed

$$r = \frac{(6000 \text{ rpm})\frac{\text{min}}{60 \text{ s}} \cdot \frac{2\pi \text{ rad}}{\text{rev}}}{\sqrt{k/m} \text{ rad/s}} = \frac{628.12}{\sqrt{1.4 \times 10^7/m}} = 1.406$$

Solving for the mass m yields

$$m = 70.15 \text{ kg}$$

Since the original design value of the mass of the disk is 55 kg, the mass must be increased by 27.5% to produce a design that has its running speed deflection limited to 2 mm.

If the compressor is to be used in an application fixed to ground (such as a building), then adding 15 kg of mass to the disk may be a perfectly reasonable solution, provided that the bearings are capable of the increased force. However, if the compressor is to be used in a vehicle where weight is a consideration such as an airplane, a 27% increase in mass may not be an acceptable design. In this case equation (5.81) can be used to examine a possible redesign by making a change in stiffness. Equation (5.81) with the appropriate parameter values yields

$$r = 1.406 = \frac{628.12 \text{ rad/s}}{\sqrt{k/55}}$$

Solving for k yields

$$k = 1.0977 \times 10^7 \text{ N/m}$$

This amounts to about a 27% change in the stiffness. Unfortunately, the stiffness of the shaft cannot be changed very easily. It is determined by geometric and material properties. The material is often determined by temperature and cost considerations as well as toughness. It can be difficult to change the stiffness by 27%.

$\square$

Note from Example 5.7.2 that the amplitude of whirling is sensitive to changes in mass and changes in stiffness. Also note from Figure 5.31 that the damping value is of little concern when choosing the design for whirl amplitude if chosen far enough from resonance (i.e., $r > 2$). Rather damping is chosen to limit the amplitude near resonance, which should occur only during startup and run down (i.e., $X = a/2\zeta$ at resonance).

5.8 ACTIVE VIBRATION SUPPRESSION

The goal of creating a design that limits vibration amplitudes and durations in the presence of both shock and vibration disturbances faces limitations in the form of constraints on the choice of mass, damping, and stiffness (static deflection) values. For instance, in the design of an isolation system, it often occurs that the desired design calls for a value of stiffness that results in a static deflection that is too large for the intended application. Sometimes, a given isolation design might be required to operate over a load range that is impossible to meet with a single choice of mass and stiffness. In addition, once the materials are fixed for a given system, it is difficult to change the mass and stiffness of the system more than of a few percent. Basically, the choice of the physical parameters m, c, and k determines the response of the system. The choice of these parameters to obtain a desired response is the design problem. This design procedure can be thought of as *passive control* (e.g., adding mass to a machine base to lower its frequency). If the constraints on m, c, and k are such that the desired response cannot be obtained by changing m, c, and k, active control may provide an attractive alternative.

Active control uses an external adjustable (or active) device, called an *actuator* (e.g., a hydraulic piston, a piezoelectric device, an electric motor) to provide a force to the device, structure, or machine whose vibration properties are to be changed. The force applied to the structure by the actuator is dependent on a measurement of the response of the system. This is called feedback control and is illustrated in Figure 5.32. If the goal of the active control system is to remove unwanted vibration, the control system is called *active vibration suppression*, which consists of measuring the output or response of the structure or machine to determine the force to apply to the mass to obtain the desired response. The device used to apply the force (i.e., the actuator), together with the sensor used to measure the response of the mass and the electronic circuit required to read the sensor's output and apply the appropriate signal to the actuator is called the *control system*. The mathematical rule used to apply the force from the sensor measurement is called the *control law*. A multitude of control systems hardware and a variety of control laws are available for use in vibration suppression.

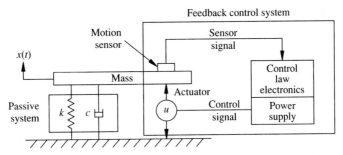

Figure 5.32   Feedback control system, illustrating the required sensor and actuator setup for active vibration suppression by measuring the response and using the response to calculate the force applied to the mass by the actuator.

A simple method of control is to use the control law referred to as state-variable feedback. In the case of a single-degree-of-freedom system, such as an isolation design, state feedback consists of using velocity and position measurements of the mass as determined by an acceleration measurement (appropriately integrated). The controller is then designed to provide a signal to the actuator proportional to measured velocity and displacement (also called PD control or position and derivative control). The actuator is then commanded to provide a force to the mass of the form

$$u = -g_1 x - g_2 \dot{x} \qquad (5.82)$$

where g_1 and g_2 are called *gains*. The gains g_1 and g_2 are determined electronically by sensor and actuator properties and by the designer's choice of control law.

The equation of motion for a single-degree-of-freedom oscillator with PD control given by equation (5.82) becomes

$$m\ddot{x} + (c + g_2)\dot{x} + (k + g_1)x = 0 \qquad (5.83)$$

The closed-loop equation (5.83) has two more parameters (i.e., g_1 and g_2) than the passive system, which can be adjusted so that the response has the desired form.

Example 5.8.1

Suppose that it is desired to build a vibration isolator with a damping ratio of 0.9 and a frequency of 10 rad/s for a system of 10 kg mass. The dashpot available produces damping coefficients between $0 \leq c \leq 100$ N·s/m. Calculate a passive system to provide the desired properties. If this does not work, calculate an active control system to provide the desired damping and stiffness properties.

Solution   Since $\omega = \sqrt{k/m}$, $\omega = 10$ rad/s, and $m = 10$ kg, the stiffness is given by

$$k = \omega^2 m = \frac{100(10\ \text{kg})}{\text{s}^2} = 1000\ \text{N/m}$$

The damping ratio is given by $\zeta = c/2\sqrt{km}$, so that

$$c = (0.9)2\sqrt{(1000)(10)} = 180\ \text{N} \cdot \text{s/m}$$

This value is larger than can be provided (i.e., $0 < c < 100$ N·s/m in this case). An active feedback control system can be introduced of the form $u = g_2\dot{x}$. In this case equation (5.83) with harmonic excitation becomes

$$m\ddot{x} + (c + g_2)\dot{x} + kx = F_0 \sin \omega_{dr} t \tag{5.84}$$

where $F_0 \sin \omega_{dr} t$ is the excitation force and $g_2\dot{x}$ represents the control force. The damping ratio ζ now must satisfy [from comparing the $\dot{x}$ coefficient with that of equation (1.48)]

$$2\zeta\omega = \frac{c + g_2}{m}$$

so that

$$c + g_2 = 2\zeta m\omega = 2(0.9)(10)(10) = 180\ \text{N} \cdot \text{s/m}$$

Since c can be chosen as a maximum of 100 N·s/m, an electronic gain of $g_2 = 180 - 100 = 80$ N·s^2/m will provide the desired damping ratio using feedback control. $\qquad \square$

Active control systems provide increased versatility and better performance in the design of vibration suppression systems. However, they do so with substantial increase in cost and potential decrease in reliability. Even in the face of increased cost and complexity, active control methods for vibration suppression are often the only alternatives. For example, active control can be used to make an unstable system stable, as illustrated in Example 3.8.2. Such devices have recently been used to improve the performance of automobile suspension systems and to protect buildings against mild earthquake excitations. An introduction to control for vibration suppression can be found in (Inman, 1989).

5.9 PRACTICAL ISOLATION DESIGN

Section 5.2 developed the use of displacement and force transmissibility for the purpose of both shock and vibration isolation. These transmissibility formulations are integrated with the design process of selecting a commercial vibration isolator for a given application in this section. In particular, terminology and design selection criteria are presented in a manner compatible to that used by sales engineers and designers.

In the selection of vibration isolation, shock isolation, or random vibration isolation components, several practical considerations need to be considered. The first consideration of importance is the size, weight, shape, and center of gravity of the equipment requiring isolation. As mentioned in Section 5.2, the static deflection and amplitude of vibration provide constraints in solving isolation problems. The center of gravity of the device is important, as its relative position determines whether or not the equipment can be modeled as a simple single-degree-of-freedom system. The next consideration of importance is the type of disturbance that characterizes the isolation problem. The disturbance could be a shock, a harmonic disturbance (called vibration), a random signal, or some combination. As pointed out in Section 5.2, a good vibration isolation design is not always a good shock isolation design. Hence it is very important to be able to characterize the disturbance in solving an isolation design problem.

In addition to knowing the input to the isolation system, it is important to know and characterize an allowable response. The equipment, device, or structure being protected by the isolation application must be studied, and a set of allowable response characterizations, such as amplitudes, must be determined in order to specify a successful design. Otherwise, the design might allow parts to bang together and fail.

An often neglected consideration is that of the ambient environment. Most commercial isolators are made of *elastomers*, a term used to describe viscoelastic material made of rubber, both synthetic and natural. As pointed out in Section 5.6, the stiffness and damping properties of elastomers depend on the operating temperature. In general, elastomers tend to stiffen and yield larger loss factors at lower temperatures. The temperature effect is quite pronounced and must be included in the design of an effective isolation system. Other effects, such as humidity and pressures, are usually ignored as they tend to be small. However, all the properties of the operating environment should be considered before risking a prototype. For example, if oil or gas is likely to spill on the elastomer or if it is to operate in a vacuum, some type of enclosure must also be designed.

The service life of an isolation design is also extremely important in choosing materials for a specific application. If the isolator only has to function a few times, such as a shock isolation system in a packaging application, or if it must function continuously for years, such as a vibration isolation system for a motor mount application, it makes a large difference in the types of materials and devices appropriate for the design.

Most, but not all, isolator problems are designed by using the single-degree-of-freedom model discussed in Section 5.2. This allows simple calculations and easy communication among the various people involved in the design process. This assumption is reasonable as long as (1) the center of gravity of the equipment isolator system coincides with the elastic center of the isolation system, (2) the material used is isoelastic (i.e., has the same translational spring rate in each direction, and (3) the equipment to be isolated is relatively light compared to the support structures (which must be rigid enough to be considered as ground). Such assumptions allow use of the formulas of Window 5.1

and Section 5.2 in the design of an isolation system. The window is repeated here for reference.

Window 5.1
A Summary of Vibration Isolation Transmissibility Formulas for both Force and Displacement.

The moving-base model on the left is used in designing isolation to protect the device from motion of its point of attachment (base). The model on the right is used to protect the point of attachment (ground) from vibration of the mass.

Displacement Transmissibility

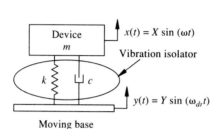

Moving base
Vibration source modeled as base motion

Force Transmissibility

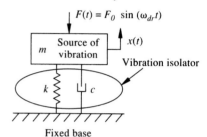

Fixed base
Vibration source mounted on isolator

Here $y(t) = Y \sin \omega_{dr} t$ is the disturbance and from equation (2.41)

$$\frac{X}{Y} = \left[\frac{1 + (2\zeta r)^2}{(1 - r^2)^2 + (2\zeta r)^2}\right]^{1/2}$$

defines the displacement transmissibility and is plotted in Figure 2.8
From equation (2.47)

$$\frac{F_T}{kY} = r^2 \left[\frac{1 + (2\zeta r)^2}{(1 - r^2)^2 + (2\zeta r)^2}\right]^{1/2}$$

defines the related force transmissibility and is plotted in Figure 2.9.

Here $F(t) = F_0 \sin \omega_{dr} t$ is the disturbance and

$$\frac{F_T}{F_0} = \left[\frac{1 + (2\zeta r)^2}{(1 - r^2)^2 + (2\zeta r)^2}\right]^{1/2}$$

defines the force transmissibility for isolating the source of vibration as derived in Section 5.2.

Several other quantities are required before a practical design can be implemented. First consider the real part of the shear modulus G', which is related to the real part elastic modulus E' by Poisson's ratio v, through the relation

$$E' = 2(1 + v)G' \tag{5.85}$$

which comes from classical elasticity theory. The shear modulus is related to the shear stress σ_s and strain ϵ_s by

$$\sigma_s = G^*\epsilon_s = G'(1 + \eta_s j)\epsilon_s \tag{5.86}$$

where G^* is the complex shear modulus and η_s is the shear loss factor. Typically for elastomeric materials in normal applications, $v = 0.05, E = 3G$, and $\eta_s = \eta$, the loss factor defined in Section 5.6 for the extensional modulus in equation (5.65). Here E' is used to denote the real part of Young's modulus, which is denoted just by E in equation (5.65). Similarly, E'' and G'' denote the imaginary part of the elastic and shear modulus, respectively (i.e., the products $E'\eta$ and $G'\eta_s$). The quantity G'' is called the *loss modulus*. Some useful approximations are

$$\frac{G''}{G'} \cong \eta \tag{5.87}$$

$$\frac{G'}{G''} \cong T_R \tag{5.88}$$

where T_R denotes the value of the transmissibility ratio at resonance.

The shear stiffness k' of an isolation material is related to the area of the elastomer to which the load is applied, A, the thickness of the elastomer, h, and the shear modulus, G', by

$$k' = \frac{AG'}{h} \tag{5.89}$$

This is in turn related to the natural frequency (in hertz) by $f = 3.13\sqrt{k_T'/m}$ where k_T' is the total system spring rate. The modulus G', and hence T_R via equation (5.88), are known to vary quite a bit with the strain due to dynamic deformation. This variation is not directly accounted for in the simple models proposed in Section 5.2 for isolator design. This effect must be corrected for by using empirical data provided by manufacturers such as illustrated in Figures 5.33 and 5.34. These figures describe the strain of the modulus and resonant transmissibility.

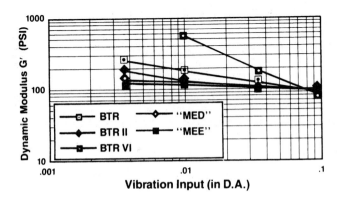

Figure 5.33   Empirical sample of the variation in dynamic modulus G' of equation (5.85) versus input amplitude for sinusoidal excitation of five different elastomers. (Courtesy of Lord Corporation, Erie, PA.)

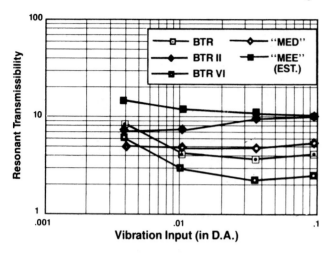

Figure 5.34 Empirical plot of T_R, the transmissibility at resonance, versus input amplitude for sinusoidal excitation of five different elastomers. (Courtesy of Lord Corporation, Erie, PA.)

Temperature sensitivity of modulus and loss factor values are also extremely important. Again, manufacturers' data for temperature dependence must be consulted. Sample data are presented in Figure 5.35, which illustrates the effect of operating temperature on dynamic stiffness. Figure 5.36 illustrates the effect of temperature on loss factor by examining the transmissibility at resonance, T.R., versus temperature. Ideally, these curves should be constant if temperature effects are to be ignored.

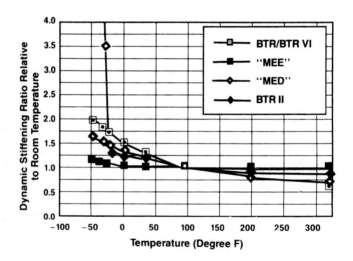

Figure 5.35 Empirical plot of the ratio of the dynamic stiffness at the indicated temperature to the dynamic stiffness at room temperature versus temperature for four different elastomers. For the equations of Window 5.1 to be completely accurate for all temperatures, the plot would have to be a constant value of 1. (Courtesy of Lord Corporation, Erie, PA.)

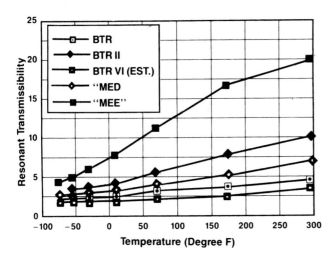

Figure 5.36 Empirical plot of T_R, the transmissibility at resonance, versus temperature for five different elastomers. (Courtesy of Lord Corporation, Erie, PA.)

Basically, the single-degree-of-freedom formulas of Window 5.1 are approximations of reality. They do not account for strain or temperature sensitivity. These effects can be accounted for by using a combination of the data plots, such as given in Figures 5.33 to 5.36 and the single-degree-of-freedom system equations given in Window 5.1. The strain changes drastically with the type of dynamic load. A rule of thumb for stiffness is

$$k_{\text{vib}} > k_{\text{shock}} > k_{\text{static}} \tag{5.90}$$

where k_{vib} is the stiffness during harmonic excitation, k_{shock} the stiffness during shock excitation, and k_{static} the usual static stiffness. Empirical data suggests that

$$k_{\text{shock}} = 1.4 k_{\text{static}} \tag{5.91}$$

as an example of how much the dynamic load affects the isolator's properties. To improve on the model for vibration isolation given in the left column of Window 5.1, the elastic modulus model of Figure 5.37 is often used. Using this model, the transmissibility

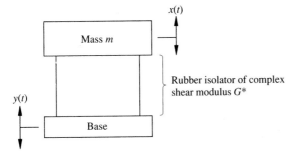

Figure 5.37 Complex shear modulus model of the vibration-isolation problem.

function in Window 5.1 becomes

$$\text{T.R.} = \frac{\sqrt{1+\eta^2}}{\sqrt{[1 - r^2(G'/G'_\omega)]^2 + \eta^2}} \tag{5.92}$$

where G'_ω denotes the dynamic shear modulus at the specified driving frequency as obtained from data plots similar to Figure 5.38, and G' is the value of the shear modulus corresponding to the undamped natural frequency obtained from $\omega^2 = k'/m$, where k' is as defined in equation (5.89). Equation (5.92) yields the same shaped curve as the traditional T.R. (see Figure 5.5) with two notable differences:

1. The crossover point between amplification and isolation is moved from $r = \sqrt{2}$, as indicated in Figure 5.5, to larger than $\sqrt{2}$. The actual value depends on temperature and load conditions (i.e., the effects of Figures 5.34 and 5.36).
2. As computed by equation (5.92), the T.R. at high frequency will actually be larger than indicated by the expression given in Window 5.1.

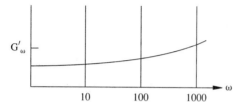

Figure 5.38 Sample plot of shear modulus G'_ω versus frequency for use in equation (5.92) at constant temperature for a sample rubber (Thiokol RD).

Using equation (5.92) along with plots of material parameters such as given in Figure 5.38, manufacturers usually provide plots of T.R. versus frequency using equation (5.92) for the various isolation products they manufacture. These plots along with load versus deflection curves, geometric sizes of the device, and the isolator's natural frequency are provided in data sheets by manufacturers. These data sheets are used by designers to choose isolation mounts for various pieces of equipment. A sample data sheet is given in Figure 5.39.

The starting point in choosing a particular elastomer or particular isolation device is to determine the amount of vibration or shock that the equipment to be isolated can withstand without breaking or malfunctioning. This is often called the *fragility* of a piece of equipment. This is determined for sinusoidal vibration input similar to the vibration criteria discussed in Section 5.1. Next, the vibration input is characterized in terms of amplitude and frequency. Both the fragility and input are plotted on the same plot. A sample plot is illustrated in Figure 5.40, which is used to start the design process. The figure indicates clearly the region in which vibration isolation must be applied.

Once the frequency range of required isolation is determined, the natural frequency of the combined isolator and equipment system is chosen. Since isolation occurs for $r > \sqrt{2}$, the frequency of the isolation system is chosen to be at least a factor of 2 less than the frequency indicated in Figure 5.40, where the equipment fragility falls below the input amplitude. By using $k' = \omega^2 m$, the stiffness of the isolator is specified. Other

AM001 Series
(Metric values in red)

Maximum static load per mount: 3 lbs. (1.4kg)

Maximum dynamic input at resonance:
.036 in. (.91mm) D.A.

Weight: .21 oz. (6.0g)

Material: Inner and outer member — aluminum alloy chromate treated per MIL-C-5541, Class 1A

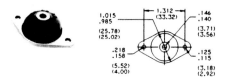

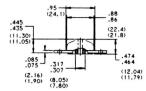

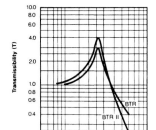

Performance Characteristics

AM001 Series Part Numbers	Axial nat. freq. f_n (Hz)*	BTR® Dynamic axial spring rate		BTR® Dynamic radial spring rate	
		lbs/in	N/mm	lbs/in	N/mm
AM001-2	17	89	16	74	13
AM001-3	19	104	18	87	15
AM001-4	20	122	21	102	18
AM001-5	22	143	25	119	21
AM001-6	23	164	29	137	24
AM001-7	25	187	33	156	27
AM001-8	27	215	38	179	31
AM001-9	29	247	43	206	36
AM001-10	31	284	50	237	41
		BTR® II			
AM001-17	15	68	12	57	10
AM001-18	17	90	16	75	13
AM001-19	20	117	20	98	17
AM001-20	22	146	26	122	21
AM001-21	25	195	34	163	28

*At .036 in. (.91mm) D.A. input and maximum static load.
To correct for loads below rated loads, use:

$$f_n = f_{nn}\sqrt{P_R/P_A}$$

where:

f_n = natural frequency at actual load
f_{nn} = nominal natural frequency
P_R = rated load
P_A = actual load

Transmissibility vs. frequency

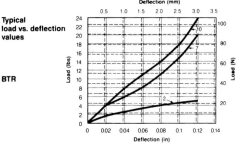

Typical load vs. deflection values

BTR

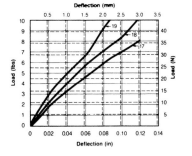

BTR II

Figure 5.39 Manufacturers' data sheet for designing and choosing isolation mounts. (Courtesy of Lord Corporation, Erie, PA.)

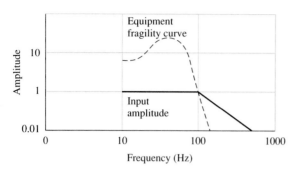

Figure 5.40 Plot of acceptable vibration amplitude versus frequency for a piece of equipment (called a fragility curve) and a plot of vibration input amplitude versus frequency, where the fragility curve falls below the input curve specifies the region in which isolation is required. For this sample it is the frequency range above 100 Hz.

facts, such as static load, geometric size, temperature, and environmental conditions, must be checked.

PROBLEMS

Section 5.1

5.1. Using the nomograph of Figure 5.1, determine the frequency range of vibration for which a machine oscillation remains at a satisfactory level under rms acceleration of 1g.

5.2. Using the nomograph of Figure 5.1, determine the frequency range of vibration for which a structure's rms acceleration will not cause wall damage if vibrating with an rms displacement of 1 mm or less.

5.3. What natural frequency must a hand drill have if its vibration must be limited to a minimum rms displacement of 10 μm and rms acceleration of 0.1 m/s^2? What rms velocity will the drill have?

5.4. A machine of mass 500 kg is mounted on a support of stiffness 197,392,000 N/m. Is the vibration of this machine acceptable (Figure 5.1) for an rms amplitude of 10 μm? If not, suggest a way to make it acceptable.

5.5. Using the expression for the amplitude of the displacement, velocity, and acceleration of an undamped single-degree-of-freedom system, calculate the velocity and acceleration amplitude of a system with a maximum displacement of 10 cm and a natural frequency of 10 Hz. If this corresponds to the vibration of the wall of a building under a wind load, is it an acceptable level?

Section 5.2

5.6. A 100-kg machine is supported on an isolator of stiffness 700×10^3 N/m. The machine contains a rotating unbalance that causes a vertical disturbance force of 350 N at a revolution of 3000 rpm. The damping ratio of the isolator is $\zeta = 0.2$. Calculate (**a**) the amplitude of motion caused by the unbalanced force, (**b**) the transmissibility ratio, and (**c**) the magnitude of the force transmitted to ground through the isolator.

5.7. Plot the T.R. of Problem 5.6 for the cases $\zeta = 0.001$, $\zeta = 0.025$, and $\zeta = 1.1$.

5.8. A simplified model of a washing machine is illustrated in Figure 5.41. A bundle of wet clothes forms a mass of 10 kg (m_b) in the machine and causes a rotating unbalance. The

Top view

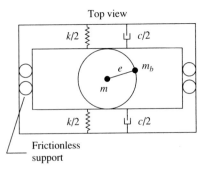

Figure 5.41 Simple model of the vibration of a washing machine induced by a rotating imbalance such as commonly caused by an uneven distribution of wet clothes during a rinse cycle.

rotating mass is 20 kg and the diameter of the washer basket ($2e$) is 50 cm. Assume that the spin cycle rotates at 300 rpm. Let k be 1000 N/m and $\zeta = 0.01$. Calculate the force transmitted to the sides of the washing machine. Discuss the assumptions made in your analysis in view of what you might know about washing machines.

5.9. Referring to Problem 5.8, let the spring constant and damping rate become variable. The quantities m, m_b, e, and ω_{dr} are all fixed by the previous design of the washing machine. Design the isolation system (i.e., decide on which value of k and c to use) so that the force transmitted to the side of the washing machine (considered as ground) is less than 100 N.

5.10. A harmonic force of maximum value 25 N and frequency of 180 cycles/min acts on a machine of 25 kg mass. Design a support system for the machine (i.e., choose c, k) so that only 10% of the force applied to the machine is transmitted to the base supporting the machine.

5.11. Consider a machine of mass 70 kg mounted to ground through an isolation system of total stiffness 30,000 N/m, with a measured damping ratio of 0.2. The machine produces a harmonic force of 450 N at 13 rad/s during steady-state operating conditions. Determine (**a**) the amplitude of motion of the machine, (**b**) the phase shift of the motion (with respect to a zero phase exciting force), (**c**) the transmissibility ratio, (**d**) the maximum dynamic force transmitted to the floor, and (**e**) the maximum velocity of the machine.

5.12. A small compressor weighs about 70 lb and runs at 900 rpm. The compressor is mounted on four supports made of metal with negligible damping.
(**a**) Design the stiffness of these supports so that only 15% of the harmonic force produced by the compressor is transmitted to the foundation.
(**b**) See if you can find a metal and geometry that should offer the required stiffness (refer to Table 1.2 for geometry).

5.13. Typically, in designing an isolation system, one cannot choose any continuous value of k and c but rather, works from a parts catalog wherein manufacturers list isolators available and their properties (and costs, details of which are ignored here). Table 5.3 lists several made up examples of available parts. Using this table to design an isolator for a 500-kg compressor running in steady state at 1500 rev/min. Keep in mind that as a rule of thumb compressors usually require a frequency ratio of $r = 3$.

5.14. An electric motor of mass 10 kg is mounted on four identical springs as indicated in Figure 5.42. The motor operates at a steady-state speed of 1750 rpm. The radius of gyration (see Example 1.4.6 for a definition) is 100 mm. Assume that the springs are undamped and choose a design (i.e., pick k) such that the transmissibility ratio in the vertical direction is

TABLE 5.3 CATALOG VALUES OF STIFFNESS AND DAMPING PROPERTIES
OF VARIOUS OFF-THE-SHELF ISOLATORS

Part No.[a]	R-1	R-2	R-3	R-4	R-5	M-1	M-2	M-3	M-4	M-5
$k(10^3 \text{N/m})$	250	500	1000	1800	2500	75	150	250	500	750
$c(\text{N} \cdot \text{s/m})$	2000	1800	1500	1000	500	110	115	140	160	200

[a]The "R" in the part number designates that the isolator is made of rubber, and the "M"
designates metal. In general, metal isolators are more expensive than rubber isolators.

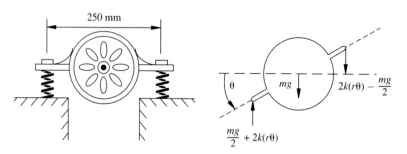

Figure 5.42   Vibration model of an electric motor mount.

0.0194. With this value of k, determine the transmissibility ratio for the torsional vibration
(i.e., using θ rather than x as the displacement coordinates).

5.15. A large industrial exhaust fan is mounted on a steel frame in a factory. The plant manager has
decided to mount a storage bin on the same platform. Adding mass to a system can change
its dynamics substantially and the plant manager wants to know if this is a safe change to
make. The original design of the fan support system is not available. Hence measurements
of the floor amplitude (horizontal motion) are made at several different motor speeds in an
attempt to measure the system dynamics. No resonance is observed in running the fan from
zero to 500 rpm. Deflection measurements are made and it is found that the amplitude is
10 mm at 500 rpm and 4.5 mm at 400 rpm. The mass of the fan is 50 kg and the plant
manager would like to store up to 50 kg on the same platform. The best operating speed
for the exhaust fan is between 400 and 500 rpm depending on environmental conditions in
the plant.

5.16. A 350-kg rotating machine operates at 800 cycles/min. It is desired to reduce the transmissibility ratio by one-fourth of its current value by adding a rubber vibration isolation pad.
How much static deflection must the pad be able to withstand?

5.17. A 68-kg electric motor is mounted on an isolator of mass 1200 kg. The natural frequency of
the entire system is 160 cycles/min and has a measured damping ratio of $\zeta = 1$. Determine
the amplitude of vibration and the force transmitted to the floor if the out-of-balance force
produced by the motor is $F(t) = 100 \sin(31.4t)$ in newtons.

5.18. The force exerted by an eccentric ($e = 0.22$ mm) flywheel of 1000 kg, is $600 \cos(52.4t)$ in
newtons. Design a mounting to reduce the amplitude of the force exerted on the floor to
1% of the force generated. Use this choice of damping to ensure that the maximum force
transmitted is never greater than twice the generated force.

5.19. A rotating machine weighing 4000 lb has an operating speed of 2000 rpm. It is desired to reduce the amplitude of the transmitted force by 80% using isolation pads. Calculate the stiffness required of the isolation pads to accomplish this design goal.

5.20. The mass of a system may be changed to improve the vibration isolation characteristics. Such isolation systems often occur when mounting heavy compressors on factory floors. This is illustrated in Figure 5.43. In this case the soil provides the stiffness of the isolation system (damping is neglected) and the design problem becomes that of choosing the value of the mass of the concrete block/compressor system. Assume that the stiffness of the soil is about $k = 2.0 \times 10^7$ N/m and design the size of the concrete block (i.e., choose m) such that the isolation system reduces the transmitted force by 75%. Assume that the density of concrete is $\rho = 23,000$ N/m^3. The surface area of the cement block in 4 m^2. The steady-state operating speed of the compressor is 1800 rpm.

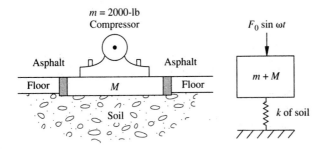

Figure 5.43 Model of a floor-mounted compressor illustrating the use of added mass to design a vibration-isolation system.

5.21. The instrument board of an aircraft is mounted on an isolation pad to protect the panel from vibration of the aircraft frame. The dominant vibration in the aircraft is measured to be at 2000 rpm. Because of size limitation in the aircraft's cabin, the isolators are only allowed to deflect $\frac{1}{8}$ in. Find the percent of motion transmitted to the instrument panel if it weighs 50 lb.

5.22. Design a base isolation system for an electronic module of mass 5 kg so that only 10% of the displacement of the base is transmitted into displacement of the module at 50 Hz. What will the transmissibility be if the frequency of the base motion changes to 100 Hz? What if it reduces to 25 Hz?

5.23. Redesign the system of Problem 5.22 such that the smallest transmissibility ratio possible is obtained over the range 50 to 75 Hz.

5.24. A 2-kg printed circuit board for a computer is to be isolated from external vibration of frequency 3 rad/s at a maximum amplitude of 1 mm as illustrated in Figure 5.44. Design an undamped isolator such that the transmitted displacement is 10% of the base motion. Also calculate the range of transmitted force.

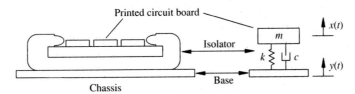

Figure 5.44 Isolation system for a printed circuit board.

5.25. Change the design of the isolator of Problem 5.24 by using a damping material with damping value ζ chosen such that the maximum T.R. at resonance is 2.

5.26. Calculate the damping ratio required to limit the displacement transmissibility to 4 at resonance for any damped isolation system.

Section 5.3

5.27. A motor is mounted on a platform that is observed to vibrate excessively at an operating speed of 6000 rpm producing a 250-N force. Design a vibration absorber (undamped) to add to the platform. Note that in this case the absorber mass will only be allowed to move 2 mm because of geometric and size constraints.

5.28. Consider an undamped vibration absorber with $\beta = 1$ and $\mu = 0.2$. Determine the operating range of frequencies for which $|Xk/F_0| \leq 0.5$.

5.29. Consider an internal combustion engine that is modeled as a lumped inertia attached to ground through a spring. Assuming that the system has a measured resonance of 100 rad/s, design an absorber so that the amplitude is 0.01 m for a (measured) force input of 10^2N.

5.30. A small rotating machine weighing 50 lb runs at a constant speed of 6000 rpm. The machine was installed in a building and it was discovered that the system was operating at resonance. Design a retrofit undamped absorber such that the nearest resonance is at least 20% away from the driving frequency.

5.31. A 3000-kg machine tool exhibits a large resonance at 120 Hz. The plant manager attaches an absorber to the machine of 600 kg tuned to 120 Hz. Calculate the range of frequencies at which the amplitude of the machine vibration is less with the absorber fitted than without the absorber.

5.32. A motor-generator set is designed with steady-state operating speed between 2000 and 4000 rpm. Unfortunately, due to an imbalance in the machine, a large violent vibration occurs at around 3000 rpm. An initial absorber design is implemented with a mass of 2 kg tuned to 3000 rpm. This, however, causes the combined system natural frequencies to occur at 2500 and 3500 rpm. Redesign the absorber so that $\omega_1 < 2000$ rpm and $\omega_2 > 4000$ rpm, rendering the system safe for operation.

5.33. A rotating machine is mounted on the floor of a building. Together, the mass of the machines and the floor is 2000 lb. The machine operates in steady state at 600 rpm and causes the floor of the building to shake. The floor-machine system can be modeled as a spring–mass system similar to the optical table of Figure 5.11. Design an undamped absorber system to correct this problem. Make sure you consider the bandwidth.

5.34. A pipe carrying steam through a section of a factory vibrates violently when the driving pump hits a speed of 300 rpm (see Figure 5.45). In an attempt to design an absorber, a trial 9-kg absorber tuned to 300 rpm was attached. By changing the pump speed it was found that the pipe–absorber system has resonances at 207 and 414 rpm. Redesign the absorber so that the natural frequencies are 40% away from the driving frequency.

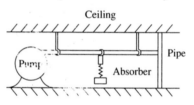

Figure 5.45 Schematic of a steam pipe system with an absorber attached.

5.35. A machine sorts bolts according to their size by moving a screen back and forth using a primary system of 2500 kg with a natural frequency of 400 cycle/min. Design a vibration absorber so that the machine–absorber system has natural frequencies below 160 cycles min and above 320 rpm. The machine is illustrated in Figure 5.46.

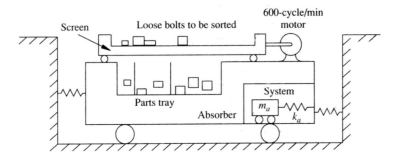

Figure 5.46   Model of a parts sorting machine.   The parts (bolts here) are placed on a screen that shakes. Parts that are small enough fall through the screen into the tray below. The larger ones remain on the screen.

5.36. A dynamic absorber is designed with $\mu = 1/4$ and $\omega_a = \omega_p$. Calculate the frequency range for which the ratio $|Xk/F_0| < 1$.

Section 5.4

5.37. A machine, largely made of aluminum, is modeled as a simple mass (of 100 kg) attached to ground through a spring of 2000 N/m. The machine is subjected to a 100-N harmonic force at 20 rad/s . Design an undamped tuned absorber system (i.e., calculate m_a and k_a) so that the machine is stationary at steady state. Aluminum, of course, is not completely undamped and has internal damping that gives rise to a damping ratio of about $\zeta = 0.001$. Similarly, the steel spring for the absorber gives rise to internal damping of about $\zeta_a = 0.0015$. Calculate how much this spoils the absorber design by determining the magnitude X using equation (5.32).

5.38. Plot the magnitude of the primary system calculated in Problem 5.36 with and without the internal damping. Discuss how the damping affects the bandwidth and performance of the absorber designed without knowledge of internal damping.

5.39. Derive equation (5.35) for the damped absorber from equations (5.34) and (5.32) along with Window 5.4. Also derive the nondimensional form of equation (5.37) from equation (5.35). Note the definition of ζ given in equation (5.36) is not the same as the ζ values used in Problems 5.37 and 5.38.

5.40. (Project) If you have a three-dimensional graphics routine available, plot equation (5.37) [i.e., plot (X/δ_{st}) versus both r and ζ for $0 < \zeta < 1$ and $0 < r < 3$, and a fixed μ and β.] Discuss the nature of your results. Does this plot indicate any obvious design choices? How does it compare to the information obtained by the series of plots given in Figures 5.16 to 5.18? (Three-dimensional plots such as these are becoming commonplace and have not yet been taken advantage of fully in vibration absorber design.)

5.41. Repeat Problem 5.40 by plotting $|X/\delta_{st}|$ versus r and β for a fixed ζ and μ.

5.42. (Project) The full damped vibration absorber equations (5.32) and (5.33) have not historically been used in absorber design because of the complicated nature of the complex arithmetic involved. However, if you have a symbolic manipulation code available to you, calculate an expression for the magnitude X by using the code to calculate the magnitude and phase of equation (5.32). Apply your results to the absorber design indicated in Problem 5.37 by using m_a, k_a, and ζ_a as design variables (i.e., design the absorber).

5.43. A machine of mass 200 kg is driven harmonically by a 100-N force at 10 rad/s. The stiffness of the machine is 20,000 N/m. Design a broadband vibration absorber [i.e., equation (5.37)] to limit the machine's motion as much as possible over the frequency range 8 to 12 rad/s. Note that other physical constraints limit the added absorber mass to be at most 50 kg.

5.44. Often absorber designs are afterthoughts such as indicated in Example 5.3.1. Add a damper to the absorber design of Figure 5.14 to increase the useful bandwidth of operation of the absorber system in the event the driving frequency drifts beyond the range indicated in Example 5.3.2.

5.45. Again consider the absorber design of Example 5.3.1. If the absorber spring is made of aluminum and introduces a damping ratio of $\zeta = 0.001$, calculate the effect of this on the deflection of the saw (primary system) with the design given in Example 5.3.1.

5.46. Consider the undamped primary system with a viscous absorber as modeled in Figure 5.19 and the rotational counterpart of Figure 5.20. Calculate the magnification factor $|Xk/M_0|$ for a 400-kg compressor having a natural frequency of 16.2 Hz if driven at resonance, for an absorber system defined by $\mu = 0.133$ and $\zeta = 0.025$.

5.47. Recalculate the magnification factor $|Xk/M_0|$ for the compressor of Problem 5.46 if the damping factor is changed to $\zeta = 0.1$. Which absorber design produces the smallest displacement of the primary system $\zeta = 0.025$ or $\zeta = 0.1$?

5.48. Consider a one-degree-of-freedom model of the nose of an aircraft (A-10) as illustrated in Figure 5.47. The nose cracked under fatigue during battle conditions. This problem has been fixed by adding a viscoelastic material to the inside of the skin to act as a damped vibration absorber as illustrated in Figure 5.47. This fixed the problem and the vibration fatigue cracking disappeared in the A-10's after they were retrofitted with viscoelastic damping treatments. While the actual values remain classified, use the following data to calculate the required damping $M = 100$ kg, $f_{dr} = 30$ Hz, and $k = 10^6$ N/m. Note that since mass always needs to be limited in an aircraft, use $\mu = 0.1$ in your design.

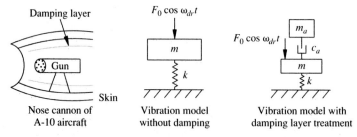

Figure 5.47 Simplified model of the A-10 nose cannon vibration problem. The nose cannon of the A-10 aircraft can be modeled as applying a harmonic force of $F_0 \cos \omega_{dr} t$ to the skin of the aircraft nose. The skin can be modeled as a spring–mass system based on the stiffness model of Figure 1.19 (i.e., $k = 3EI/l^3$).

5.49. Plot an amplification curve such as Figure 5.21 by using equation (5.39) for $\zeta = 0.02$ after several values of μ($\mu = 0.1, 0.25, 0.5$, and 1). Can you form any conclusions about the effect of the mass ratio on the response of the primary system? Note that as μ gets large $|(Xk/M_0)|$ gets very small. What is wrong with using very large μ in absorber design?

5.50. A Houdaille damper is to be designed for an automobile engine. Choose a value for ζ and μ if the magnification $|(Xk/M_0)|$ is to be limited to 4 at resonance. (One solution is $\mu = 1, \zeta = 0.129$.)

5.51. Determine the amplitude of vibration for the various dampers of Problem 5.46 if $\zeta = 0.1$, and $F_0 = 100$ N.

5.52. (Project) Use your knowledge of absorbers and isolation to design a device that will protect a mass from both shock inputs and harmonic inputs. It may help to have a particular device in mind such as the module discussed in Figure 5.6.

Section 5.5

5.53. Design a Houdaille damper for an engine modeled as having an inertia of 1.5 kg·m^2 and a natural frequency of 33 Hz. Choose a design such that the maximum dynamic magnification is less than 6:

$$|\frac{Xk}{M}_0| < 6$$

The design consists of choosing m_a and c_a, the required optimal damping.

5.54. Consider the damped vibration absorber of equation (5.37) with β fixed at $\beta = 1/2$ and μ fixed at $\mu = 0.25$. Calculate the value of ζ that minimizes $|X/\delta_{st}|$. Plot this function for several values of $0 < \zeta < 1$ to check your design. If you cannot solve this analytically, consider using a three-dimensional plot of $|X/\delta_{st}|$ versus r and ζ to determine your design.

5.55. For a Houdaille damper with mass ratio $\mu = 0.25$, calculate the optimum damping ratio and the frequency at which the damper is most effective at reducing the amplitude of vibration of the primary system.

5.56. Consider again the system of Problem 5.53. If the damping ratio is changed to $\zeta = 0.1$, what happens to $|Xk/M_0|$?

5.57. Derive equation (5.42) from equation (5.35) and derive equation (5.49) for the optimal damping ratio.

5.58. Consider the design suggested in Example 5.5.1. Calculate the percent change in the maximum deflection if the damping constant changes 10% from its optimal value. If the optimal damping is fixed but the mass of the absorber changes by 10%, what percent change in $|Xk/M_0|_{max}$ results? Is the optimal absorber design more sensitive to changes in c_a or m_a?

5.59. Consider the elastic isolation problem described in Figure 5.23. Derive equations (5.54) and (5.55) from equation (5.53).

5.60. Use the derivative calculation for finding maximum and minimum to derive equations (5.57) and (5.58) for the elastic damper system.

5.61. A 1000-kg mass is isolated from ground by a 40,000-N/m spring. A viscoelastic damper is added, as indicated in Figure 5.23. Design the isolator (choose k_2 and c) such that when a 70-N sinusoidal force is applied to the mass, no more than 100 N is transmitted to ground.

5.62. Consider the isolation design of Example 5.5.2. If the value of the damping coefficient changes 10% from the optimal value (of 188.56 kg/s), what percent change occurs in (T.R.)$_{max}$? If c remains at its optimal value and k_2 changes by 10%, what percent change occurs in (T.R.)$_{max}$? Is the design of this type of isolation more sensitive to changes in damping or stiffness?

5.63. A 3000-kg machine is mounted on an isolator with an elastically coupled viscous damper such as indicated in Figure 5.23. The machine stiffness (k_1) is 2.943×10^6 N/m, $\gamma = 0.5$, and $c = 56.4 \times 10^3$ N·s/m. The machine, a large compressor, develops a harmonic force of 1000 N at 7 Hz. Determine the amplitude of vibration of the machine.

5.64. Again consider the compressor isolation design given in Problem 5.61. If the isolation material is changed so that the damping in the isolator is changing so that $\zeta = 0.15$, what is the force transmitted? Next determine the optimal value for the damping ratio and calculate the resulting transmitting force.

5.65. Consider the optimal vibration isolation design of Problem 5.64. Calculate the optimal design if the compressor's steady-state driving frequency changes to 24.7 Hz. If the wrong optimal point is used (i.e., if the optimal damping for the 7-Hz driving frequency is used), what happens to the transmissibility ratio?

5.66. Recall the optimal vibration absorber of Problem 5.53. This design is based on a steady-state response. Calculate the response of the primary system to an impulse of magnitude M_0 applied to the primary inertia J_1. How does the maximum amplitude of the transient compare to that in steady state?

Section 5.6

5.67. Compare the resonant amplitude at steady state (assume a driving frequency of 100 Hz) of a piece of nitrite rubber at 50°F versus the value at 75°F. Use the values for η from Table 5.1.

5.68. Using equation (5.67), calculate the new modulus of a $0.05 \times 0.01 \times 1$m piece of pinned–pinned aluminum covered with a 1-cm-thick piece of nitrite rubber at 75°F driven at 100 Hz.

5.69. Calculate Problem 5.68 again at 50°F. What percent effect does this change in temperature have on the modulus of the layered material?

5.70. Repeat the design of Example 5.6.1 by
   (a) changing the operating frequency to 1000 Hz, and
   (b) changing the operating temperature to 50°F.
   Discuss which of these designs yields the most favorable system.

5.71. Reconsider Example 5.6.2. Make a plot of thickness of the damping treatment versus loss factor.

5.72. Calculate the maximum transmissibility coefficient of the center of the shelf of Example 5.6.1. Make a plot of the maximum transmissibility ratio for this system frequency using Table 5.1 for each temperature.

5.73. The damping ratio associated with steel is about $\zeta = 0.001$. Does it make any difference whether the shelf in Example 5.6.1 is made out of aluminum or steel? What percent improvement in damping ratio at resonance does the rubber layer provide the steel shelf?

Section 5.7

5.74. A 100-kg compressor rotor has a shaft stiffness of 1.4×10^7 N/m. The compressor is designed to operate at a speed of 6000 rpm. The internal damping of the rotor shaft system is measured to be $\zeta = 0.01$.
   (a) If the rotor has an eccentric radius of 1 cm, what is the rotor system's critical speed?

(b) Calculate the whirl amplitude at critical speed. Compare your results to those of Example 5.7.1.

5.75. Redesign the rotor system of Problem 5.74 such that the whirl amplitude at critical speed is less than 1 cm by changing the mass of the rotor.

5.76. Determine the effect of the rotor system's damping ratio on the design of the whirl amplitude at critical speed for the system of Example 5.7.1 by plotting X at $r = 1$ for ζ between $0 < \zeta < 1$.

5.77. The flywheel of an automobile engine has a mass of about 50 kg and an eccentricity of about 1 cm. The operating speed ranges from 1200 rpm (idle) to 5000 rpm (red line). Choose the remaining parameters so that the whirling amplitude is never more than 1 mm.

5.78. Consider the design of the compressor rotor system of Example 5.7.1. The amplitude of the whirling motion depends on the parameters a, ζ, m, k and the driving frequency. Which parameter has the greatest affect on the amplitude? Discuss your results.

5.79. At critical speed the amplitude is determined entirely by the damping ratio and the eccentricity. If a rotor has an eccentricity of 1 cm, what value of damping ratio is required to limit the deflection to 1 cm?

5.80. A rotor system has damping limited by $\zeta < 0.05$. What is the maximum value of eccentricity allowable in the rotor design if the maximum amplitude at critical speed must be less than 1 cm?

Section 5.8

5.81. Recall the definitions of settling time, time to peak, and overshoot given in Example 3.2.1 and illustrated in Figure 3.5. Consider a single-degree-of-freedom system with mass $m = 2$ kg, damping coefficient $c = 0.8$ N·s/m, and stiffness 8 N/m. Design a PD controller such that the settling time of the closed-loop system is less than 10 s.

5.82. Redesign the control system given in Example 5.8.1 if the available internal damping is reduced to 50 N·s/m.

5.83. Consider the compressor rotor–shaft system discussed in Problem 5.74. Modern designers have considered using electromagnetic bearings in such rotor systems to improve their design. Use a derivative feedback control law on the design of this compressor to increase the effective damping ratio to $\zeta = 0.5$. Calculate the required gain. How does this affect the answer to parts (a) and (b) of Problem 5.74?

5.84. Calculate the magnitude of the force required of the actuator used in the feedback control system of Example 5.8.1. See if you can find a device that provides this much force.

5.85. In some cases the force actuator used in a control system also introduces dynamics. In this case a system of the form given in equation (5.27) may result where m_a, c_a, and k_a are values associated with the actuator (rather than an absorber). In this case the absorber system indicated in Figure 5.15 can be considered as the control system and the motion of the mass m is the object of the control system. Let $m = 10$ kg, $k = 100$ N/m, and $c = 0$. Choose the feedback control law to be

$$u = -g_1 x - g_2 \dot{x}$$

and assume that $c_a = 20$ N · s/m, $k_a = 100$ N/m, and $m_a = 1$ kg. Calculate g_1 and g_2 so that x is as small as possible for a driving frequency of 5 rad/s.

Section 5.9

5.86. Reconsider Example 5.2.1, which describes the use of a vibration isolation to protect an electronic module. Recalculate the solution to this example using equation (5.92).

5.87. A machine part is driven at 40 Hz at room temperature. The machine has a mass of 100 kg. Use Figure 5.39 to determine an appropriate isolator so that the transmissibility is less than 1.

5.88. Make a comparison between the transmissibility ratio of Window 5.1 and that of equation (5.92).

MATLAB VIBRATION TOOLBOX

If you have not yet used the *Vibration Toolbox* program, return to the end of Chapter 1 for a brief introduction to using MATLAB files or open the file/directory "INTRO.TXT" on the disk on the inside of the back cover. The files contained in folder VTB5 may be used to help solve the problems listed above. The M-files from earlier chapters (VTBX.X.M, etc.) may also be useful. The exercises listed below are intended to help you gain some experience with the concepts introduced in this chapter for designing vibrating systems and to build experience with the various formulas.

TOOLBOX PROBLEMS

TB5.1. Use file VTB5_1 to reproduce the plots of Figure 5.5 by inputting various values of m, c, and k. First fix m and k and vary c. Then fix m and c and vary k.

TB5.2. Use file VTB5_2 to verify the solution of Example 5.2.1 for the magnitude of the force transmitted to the electronic module.

TB5.3. Use file VTB5_3 to examine what happens to the shaded region in Figure 5.12 as μ is varied. Do this by increasing the absorber mass ($\mu = m_a/m$) so that μ varies in increments of 0.1 from 0.1 to 1 for a fixed value of β.

TB5.4. Examine the effect of β on Figure 5.13 by using file VTB5_4 to plot Figure 5.13 over again for several different values of $\beta(\beta = 0.1, 0.5, 1, 1.5)$. What do you notice? What does changing β correspond to in terms of choosing the values of the absorber design?

TB5.5. Consider the amplitude plot of the damped absorber given in Figure 5.16. Use file VTB5_5 to see the effect of changing the primary mass m has on the design. Choose $m_a = 10, c_a = 1, k = 1000, k_a = 1000$. Plot $|Xk/F_0|$ for various values of the primary mass m (i.e., $m = 1, 10, 100, 1000$) versus r. Can you draw any conclusions?

TB5.6. File VTB5_6 plots the three-dimensional mesh used to generate Figure 5.22, which illustrate the optimal damped absorber design. Use this program to note the effects of changing the spring–mass on the values of ζ corresponding to the absolute minimum of the maximum response.

6

Distributed-Parameter Systems

This chapter presents methods of accounting for flexibility in standard components by modeling the mass and stiffness as distributed throughout the spatial definition of a structural component rather than lumping these properties at discrete points as is examined in Chapter 4. Such distributed-parameter models are useful in providing understanding of the vibration of the wing in the stealth fighter bomber and of the cello strings pictured here. While distributed models are limited to basic elements such as strings, beams, and plates, they form a sound basis for understanding the vibrations of complex structures such as the aircraft and cello pictured here, and they form the basis of the generally applicable finite element method discussed in Chapter 8.

In previous chapters all systems considered are modeled as lumped-parameter systems; that is, the motion of each point in the system under consideration is modeled as if the mass were concentrated at that point. Multiple-degree-of-freedom systems are considered as arrangements of various lumped masses separated by springs and dampers. In this sense, the parameters of the system are discrete sets of finite numbers. Hence such systems are also called discrete systems or finite dimensional systems. In this chapter the flexibility of structures is considered. Here the mass of an object is considered to be distributed throughout the structure as a series of infinitely small elements. When a structure vibrates, each of these infinite number of elements move relative to each other in a continuous fashion. Hence these systems are called *infinite-dimensional systems*, *continuous systems*, or *distributed-parameter systems*. The choice of modeling a given mechanical system as a lumped parameter system or a distributed-parameter system depends on the purpose at hand as well as the nature of the object. There are only a few distributed-parameter models that have closed-form solutions. However, these solutions provide insight into a large number of problems that cannot be solved in closed form.

The time response of a distributed-parameter system is described spatially by a continuous function of the relative position along the system. In contrast, the time response of a lumped parameter system is described spatially by labeling a discrete number of points throughout the system in the form of a vector. Here the terms *lumped parameter* and *distributed parameter* are used rather than discrete and continuous to avoid confusion with discrete-time systems (used in numerical integration and measurement). The specific cases considered here are the vibrations of strings, rods, beams, membranes, and plates. Common examples of such structures are a vibrating guitar string and the swaying motion of a bridge or tall building. In addition, systems having both lumped parameters and distributed parameters are considered.

6.1 VIBRATION OF A STRING OR CABLE

String instruments (guitars, violins, etc.) provide an excellent and intuitive example of the vibration of a distributed-parameter object. Consider the string of Figure 6.1 with mass density ρ, fixed at both ends and under a tension denoted by τ. The string moves up and down in the y direction. The motion at any point on the string must be a function of both the time t and the position along the string, x. The deflection of the string is thus denoted by $w(x, t)$. Let $f(x,t)$ be an external force per unit length also distributed

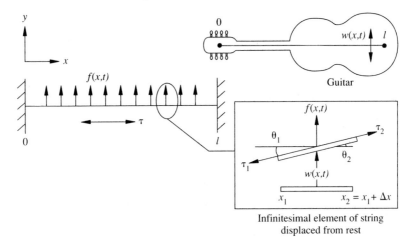

Figure 6.1   Geometry of a vibrating string with applied force $f(x, t)$ and displacement $w(x, t)$.

along the string and consider the infinitesimal element (Δx long) of the displaced string indicated in Figure 6.1.

The net force acting on the infinitesimal element in the y direction must be equal to the inertial force in the y direction, $\rho \Delta x (\partial^2 w / \partial t^2)$, so that

$$-\tau_1 \sin \theta_1 + \tau_2 \sin \theta_2 + f(x, t)\Delta x = \rho \Delta x \frac{\partial^2 w(x, t)}{\partial t^2} \qquad (6.1)$$

Note that the acceleration is stated in terms of partial derivatives ($\partial^2 / \partial t^2$) because w is a function of two variables. The expressions in equation (6.1) can be approximated in the case of small deflections so that θ_1 and θ_2 are small. In this case τ_1 and τ_2 can easily be related to the initial tension in the string τ by noting that the horizontal component of the deflected string tension is $\tau_1 \cos \theta_1$ at end 1, and $\tau_2 \cos \theta_2$ at end 2. In the small-angle approximation $\cos \theta_1 \simeq 1$ and $\cos \theta_2 \simeq 1$, so it is reasonable to set $\tau_1 = \tau_2 = \tau$. Also, for small θ_1,

$$\sin \theta_1 \simeq \tan \theta_1 = \left. \frac{\partial w(x, t)}{\partial x} \right|_{x_1} \qquad (6.2)$$

and

$$\sin \theta_2 \simeq \tan \theta_2 = \left. \frac{\partial w(x, t)}{\partial x} \right|_{x_2} \qquad (6.3)$$

where $(\partial w / \partial x)|_{x_1}$ is the slope of the string at point x_1 and $(\partial w / \partial x)|_{x_2}$ is the slope of the string at point $x_2 = x_1 + \Delta x$. With these approximations equation (6.1) now becomes

$$\left. \left(\tau \frac{\partial w(x, t)}{\partial x} \right) \right|_{x_2} - \left. \left(\tau \frac{\partial w(x, t)}{\partial x} \right) \right|_{x_1} + f(x, t)\Delta x = \rho \frac{\partial^2 w(x, t)}{\partial x^2} \Delta x \qquad (6.4)$$

The slopes can be evaluated further by recalling the Taylor series expansion for the function $\tau(\partial w/\partial x)$ around the point x_1 from calculus. This yields

$$\left(\tau\frac{\partial w}{\partial x}\right)\bigg|_{x_2} = \left(\tau\frac{\partial w}{\partial x}\right)\bigg|_{x_1} + \Delta x\frac{\partial}{\partial x}\left(\tau\frac{\partial w}{\partial x}\right)\bigg|_{x_1} + O(\Delta x^2) \tag{6.5}$$

where $O(\Delta x^2)$ denotes the rest of the Taylor series, which consists of terms of order Δx^2 and higher. Since Δx is small, $O(\Delta x^2)$ is even smaller and hence neglected. Substitution of expression (6.5) into equation (6.4) yields

$$\frac{\partial}{\partial x}\left(\tau\frac{\partial w}{\partial x}\right)\bigg|_{x_1} \Delta x + f(x,t)\Delta x = \rho\frac{\partial^2 w(x,t)}{\partial t^2}\Delta x \tag{6.6}$$

Dividing by Δx and realizing that since Δx is infinitesimal, the designation of point 1 becomes unnecessary, the equation of motion for the string becomes

$$\frac{\partial}{\partial x}\left(\tau\frac{\partial w(x,t)}{\partial x}\right) + f(x,t) = \rho\frac{\partial^2 w(x,t)}{\partial t^2} \tag{6.7}$$

Since the tension τ is constant, and if the external force is zero this becomes

$$c^2\frac{\partial^2 w(x,t)}{\partial x^2} = \frac{\partial^2 w(x,t)}{\partial t^2} \tag{6.8}$$

where $c = \sqrt{\tau/\rho}$ depends only on the physical properties of the string (called the *wave speed* and is not to be confused with the symbol used for damping coefficient in earlier chapters). Equation (6.8) is the one-dimensional wave equation, also called the *string equation*, and is subject to two initial conditions in time because of the dependence on the second time derivative. These are written as $w(x,0) = w_0(x)$ and $w_t(x,0) = \dot{w}_0(x)$, where the subscript t is an alternative notation for the partial derivative $\partial/\partial t$ and where $w_0(x)$ and $\dot{w}_0(x)$ are the initial displacement and velocity distributions of the string, respectively. The second spatial derivative in equation (6.8) implies that two other conditions must be applied to the solution $w(x,t)$ in order to calculate the two constants of integration arising from integrating these spatial derivatives. These conditions come from examining the boundaries of the string. In the configuration of Figure 6.1 the string is fixed at both ends (i.e., at $x = 0$ and $x = l$). This means that the deflection $w(x,t)$ must be zero at these points so that

$$w(0,t) = w(l,t) = 0 \qquad t > 0 \tag{6.9}$$

These two conditions at the boundary provide the other two constants of integration resulting from the two spatial derivatives of $w(x,t)$. Because these conditions occur at the boundaries, the problem described by equations (6.8) and (6.9) and the initial conditions is called a *boundary value problem*.

Using the subscript notation for partial differentiation, the various derivatives of the deflection, $w(x,t)$, have the following physical interpretations. The quantity $w_x(x,t)$ denotes the slope of the string, while $\tau w_{xx}(x,t)$ corresponds to the restoring force of the string (i.e., the string's stiffness or elastic property). The quantity $w_t(x, t)$ is the velocity and $w_{tt}(x, t)$ is the acceleration of the string at any point x and time t.

The string vibration problem described above forms a convenient and simple model to study the vibration of distributed-parameter systems. This is analogous to the spring–mass model of Section 1.1, which provided a building block for the study of lumped-parameter systems. To that end, the string with fixed endpoints is used in the next section to develop general techniques of solving for the vibration response of distributed-parameter systems. More about the string equation and its use in wave propagation can be found in introductory physics texts. Note that these developments apply to cables as well as strings.

Example 6.1.1

Consider the cable of Figure 6.2, which is pinned at one end and attached to a spring at the other end. Determine the governing equation for the vibration of the system.

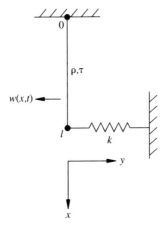

Figure 6.2 Cable fixed at one end and attached to a spring at the other end. Note that the motion $w(x, t)$ is in the y direction.

Solution The equation of motion for the string and the cable are the same and are given by equation (6.8). The initial conditions are also unaffected. However, the boundary condition at $x = l$ changes. Writing a force balance in the y direction at point $x = l$ yields

$$\sum_y F\,|_{x=l} = \tau \sin\theta + kw(l, t) = 0$$

where k is the stiffness of the (lumped) spring. Again enforcing the small-angle approximation, this becomes

$$\tau \left.\frac{\partial w(x, t)}{\partial x}\right|_{x=l} = -kw(x, t)|_{x=l}$$

The boundary condition at $x = 0$ remains unchanged.

□

6.2 MODES AND NATURAL FREQUENCIES

In this section the string equation is solved using the technique of *separation of variables*. This method leads in a natural way to modal analysis and the concepts of mode shapes and natural frequencies for distributed-parameter systems that are used extensively for lumped-parameter systems in Chapter 4. The solution procedures are described in detail in introductory differential equations (see Boyce and DiPrima, 1986, for instance) and reviewed here. First, it is assumed that the displacement $w(x, t)$ can be written as the product of two functions, one depending only on x and the other depending only on t (hence separation of variables). Thus

$$w(x, t) = X(x)T(t) \tag{6.10}$$

Substitution of this separated form into the string equation (6.8) yields

$$c^2 X''(x)T(t) = X(x)\ddot{T}(t) \tag{6.11}$$

where the primes on $X''(x)$ denote total differentiation (twice in this case) with respect to x and the overdots indicate total differentiation (twice in this case) with respect to the time, t. These derivatives become total derivatives, instead of partial derivatives, because the functions $X(x)$ and $T(t)$ are each a function of a single variable. A simple rearrangement of equation (6.11) yields

$$\frac{X''(x)}{X(x)} = \frac{\ddot{T}(t)}{c^2 T(t)} \tag{6.12}$$

Since each side of the equation is a function of a different variable, it is argued that each side must be constant. To see this, differentiate with respect to x. This yields

$$\frac{d}{dx}\left(\frac{X''}{X}\right) = 0 \tag{6.13}$$

which becomes upon integration

$$\frac{X''}{X} = \text{constant} = -\sigma^2 \tag{6.14}$$

In this case $-\sigma^2$ is the constant chosen to ensure that the quantity on the right side of equation (6.14) is negative. Actually, all possible choices (negative, positive, and zero) for this constant need to be considered. The other two choices (positive and zero) lead to physically unacceptable results as discussed in Example 6.2.1 to follow. Equation (6.14) also requires that

$$\frac{\ddot{T}(t)}{c^2 T(t)} = -\sigma^2 \tag{6.15}$$

in order to satisfy equation (6.12).

Rearranging equation (6.14) yields the result that the function $X(x)$ must satisfy

$$X''(x) + \sigma^2 X(x) = 0 \tag{6.16}$$

Equation (6.16) has the solution (see Example 6.2.1)

$$X(x) = a_1 \sin \sigma x + a_2 \cos \sigma x \qquad (6.17)$$

where a_1 and a_2 are constants of integration. To solve for these constants of integration, consider the boundary conditions of equation (6.9) in the separated form implied by equation (6.10). They become

$$X(0)T(t) = 0 \qquad \text{and} \qquad X(l)T(t) = 0 \qquad t > 0 \qquad (6.18)$$

Since it is as assumed that $T(t)$ cannot be zero for all t [this would yield only the uninteresting solution $w(x, t) = 0$ for all time], equation (6.18) reduces to

$$X(0) = 0 \qquad X(l) = 0 \qquad (6.19)$$

Applying these two conditions to the solution, equation (6.17), yields the two simultaneous equations

$$X(0) = a_2 = 0$$
$$\qquad\qquad\qquad\qquad (6.20)$$
$$X(l) = a_1 \sin \sigma l = 0$$

The first of these two expressions eliminates the cosine term in the solution, and the last of these two expressions yields values of σ by requiring that $\sin \sigma l = 0$. This is called the *characteristic equation*, which has solutions $\sigma l = n\pi$. Since this must be true for any integer value of n, there exists an infinite number of σ values that satisfy the condition $\sigma l = n\pi$. Hence σ is indexed to be σ_n and takes on the values

$$\sigma_n = \frac{n\pi}{l} \qquad n = 1, 2, 3, \ldots \qquad (6.21)$$

The two simultaneous equations resulting from the boundary conditions given by equation (6.20) can also be written in matrix form as

$$\begin{bmatrix} \sin \sigma l & 0 \\ 0 & 1 \end{bmatrix} \begin{bmatrix} a_1 \\ a_2 \end{bmatrix} = \begin{bmatrix} 0 \\ 0 \end{bmatrix}$$

Recall from Section 4.1, Equation (4.20), that this vector equation has a nonzero solution (for a_1 and a_2) as long as the coefficient matrix has a zero determinant. Thus this alternative formulation yields the characteristic equation

$$\det \begin{bmatrix} \sin \sigma l & 0 \\ 0 & 1 \end{bmatrix} = \sin \sigma l = 0$$

This provides a more systematic calculation of the characteristic equation from the statement of the boundary conditions.

Since there are an infinite number of values of σ_n, the solution (6.17) then becomes the infinite number of solutions

$$X_n(x) = a_n \sin \left(\frac{n\pi}{l} x \right) \qquad n = 1, 2, \ldots \qquad (6.22)$$

Here X is now indexed by n because of its dependence on σ_n, and the a_n are arbitrary constants, potentially depending on the index n as well, which still need to be determined.

Equation (6.22) forms the spatial solution of the vibrating string problem. The functions $X_n(x)$ satisfy the boundary value problem

$$-\frac{\partial^2}{\partial x^2}(X_n) = \lambda_n X_n$$

$$X_n(0) = X_n(l) = 0$$

(6.23)

where λ_n are constants ($\lambda_n = \sigma_n^2$) and where the function X_n is never identically zero over all values of x. Comparison of this with the definition of the matrix eigenvalue and eigenvector problem of Section 4.2 yields some very strong similarities. Instead of an eigenvector consisting of a column of constants, equation (6.23) defines the *eigenfunctions*, $X_n(x)$, which are nonzero functions of x satisfying boundary conditions as well as a differential equation. The constants λ_n are called *eigenvalues* just as in the matrix case. The differential operator $-\partial^2/\partial x^2$ takes the place of the matrix in this eigenproblem. Similar to the eigenvector of Chapter 4, the eigenfunction is only known to within a constant. That is, if $X_n(x)$ is an eigenfunction, so is $aX_n(x)$, where a is any constant. In fact, the concept of eigenvector and that of eigenfunction are mathematically identical. As is illustrated in the following, the eigenvalues, as determined by the characteristic equation given in this case by expression (6.20) and the eigenfunctions described in (6.22), will determine the natural frequencies and mode shapes of the vibrating string.

To this end, consider next the temporal equation given by (6.15) with the quantities (6.21) substituted for σ_n:

$$\ddot{T}_n(t) + \sigma_n^2 c^2 T_n(t) = 0 \qquad n = 1, 2 \ldots$$

(6.24)

where $T(t)$ is now indexed because there is one solution for each value of σ_n. The general form of the solution of (6.24) is given in Section 2.3 as

$$T_n(t) = A_n \sin \sigma_n ct + B_n \cos \sigma_n ct$$

(6.25)

where A_n and B_n are constants of integration. Since each of the functions $X_n(x)$ and $T_n(t)$ are found to be dependent on n, the solution $w(x, t) = X_n(x)T_n(t)$ must also be a function of n, so that

$$w_n(x, t) = c_n \sin\left(\frac{n\pi}{l}x\right) \sin\left(\frac{n\pi c}{l}t\right) + d_n \sin\left(\frac{n\pi}{l}x\right) \cos\left(\frac{n\pi c}{l}t\right) \qquad n = 1, 2, \ldots$$

(6.26)

where c_n and d_n are new constants to be determined. Note that an unknown constant a_n times another unknown constant A_n is the unknown constant c_n (similarly, $d_n = a_n B_n$). Since the string equation is linear, any linear combination of solutions is a solution. Hence the general solution is of the form

$$w(x, t) = \sum_{n=1}^{\infty} (c_n \sin \sigma_n x \sin \sigma_n ct + d_n \sin \sigma_n x \cos \sigma_n ct)$$

(6.27)

The set of constants $\{c_n\}$ and $\{d_n\}$ can be determined by applying the initial conditions on $w(x, t)$ and the orthogonality of the set of functions $\sin(n\pi x/l)$. The orthogonality of the set of functions $\sin(n\pi x/l)$ states that

$$\int_0^l \sin\frac{n\pi x}{l} \sin\frac{m\pi x}{l} dx = \begin{cases} \frac{l}{2} & n = m \\ 0 & n \neq m \end{cases} \tag{6.28}$$

which is identical to the orthogonality of mode shape vectors discussed in Section 4.2. This orthogonality condition can be derived by simple trigonometric identities and integration (as suggested in Section 3.3).

Consider the initial condition on the displacement applied to equation (6.27):

$$w(x,0) = w_0(x) = \sum_{n=1}^{\infty} d_n \sin\frac{n\pi x}{l} \cos(0) \tag{6.29}$$

Multiplying both sides of this equality by $\sin m\pi x/l$ and integrating over the length of the string yields

$$\int_0^l w_0(x) \sin\frac{m\pi x}{l} dx = \sum_{n=1}^{\infty} d_n \int_0^l \sin\frac{n\pi x}{l} \sin\frac{m\pi x}{l} dx = d_m \left(\frac{l}{2}\right) \tag{6.30}$$

where each term in the summation on the right-hand side of equation (6.29) is zero except for the mth term, by direct application of the orthogonality condition of (6.28). Equation (6.30) must hold for each value of m, so that

$$d_m = \frac{2}{l} \int_0^l w_0(x) \sin\frac{m\pi x}{l} dx \qquad m = 1, 2, 3, \ldots \tag{6.31}$$

It is customary to replace m by n in the above, since the index runs over all positive values [i.e., the index in equation (6.31) is a free index and it does not matter what it is named]. It is most convenient to rename it d_n.

A similar expression for the constants $\{c_n\}$ is obtained by using the initial velocity condition. Time differentiation of the summation of equation (6.27) yields

$$\dot{w}_0(x) = w_t(x, 0) = \sum_{n=1}^{\infty} c_n \sigma_n c \sin\frac{n\pi x}{l} \cos(0) \tag{6.32}$$

Again, multiplying by $\sin m\pi x/l$, integrating over the length of the string, and applying the orthogonality condition of (6.28) yields

$$c_n = \frac{2}{n\pi c} \int_0^l \dot{w}_0(x) \sin\frac{n\pi x}{l} dx \qquad n = 1, 2, 3, \ldots \tag{6.33}$$

where the index has been renamed n. Equations (6.31) and (6.33) combined with equation (6.27) form the complete solution for the vibrating string (i.e., they describe the vibration response of the string at any point x and any time t).

Example 6.2.1

The solution of a second-order ordinary differential equation with constant coefficients subject to boundary conditions is used throughout this section. This example clarifies the choice of a negative constant for the separation of variables procedure and reviews the solution technique for second-order differential equations with constant coefficients (see e.g., Boyce and DiPrima, 1986). Consider again equation (6.14), where this time the separation constant is chosen to be $-\beta$, where β is of arbitrary sign. This yields

$$X'' + \beta X = 0 \tag{6.34}$$

Assume that the solution of (6.34) is of the form $X(x) = e^{\lambda x}$, where λ is to be determined. Substitution into (6.34) yields

$$(\lambda^2 + \beta)e^{\lambda x} = 0 \tag{6.35}$$

Since $e^{\lambda x}$ is never zero, this requires that

$$\lambda = \pm\sqrt{-\beta} \tag{6.36}$$

Here λ will be purely imaginary or real, depending on the sign of the separation constant β. Thus there are two solutions and the general solution is the sum

$$X(x) = Ae^{-\sqrt{-\beta}x} + Be^{+\sqrt{-\beta}x} \tag{6.37}$$

where A and B are constants of integration to be determined by the boundary conditions.

Applying the boundary conditions to (6.37) yields

$$X(0) = A + B = 0$$
$$\tag{6.38}$$
$$X(l) = Ae^{-\sqrt{-\beta}l} + Be^{\sqrt{-\beta}l} = 0$$

This system of equations can be written in the matrix form

$$\begin{bmatrix} 1 & 1 \\ e^{-\sqrt{-\beta}l} & e^{\sqrt{-\beta}l} \end{bmatrix} \begin{bmatrix} A \\ B \end{bmatrix} = \begin{bmatrix} 0 \\ 0 \end{bmatrix}$$

which has a nonzero solution for A and B if and only if (recall Section 4.1) the determinant of the matrix coefficient is zero. This yields

$$e^{+\sqrt{-\beta}l} - e^{-\sqrt{-\beta}l} = 0 \tag{6.39}$$

must be satisfied. For negative real values of β, say $\beta = -\sigma^2$, this becomes [recall the definition $\sinh u = (e^u - e^{-u})/2$]

$$e^{\sigma l} - e^{-\sigma l} = 2\sinh \sigma l = 0 \tag{6.40}$$

which has only the trivial solution and hence $\beta \neq -\sigma^2$. This means that the separation constant σ in the development of equation (6.14) cannot be positive. Thus $\sqrt{-\beta}$ must be complex (i.e., $\beta = \sigma^2$), so that equation (6.39) becomes

$$e^{\sigma jl} - e^{-\sigma jl} = 0 \tag{6.41}$$

where $j = \sqrt{-1}$. Euler's formula for the sine function is $\sin u = (e^{uj} - e^{-uj})/2j$, so that (6.41) becomes

$$\sin \sigma l = 0 \tag{6.42}$$

which has the solution $\sigma = n\pi/l$ as used in equation (6.20).

The only possibility that remains to check is the case $\beta = 0$, which yields $X'' = 0$, or upon integrating twice,

$$X(x) = a + bx \tag{6.43}$$

Applying the boundary conditions results in

$$X(0) = a = 0$$

$$X(l) = bl = 0$$

(6.44)

which yields only the trivial solution $a = b = 0$. Hence the choice of separation constant in equation (6.14) as $-\sigma^2$ is fully justified, and the solution of the spatial equation for the string fixed at both ends is of the form

$$X_n(x) = a_n \sin \frac{n\pi x}{l}$$

Here the subscript n has been added to $X(x)$ to denote its dependence on the index n and to indicate that more than one solution results. In this case an infinite number of solutions result, one for each integer n.

$\square$

Note that the solution of this spatial equation is very much like the solution of the equation of the single-degree-of-freedom oscillator in Section 1.1 and reviewed in Window 6.1. There are, however, two main differences. The sign of the coefficient in the single-degree-of-freedom oscillator is determined on physical grounds (i.e., the ratio of stiffness to mass is positive), and second, the constants of integration were both evaluated at the beginning of the interval instead of one at each end. The spatial string equation is a boundary value problem, while the single-degree-of-freedom oscillator equation is an initial value problem. Note that the temporal equation for the string, however, is identical to the equation of a single-degree-of-freedom oscillator. The temporal equation is also an initial value problem.

Window 6.1
Review of the Solution of a Single-Degree-of-Freedom System

The solution to $m\ddot{x} + kx = 0$, $x(0) = x_0$, $\dot{x}(0) = v_0$ for $m, k > 0$ is

$$x(t) = \frac{\sqrt{\omega^2 x_0^2 + v_0^2}}{\omega} \sin\left(\omega t + \tan^{-1} \frac{\omega x_0}{v_0}\right)$$

(1.10)

where $\omega = \sqrt{k/m}$.

Now that the mathematical solution of the vibrating string is established consider the physical interpretation of the various terms in equation (6.27). Consider giving the string an initial displacement of the form

$$w_0(x) = \sin \frac{\pi x}{l}$$

(6.45)

and an initial velocity of $\dot{w}_0(x) = 0$. With these values of the initial conditions, the computation of the coefficients in (6.27) by using equations (6.31) and (6.33) yields

$c_n = 0$ for $n = 1, 2, \ldots, d_n = 0$ for $n = 2, 3, \ldots$, and $d_1 = 1$. Substitution of these coefficients back into equation (6.27) yields the solution

$$w(x, t) = \sin \frac{\pi x}{l} \cos \frac{\pi c}{l} t \tag{6.46}$$

which is the first term of the series. Comparing the second factor of equation (6.46) to $\cos \omega t$ states that the string is oscillating in time at a frequency of

$$\frac{\pi c}{l} = \frac{\pi}{l} \sqrt{\frac{\tau}{\rho}} \tag{6.47}$$

in radians per second, where $c = \sqrt{\tau/\rho}$, in the spatial shape of $\sin(\pi x/l)$. Hence for a fixed time t, the string will be deformed in the shape of a sinusoid. Each point of the string is moving up and down in time (i.e., vibrating at frequency $\pi c/l$). Borrowing the jargon of Chapter 4, the function $X_1(x) = \sin(\pi x/l)$ is called a *mode shape*, or *mode*, of the string and the quantity $(\pi c/l)$ is called a *natural frequency* of the string.

This procedure can be repeated for each value of the index n by choosing the initial displacement to be $\sin(n\pi x/l)$ and the velocity to be zero. This gives rise to an infinite number of mode shapes, $\sin(n\pi x/l)$, and natural frequencies $(n\pi c/l)$ and is the reason distributed-parameter systems are called *infinite-dimensional systems*. Figure 6.3 illustrates the first two mode shapes of the fixed endpoint string. If the string is excited in the second mode by using the initial conditions $w_0 = \sin(2\pi x/l)$, $\dot{w}_0 = 0$, and the resulting motion is viewed by a stroboscope flashing with frequency $2\pi c/l$, the curve labeled $n = 2$ in Figure 6.3 would be visible.

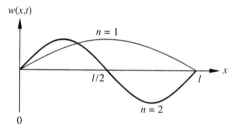

Figure 6.3 Plot of the deflection $w(x, t)$ versus the position x for a fixed time illustrating the first two mode shapes of a vibrating string fixed at both ends.

An interesting property of the modes of a string are the points where the modes $\sin(n\pi x/l)$ are zero. These points are called *nodes*. Note from Figure 6.3 that the node of the second mode is at the point $l/2$. If the string is given an initial displacement equal to the second mode, there will be no motion at the point $l/2$ for any value of t. This node, as in the case of lumped-parameter systems, is a point on the string that does not move if the structure is excited only in that mode.

The modes and natural frequencies for a string and those for a multiple-degree-of-freedom system are very similar. For a multiple-degree-of-freedom system the mode shape is a vector, the elements of which yield the relative amplitude of vibration of each coordinate of the system. The same is true for an eigenfunction of the string, the difference being that instead of elements of a vector, the eigenfunction gives the modal response magnitude at each value of the position x along the string.

Equation (6.27) is the distributed-parameter equivalent of the modal expansion given by equation (4.103) for lumped-parameter systems. Instead of an expansion of the solution vector $\mathbf{x}(t)$ in terms of the modal vectors $\mathbf{u}_i$ indicated in equation (4.103) and reviewed in Window 6.2, the distributed parameter equivalent consists of an expansion in terms of the mode shape functions $\sin \sigma_n x$. Just as the mode shape vectors determine the relative magnitude of the motion of various masses of the lumped system, the mode shape functions determine the magnitude of the motion of the mass distribution of the distributed parameter system. The procedures outlined above in this section constitute the modal analysis of a distributed parameter system. This basic procedure is repeated in the following sections to determine the vibration response of a variety of different distributed-parameter systems. Modal analysis for damped systems is given in Section 6.7 and for the forced response in Section 6.8.

Window 6.2
A Review of the Modal Expansion Solution of a Multiple-Degree-of-Freedom System

The solution of $M \ddot{\mathbf{x}} + K \mathbf{x} = \mathbf{0}$, $\mathbf{x}(0) = \mathbf{x}_0$, $\dot{\mathbf{x}}(0) = \dot{\mathbf{x}}_0$ is

$$\mathbf{x}(t) = \sum_{i=1}^{n} c_i \sin(\omega_i t + \phi_i) \mathbf{u}_i \qquad (4.103)$$

where $\mathbf{u}_i$ are the eigenvectors of the matrix $M^{-1/2} K M^{-1/2}$ and ω_i^2 are the associated eigenvalues of $M^{-1/2} K M^{-1/2}$. The constants c_i and ϕ_i are determined from the initial conditions by using the orthogonality of $\mathbf{u}_i$ to obtain

$$c_i = \frac{\mathbf{u}_i^T M \mathbf{x}(0)}{\sin \phi_i}$$

$$\phi_i = \tan^{-1} \frac{\omega_i \mathbf{u}_i^T M \mathbf{x}(0)}{\mathbf{u}_i^T M \dot{\mathbf{x}}(0)}$$

Example 6.2.2

For the sake of obtaining a feeling for the units of a vibrating string, consider a piano wire. A reasonable model of a piano wire is the string fixed at both ends of Figure 6.1. A piano wire is 1.4 m long, has a mass of about 110 g, and has a tension of about $\tau = 11.1 \times 10^4$ N. Its first natural frequency is then

$$\omega_1 = \frac{\pi c}{l} = \frac{\pi}{1.4 \text{ m}} \sqrt{\frac{\tau}{\rho}}$$

Since $\rho = 110$ g per 1.4 m = 0.0786 kg/m, this yields

$$\omega_1 = \frac{\pi}{1.4 \text{ m}} \left(\frac{11.1 \times 10^4 \text{ N}}{0.0786 \text{ kg/m}} \right)^{1/2} = 2666.69 \text{ rad/s} \quad \text{or} \quad 424 \text{ Hz}$$

$\square$

Example 6.2.3

Calculate the mode shapes and natural frequencies of the spring–cable system of Example 6.1.1. The solution of the string equation will be the same as that presented for the string fixed at both ends except for the evaluation of the constants by using the boundary conditions.

Solution   Referring to equation (6.17) and applying the boundary conditions given in Example 6.1.1 yields

$$X(0) = a_2 = 0 \qquad (6.48)$$

at one end. At the other end substitution of $w(x, t) = X(x)T(t)$ into the boundary condition yields

$$\tau X'(l)T(t) = -kX(l)T(t) \qquad (6.49)$$

Upon further substitution of $X = a_1 \sin \sigma x$ this becomes

$$\tau \sigma \cos \sigma l = -k \sin \sigma l \qquad (6.50)$$

so that

$$\tan \sigma l = -\frac{\tau \sigma}{k} \qquad (6.51)$$

As in the case of the string fixed at both ends, satisfying the second boundary condition yields an infinite number of values of σ. Hence σ becomes σ_n, where σ_n are the solutions of (6.51). These values of σ_n can be visualized (and computed) by plotting $\tan \sigma l$ and $-\tau\sigma/k$ versus σ on the same plot as illustrated in Figure 6.4. Thus the eigenfunctions or mode shapes of a cable connected to a spring are

$$a_n \sin \sigma_n x \qquad (6.52)$$

where the σ_n must be calculated numerically for given values of k, τ, and l from equation (6.51). Note that by examining the curves of Figure 6.4, the values of σ_n (the point where the two curves cross) for large values of n become very close to $(2n - 1)\pi/(2l)$.

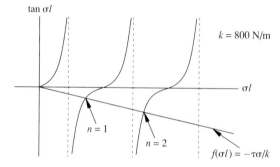

Figure 6.4   Graphical solution of the transcendental equation $\tan \sigma l = -\tau\sigma/k$.

The separation-of-variables modal analysis method illustrated in this section is summarized in Window 6.3. Note that the procedure is analogous to that summarized in Window 6.2 for lumped-parameter systems.

Window 6.3
Summary of the Separation-of-Variables Solutions Method

1. A solution to a partial differential equation is assumed to be of the form $w(x, t) = X(x)T(t)$ (i.e., separated).

2. The separated form is next substituted into the partial differential equation of motion and separated by trying to position all of the terms containing $X(x)$ on one side of the equality, and all of the terms containing $T(t)$ on the other side. (Note that this cannot always be done.)

3. Once written in separated form, equation (6.12), for example, each side of the equality is set equal to the same constant to obtain two new equations: a spatial equation in the single-valued variable $X(x)$ and a temporal equation in the single-valued variable $T(t)$. This constant is called the separation constant and is denoted σ^2.

4. The boundary conditions are applied to the spatial equation that results in an algebraic eigenvector problem which yields an infinite number of values of the separation constant, σ_n, and the mode shapes $X_n(x)$.

5. The separation constants obtained in step 4 are substituted into the temporal equation and the solutions $T_n(t)$ are obtained in terms of two arbitrary constants.

6. The infinite number of solutions $w_n(x, t) = X_n(x)T_n(t)$ are then formed and summed to produce the series solution

$$w(x, t) = \sum_{n=1}^{\infty} X_n(x)T_n(t).$$

which contains two sets of undetermined constants.

7. The unknown constants are determined by applying the initial conditions $w(x, 0)$ and $w_t(x, 0)$ to the series solution, multiplying by a mode $X_m(x)$, and integrating over x. The orthogonality condition then yields simple equations for the unknown constants in terms of the initial conditions. See equations (6.31) and (6.33), for example.

8. The series solution is written with the constants evaluated from step 7 and the solution is complete.

6.3 VIBRATION OF RODS AND BARS

Next consider the vibration of an elastic bar (or rod) of length l and of varying cross-sectional area in the direction indicated in Figure 6.5. The density of the bar is denoted by ρ (not to be confused with mass per unit length used for the string) and the cross-sectional area by $A(x)$. Using the coordinate system indicated in the figure, the forces on the infinitesimal element summed in the x direction are

$$F + dF - F = \rho A \, dx \frac{\partial^2 w(x, t)}{\partial t^2} \tag{6.53}$$

where $w(x, t)$ is the deflection of the rod in the x direction, F denotes the force acting on the infinitesimal element to the left, and $F + dF$ denotes the force to the right, as the element is displaced. From introductory strength of materials the force F is given by $F = \sigma_s A$, where σ_s is the unit stress in the x direction and has the value $E w_x(x, t)$, where E is the Young's modulus (or modulus of elasticity) and w_x is the unit strain. This yields

$$F = EA(x) \frac{\partial w(x, t)}{\partial x} \tag{6.54}$$

The differential force becomes $dF = (\partial F / \partial x) dx$, from the chain rule for partial derivatives. Substitution of these quantities into equation (6.53) and dividing by dx yields

$$\frac{\partial}{\partial x} \left(E A(x) \frac{\partial w(x, t)}{\partial x} \right) = \rho A(x) \frac{\partial^2 w(x, t)}{\partial t^2} \tag{6.55}$$

In those cases where $A(x)$ is a constant this becomes

$$\left(\frac{E}{\rho} \right) \frac{\partial^2 w(x, t)}{\partial x^2} = \frac{\partial^2 w(x, t)}{\partial t^2} \tag{6.56}$$

which has exactly the same form as the string equation of Section 6.2. The quantity $c = \sqrt{E/\rho}$ defines the velocity of propagation of the displacement (or stress wave) in the bar. Hence the solution technique used in Section 6.2 is exactly applicable here.

Because the bar can support its own weight, a variety of boundary conditions are possible. Various springs, masses, or dashpots can be attached to one end of the bar or the other, to model a variety of situations. Consider the boundary conditions for the clamped–free configuration (called cantilevered) of Figure 6.5. At the clamped end, $x = 0$, the displacement must be zero, so that

$$w(0, t) = 0 \qquad t > 0 \tag{6.57}$$

At the free end, $x = l$, the stress in the bar must be zero, or

$$EA \left. \frac{\partial w(x, t)}{\partial x} \right|_{x=l} = 0 \tag{6.58}$$

These are the simplest boundary conditions.

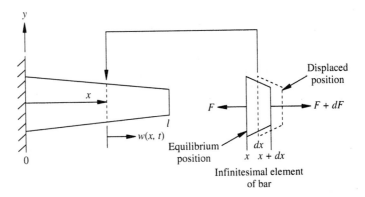

Figure 6.5   Cantilevered bar in longitudinal vibration along x.

Example 6.3.1

Calculate the mode shapes and natural frequencies of the cantilevered (clamped–free) bar of Figure 6.5 for the case of constant cross section.

Solution   Following the solution details of the string equation and the method outlined in Window 6.3, the spatial response of the bar will be of the form

$$X(x) = a \sin \sigma x + b \cos \sigma x \tag{6.59}$$

Applying the condition at the fixed end, $x = 0$, yields the result that $b = 0$, so that $X(x)$ has the form

$$X(x) = a \sin \sigma x$$

At $x = l$, the boundary condition yields

$$AEX'(l) = 0 = AE\sigma \, a \cos \sigma l \tag{6.60}$$

Since A, E, a, and σ are nonzero, this requires that $\cos \sigma l = 0$, or that

$$\sigma_n = \frac{2n - 1}{2l}\pi \qquad n = 1, 2, 3, \dots \tag{6.61}$$

The mode shapes are thus of the form

$$a_n \sin \frac{(2n - 1)\pi x}{2l} \tag{6.62}$$

and the natural frequencies are of the form

$$\omega_n = \frac{(2n - 1)\pi c}{2l} = \frac{(2n - 1)\pi}{2l}\sqrt{\frac{E}{\rho}} \qquad n = 1, 2, 3, \dots \tag{6.63}$$

The solution is given by equation (6.27) with $c = \sqrt{E/\rho}$ and σ_n as given by expression (6.61). $\square$

In distributed-parameter systems, the form of the spatial differential equation is often determined by the stiffness term. The differential form of the stiffness term along with the boundary conditions determine whether or not the eigenfunctions, and hence the mode shapes, are orthogonal. The eigenfunction orthogonality condition for distributed-parameter systems is an exact replica of the eigenvector and mode shape orthogonality condition for lumped-parameter systems. Precise orthogonality conditions are beyond the scope of this chapter and can be found in Inman (1989), for instance. Note in Example 6.3.1 that the mode shapes given by expression (6.62) are orthogonal [recall equation (6.28)]. Just as in the lumped-parameter case, the orthogonality of mode shapes for a distributed-parameter system is extremely useful in calculating the vibration response of the system. The following example illustrates the use of modes to calculate the response to a given set of initial conditions.

Example 6.3.2

Calculate the response of the bar of Example 6.3.1 to an initial velocity of 3 cm/s at the free end and a zero initial displacement. Assume that the bar is 5 m long, has a density of $\rho = 8 \times 10^3$ kg/m^3, and has a modulus of $E = 20 \times 10^{10}$ N/m^2. Plot the response at $x = l$ and $x = l/2$.

Solution The solution $w(x, t)$ for the longitudinal vibration of the bar is given by equation (6.27) as

$$w(x, t) = \sum_{n=1}^{\infty} (c_n \sin \sigma_n ct + d_n \cos \sigma_n ct) \sin \frac{(2n - 1)\pi x}{2l}$$

where σ_n is calculated in Example 6.3.1 and given in equation (6.61). To calculate the coefficients c_n and d_n, note that $w(x, 0) = 0$ and $w_t(l, 0) = 0.03$ m/s. Thus from equation (6.31) $d_n = 0$, and from equation (6.32) the c_n are determined from

$$w_t(x, 0) = \sum_{n=1}^{\infty} c_n \sigma_n c \sin \sigma_n x \cos(0)$$

Multiplying this last expression by $\sin \sigma_m x$ and integrating yields

$$(0.03) \int_0^l \sin(\sigma_m x)dx = c_m \sigma_m c \int_0^l \sin^2(\sigma_m x)dx$$

since the set of functions $\sin \sigma_m x$ are orthogonal. Evaluating the integrals yields

$$\frac{0.03}{\sigma_m} = \frac{c\sigma_m l}{2} c_m$$

for each m. Solving this, reindexing, and substituting the appropriate physical parameters yields

$$c_n = \frac{(0.03)2}{lc\sigma_n^2} = \frac{(0.06)4l^2}{\pi^2(5)\sqrt{E/\rho}} \frac{1}{(2n - 1)^2}$$

Thus the total solution becomes

$$w(x, t) = 2.43 \times 10^{-5} \sum_{n=1}^{\infty} \frac{1}{(2n - 1)^2} \sin \frac{(2n - 1)\pi x}{10} \sin[500(2n - 1)\pi t]$$

At $x = l$ this becomes

$$w(l, t) = 2.43 \times 10^{-5} \sum_{n=1}^{\infty} \frac{(-1)^{n+1}}{(2n - 1)^2} \sin[500(2n - 1)\pi t]$$

and at $x = l/2$ the response is given by

$$w(l/2, t) = 2.43 \times 10^{-5} \sum_{n=1}^{\infty} \frac{(-1)^{n+1}}{(2n-1)^2} \sin(500)(2n-1)\pi t$$

To plot these, the infinite series must be approximated by truncating the series. In Figure 6.6, the response of $w(l/2, t)$ is plotted for one, two, five, and ten terms. Note that the only noticeable difference in the plot of the first term in the sum, and that of the first two terms is the increase in amplitude. After five terms, not much change in the plots occur. The toolbox problems at the end of this chapter investigate this further.

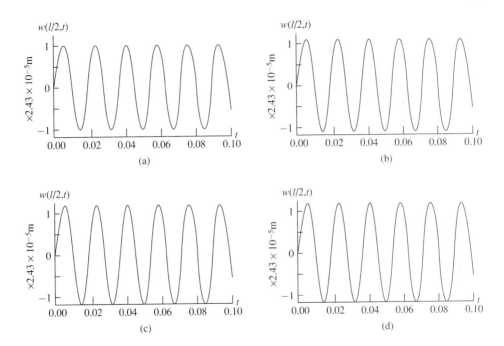

Figure 6.6   Response of the bar of Example 6.3.2 plotted for $x = l/2$ for (a) one, (b) two, (c) five, and (d) 10 terms of the series solution.

The analysis of a lumped-parameter system and that of a distributed-parameter system are similar in as much as they both require the computation of natural frequencies and mode shapes (eigenvectors in the lumped case, eigenfunctions in the distributed case). However, the time response of the distributed-parameter system is distributed spatially in a continuous manner, whereas that of a lumped-parameter system is spatially discrete. Thus for a lumped system it is useful to plot the time response of a single coordinate $x_i(t)$, and for a distributed-parameter system the response plot is at a single point [i.e, $w(l/2, t)$] as in Figure 6.6. For the lumped case there is a finite number of degrees of freedom and hence a finite number of responses $x_i(t)$ ($i = 1, 2, \ldots n$) to consider. In the distributed-parameter case, there is an infinite number of responses $w(x, t)$ to consider, as the spatial variable x can take on any value between 0 and l. Boundary conditions for various configurations of the bar are listed in Table 6.1 along with appropriate frequencies and mode shapes. The boundary, or end, of the beam can be free, requiring the longitudinal stress to vanish: $\partial w / \partial x = 0$, fixed requiring the displacement to vanish $w = 0$, attached to a spring of stiffness k requiring $kw = AE \, \partial w / \partial x$, or attached to a mass, m, requiring $-m \partial^2 w / \partial t^2 = AE \, \partial w / \partial x$. The results of using some of these boundary conditions are listed in Table 6.1.

TABLE 6.1   VARIOUS CONFIGURATIONS OF A UNIFORM BAR OF LENGTH l IN LONGITUDINAL VIBRATION ILLUSTRATING THE NATURAL FREQUENCIES AND MODE SHAPES[a]

Configuration	Frequency (rad/s) or characteristic equation	Mode shape
Free–free	$\omega_n = \dfrac{n \pi c}{l}, \quad n = 0, 1, 2, \ldots$	$\cos \dfrac{n \pi x}{l}$
Fixed–free	$\omega_n = \dfrac{(2n - 1) \pi c}{2l}, \quad n = 1, 2, \ldots$	$\sin \dfrac{(2n - 1) \pi x}{2l}$
Fixed–fixed	$\omega_n = \dfrac{n \pi c}{l}, \quad n = 1, 2, \ldots$	$\sin \dfrac{n \pi x}{l}$
Fixed–spring	$\lambda_n \cot \lambda_n = -\left(\dfrac{kl}{EA} \right)$ $\omega_n = \dfrac{\lambda_n c}{l}$	$\sin \dfrac{\lambda_n x}{l}$
Fixed–mass	$\cot \lambda_n = \left(\dfrac{m}{\rho Al} \right) \lambda_n$ $\omega_n = \dfrac{\lambda_n c}{l}$	$\sin \dfrac{\lambda_n x}{l}$

[a] Note that the last two examples require a numerical procedure (suggested in the toolbox problems at the end of the chapter) in order to calculate the values of the natural frequencies. Here $c = \sqrt{E/\rho}$.

6.4 TORSIONAL VIBRATION

The bar of Section 6.3 may also vibrate in the torsional direction as indicated by the circular shaft of Figure 6.7. In this case the vibration occurs in an angular direction around the center axis of the shaft in a plane perpendicular to the cross section of the shaft or rod. The rotation of the shaft, θ, about the center axis is a function of both the position along the length of the rod, x, and the time, t. Thus θ is a function of two variables denoted $\theta(x, t)$. The equation of motion can be determined by considering a moment balance of an infinitesimal element of the rod of length dx illustrated in the figure. Referring to Figure 6.7, the torque at the right face of the element dx (i.e., at position x) is τ (here τ is used as a torque, not a tension as in Section 6.1), while that at the left end, at position $x + dx$, is $\tau + \frac{\partial \tau}{\partial x} dx$. From solid mechanics the applied torque is related to the torsional deflection by (Shames, 1989)

$$\tau = GJ \frac{\partial \theta(x, t)}{\partial x} \tag{6.64}$$

where GJ is the torsional stiffness composed of the shear modulus G and the polar moment of area J of the cross section. Note that J could be a function of x as well, but is considered constant here. The total torque acting on dx becomes

$$\tau + \frac{\partial \tau}{\partial x} dx - \tau = J_0 \frac{\partial^2 \theta}{\partial t^2} dx \tag{6.65}$$

where J_0 is the polar moment of inertia of the shaft per unit length and $\partial^2 \theta / \partial t^2$ is the angular acceleration. If the shaft is of uniform circular cross section, J_0 becomes simply $J_0 = \rho J$, where ρ is the shaft's material mass density. Substitution of the expression for the torque given by equation (6.64) into equation (6.65) yields

$$\frac{\partial}{\partial x} \left(GJ \frac{\partial \theta}{\partial x} \right) = \rho J \frac{\partial^2 \theta}{\partial t^2}$$

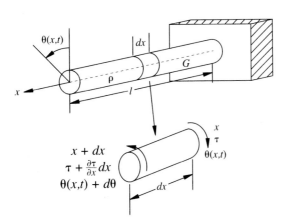

Figure 6.7   Circular shaft illustrating an angular motion, $\theta(x, t)$, as the result of a moment acting on a differential element dx of the shaft of density ρ, length l, and given modulus G. The function $\theta(x, t)$ denotes the angle of twist.

Simplifying for the case of constant stiffness GJ yields

$$\frac{\partial^2\theta(x,t)}{\partial t^2} = \left(\frac{G}{\rho}\right)\frac{\partial^2\theta(x,t)}{\partial x^2} \tag{6.66}$$

for the equation of twisting vibration of a rod. This is, again, identical in mathematical form to the formula of equation (6.8) for the string or cable, and of equation (6.56) for the longitudinal vibrations of a rod or bar.

For other types of cross sections, the torsional equation of a shaft can still be used to approximate the torsional motion by replacing J in equation (6.64) with a *torsional constant* γ defined to be the moment required to produce a torsional rotation of 1 rad on a unit length of shaft divided by the shear modulus. Thus a shaft with noncircular cross section can be approximated by the equation

$$\frac{\partial^2\theta(x,t)}{\partial t^2} = \left(\frac{G\gamma}{\rho J}\right)\frac{\partial^2\theta(x,t)}{\partial x^2} \tag{6.67}$$

Some values of γ are presented in Table 6.2 for several common cross sections. Keep in mind that equation (6.67) is approximate and assumes in particular that the center of mass and center of rotation coincide so that torsional and flexural vibrations do not couple.

The solution of equation (6.66) depends on two initial conditions in time [i.e., $\theta(x,0)$ and $\theta_t(x,0)$] and two boundary conditions, one at each end of the rod. The possible choices of boundary conditions are similar to those for the string and the bar (i.e., either the deflection is zero if the rod is fixed at a boundary or the torque is zero if the rod is free at a boundary). For example, the clamped–free rod of Figure 6.7 has boundary conditions

$$\text{(deflection at 0)}\quad \theta(0,t) = 0 \tag{6.68}$$

$$\text{(torque at } l)\quad \gamma G\theta_x(l,t) = 0 \tag{6.69}$$

Some other common boundary conditions are that an end could be connected to a spring of stiffness k in which case the boundary condition becomes $k\theta = G\gamma\partial\theta/\partial x$ at the boundary, or an end could be connected to a lumped mass of polar mass moment of inertia J_1, in which case the boundary condition becomes $-J_1\partial^2\theta/\partial t^2 = G\gamma\partial\theta/\partial x$. The solution technique is identical to the procedure used in Example 6.3.1 for a clamped-free bar with a different interpretation given to the displacement and to the physical constants.

Example 6.4.1

The vibration of a grinding tool can be modeled as a shaft or rod with a disk at one end as illustrated in Figure 6.8. The top end of the shaft, at $x = 0$, is connected to a pulley. The effects of the drive belt and motor are accounted for by including their collective inertia in with the mass moment of inertia of the pulley denoted J_1. Determine the natural frequencies of the system.

Solution The frequencies of vibration are determined by equation (6.66) subject to the appropriate boundary conditions. The boundary conditions are determined by examining either the deflections or torques at the boundaries. In this example the deflection is unspecified at

TABLE 6.2 SOME VALUES OF THE TORSIONAL
CONSTANT FOR VARIOUS CROSS-SECTIONAL SHAPES
USED IN APPROXIMATING TORSIONAL VIBRATION FOR
NONCIRCULAR CROSS SECTIONS

Cross section	Torsional Constant, γ
	Circular shaft $\dfrac{\pi R^4}{2}$
	Hollow circular shaft $\dfrac{\pi}{2}(R_2^4 - R_1^4)$
	Square shaft $0.1406a^4$
	Hollow rectangular shaft $\dfrac{2AB(a - A)^2(b - B)^2}{aA + bB - A^2 - B^2}$

$x = 0$, but the torque must match that supplied by the pulley or

$$GJ \frac{\partial \theta(0, t)}{\partial x} = J_1 \frac{\partial^2 \theta(0, t)}{\partial t^2} \tag{6.70}$$

Similarly, at $x = l$ a balance of torques yields

$$GJ \frac{\partial \theta(l, t)}{\partial x} = -J_2 \frac{\partial^2 \theta(l, t)}{\partial t^2} \tag{6.71}$$

where the minus sign arises from the right-hand rule. Following the method of Section 6.2, the solution for $\theta(x, t)$ is assumed to separate and be of the form $\theta(x, t) = \Theta(x)T(t)$.

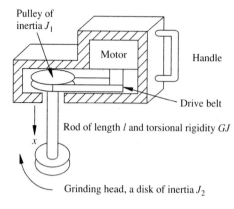

Pulley of
inertia J_1

Motor

Handle

Drive belt

Rod of length l and torsional rigidity GJ

x

Grinding head, a disk of inertia J_2

Figure 6.8 Simple model of a grinding tool or disk sander.

Substitution into equation (6.66) and rearranging yields

$$\frac{\Theta''(x)}{\Theta(x)} = \left(\frac{\rho}{G}\right)\frac{\ddot{T}(t)}{T(t)} = -\sigma^2 \qquad (6.72)$$

Defining $(\rho/G) = 1/c^2$ and realizing that equation (6.72) splits into a temporal equation and a spatial equation, the spatial equation becomes

$$\Theta''(x) + \sigma^2\Theta(x) = 0 \qquad (6.73)$$

where σ is the separation constant and is related to the natural frequencies of the system by

$$\omega = \sigma c = \sigma\sqrt{\frac{G}{\rho}} \qquad (6.74)$$

From equation (6.70) the boundary condition at $x = 0$ becomes

$$GJ\,\Theta'(0)T(t) = J_1\Theta(0)\ddot{T}(t) \qquad (6.75)$$

or

$$\frac{GJ\,\Theta'(0)}{J_1\Theta(0)} = \frac{\ddot{T}(t)}{T(t)} = -c^2\sigma^2 \qquad (6.76)$$

where the last equality follows from the right-hand side of equation (6.72). Upon further manipulation, equation (6.76) yields

$$\Theta'(0) = -\frac{\sigma^2 J_1}{\rho J}\Theta(0) \qquad (6.77)$$

Similarly, the boundary conditions at $x = l$ given by equation (6.71) yields

$$\Theta'(l) = \frac{\sigma^2 J_2}{\rho J}\Theta(l) \qquad (6.78)$$

The general solution of equation (6.73) is

$$\Theta(x) = a_1 \sin \sigma x + a_2 \cos \sigma x \qquad (6.79)$$

so that

$$\Theta'(x) = a_1\sigma\cos\sigma x - a_2\sigma\sin\sigma x \tag{6.80}$$

Substitution of these expressions into equation (6.77) for the boundary condition at $x = 0$ yields

$$a_1 = -\frac{\sigma J_1}{\rho J}a_2 \tag{6.81}$$

Application of the boundary conditions at $x = l$ by substitution of equations (6.79), (6.80), and (6.81) into equation (6.78) yields the characteristic equation given by

$$\tan\sigma l = \frac{\rho J l(J_1 + J_2)(\sigma l)}{J_1 J_2(\sigma l)^2 - \rho^2 J^2 l^2} \tag{6.82}$$

This expression is a transcendental equation in the quantity σl, which must be solved numerically, similar to Example 6.2.3. Because of the tangent term, equation (6.82) has an infinite number of solutions which can be denoted $\sigma_n l, n = 1, 2, 3, \ldots \infty$. The numerical solutions σ_n determine the natural frequencies of vibrations according to equation (6.74):

$$\omega_n = \sigma_n\sqrt{\frac{G}{\rho}} \tag{6.83}$$

Note that the first solution of equation (6.82) corresponding to the first eigenvalue yields

$$\omega_1 = 0$$

Since $\omega_1 = \sigma_1 = 0$, the time equation (6.72) becomes $\ddot{T}_1(t) = 0$, or $T_1(t) = a + bt$, where a and b are constants determined by the initial conditions. This solution corresponds to the rigid-body mode of the system which in this case is a constant shaft rotation. The rigid-body mode shape is calculated from equation (6.73) with $\sigma = 0$. This yields $\Theta_1''(x) = 0$, so that $\Theta_1(x) = a_1 + b_1 x$. Applying the boundary condition at $x = 0$ yields $GJ(b_1) = 0$, so that $b_1 = 0$. The boundary condition at $x = l$ also results in $b_1 = 0$, so that the first mode shape, or first eigenfunction, is simply $\Theta_1(x) = a_1$, a nonzero constant. This defines the rigid-body mode shape.

For very large values of σl, equation (6.82) reduces to $\tan\sigma l = 0$, so that the high frequencies of vibration approach $\omega_n = n\pi c$. Examining the form of the characteristic equation indicates that it can be written as

$$(bx^2 - a)\tan x = x \tag{6.84}$$

where $x = \sigma l, a = \rho J l/(J_1 + J_2)$, and $b = J_1 J_2/[(J_1 + J_2)\rho J l]$. The solution of this expression for the case $J_1 = 10 \text{ kg} \cdot \text{m}^2/\text{rad}, J_2 = 10 \text{ kg} \cdot \text{m}^2/\text{rad}, \rho = 2700 \text{ kg/m}^3, J = 5 \text{ kg} \cdot \text{m}^4/\text{rad}, l = 0.25 \text{ m}$ is illustrated by the points of intersection of the two curves in Figure 6.9. These points of intersection are determined by using MATLAB to solve for the roots of equation (6.84). For $G = 25 \times 10^9$ Pa, this results in

$$f_1 = 0 \qquad\qquad f_2 = 38{,}013 \text{ Hz} \qquad f_3 = 76{,}026 \text{ Hz}$$

$$f_4 = 114{,}039 \text{ Hz} \qquad f_5 = 152{,}052 \text{ Hz} \qquad f_6 = 190{,}066 \text{ Hz}$$

with the higher frequencies approaching $n\pi$ in radians.
□

The natural frequencies for various configurations of torsional systems are listed in Table 6.3. Note the similarity to Table 6.1 for longitudinal vibration. The only difference

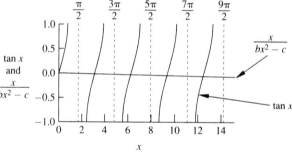

Figure 6.9 Plot of $\tan x$ versus x and $x/(bx^2 - a)$ versus x. These points of intersection are determined using the `fzero` command in MATLAB, as discussed at the end of the problem section. This yields the roots of equation (6.84).

TABLE 6.3 SAMPLE OF VARIOUS CONFIGURATIONS OF A UNIFORM SHAFT IN TORSIONAL VIBRATION, OF LENGTH l, ILLUSTRATING THE NATURAL FREQUENCIES AND MODE SHAPES[a]

Configuration	Frequency (rad/s) or characteristic equation	Mode shape
Free–free	$\omega_n = \dfrac{n\pi c}{l}, \quad n = 0, 1, 2, \ldots$	$\cos\dfrac{n\pi x}{l}$
Fixed–free	$\omega_n = \dfrac{(2n-1)\pi c}{2l}, \quad n = 1, 2, \ldots$	$\sin\dfrac{(2n-1)\pi x}{2l}$
Fixed–fixed	$\omega_n = \dfrac{n\pi c}{l}, \quad n = 1, 2, \ldots$	$\sin\dfrac{n\pi x}{l}$
Fixed–spring (rotational)	$\lambda_n \cot \lambda_n = -\dfrac{kl}{G\gamma}$ $\omega_n = \dfrac{\lambda_n c}{l}$	$\sin\dfrac{\lambda_n x}{l}$
Fixed–inertia	$\cot \lambda_n = \dfrac{J_0}{\rho l \gamma}\lambda_n$ $\omega_n = \dfrac{\lambda_n c}{l}$	$\sin\dfrac{\lambda_n x}{l}$

[a] Here $c = \sqrt{G\gamma/\rho J}$. The values of γ can be found in Table 6.2. Note that a numerical procedure is required for the last two cases, as illustrated in Example 6.4.1.

is the physical interpretation of the motion. Table 6.3 can be combined with Table 6.2 to approximate the natural frequencies of torsional vibration for noncircular cross sections as well. A more complete tabulation can be found in Blevins (1987).

6.5 BENDING VIBRATION OF A BEAM

This section again considers the vibration of the bar or beam of Figure 6.5. In this case, however, vibration of the beam in the direction perpendicular to its length is considered. Such vibrations are often called *transverse vibrations*, or *flexural vibrations*, because they move across the length of the beam. Transverse vibration is easily felt by humans when walking over bridges, for example. Figure 6.10 illustrates a cantilevered beam with the transverse direction of vibration indicated [i.e., the deflection, $w(x, t)$, is in the y direction]. The beam is of rectangular cross section $A(x)$ with width h_y, thickness h_z, and length l. Also associated with the beam is a flexural (bending) stiffness $EI(x)$, where E is Young's elastic modulus for the beam and $I(x)$ is the cross-sectional area moment of inertia about the "z axis." From mechanics of materials, the beam sustains a bending moment $M(x, t)$ which is related to the beam deflection, or bending deformation, $w(x, t)$, by

$$M(x, t) = EI(x)\frac{\partial^2 w(x, t)}{\partial x^2} \tag{6.85}$$

A model of bending vibration may be derived from examining the force diagram of an infinitesimal element of the beam as indicated in Figure 6.10. Assuming the deformation to be small enough such that the shear deformation is much smaller than $w(x, t)$ (i.e., so that the sides of the element dx do not bend), a summation of forces in the y direction yields

$$\left(V(x, t) + \frac{\partial V(x, t)}{\partial x}dx\right) - V(x, t) + f(x, t)\, dx = \rho A(x)dx\frac{\partial^2 w(x, t)}{\partial t^2} \tag{6.86}$$

Here $V(x,t)$ is the shear force at the left end of the element dx, $V(x, t) + V_x(x, t)dx$ is the shear force at the right end of the element dx, $f(x,t)$ is the total external force applied to the element per unit length and the term on the right side of the equality is the inertial force of the element. The assumption of small shear deformation used in the force balance of equation (6.86) is true if $l/h_z \geq 10$ and $l/h_y \geq 10$ (i.e., for long slender beams).

Next the moments acting on the element dx about the z axis through point Q are summed. This yields

$$\left[M(x, t) + \frac{\partial M(x, t)}{\partial x}dx\right] - M(x, t) + \left[V(x, t) + \frac{\partial V(x, t)}{\partial x}dx\right] dx + [f(x, t)dx]\frac{dx}{2} = 0 \tag{6.87}$$

Here the left-hand side of the equation is zero since it is also assumed that the rotary inertia of the element dx is negligible. Simplifying this expression yields

$$\left[\frac{\partial M(x, t)}{\partial x} + V(x, t)\right] dx + \left[\frac{\partial V(x, t)}{\partial x} + \frac{f(x, t)}{2}\right] (dx)^2 = 0 \tag{6.88}$$

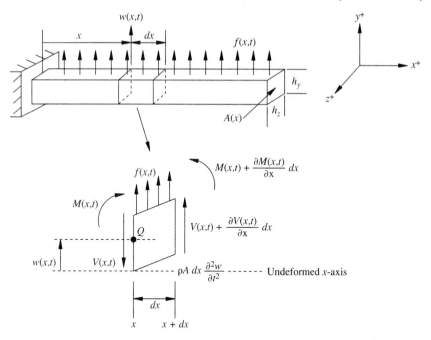

Figure 6.10 Simple beam in transverse vibration and a free-body diagram of a small element of the beam as it is deformed by a distributed force per unit length, denoted $f(x, t)$.

Since dx is assumed to be very small, $(dx)^2$ is assumed to be almost zero, so that this moment expression yields (dx is small, but not zero)

$$V(x, t) = -\frac{\partial M(x, t)}{\partial x} \tag{6.89}$$

This states that the shear force is proportional to the spatial change in the bending moment. Substitution of this expression for the shear force into equation (6.86) yields

$$-\frac{\partial^2}{\partial x^2}[M(x, t)]\,dx + f(x, t)\,dx = \rho A(x)\,dx\,\frac{\partial^2 w(x, t)}{\partial t^2} \tag{6.90}$$

Further substitution of equation (6.85) into (6.90) and dividing by dx yields

$$\rho A(x)\frac{\partial^2 w(x, t)}{\partial t^2} + \frac{\partial^2}{\partial x^2}\left[EI(x)\frac{\partial^2 w(x, t)}{\partial x^2}\right] = f(x, t) \tag{6.91}$$

If no external force is applied so that $f(x, t) = 0$ and if $EI(x)$ and $A(x)$ are assumed to be constant, equation (6.91) simplifies so that free vibration is governed by

$$\frac{\partial^2 w(x, t)}{\partial t^2} + c^2\frac{\partial^4 w(x, t)}{\partial x^4} = 0 \qquad c = \sqrt{\frac{EI}{\rho A}} \tag{6.92}$$

Note that unlike the previous equations, the free vibration equation (6.92) contains four spatial derivatives and hence requires four (instead of two) boundary conditions in calculating a solution. The presence of the two time derivatives again requires that two initial conditions, one for the displacement and one for the velocity, be specified.

The boundary conditions required to solve the spatial equation in a separation-of-variables solution of equation (6.92) are obtained by examining the deflection $w(x, t)$, the slope of the deflection $\partial w(x, t)/\partial x$, the bending moment $EI \partial^2 w(x, t)/\partial x^2$, and the shear force $\partial[EI \partial^2 w(x, t)/\partial x^2]/\partial x$ at each end of the beam. A common configuration is *clamped–free* or *cantilevered* as illustrated in Figure 6.10. In addition to a boundary being clamped or free, the end of a beam could be resting on a support restrained from bending or deflecting. The situation is called *simply supported* or *pinned*. A *sliding* boundary is one in which displacement is allowed but rotation is not. The shear load at a sliding boundary is zero.

If a beam in transverse vibration is free at one end, the deflection and slope at that end are unrestricted, but the bending moment and shear force must vanish:

$$\text{bending moment} = EI \frac{\partial^2 w}{\partial x^2} = 0$$

$$\text{shear force} = \frac{\partial}{\partial x} \left[EI \frac{\partial^2 w}{\partial x^2} \right] = 0$$

$$(6.93)$$

If, on the other hand, the end of a beam is clamped (or fixed), the bending moment and shear force are unrestricted, but the deflection and slope must vanish at that end:

$$\text{deflection} = w = 0$$

$$\text{slope} = \frac{\partial w}{\partial x} = 0$$

$$(6.94)$$

At a simply supported or pinned end, the slope and shear force are unrestricted and the deflection and bending moment must vanish:

$$\text{deflection} = w = 0$$

$$\text{bending moment} = EI \frac{\partial^2 w}{\partial x^2} = 0$$

$$(6.95)$$

At a sliding end, the slope or rotation is zero and no shear force is allowed. On the other hand, the deflection and bending moment are unrestricted. Hence, at a sliding boundary,

$$\text{slope} = \frac{\partial w}{\partial x} = 0$$

$$\text{shear force} = \frac{\partial}{\partial x} \left(EI \frac{\partial^2 w}{\partial x^2} \right) = 0$$

$$(6.96)$$

Other boundary conditions are possible by connecting the ends of a beam to a variety of devices such as lumped masses, springs, and so on. These boundary conditions can be determined by force and moment balances.

In addition to satisfying four boundary conditions, the solution of equation (6.92) for free vibration can be calculated only if two initial conditions (in time) are specified. As in the case of the rod, string, and bar, these initial conditions are specified initial deflection and velocity profiles:

$$w(x, 0) = w_0(x) \qquad \text{and} \qquad w_t(x, 0) = \dot{w}_0(x)$$

assuming that $t = 0$ is the initial time. Note that w_0 and $\dot{w}_0$ cannot both be zero or no motion will result.

The solution of equation (6.92) subject to four boundary conditions and two initial conditions proceeds following exactly the same steps used in previous sections. A separation-of-variables solution is assumed of the form $w(x, t) = X(x)T(t)$. This is substituted into the equation of motion, equation (6.92), to yield (after rearrangement)

$$c^2 \frac{X''''(x)}{X(x)} = -\frac{\ddot{T}(t)}{T(t)} = \omega^2 \tag{6.97}$$

where the partial derivatives have been replaced with total derivatives as before (*note:* $X'''' = d^4X/dx^4$, $\ddot{T} = d^2T/dt^2$). Here the choice of separation constant, ω^2, is made, based on experience with the systems of Section 6.4, that the natural frequency comes from the temporal equation:

$$\ddot{T}(t) + \omega^2 T(t) = 0 \tag{6.98}$$

which is the right side of equation (6.97). This temporal equation has a solution of the form

$$T(t) = A \sin \omega t + B \cos \omega t \tag{6.99}$$

where the constants A and B will eventually be determined by the specified initial conditions after being combined with the spatial solution.

The spatial equation comes from rearranging equation (6.97) which yields:

$$X''''(x) - \left(\frac{\omega}{c}\right)^2 X(x) = 0 \tag{6.100}$$

By defining [recall equation (6.92)]

$$\beta^4 = \frac{\omega^2}{c^2} = \frac{\rho A \omega^2}{EI} \tag{6.101}$$

and assuming a solution to (6.100) of the form $Ae^{\sigma x}$, the general solution of equation (6.100) can be calculated to be of the form (see Problem 6.42)

$$X(x) = a_1 \sin \beta x + a_2 \cos \beta x + a_3 \sinh \beta x + a_4 \cosh \beta x \tag{6.102}$$

Here the value for β and three of the four constants of integration a_1, a_2, a_3, and a_4 will be determined from the four boundary conditions. The fourth constant becomes combined with the constants A and B from the temporal equation, which are then determined from the initial conditions. The following example illustrates the solution procedure for a beam fixed at one end and simply supported at the other end.

Example 6.5.1

Calculate the natural frequencies and mode shapes for the transverse vibration of a beam of length l that is fixed at one end and pinned at the other end.

Solution   The boundary conditions in this case are given by equation (6.94) at the fixed end and equation (6.95) at the pinned end. Substitution of the general solution given by equation (6.102) into equation (6.94) at $x = 0$ yields

$$X(0) = 0 \Rightarrow a_2 + a_4 = 0$$

$$X'(0) = 0 \Rightarrow \beta(a_1 + a_3) = 0$$

Similarly, at $x = l$ the boundary conditions result in

$$X(l) = 0 \Rightarrow a_1 \sin \beta l + a_2 \cos \beta l + a_3 \sinh \beta l + a_4 \cosh \beta l = 0$$

$$EIX''(l) = 0 \Rightarrow \beta^2(-a_1 \sin \beta l - a_2 \cos \beta l + a_3 \sinh \beta l + a_4 \cosh \beta l) = 0$$

These four boundary conditions thus yield four equations in the four unknown coefficients a_1, a_2, a_3, and a_4. These can be written as the single vector equation

$$\begin{bmatrix} 0 & 1 & 0 & 1 \\ \beta & 0 & \beta & 0 \\ \sin \beta l & \cos \beta l & \sinh \beta l & \cosh \beta l \\ -\beta^2 \sin \beta l & -\beta^2 \cos \beta l & \beta^2 \sinh \beta l & \beta^2 \cosh \beta l \end{bmatrix} \begin{bmatrix} a_1 \\ a_2 \\ a_3 \\ a_4 \end{bmatrix} = \begin{bmatrix} 0 \\ 0 \\ 0 \\ 0 \end{bmatrix}$$

Recall from Chapter 4 that this vector equation can have a nonzero solution for the vector $\mathbf{a} = [a_1 \quad a_2 \quad a_3 \quad a_4]^T$ only if the determinant of the coefficient matrix vanishes (i.e., if the coefficient matrix is singular). Furthermore, recall that since the coefficient matrix is singular, not all of the elements of the vector $\mathbf{a}$ can be calculated.

Setting the determinant above equal to zero yields the characteristic equation

$$\tan \beta l = \tanh \beta l$$

This equality is satisfied for an infinite number of choices for β, denoted β_n. The solution can be visualized by plotting both $\tan \beta l$ and $\tanh \beta l$ versus βl on the same plot. This is similar to the solution technique used in Example 6.4.1 and illustrated in Figure 6.9. The first five solutions are

$$\beta_1 l = 3.926602 \qquad \beta_2 l = 7.068583 \qquad \beta_3 l = 10.210176$$

$$\beta_4 l = 13.351768 \qquad \beta_5 l = 16.49336143$$

For the rest of the modes (i.e., for values of the index $n > 5$), the solutions to the characteristic equation are well approximated by

$$\beta_n l = \frac{(4n + 1)\pi}{4}$$

With these values of the weighted frequencies $\beta_n l$, the individual modes of vibration can be calculated. Solving the matrix equation above for the individual coefficients a_i yields $a_1 = -a_3$, $a_2 = -a_4$, and

$$(\sinh \beta_n l - \sin \beta_n l)a_3 + (\cosh \beta_n l - \cos \beta_n l)a_4 = 0$$

Thus

$$a_3 = -\frac{\cosh \beta_n l - \cos \beta_n l}{\sinh \beta_n l - \sin \beta_n l} a_4$$

for each n. The fourth coefficient a_4 cannot be determined by this set of equations, because the coefficient matrix is singular (otherwise, each a_i would be zero). This remaining coefficient becomes the arbitrary magnitude of the eigenfunction. As this constant depends on n, denote it by $(a_4)_n$. Substitution of these values of a_i in the expression $X(x)$ for the spatial solution yields the result that the eigenfunctions or mode shapes have the form

$$X_n(x) = (a_4)_n \left[\frac{\cosh \beta_n l - \cos \beta_n l}{\sinh \beta_n l - \sin \beta_n l} (\sinh \beta_n x - \sin \beta_n x) - \cosh \beta_n x + \cos \beta_n x \right],$$

$$n = 1, 2, 3, \ldots$$

The first three mode shapes are plotted in Figure 6.11 for $(a_2)_n = 1$, $n = 1, 2, 3$, and $l = 1$.

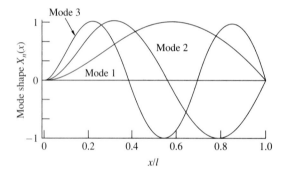

Figure 6.11 Plot of the first three mode shapes of the clamped–pinned beam of Example 6.5.1, arbitrarily normalized to unity.

These mode shapes can be shown to be orthogonal, so that

$$\int_0^l X_n(x) X_m(x) \, dx = 0$$

for $n \neq m$ (see Exercise 6.45). As in Example 6.3.2, this orthogonality, along with the initial conditions, can be used to calculate the constants A_n and B_n in the series solution for the displacement

$$w(x, t) = \sum_{n=1}^{\infty} (A_n \sin \omega_n t + B_n \cos \omega_n t) X_n(x)$$

□

Table 6.4 summarizes a number of different boundary configurations for the slender beam. The slender beam model given in equation (6.91) is often referred to as the Euler–

TABLE 6.4 SAMPLE OF VARIOUS BOUNDARY CONFIGURATIONS OF A SLENDER BEAM IN TRANSVERSE VIBRATION OF LENGTH l ILLUSTRATING WEIGHTED NATURAL FREQUENCIES AND MODE SHAPES[a]

Configuration	Weighted frequencies $\beta_n l$ and characteristic equation	Mode shape	σ_n
Free–free	0 (rigid-body mode)	$\cosh \beta_n x + \cos \beta_n x$	
	4.73004074		0.9825
	7.85320462	$-\sigma_n (\sinh \beta_n x + \sin \beta_n x)$	1.0008
	10.9956078		0.9999
	14.1371655		1.0000
	17.2787597		0.9999
	$\dfrac{(2n+1)\pi}{2}$ for $n > 5$		1 for $n > 5$
	$\cos \beta l \cosh \beta l = 1$		
Clamped–free	1.87510407	$\cosh \beta_n x - \cos \beta_n x$	0.7341
	4.69409113		1.0185
	7.85475744	$-\sigma_n (\sinh \beta_n x - \sin \beta_n x)$	0.9992
	10.99554073		1.0000
	14.13716839		1.0000
	$\dfrac{(2n-1)\pi}{2}$ for $n > 5$		1 for $n > 5$
	$\cos \beta l \cosh \beta l = -1$		
Clamped–pinned	3.92660231	$\cosh \beta_n x - \cos \beta_n x$	1.0008
	7.06858275		1 for $n > 1$
	10.21017612	$-\sigma_n (\sinh \beta_n x - \sin \beta_n x)$	
	13.35176878		
	16.49336143		
	$\dfrac{(4n+1)\pi}{4}$ for $n > 5$		
	$\tan \beta l = \tanh \beta l$		
Clamped–sliding	2.36502037	$\cosh \beta_n x - \cos \beta_n x$	0.9825
	5.49780392		1 for $n > 1$
	8.63937983	$-\sigma_n (\sinh \beta_n x - \sin \beta_n x)$	
	11.78097245		
	14.92256510		
	$\dfrac{(4n-1)\pi}{4}$ for $n > 5$		
	$\tan \beta l + \tanh \beta l = 0$		
Clamped–clamped	4.73004074	$\cosh \beta_n x - \cos \beta_n x$	0.9825
	7.85320462		1.0008
	10.9956079	$-\sigma_n (\sinh \beta_n x - \sin \beta_n x)$	0.9999
	14.1371655		1.0000
	17.2787597		0.9999
	$\dfrac{(2n+1)\pi}{2}$ for $n > 5$		1 for $n > 5$
	$\cos \beta l \cosh \beta l = 1$		
Pinned–pinned	$n\pi$	$\sin \dfrac{n\pi x}{l}$	none
	$\sin \beta l = 0$		

[a] Here the weighted natural frequencies $\beta_n l$ are related to the natural frequencies by equation (6.101) or $\omega_n = \beta_n^2 \sqrt{EI/\rho A}$, as used in Example 6.5.1. The values of σ_i for the mode shapes are computed from the formulas given in Table 6.5.

Bernoulli or Bernoulli–Euler beam equation. The assumptions used in formulating this model are that the beam be

- Uniform along its span, or length, and slender
- Composed of a linear, homogeneous, isotropic elastic material without axial loads
- Such that plane sections remain plane
- Such that the plane of symmetry of the beam is also the plane of vibration so that rotation and translation are decoupled
- Such that rotary inertia and shear deformation can be neglected

The key to solving for the time response of distributed parameter systems is the orthogonality of the mode shapes. Note from Table 6.5 that the mode shapes are quite complicated in many configurations. This does not mean that orthogonality is necessarily violated, just that evaluating the integrals in the modal analysis procedure becomes more difficult.

TABLE 6.5 EQUATIONS FOR THE MODE
SHAPE COEFFICIENTS σ_n FOR USE IN
TABLE 6.4[a]

Boundary condition	Formula for σ_n
Free–free	$\sigma_n = \dfrac{\cosh \beta_n l - \cos \beta_n l}{\sinh \beta_n l - \sin \beta_n l}$
Clamped–free	$\sigma_n = \dfrac{\sinh \beta_n l - \sin \beta_n l}{\cosh \beta_n l + \cos \beta_n l}$
Clamped–pinned	$\sigma_n = \dfrac{\cosh \beta_n l - \cos \beta_n l}{\sinh \beta_n l - \sin \beta_n l}$
Clamped–sliding	$\sigma_n = \dfrac{\sinh \beta_n l - \sin \beta_n l}{\cosh \beta_n l + \cos \beta_n l}$
Clamped–clamped	Same as free–free

[a] These coefficients are used in the calculations for the mode shapes, as illustrated in Example 6.5.1.

The model of the transverse vibration of the beam presented in equation (6.91) ignores the effects of shear deformation and rotary inertia. These effects are considered next. As mentioned above, it is safe to ignore the shear deformation as long as the h_z and h_y illustrated in Figure 6.10 are small compared with the length of the beam. As the beam becomes shorter, the effect of shear deformation becomes evident. This is illustrated in Figure 6.12, which is a repeat of the element dx of Figure 6.10 with shear deformation included.

Referring to the figure, the line OA is a line through the center of the element dx perpendicular to the face at the right side. The line OB, on the other hand, is the

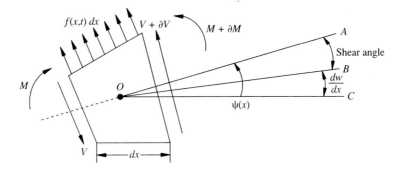

Figure 6.12   Effects of shear deformation on an element of a bending beam.

line through the center tangent to the centerline of the beam, while the line OC is the centerline of the beam while at rest. As the beam bends, the shear angle appears as the length l is decreased relative to the beam width. For the case of a long beam, the lines OB and OA coincide, as in Figure 6.10. The presence of significant shear causes the rectangular infinitesimal element of Figure 6.10 to deform into the distorted almost-diamond shape of Figure 6.12. Referring to Figure 6.12, note that the shear angle given by $\psi - dw/dx$ (i.e., the difference between the total angle due to bending, ψ and the slope of the centerline of the beam, dw/dx), represents the effect of shear deformation. From elastic considerations (see, e.g., Reismann and Pawlik, 1974) the bending moment equation becomes

$$EI\frac{d\psi(x,t)}{dx} = M(x,t) \tag{6.103}$$

and the shear force equation becomes

$$\kappa^2 AG\left[\psi(x,t) - \frac{dw(x,t)}{dx}\right] = V(x,t) \tag{6.104}$$

where E, I, A, ψ, V and M are as defined previously, G is the *shear modulus* and κ^2 is a dimensionless factor that depends on the shape of the cross-sectional area. The constant κ^2 is called a *shear coefficient* and has been tabulated by Cowper (1966). As in the case of equation (6.86), a dynamic force balance yields

$$\rho A(x)dx\frac{\partial^2 w(x,t)}{\partial t^2} = -\left[V(x,t) + \frac{\partial V(x,t)}{\partial x}dx\right] + V(x,t) + f(x,t)dx \tag{6.105}$$

If the rotary inertia is included, then the moment balance on dx previously given by equation (6.87) becomes

$$\rho I(x)dx\frac{\partial^2 \psi(x,t)}{\partial t^2} = \left[M(x,t) + \frac{\partial M(x,t)}{\partial x}dx\right] - M(x,t)$$

$$+ \left[V(x,t) + \frac{\partial V(x,t)}{\partial x}dx\right]dx + f(x,t)\frac{dx^2}{2} \tag{6.106}$$

Substitution of equations (6.103) and (6.104) into (6.105) and (6.106) yields the two coupled equations

$$\frac{\partial}{\partial x}\left[EI\frac{\partial\psi}{\partial x}\right] + \kappa^2 AG\left(\frac{\partial w}{\partial x} - \psi\right) = \rho I \frac{\partial^2\psi}{\partial t^2} \tag{6.107}$$

and

$$\frac{\partial}{\partial x}\left[\kappa^2 AG\left(\frac{\partial w}{\partial x} - \psi\right)\right] + f(x,t) = \rho A \frac{\partial^2 w}{\partial t^2} \tag{6.108}$$

governing the vibration of a beam including the effects of rotary inertia and shear deformation. Assuming that the coefficients are all constant and that no external force is applied, $\psi(x,t)$ can be eliminated and the coupled equations can be reduced to one single equation for free vibration of uniform beams. This is

$$EI\frac{\partial^4 w}{\partial x^4} + \rho A\frac{\partial^2 w}{\partial t^2} - \rho I\left(1 + \frac{E}{\kappa^2 G}\right)\frac{\partial^4 w}{\partial x^2 \partial t^2} + \frac{\rho^2 I}{\kappa^2 G}\frac{\partial^4 w}{\partial t^4} = 0 \tag{6.109}$$

Equation (6.109) is now subject to four initial conditions and four boundary conditions. For a clamped end, the boundary conditions become (at $x = 0$, say)

$$\psi(0,t) = w(0,t) = 0 \tag{6.110}$$

At a simply supported end, they become

$$EI\frac{\partial\psi(0,t)}{\partial x} = w(0,t) = 0 \tag{6.111}$$

and at a free end they become

$$\kappa^2 AG\left(\frac{\partial w}{\partial x} - \psi\right) = EI\frac{\partial\psi}{\partial x} = 0 \tag{6.112}$$

These equations can be solved by the methods suggested for the beam model given by equation (6.91). Equation (6.91) is called the *Euler–Bernoulli beam model* or classical beam model and equation (6.109) is called the *Timoshenko beam model*.

Which of these two beam models to use is largely dependent on the beam geometry, which modes are of interest, and how many modes are important. A steel beam 12 m long, 15 cm wide, and 0.6 m deep, shows a difference in only 0.4% between the first natural frequency of the Euler–Bernoulli model and that of the Timoshenko model. This grows to a 10% difference in the fifth natural frequency. Hence if only the first mode is of interest, an Euler–Bernoulli model for this system would be good enough. On the other hand, if the fifth mode is of interest, the Timoshenko model might be worth the extra complexity. For a beam of typical metal with rectangular cross section, the shear deformation is about three times more significant than rotary inertia effects. This is examined in the next example.

Example 6.5.2

To obtain a feeling for the differences between the Euler–Bernoulli beam model and the more complicated Timoshenko model, calculate the natural frequencies of a pinned–pinned beam using the Timoshenko equation and compare them to the natural frequencies predicted by the beam without shear deformation or rotary inertia as given in Table 6.4.

Solution   Equation (6.109) cannot easily be solved by separation of variables as suggested in Window 6.3. This results because substitution of $w(x, t) = X(x)T(t)$ into equation (6.3) yields an equation of the form (a_i are constants)

$$a_1 X''''T + (a_2 X + a_3 X'')\ddot{T} + a_4 X \ddddot{T} = 0$$

which does not readily separate because of the middle term and the fourth time derivative. This expression can be solved by assuming that the mode shapes of pinned–pinned Euler–Bernoulli beam and that of the Timoshenko beam are the same (a reasonable assumption, as one is a more complete account of the other) and that the temporal response is periodic (also reasonable, as the system is undamped). Preceding with these assumptions a solution of equation (6.109) is *assumed* to be of the specific separated form

$$w_n(x, t) = \sin \frac{n\pi x}{l} \cos \omega_n t$$

where $\sin(n\pi x/l)$ is the assumed nth, mode shape of the pinned–pinned configuration and ω_n is, at this point, the unknown natural frequency. Substitution of this form into equation (6.109) yields

$$EI \left(\frac{n\pi}{l}\right)^4 \sin \frac{n\pi x}{l} \cos \omega_n t - \rho I \left(1 + \frac{E}{\kappa^2 G}\right) \left(\frac{n\pi}{l}\right)^2 \omega_n^2 \sin \frac{n\pi}{l} \cos \omega_n t$$

$$= -\frac{\rho^2 I}{\kappa^2 G} \omega_n^4 \sin \frac{n\pi x}{l} \cos \omega_n t + \rho A \omega_n^2 \sin \frac{n\pi x}{l} \cos \omega_n t$$

Each term contains the factor $\sin(n\pi x/l) \cos(\omega_n t)$, which can be factored out to reveal the characteristic equation

$$\omega_n^4 \frac{\rho r^2}{\kappa^2 G} - \left(1 + \frac{n^2\pi^2 r^2}{l^2} + \frac{n^2\pi^2 r^2}{l^2} \frac{E}{\kappa^2 G}\right) \omega_n^2 + \frac{\alpha^2 n^4 \pi^4}{l^4} = 0$$

where r and α are defined by

$$\alpha^2 = \frac{EI}{\rho A} \qquad r^2 = \frac{I}{A}$$

The characteristic equation for the frequencies ω_n is quadratic in ω_n^2, and hence easily solved. This expression for ω_n provides a mechanism for observing the effects of shear deformation and rotary inertia on the natural frequencies of a pinned–pinned beam. Note the following comparisons:

1. Of the two roots for each value of n determined by the frequency equation, the smaller value is associated with bending deformation, and the larger root is associated with shear deformation.

2. The natural frequencies for just the Euler–Bernoulli beam are

$$\omega_n^2 = \frac{\alpha^2 n^4 \pi^4}{l^4}$$

3. The natural frequencies for including just rotary inertia (i.e., no shear, so that terms involving κ are eliminated) are

$$\omega_n^2 = \frac{\alpha^2 n^4 \pi^4}{l^4 \left(1 + n^2 \pi^2 r^2 / l^2\right)}$$

4. The natural frequencies for neglecting the rotary inertia and including the shear deformation are

$$\omega_n^2 = \frac{\alpha^2 n^4 \pi^4}{l^4 \left[1 + (n^2 \pi^2 r^2 / l^2)(E / \kappa G)\right]}$$

These expressions can be used to investigate the various effects on the natural frequencies of prismatic beams in the pinned–pinned configuration. These expressions can also be used to gain insight into the effects of rotary inertia and shear deformation for other configurations. By comparing notes 2, 3, and 4 above, the general effect of shear deformation and rotary inertia is to reduce the value of the natural frequencies. Also note that for high frequencies (large n) the effects of shear deformation and rotary inertia is more pronounced because of the $1 + (n \pi r / l)^2$ term in the denominator. This is investigated further in Problem TB.6.4 at the end of the chapter.

$\square$

6.6 VIBRATION OF MEMBRANES AND PLATES

The string, rod, and beam models considered in the previous five sections have displacements that are a function of a single direction, x, along the mass. In this sense they are one-dimensional problems. In this section membranes and plates are considered having displacements, which are functions of two dimensions (i.e., they are defined in a plane region in space as illustrated in Figure 6.13). A membrane is essentially a two-dimensional string and a plate is essentially a two-dimensional beam. The equation for a membrane and plate are not derived here but follow similar arguments used in developing the string and Euler–Bernoulli beam equations respectively.

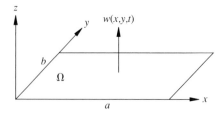

Figure 6.13   Schematic of a rectangular membrane illustrating the vibration perpendicular to its surface. The boundary conditions are specified around the edge of the membrane (i.e., along the lines $x = a$, $x = 0$, $y = b$, and $y = 0$).

First, consider the vibrational equations for a membrane. A membrane is basically a two-dimensional system that lies in a plane when in equilibrium. A drumhead and a soap film are physical examples of objects that may be modeled as a membrane. The structure itself provides no resistance to bending, so that the restoring force is due only to the tension in the membrane. Thus a membrane is similar to a string. The reader is referred to Weaver, Timoshenko, and Young (1990) for the derivation of the membrane equation.

Let $w(x, y, t)$ represent the displacement in the z direction of a membrane, lying in the xy plane at the point (x,y) and time t. The displacement is assumed to be small, with small slopes, and is perpendicular to the xy plane. Let τ be the constant tension per unit length of the membrane and ρ be the mass per unit area of the membrane. Then the equation for free vibration is given by

$$\tau \nabla^2 w(x, y, t) = \rho w_{tt}(x, y, t) \tag{6.113}$$

where x and y lie inside the region, Ω, occupied by the membrane as indicated in Figure 6.13. Here ∇^2 is the *Laplace operator*. In rectangular coordinates this operator has the form

$$\nabla^2 = \frac{\partial^2}{\partial x^2} + \frac{\partial^2}{\partial y^2} \tag{6.114}$$

The Laplace operator takes on other forms if a different coordinate system is used. For example a drumhead is best written in a circular coordinate system. For the rectangular system considered here, equation (6.113) becomes

$$\frac{\partial^2 w(x, y, t)}{\partial x^2} + \frac{\partial^2 w(x, y, t)}{\partial y^2} = \frac{1}{c^2} \frac{\partial^2 w(x, y, t)}{\partial t^2} \tag{6.115}$$

where $c = \sqrt{\tau/\rho}$. The boundary conditions for the membrane must be specified along the shape of the boundary, not just at points, as in the case of the string. If the membrane is fixed or clamped at a segment of the boundary, the defection must be zero along that segment. If $\partial\Omega$ is the curve in the xy plane corresponding to the edge of the membrane (i.e., the boundary of Ω), the clamped boundary condition is denoted by

$$w(x, y, t) = 0 \qquad \text{for} \quad x, y \in \partial\Omega \tag{6.116}$$

If for some segment of $\partial\Omega$, denoted by $\partial\Omega_1$, the membrane is free to deflect transversely, there can be no force component in the transverse direction, and the boundary condition becomes

$$\frac{\partial w(x, y, t)}{\partial n} = 0 \qquad \text{for} \quad x, y \in \partial\Omega_1 \tag{6.117}$$

Here, $\partial/\partial n$ denotes the derivative of $w(x, y, t)$ normal to the boundary in the reference plane of the membrane. The following example illustrates the procedure for calculating the solution for a vibrating membrane.

Example 6.6.1

Consider the vibration of a square membrane, as indicated in Figure 6.13. The membrane is clamped at all the edges. The equation of motion is given by equation (6.115). Compute the natural frequencies and mode shapes for the case $a = b = 1$.

Solution   Assuming that the solution separates [i.e., that $w(x, y, t) = X(x)Y(y)T(t)$], equation (6.115) becomes

$$\frac{1}{c^2}\frac{\ddot{T}}{T} = \frac{X''}{X} + \frac{Y''}{Y} \tag{6.118}$$

This implies that $\ddot{T}/(Tc^2)$ is a constant (recall the argument used in Example 6.2.1). Denote the constant by ω^2, so that

$$\frac{\ddot{T}}{Tc^2} = -\omega^2 \tag{6.119}$$

This assumption leads to the appropriate time solution as before. Then equation (6.118) implies that

$$\frac{X''}{X} = -\omega^2 - \frac{Y''}{Y} \tag{6.120}$$

By the same argument as that used before, both X''/X and Y''/Y must be constant (i.e., independent of t and x or y). Hence

$$\frac{X''}{X} = -\alpha^2 \tag{6.121}$$

and

$$\frac{Y''}{Y} = -\gamma^2 \tag{6.122}$$

where α^2 and γ^2 are constants. Equation (6.120) then yields

$$\omega^2 = \alpha^2 + \gamma^2 \tag{6.123}$$

This results in two spatial equations to be solved,

$$X'' + \alpha^2 X = 0 \tag{6.124}$$

which has a solution (A and B constants of integration) of the form

$$X(x) = A \sin \alpha x + B \cos \alpha x \tag{6.125}$$

and

$$Y'' + \gamma^2 Y = 0 \tag{6.126}$$

which yields a solution of the form (C and D are constants of integration)

$$Y(y) = C \sin \gamma y + D \cos \gamma y \tag{6.127}$$

The total spatial solution is the product $X(x)Y(y)$, or

$$X(x)Y(y) = A_1 \sin \alpha x \sin \gamma y + A_2 \sin \alpha x \cos \gamma y$$
$$+ A_3 \cos \alpha x \sin \gamma y + A_4 \cos \alpha x \cos \gamma y \tag{6.128}$$

Here the constants A_i consist of the products of the constants in equations (6.125) and (6.127) and are to be determined by the boundary and initial conditions.

Equation (6.128) can now be used with the boundary conditions to calculate the eigenvalues and eigenfunctions of the system. The clamped boundary condition along $x = 0$ in Figure 6.13, yields

$$T(t)X(0)Y(y) = T(t)(A_3 \sin \gamma y + A_4 \cos \gamma y) = 0$$

or

$$A_3 \sin \gamma y + A_4 \cos \gamma y = 0 \qquad (6.129)$$

Now, equation (6.129) must hold for any value of y. Thus, as long as γ is not zero (a reasonable assumption, since if it is zero the system has a rigid-body motion), A_3 and A_4 must be zero. Hence the solution must have the form

$$X(x)Y(y) = A_1 \sin \alpha x \sin \gamma y + A_2 \sin \alpha x \cos \gamma y \qquad (6.130)$$

Next, application of the boundary condition $w = 0$ along the line $x = 1$ yields

$$A_1 \sin \alpha \sin \gamma y + A_2 \sin \alpha \cos \gamma y = 0 \qquad (6.131)$$

Factoring this expression yields

$$\sin \alpha (A_1 \sin \gamma y + A_2 \cos \gamma y) = 0 \qquad (6.132)$$

Now either $\sin \alpha = 0$ or, by the preceding argument, A_1 and A_2 must be zero. However, if A_1 and A_2 are both zero, the solution is trivial. Hence for a nontrivial solution to exist, $\sin \alpha = 0$. This yields

$$\alpha = n\pi \qquad n = 1, 2, \ldots, \infty \qquad (6.133)$$

Using the boundary condition $w = 0$ along the lines $y = 1$ and $y = 0$ results in a similar procedure and yields

$$\gamma = m\pi \qquad m = 1, 2, \ldots, \infty \qquad (6.134)$$

Note that the possibility of $\gamma = \alpha = 0$ is not used because it was necessary to assume that $\gamma \neq 0$ in order to derive equation (6.130). Equation (6.123) yields that the constant ω in the temporal equation must have the form

$$\omega_{mn} = \sqrt{\alpha_n^2 + \gamma_m^2}$$

$$= \pi\sqrt{m^2 + n^2} \qquad m, n = 1, 2, 3, \ldots, \infty$$

Thus the eigenvalues and eigenfunctions for the clamped membrane are, respectively, $\pi\sqrt{m^2 + n^2}$ and $\{\sin n\pi x \sin m\pi y\}$. The solution of equation (6.115) becomes

$$w(x, y, t) = \sum_{m=1}^{\infty}\sum_{n=1}^{\infty}(\sin m\pi x \sin n\pi y)\{A_{mn} \sin \sqrt{n^2 + m^2}c\pi t$$

$$+ B_{mn} \cos \sqrt{n^2 + m^2}c\pi t\} \qquad (6.135)$$

where A_{mn} and B_{mn} are determined by the initial conditions. To see this, multiply equation (6.135) by the mode shape $(\sin m\pi x \sin n\pi y)$ and integrate the resulting equation over

dx and dy along the edges of the membrane. Because the set of functions $\{\sin m\pi x \sin n\pi y\}$ is orthogonal, the summations are reduced to the single term

$$\int_0^1 \int_0^1 w(x, y, t) \sin m\pi x \sin n\pi y\, dx\, dy = \tfrac{1}{4}(A_{mn} \cos \omega_{mn} ct + B_{mn} \sin \omega_{mn} ct) \qquad (6.136)$$

Setting $t = 0$ to obtain the initial conditions in equation (6.136) yields

$$A_{mn} = 4 \int_0^1 \int_0^1 w(x, y, 0) \sin m\pi x \sin n\pi y\, dx\, dy$$

Differentiating equation (6.136) and setting $t = 0$ yields

$$B_{mn} = \frac{4}{\omega_{nm} c} \int_0^1 \int_0^1 w_t(x, y, 0) \sin m\pi x \sin n\pi y\, dx\, dy$$

These last two expressions yield the expansion coefficients in terms of the initial displacement $w(x, y, 0)$ and the initial velocity $w_t(x, y, 0)$.

The mode shapes of the membrane are the set of functions

$$w_{nm}(x, y) = \sin m\pi x \sin n\pi y, \qquad m = 1, 2, \ldots, \qquad n = 1, 2, \ldots$$

The first mode corresponds to the index $m = n = 1$. If the initial velocity is zero [i.e., $w_t(x, y, 0) = 0$], the coefficients B_{mn} are all zero. If, in addition, the initial displacement $w(x, y, 0)$ is chosen such that all the coefficients A_{mn}, are zero except for A_{11}, the solution becomes the single term

$$w_{11}c(x, y, z) = (A_{11} \sin \omega_{11} ct) \sin \pi x \sin \pi y$$

where $\omega_{11} = \pi\sqrt{2}$. This last expression describes the fundamental mode shape of the membrane vibrating at the single frequency $\pi\sqrt{2}$ rad/s.

Note from the expression for w_{nm} that the frequency corresponding to $m = 1, n = 2$ will be the same as the corresponding to $m = 2, n = 1$ (i.e., $\omega_{12} = \omega_{21} = \pi\sqrt{5}$). However, the mode shapes are different:

$$w_{12}(x, y, t) = (A_{12} \cos c\pi\sqrt{5}t) \sin \pi x \sin 2\pi y$$

$$w_{21}(x, y, t) = (A_{21} \cos c\pi\sqrt{5}t) \sin 2\pi x \sin \pi y$$

Thus the membrane can vibrate the single frequency $c\pi\sqrt{5}$ in two different ways, exhibiting different nodal lines.

☐

Note from this example that the mode shapes are functions of both x and y, so that the nodes of the modes of a membrane form a line of no motion along the membrane when excited at a particular natural frequency. This turns out to be relatively simple to verify experimentally, adding creditability to the analysis presented here.

In progressing from the vibration of a string to considering the transverse vibration of a beam, the beam equation allowed for bending stiffness. In somewhat the same manner, a plate differs from a membrane because plates have bending stiffness. The reader is referred to Reismann (1988) for a more detailed explanation and for a precise derivation of the plate equation. Basically, the plate, like the membrane, is defined in a plane (xy) with the deflection $w(x, y, t)$ taking place along the z axis perpendicular to the xy plane. The basic assumption is again small deflections with respect to the thickness h. Thus the plane running through the middle of the plate is assumed not to deform during bending (called a *neutral plane or surface*). In addition, normal stresses in the direction transverse to the plate are assumed to be negligible. Again there is no thickness stretch. The displacement equation of motion for the free vibration of the plate is

$$D_E \nabla^4 w(x, y, t) = \rho w_{tt}(x, y, t) \tag{6.137}$$

where E again denotes the elastic modulus, ρ is the mass density, and the constant D_E, the plate flexural rigidity, is defined in terms of Poison's ratio v and the plate thickness h as

$$D_E = \frac{Eh^3}{12(1 - v^2)} \tag{6.138}$$

The operator ∇^4, called the *biharmonic operator*, is a fourth-order operator, the exact form of which depends on the choice of coordinate systems. In rectangular coordinates the biharmonic operator becomes

$$\nabla^4 = \frac{\partial^4}{\partial x^4} + 2\frac{\partial^4}{\partial x^2 \partial y^2} + \frac{\partial^4}{\partial y^4} \tag{6.139}$$

The boundary conditions for a plate are a little more difficult to write, as their form, in some cases, also depends on the coordinate system in use.

For a clamped edge the deflection and normal derivative, $\partial/\partial n$, are both zero along the edge:

$$w(x, y, t) = 0 \quad \text{and} \quad \frac{\partial w(x, y, t)}{\partial n} = 0 \quad \text{for} \quad x, y \text{ along } \partial\Omega \tag{6.140}$$

Here the normal derivative is the derivative of w normal to the plate boundary and in the neutral plane. For a rectangular plate, the simply supported boundary conditions become

$$w(x, y, t) = 0 \qquad \text{along all edges} \tag{6.141}$$

$$\frac{\partial^2 w(x, y, t)}{\partial x^2} = 0 \qquad \text{along the edges } x = 0, \quad x = l_1$$
$$\tag{6.142}$$

$$\frac{\partial^2 w(x, y, t)}{\partial y^2} = 0 \qquad \text{along the edges } y = 0, \quad y = l_2$$

where l_1 and l_2 are the lengths of the plate edges and the second partial derivatives reflect the normal strains along these edges.

This plate model is basically a two-dimensional analog of the Euler–Bernoulli beam and is referred to as *thin plate theory*. Hence it is limited to cases where shear deformation and rotary inertia are negligible. The plate equation can be improved by adding shear deformation and rotary inertia to produce a two-dimensional analog of the Timoshenko beam. This is called Mindlin–Timoshenko theory and is not discussed here (see, e.g., Magrab, 1979).

6.7 MODELS OF DAMPING

The models discussed in the preceding six sections do not account for energy dissipation. As in Section 4.5 for lumped-mass systems, damping can be introduced in two ways: either as modal damping or as a physical damping model. In modeling single-degree-of-freedom systems, viscous damping was used as much for mathematical convenience as for physical truth. This is the case here as well.

A simple procedure for including damping is to add it to the temporal equation after separation of variables. For example, consider the temporal equation for the string as given by equation (6.24):

$$\ddot{T}_n(t) + \sigma_n^2 c^2 T_n(t) = 0 \qquad n = 1, 2, \ldots$$

These expressions yield the distributed parameter analog of equations (4.85) for a lumped-mass system and can be called *modal equations*. Modal damping can be added to equation (6.24) by including the term

$$2\zeta_n \omega_n \dot{T}_n(t) \qquad n = 1, 2, 3, \ldots \tag{6.143}$$

where ω_n is the nth natural frequency and ζ_n is the nth modal damping ratio. The damping ratios ζ_n are chosen, like those of equation (4.123), based on experience or on experimental measurements. Usually, ζ_n is a small positive number between 0 and 1, with most common values of $\zeta_n \leq 0.05$.

Once the modal damping ratios are assigned, the damping term of equation (6.143) is added to equation (6.24) to yield

$$\ddot{T}_n(t) + 2\zeta_n \omega_n \dot{T}_n(t) + \omega_n^2 T_n(t) = 0 \qquad n = 1, 2, 3, \ldots \tag{6.144}$$

where $\omega_n = \sigma_n c$. The solution, for an underdamped mode, becomes (see Window 6.4)

$$T_n(t) = A_n e^{-\zeta_n \omega_n t} \sin(\omega_{dn} t + \phi_n) \qquad n = 1, 2, 3, \ldots \tag{6.145}$$

Window 6.4
Review of a Damped Single-Degree-of-Freedom System

The solution of $m\ddot{x} + c\dot{x} + kx = 0$, $x(0) = x_0$, $\dot{x}(0) = \dot{x}_0$ or $\ddot{x} + 2\zeta\omega\dot{x} + \omega^2 x = 0$ is (for the underdamped case $0 < \zeta < 1$)

$$x(t) = A e^{-\zeta\omega t} \sin(\omega_d t + \theta)$$

where $\omega = \sqrt{k/m}$, $\zeta = c/2\,m\omega$, and

$$A = \left[\frac{(\dot{x}_0 + \zeta\omega x_0)^2 + (x_0\omega_d)^2}{\omega_d^2} \right]^{1/2} \qquad \theta = \tan^{-1} \frac{x_0\omega_d}{\dot{x}_0 + \zeta\omega x_0}$$

from equations (1.36), (1.39), and (1.38).

where $\omega_{dn} = \omega_n \sqrt{1 - \zeta_n^2}$ and where A_n and ϕ_n are constants to be determined by the initial condition. Once the temporal coefficients are determined, the rest of the solution procedure is the same as given in Section 6.2.

Example 6.7.1

Calculate the response of the bar of Example 6.3.1 to an initial displacement of $w(x, 0) = (x/l)$ cm and an initial velocity of $w_t(x, 0) = 0$. Assume that the bar exhibits modal damping of $\zeta_n = 0.01$ in each mode.

Solution   From Example 6.3.1, the mode shapes are $a_n \sin[(2n - 1)\pi x / 2l]$ and the undamped natural frequencies are

$$\omega_n = \sigma_n \sqrt{\frac{E}{\rho}} = \frac{(2n - 1)\pi}{2l} \sqrt{\frac{E}{\rho}}$$

Since the modal damping ratio is chosen to be 0.01, the damped natural frequencies become

$$\omega_{dn} = \omega_n \sqrt{1 - \zeta_n^2} = 0.9999 \frac{2n - 1}{2l} \pi \sqrt{\frac{E}{\rho}}$$

From equation (6.145), the temporal solution becomes

$$T_n(t) = A_n e^{-0.01\omega_n t} \sin(\omega_{dn} t + \phi_n)$$

The total solution is then of the form

$$w(x, t) = \sum_{n=1}^{\infty} A_n e^{-0.01\omega_n t} \sin(\omega_{dn} t + \phi_n) \sin \frac{2n - 1}{2l} \pi x \tag{6.146}$$

Applying the initial condition yields

$$0.01 \left(\frac{x}{l} \right) = \sum_{n=1}^{\infty} A_n \sin \phi_n \sin \sigma_n x$$

Multiplying the last expression by $\sin \sigma_m x$ and integrating over the length of the bar yields

$$\frac{0.01}{l} \int_0^l x \sin \sigma_m x \, dx = \frac{0.01}{l \sigma_m^2} (-1)^{m+1} = \sum_{n=1}^{\infty} A_n \sin \phi_n \int_0^l \sin \sigma_n x \sin \sigma_m x \, dx \tag{6.147}$$

The integral on the right side is the orthogonality condition for the modes:

$$\int_0^l \sin \sigma_n x \sin \sigma_m x \, dx = \left(\frac{l}{2} \right) \delta_{mn} \tag{6.148}$$

Substitution of the orthogonality conditions into the summation of (6.147) yields

$$\frac{0.01}{l \sigma_m^2} (-1)^{m+1} = (A_m \sin \phi_m) \left(\frac{l}{2} \right) \qquad m = 1, 2, \dots \tag{6.149}$$

which provides one equation in the two sets of unknown coefficients A_m and ϕ_m. A second equation for the unknown coefficients A_m and ϕ_m is obtained from the second initial condition. This yields

$$w_t(x, t) = 0$$

$$= \sum_{n=1}^{\infty} A_n[-0.01\omega_n e^{-0.01\omega_n t} \sin(\omega_{dn}t + \phi_n) + e^{-0.01\omega_n t}\omega_{dn}\cos(\omega_{dn}t + \phi_n)]\sin\sigma_n x$$

Again multiplying by $\sin\sigma_m x$ integrating over the length of the beam and using the orthogonality condition results in

$$0 = A_m(-0.01\omega_n \sin\phi_m + \omega_{dm}\cos\phi_m) \qquad m = 1, 2, \ldots$$

Since $A_m \neq 0$, the term in parentheses must be zero so that

$$\tan\phi_m = \frac{\sqrt{1 - \zeta^2}}{0.01} = 99.9949 \qquad m = 1, 2, \ldots$$

and hence $\phi_m = 1.5607$, which is almost $\pi/2$ radians or 90°. Substitution of this value back into equation (6.149) yields

$$A_m = \frac{0.02}{l^2\sigma_m^2}(-1)^{m+1} \qquad m = 1, 2, \ldots$$

Hence the general solution, from equation (6.146), becomes

$$w(x, t) = \sum_{n=1}^{\infty} \left(\frac{0.02}{l^2\sigma_n^2}(-1)^{n+1}\right)e^{-0.01\omega_n t}\cos\omega_{dn}t \sin\sigma_n x \qquad (6.150)$$

where $\sigma_n = (2n - 1)\pi/2l$, $\omega_n = \sigma_n\sqrt{E/\rho}$, and $\omega_{dn} = 0.9999\omega_n$.

☐

Next consider some physical models of damping. Again, as was the case for single- and lumped-parameter models, physical damping mechanisms are illusive and difficult to derive. Hence some common examples are presented here. First consider the effects of an external damping mechanism, such as air. As a string, membrane or beam in transverse vibration oscillates it pushes air around, causing an energy loss from a nonconservative force proportional to velocity (see Blevins, 1977, for a more complete explanation).

In some circumstances, this energy dissipation can be approximated as linear viscous damping of the form $\gamma w_t(x, t)$, where γ is a constant damping parameter. In this case the equation of a clamped–clamped string becomes

$$\rho w_{tt}(x, t) + \gamma w_t(x, t) - \tau w_{xx}(x, t) = 0$$
$$w(0, t) = 0 = w(l, t) \qquad (6.151)$$

where ρ and τ are the density and tension as defined in equation (6.4) and l is the length of the string. The equation for a square membrane moving in a fluid and clamped around its edges becomes

$$\rho w_{tt}(x, y, t) + \gamma w_t(x, y, t) - \tau[w_{xx}(x, y, t) + w_{yy}(x, y, t)] = 0$$
$$w(0, y, t) = w(l, y, t) = w(x, 0, t) = w(x, l, t) = 0 \qquad (6.152)$$

where ρ and τ are as defined in equation (6.113) and l is the length of the sides of the membrane.

Internal damping can be modeled by examining the various forces and moments involved in deriving the equations of motion. For example consider the Euler–Bernoulli bending vibration model given by equation (6.91), developed from Figure 6.10. One possible choice for internal damping is to assign a viscous damping proportional to the rate of strain in the beam. The equation of motion for a beam with viscous air damping (external) and strain rate damping (internal) is

$$\rho A w_{tt}(x,t) + \gamma w_t(x,t) + \beta \frac{\partial^2}{\partial x^2}\left[I\frac{\partial^3 w(x,t)}{\partial x^2 \partial t}\right] + \frac{\partial^2}{\partial x^2}\left[EI\frac{\partial^2 w(x,t)}{\partial x^2}\right] = 0 \quad (6.153)$$

For the clamped–free case the boundary conditions become

$$w(0,t) = w_x(0,t) = 0$$

$$EIw_{xx}(l,t) + \beta I w_{xxt}(l,t) = 0 \qquad\qquad (6.154)$$

$$\frac{\partial}{\partial x}[EIw_{xx}(l,t) + \beta I w_{xxt}(l,t)] = 0$$

Here E, I, ρ, and A are as defined in equation (6.91) and γ and β are constant damping parameters. If $I(x)$ is constant, the boundary conditions and equation of motion can be somewhat simplified. Note that the inclusion of strain rate damping changes the boundary conditions as well as the equation of motion. For constant $I(x)$, the change in the boundary condition does not affect the solution. See Cudney and Inman (1989) for an experimental verification of these equations of motion. Strain rate damping is also called Kelvin–Voigt *damping*.

The solution technique for systems with the physical damping models above remains the same for constant parameters (i.e., E, I, ρ constant) as for the undamped case. This is so because these damping terms are all proportional to the effective mass and stiffness terms as in the lumped-parameter case. Caughey and O'Kelly (1965) present a more detailed explanation. The following example illustrates the use of modal analysis to solve the proportionally damped case.

Example 6.7.2

Calculate the solution of the damped string, equation (6.151) by modal analysis. Equation (6.151) is first restated by using separation of variables [i.e., by substituting $w(x,t) = T(t)X(x)$]. This yields

$$\frac{\rho \ddot{T} + \gamma \dot{T}}{\tau T} = \frac{X''}{X} = \text{constant} = -\sigma^2$$

This results in two equations: one in time and one in space. The spatial equation $X''(x) + \sigma^2 X(x) = 0$ is subject to the two boundary conditions. This problem was solved in Section 6.2, which resulted in $X_n(x) = a_n \sin(n\pi x/l)$ and $\sigma_n = n\pi/l$, for $n = 1, 2, \dots$. Substitution of these values into the temporal equation above yields

$$\ddot{T}_n(t) + \frac{\gamma}{\rho}\dot{T}_n(t) + \frac{\tau}{\rho}\left(\frac{n\pi}{l}\right)^2 T_n(t) = 0 \qquad\qquad (6.155)$$

Comparing the coefficient of $\dot{T}_n(t)$ with $2\zeta_n\omega_n$ yields

$$\zeta_n = \frac{1}{2\omega_n}\frac{\gamma}{\rho} = \frac{\gamma l}{2n\pi\sqrt{\tau\rho}} \tag{6.156}$$

The solution of the temporal equation (6.155) yields (see Window 6.3)

$$T_n(t) = A_n e^{-\zeta_n\omega_n t}\sin(\omega_{dn}t + \phi_n)$$

where $\omega_{dn} = \omega_n\sqrt{1 - \zeta_n^2}$. The total solution becomes

$$w(x, t) = \sum_{n=1}^{\infty} A_n e^{-\zeta_n\omega_n t}\sin(\omega_{dn}t + \phi_n)\sin\frac{n\pi x}{l}$$

where A_n and ϕ_n are constants to be determined by the initial conditions using the orthogonality relationship.

$\square$

Damping can also be modeled at the boundary of a structure. In fact, in many cases more energy is dissipated at joints or points of connection than through internal mechanisms such as strain rate damping. For example, the longitudinal vibration of the bar of Section 6.5 could be modeled as being attached to a lumped spring–damper system as indicated in Figure 6.14.

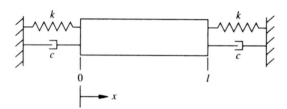

Figure 6.14 Clamped–clamped bar in longitudinal vibration with the point of attachment to ground modeled as providing viscous damping c and stiffness k.

The equation of motion remains as given in equation (6.56). However, summing forces in the x direction at the boundary yields

$$AE\frac{\partial w(0, t)}{\partial x} = kw(0, t) + c\frac{\partial w(0, t)}{\partial t}$$
$$AE\frac{\partial w(l, t)}{\partial x} = -kw(l, t) - c\frac{\partial w(l, t)}{\partial t} \tag{6.157}$$

These new boundary conditions affect both the orthogonality conditions and the system's temporal solution.

6.8 MODAL ANALYSIS OF THE FORCED RESPONSE

The forced response of a distributed-parameter system can be calculated using modal analysis just as in the lumped-parameter case of Section 4.6. The approach again uses the orthogonality condition of the unforced system's eigenfunctions to reduce the calculation of the response to a system of decoupled modal equations for the time response. The procedure is best illustrated by a simple example.

Example 6.8.1

Calculate the forced response of the string fixed at both ends (discussed in Section 6.2) with external damping coefficient γ, subject to unit impulse applied at $l/4$, where l is the length of the string. Assume that the initial conditions are both zero.

Solution   The equation of motion is

$$\rho w_{tt}(x,t) + \gamma w_t(x,t) - \tau w_{xx}(x,t) = f(x,t) \tag{6.158}$$

where $f(x,t) = \delta(x - l/4)\delta(t)$. Here $\delta(x - l/4)$ is a Dirac delta function indicating that the unit force is applied at $x = l/4$ and $\delta(t)$ indicates that the force is applied at time $t = 0$. The boundary conditions are $w(0,t) = w(l,t) = 0$. The eigenfunctions of the undamped, unforced clamped–clamped string are given by equation (6.22) to be of the form

$$X_n(x) = a_n \sin \frac{n\pi x}{l}$$

obtained by solving using the method of separation of variables.

The modal analysis procedure continues by assuming that the solution is of the form $w_n(x,t) = T_n(t)X_n(x)$, substituting this form into equation (6.158), multiplying by $X_n(x)$, integrating over the length of the string, and solving for $T_n(t)$. Following this procedure, equation (6.158) becomes

$$\left\{ \rho \ddot{T}_n(t) + \gamma \dot{T}_n(t) - \tau \left[-\left(\frac{n\pi}{l} \right)^2 \right] T_n(t) \right\} \sin \frac{n\pi x}{l} = \delta \left(x - \frac{l}{4} \right) \delta(t)$$

where the constant coefficient a_n of X_n has been arbitrarily set equal to unity. Multiplying by $\sin(n\pi x/l)$ and integrating yields

$$\left[\rho \ddot{T}_n(t) + \gamma \dot{T}_n(t) + \tau \left(\frac{n\pi}{l} \right)^2 T_n(t) \right] \frac{l}{2}$$

$$= \delta(t) \int_0^l \delta(x - l/4) \sin \frac{n\pi x}{l} dx = \delta(t) \left(\sin \frac{n\pi}{4} \right)$$

Rewriting this expression in the form of single-degree-of-freedom oscillator yields

$$\ddot{T}_n(t) + \frac{\gamma}{\rho} \dot{T}_n(t) + \left(\frac{cn\pi}{l} \right)^2 T_n(t) = \left(\frac{2}{l\rho} \sin \frac{n\pi}{4} \right) \delta(t) \qquad n = 1, 2, \ldots \tag{6.159}$$

where $c = \sqrt{\tau/\rho}$ as before. Equation (6.159) can be written in terms of the modal damping ratio, natural frequency, and input magnitude by comparing the coefficients to

$$\ddot{T}_n(t) + 2\zeta_n \omega_n \dot{T}_n(t) + \omega_n^2 T_n(t) = \hat{F}_n \delta(t) \qquad n = 1, 2, \ldots$$

Window 6.5
Solution of the Single-Degree-of-Freedom Forced Response

The response of an underdamped single degree of freedom system to an impulse modeled by

$$m\ddot{x} + c\dot{x} + kx = \hat{F}\delta(t)$$

where $\delta(t)$ is a unit impulse, is given by equations (3.7) and (3.8) as

$$x(t) = \frac{\hat{F}}{m\omega_d} e^{-\zeta\omega t} \sin \omega_d t$$

where $\omega = \sqrt{k/m}$, $\zeta = c/2\ m\omega$, $\omega_d = \omega\sqrt{1 - \zeta^2}$, and $0 < \zeta < 1$ must hold.

---

which has solution given by equations (3.7) and (3.8) and repeated in Window 6.5. Comparing coefficients between the expressions in Window 6.4 and equation (6.159) yields (for $n = 1, 2, 3, \ldots$)

$$\omega_n = \frac{cn\pi}{l}$$

$$\hat{F}_n = \frac{2}{l\rho} \sin \frac{n\pi}{4}$$

$$\zeta_n = \frac{\gamma l}{2cn\pi\rho}$$

Similarly, the nth damped natural frequency becomes

$$\omega_{dn} = \omega_n\sqrt{1 - \zeta_n^2} = \frac{cn\pi}{l}\sqrt{1 - \frac{\gamma^2 l^2}{4c^2 n^2 \pi^2 \rho^2}} \qquad n = 1, 2, 3, \ldots$$

Substitution of these values for ω_n, ω_{dn}, ζ_n and $\hat{F}_n$ into the expression of Window 6.5 yields that the solution for the nth temporal equation becomes

$$T_n(t) = \frac{\hat{F}_n}{\omega_{dn}} e^{-\zeta_n \omega_n t} \sin \omega_{dn} t$$

$$= \frac{4 \sin (n\pi/4)}{\sqrt{(2\rho cn\pi)^2 - (\gamma l)^2}} e^{-\gamma t/2\rho} \sin \left[\frac{1}{2\rho l} \sqrt{(2\rho cn\pi)^2 - (\gamma l)^2} t \right]$$

Combining this with $X_n = \sin(n\pi x/l)$ and forming the summation over all modes, the total solution becomes

$$w(x, t) = \sum_{n=1}^{\infty} \frac{4 \sin (n\pi/4)}{\sqrt{(2\rho cn\pi)^2 - (\gamma l)^2}} e^{-\gamma t/2\rho} \sin \left[\left(\frac{1}{2\rho l} \sqrt{(2\rho cn\pi)^2 - (\gamma l)^2} \right) t \right] \sin \frac{n\pi x}{l} \quad (1.160)$$

This represents the solution of the damped string subject to a unit impulse applied at $l/4$ units from one end. Similar to the series solutions for the free response, the series must converge and hence not all terms need to be calculated to obtain a reasonable solution. In fact, usually only the first few terms need be calculated, as illustrated in Figure 6.6 for a similar example.

$\square$

The response of a single-degree-of-freedom system to an impulse can be used to calculate the response of any general force input by use of the impulse response functions. This was illustrated in Section 3.2. The response of a distributed parameter system to an arbitrary force input can also be calculated using the concept of an impulse response function by defining a modal impulse response function.

The solution for the general forced response of an underdamped single-degree-of-freedom system as detailed in Section 3.2 is summarized in Window 6.6. Following the reasoning used in Example 6.8.1 for the impulse response and referring to Window 6.6, the response of a damped distributed parameter system to any force can be calculated. As in the case of the impulse response, the method is best illustrated by example.

Window 6.6
Forced Response of an Underdamped System from Section 3.2

The response of an underdamped system

$$m\ddot{x} + c\dot{x} + kx = F(t)$$

(with zero initial conditions) is given by (for $0 < \zeta < 1$)

$$x(t) = \frac{1}{m\omega_d} e^{-\zeta\omega t} \int_0^t F(\tau) e^{\zeta\omega\tau} \sin\omega_d(t - \tau)d\tau$$

where $\omega = \sqrt{k/m}$, $\zeta = c/2m\omega$, and $\omega_d = \omega\sqrt{1 - \zeta^2}$. With non zero initial conditions this becomes

$$x(t) = Ae^{-\zeta\omega t} \sin(\omega_d t + \theta) + \frac{1}{\omega_d} e^{-\zeta\omega t} \int_0^t f(\tau) e^{\zeta\omega\tau} \sin\omega_d(t - \tau)d\tau$$

where $f = F/m$ and A and θ are constants determined by the initial conditions.

Example 6.8.2

A rotating machine is mounted on the second floor of a building as illustrated in Figure 6.15. The machine excites the floor support beam with a force of 100 N at 3 rad/s. Model the floor support beam as an undamped simply supported Euler–Bernoulli beam and calculate the forced response.

Solution   The sketch on the right side of Figure 6.15 suggests that a reasonable model for the vibration is given by equation (6.92) and boundary conditions given by equation (6.95)

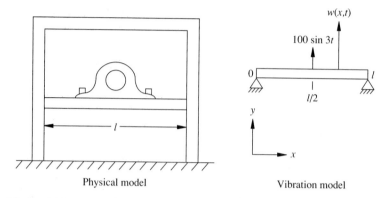

Physical model   Vibration model

Figure 6.15 A model of a building with an out-of-balance rotating machine mounted in the middle of the second floor over a support beam. The vibration model is simplified to be that of a harmonic force applied to the center of a simply supported beam.

with a driving force of $100 \sin 3t$. Hence the problem is to solve

$$w_{tt}(x, t) + c^2 w_{xxxx}(x, t) = 100 \sin 3t \delta \left(x - \frac{l}{2} \right) \tag{6.161}$$

subject to $w(0, t) = w(l, t) = w_{xx}(0, t) = w_{xx}(l, t) = 0$, where $c = \sqrt{EI/\rho A}$. First separation of variables is used to calculate the spatial mode shapes using the homogeneous version of equation (6.161). To this end, let $w(x, t) = T(t)X(x)$ in equation (6.161) so that

$$\frac{\ddot{T}(t)}{T(t)} = -c^2 \frac{X''''(x)}{X(x)} = -\omega^2$$

following equation (6.97). This leads to equations (6.98) and (6.102), so that the solution of the spatial equation is given by equation (6.102):

$$X(x) = a_1 \sin \beta x + a_2 \cos \beta x + a_3 \sinh \beta x + a_4 \cosh \beta x \tag{6.162}$$

Here $\beta^4 = \rho A \omega^2 / EI$. Applying the four boundary conditions given in equation (6.95) yields the desired eigenfunctions. The deflection must be zero at $x = 0$ [i.e., $X(0) = 0$] so that $a_2 + a_4 = 0$. Similarly, the bending moment must vanish at $x = 0$ [i.e.,. $X''(0) = 0$] so that $-a_2 + a_4 = 0$. Hence $a_2 = a_4 = 0$ is required to satisfy the boundary conditions at $x = 0$. Thus equation (6.162) for the spatial solution reduces to

$$X(x) = a_1 \sin \beta x + a_3 \sinh \beta x \tag{6.163}$$

Applying the boundary condition at $x = l$ yields

$$a_1 \sin \beta l + a_3 \sinh \beta l = 0 \tag{6.164}$$

$$-a_1 \sin \beta l + a_3 \sinh \beta l = 0 \tag{6.165}$$

which has the solution $a_3 = 0$ and $\sin \beta l = 0$. Hence the characteristic equation is $\sin \beta l = 0$, which yields $\beta l = n\pi$ and the eigenfunction becomes

$$X_n(x) = A_n \sin \frac{n\pi x}{l} \qquad n = 1, 2, \ldots \tag{6.166}$$

Recalling that $\beta^4 = \rho A \omega^2 / EI$ yields

$$\omega = \omega_n = \sqrt{\frac{EI}{\rho A}} \left(\frac{n\pi}{l}\right)^2 \qquad n = 1, 2, \ldots \qquad (6.167)$$

It is convenient at this point to normalize the eigenfunction given by equation (6.166). Following the definition of Chapter 4 for vectors, a set of eigenfunctions $X_n(x)$ is said to be *normal* if for each value of n

$$\int_0^l X_n(x) X_n(x) dx = 1 \qquad (6.168)$$

As in the case of eigenvectors, the normalization condition fixes the arbitrary constant associated with the eigenfunction. The constant A_n in equation (6.166) can be determined by substitution of the eigenfunction $X_n(x) = A_n \sin(n\pi x/l)$ into equation (6.168). This yields

$$A_n^2 \int_0^l \sin^2 \frac{n\pi x}{l} dx = 1$$

Performing the indicated integration yields

$$A_n^2 \frac{l}{2} = 1 \qquad \text{or} \qquad A_n = \sqrt{2/l}$$

so that the normalized eigenfunction becomes

$$X_n(x) = \sqrt{\frac{2}{l}} \sin \frac{n\pi x}{l} \qquad n = 1, 2, \ldots \qquad (6.169)$$

These are also referred to as normalized mode shapes. These functions also have the property that

$$\int_0^l X_n(x) X_m(x) dx = 0 \qquad m \neq n \qquad (6.170)$$

which is an orthogonality condition similar to that for eigenvectors. If a set of functions $X_n(x)$ satisfied both equations (6.170) and (6.168) for all combinations of the index n, they are said to be an *orthonormal set*.

Continuing with the modal analysis solution of equation (6.161), substitution of the form $w(x, t) = T_n(t) X_n(x)$ into the equation of motion yields

$$\ddot{T}_n(t) X_n(x) + c^2 T_n(t) X_n''''(x) = 100 \sin 3t \, \delta\left(x - \frac{l}{2}\right) \qquad (6.171)$$

From the equation below (6.161), $X_n'''' = (\omega_n^2/c^2) X_n(x)$. Substitution of this into (6.171) yields

$$[\ddot{T}_n(t) + \omega_n^2 T_n(t)] X_n(x) = (100 \sin 3t) \delta\left(x - \frac{l}{2}\right) \qquad (6.172)$$

Multiplying equation (6.172) by $X_n(x)$ and integrating over the length of the beam yields

$$\ddot{T}_n(t) + \omega_n^2 T_n(t) = 100 \sin 3t \sqrt{\frac{2}{l}} \int_0^l \delta\left(x - \frac{l}{2}\right) \sin \frac{n\pi x}{l} dx \qquad (6.173)$$

where the normalization condition is used to evaluate the integral on the left side of equation (6.173). Evaluating the integral on the left side yields

$$\ddot{T}_n(t) + \omega_n^2 T_n(t) = \sqrt{\frac{2}{l}} 100 \sin 3t \sin \frac{n\pi}{2} \qquad n = 1, 2, 3, \ldots \tag{6.174}$$

or

$$\ddot{T}_n(t) + \frac{EI}{\rho A} \left(\frac{n\pi}{l}\right)^4 T_n(t) = 0, \qquad n = 2, 4, 6, \ldots \tag{6.175}$$

$$\ddot{T}_n(t) + \frac{EI}{\rho A} \left(\frac{n\pi}{l}\right)^4 T_n(t) = \sqrt{\frac{2}{l}} 100 \sin 3t \qquad n = 1, 5, 9, \ldots \tag{6.176}$$

$$\ddot{T}_n(t) + \frac{EI}{\rho A} \left(\frac{n\pi}{l}\right)^4 T_n(t) = -\sqrt{\frac{2}{l}} 100 \sin 3t \qquad n = 3, 7, \ldots \tag{6.177}$$

where equation (6.167) has been used to evaluate ω_n^2.

Since the forced response is of interest here, the solution for $T_n(t)$ for even values of the index n determined by equation (6.175) is zero (zero force input and zero initial conditions). The solution to equations (6.176) and (6.177) is given by equation (2.7) as

$$T_n(t) = \frac{100\sqrt{2/l}}{(EI/\rho A)(n\pi/l)^4 - 9} \sin 3t \qquad n = 1, 5, 9, \ldots \tag{6.178}$$

and

$$T_n(t) = \frac{-100\sqrt{2/l}}{(EI/\rho A)(n\pi/l)^4 - 9} \sin 3t \qquad n = 3, 7, 11 \ldots \tag{6.179}$$

The forced response of the floor beam is then given by the series

$$w(x, t) = \sum_{n=1}^{\infty} T_n(t) X_n(x) \tag{6.180}$$

with T_n and X_n as indicated in equations (6.169), (6.178), and (6.179). Writing out the first few nonzero terms of this solution yields

$$w(x, t) =$$

$$\frac{200}{l} \left[\frac{\sin(\pi x/l)}{(\pi^4 EI/l^4 \rho A) - 9} - \frac{\sin(3\pi x/l)}{(81\pi^4 EI/l^4 \rho A) - 9} + \frac{\sin(5\pi x/l)}{(625\pi^4 EI/l^4 \rho A) - 9} \cdots \right] \sin 3t \tag{6.181}$$

Given values of the material parameters ρ, E, and the dimension of the support beam l, I, and A, equation (6.181) describes the forced response of the beam in the model of the vibration of a building floor due to a rotating machine mounted above it as sketched in Figure 6.15.

                                                               □

The modal analysis calculations used in Examples 6.8.1 and 6.8.2 as well as the example in Section 6.7 can be outlined for a general case just as for lumped-parameter systems in Chapter 4. This is summarized in Window 6.3 and again here. Modal analysis proceeds by substitution of the separation-of-variables form

$$w(x, t) = X_n(x)T_n(t)$$

into the equation of motion. This leads to two equations: one a boundary value problem in $X_n(x)$ and the other an initial value problem in $T_n(t)$. The boundary conditions are applied to the solution of the spatial equation for $X_n(x)$. This yields the eigenvalues (natural frequencies) and eigenfunctions (mode shapes) of the system. This step yields the same type of information as solving the eigenvalue problems for the matrix $M^{-1/2}KM^{-1/2}$ for lumped-mass systems. Next the eigenfunctions are normalized by using equation (1.168).

With the functions $X_n(x)$ completely determined and the natural frequencies known, the temporal equation for the function $T_n(t)$ can be solved by substitution of $w(x, t) = T_n(t)X_n(x)$ into the equation of motion and the initial conditions. With $T_n(t)$ known for all n the total solution is assembled as the sum

$$w(x, t) = \sum_{n=1}^{\infty} T_n(t)X_n(x) \tag{6.182}$$

This equation is also called the *expansion theorem*. Certain mathematical arguments need to be made to check the convergence of this infinite series (see, e.g., Inman, 1989), but these are beyond the scope of presentation here. In many cases it suffices to use just a few of the first terms in the series to compute a meaningful approximation to the solution. Chapter 8 and 9 discuss more about approximating the solution given in equation (6.182).

As mentioned in Window 6.3 and confirmed in Example 6.5.2, the separation of variables/modal analysis approach does not always work. That is, it is often difficult or even impossible to separate the governing partial differential equations into a separate spatial equation and a separate temporal equation. This happens, for instance, in attempting to solve the Timoshenko beam equation because of the cross term

$$\frac{\partial^4 w(x, t)}{\partial x^2 \partial t^2}$$

which involves derivatives of both variables, and because of the existence of a fourth time derivative. However, because the Euler–Bernoulli beam mode shapes are assumed to be satisfactory for use with the Timoshenko method, a separated solution is assumed by writing $w(x, t)$ as a product of $T(t)$ and the assumed mode shape. This works in Example 6.5.2. Many other distributed-parameter vibration problems cannot be directly solved by the analytical approach of separation of variables. Hence a number of approximate approaches have been developed, such as assuming the mode shape (called the *assumed mode method*), as is done in Example 6.5.2. Other approximate methods are developed in detail in Chapter 8. For distributed-parameter structures complicated beyond the simple configurations considered in this chapter, the finite element of Chapter 8 or similar approximation method must be used for vibration analysis.

PROBLEMS

Section 6.2

6.1. Prove the orthogonality condition of equation (6.28).

6.2. Calculate the orthogonality of the modes in Example 6.2.3.

6.3. Plot the first four modes of Example 6.2.3.

6.4. Consider a cable that has one end fixed and one end free. The free end cannot support a transverse force, so that $w_x(l, t) = 0$. Calculate the natural frequencies and mode shapes.

6.5. Calculate the coefficients c_n and d_n of equation (6.27) for the system of a clamped–clamped string to the initial displacement given in Figure 6.16 and an initial velocity of $w_t(x, 0) = 0$.

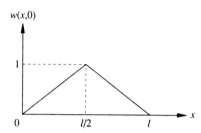

Figure 6.16   Initial displacement for the string of Problem 6.5.

6.6. Plot the response of the string in Problem 6.5 for the piano string of Example 6.2.2 at $x = l/4$ and $x = l/2$, using 3, 5, and 10 terms in the solution.

6.7. Consider the clamped string of Problem 6.5. Calculate the response of the string to the initial condition

$$w(x, 0) = \sin \frac{3\pi x}{l} \qquad w_t(x, 0) = 0$$

Plot the response at $x = l/2$ and $x = l/4$, for the parameters of Example 6.2.2.

Section 6.3

6.8. Calculate the natural frequencies and mode shapes for a free–free bar. Calculate the temporal solution of the first mode.

6.9. Calculate the natural frequencies and mode shapes of a clamped–clamped bar.

6.10. It is desired to design a 1 m clamped–free bar such that its first natural frequency is 100 Hz. Of what material should it be made?

6.11. Compare the natural frequencies of a clamped–free 1-m aluminum bar to that of a 1-m bar made of steel, a carbon composite, and a piece of wood.

6.12. Derive the boundary conditions for a clamped–free bar with a solid lumped mass, of mass M attached to the free end.

6.13. Calculate the mode shapes and natural frequencies of the bar of Problem 6.12. State how the lumped mass affects the natural frequencies and the mode shapes.

6.14. Calculate and plot the first three mode shapes of a clamped–free cable.

6.15. Calculate and plot the first three mode shapes of a clamped–clamped cable and compare them to the plots of Problem 6.14.

6.16. Calculate and compare the eigenvalues of the free–free, clamped–free, and clamped–clamped cable. Are they related? What does this state about the system's natural frequencies?

6.17. Consider the nonuniform bar of Figure 6.17 which changes cross-sectional area as indicated in the figure. In the Figure A_1, E_1, ρ_1, and l_1 are the cross-sectional area, modulus, density, and length of the first segment, respectively, and A_2, E_2, ρ_2, and l_2 are the corresponding physical parameters of the second segment. Calculate the vibrational response for the indicated cantilevered configuration to the initial conditions $w(x, 0) = 0$ and $w_t(x, 0) = \delta(x - l)$, where $l = l_1 + l_2$.

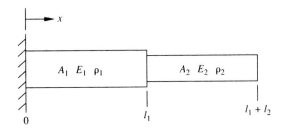

Figure 6.17   Bar with two separate cross sectional areas made of two different materials.

6.18. Show that the solution obtained to Problem 6.17 is consistent with that of a uniform bar.

6.19. Calculate the first three natural frequencies for the cable and spring system of Example 6.2.3 for $l = 1$, $k = 100$, $\tau = 100$ (SI units).

6.20. Calculate the first three natural frequencies of a clamped–free cable with a mass of value m attached to the free end. Compare these to the frequencies obtained in Problem 6.17.

6.21. Calculate the boundary conditions of a bar fixed at $x = 0$ and connected to ground through a mass and a spring as illustrated in Figure 6.18.

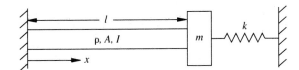

Figure 6.18   Beam with a tip mass connected to ground via a spring.

6.22. Calculate the natural frequency equation for the system of problem 6.21.

6.23. Estimate the natural frequencies of an automobile frame for vibration in its longitudinal direction (i.e., along the length of the car) by modeling the frame as a (one-dimensional) steel bar.

6.24. Consider the first natural frequency of the bar of Problem 6.1.8 with $k = 0$ and Table 6.1, which is fixed at one end and has a lumped-mass, M, attached at the free end. Compare this to the natural frequency of the same system modeled as a single-degree-of-freedom spring–mass system given in Figure 1.20. What happens to the comparison as M becomes small and goes to zero?

6.25. Following the line of thought suggested in Problem 6.24, model the system of Problem 6.19 as a lumped-mass single-degree-of-freedom system and compare this frequency to the frequencies obtained in Problem 6.19.

6.26. Calculate the response of a clamped–free bar to an initial displacement 1 cm at the free end and a zero initial velocity. Assume that $\rho = 7.8 \times 10^3$ kg/m^3, $A = 0.001$ m^2, $E = 10^{10}$ N/m^2, and $l = 0.5$ m. Plot the response at $x = l$ and $x = l/2$ using the first three modes.

6.27. Repeat the plots of Problem 6.26 for 5 modes, 10 modes, 15 modes, and so on, to answer the question of how many modes are needed in the summation of equation (6.27) in order to yield an accurate plot of the response for this system.

6.28. A moving bar is traveling along the x axis with constant velocity and is suddenly stopped at the end at $x = 0$, so that the initial conditions are $(x, 0) = 0$ and $w_t(x, 0) = v$. Calculate the vibration response.

6.29. Calculate the response of the clamped–clamped string of Section 6.2 to a zero initial velocity and an initial displacement of $w_0(x) = \sin(2\pi x/l)$. Plot the response at $x = l/2$.

Section 6.4

6.30. Calculate the first three natural frequencies of torsional vibration of a shaft of Figure 6.7 clamped at $x = 0$, if a disk of inertia $J_0 = 10$ kg $\cdot$ m^2/rad is attached to the end of the shaft at $x = l$. Assume that $l = 0.5$ m, $J = 5$ m^4, $G = 2.5 \times 10^9$ Pa, $\rho = 2700$ kg/m^3.

6.31. Compare the frequencies calculated in the previous problem to the frequencies of the lumped-mass single-degree-of-freedom approximation of the same system.

6.32. Calculate the natural frequencies and mode shapes of a shaft in torsion of shear modulus G, length l, polar inertia, J and density ρ which is free at $x = 0$ and connected to a disk of inertia J_0 at $x = l$.

6.33. Consider the lumped-mass model of Figure 4.18 and the corresponding three degree-of-freedom model of Example 4.8.1. Let $J_1 = k_1 = 0$ in this model and collapse it to a two-degree-of-freedom model. Comparing this to Example 6.4.1 it is seen that they are a lumped-mass model and a distributed-mass model of the same physical device. Referring to Chapter 1 for the effects on lumped stiffness of a rod in torsion (k_2), compare the frequencies of the lumped-mass two-degree-of-freedom model with those of Example 6.4.1.

6.34. The modulus and density of a 1-m aluminum rod are $E = 7.1 \times 10^{10}$N/m^2, $G = 2.7 \times 10^{10}$ N/m^2, and $\rho = 2.7 \times 10^3$ kg/m^3. Compare the torsional natural frequencies with the longitudinal natural frequencies for a free-clamped rod.

6.35. Consider the aluminum shaft of Problem 6.32. Add a disk of inertia J_0 to the free end of the shaft. Plot the torsional natural frequencies versus increasing the tip inertia J_0 of a single-degree-of-freedom model and for the first natural frequency of the distributed parameter model in the same plot. Are there any values of J_0 for which the single-degree-of-freedom model gives the same frequency as the fully distributed model?

6.36. Calculate the mode shapes and natural frequencies of a bar with circular cross section in torsional vibration with free–free boundary conditions. Express your answer in terms of G, l, and ρ.

6.37. Calculate the mode shapes and natural frequencies of a bar with circular cross section in torsional vibration with fixed boundary conditions. Express your answer in terms of G, l, and ρ.

6.38. Calculate the eigenfunctions of Example 6.4.1.

6.39. Show that the eigenfunctions of Problem 6.38 are orthogonal.

Section 6.5

6.40. Calculate the natural frequencies and mode shapes of a clamped-free beam. Express your solution in terms of E, I, ρ, and l. This is called the cantilevered beam problem.

6.41. Plot the first three mode shapes calculated in Problem 6.40. Next calculate the strain mode shape [i.e., $X'(x)$], and plot these next to the displacement mode shapes $X(x)$. Where is the strain the largest?

6.42. Derive the general solution to a fourth-order ordinary differential equation with constant coefficients of equation (6.100) given by equation (6.102).

6.43. Calculate the natural frequencies and mode shapes of a pinned–pinned beam in transverse vibration. Calculate the solution for $w_0(x) = \sin 2\pi x/l$ and $\dot{w}_0(x) = 0$.

6.44. Calculate the natural frequencies and mode shapes of a fixed–fixed beam in transverse vibration.

6.45. Show that the eigenfunctions or mode shapes of Example 6.5.1 are orthogonal. Make them normal.

6.46. Derive equation (6.109) from equations (6.107) and (6.108).

6.47. Show that if shear deformation and rotary inertia are neglected, the Timoshenko equation reduces to the Euler–Bernoulli equation and the boundary conditions for each model become the same.

Section 6.6

6.48. Calculate the natural frequencies of the membrane of Example 6.6.1 for the case that one edge $x = 1$ is free.

6.49. Repeat Example 6.6.1 for a rectangular membrane of size a by b. What is the effect of a and b on the natural frequencies?

6.50. Plot the first three mode shapes of Example 6.6.1.

6.51. The lateral vibrations of a circular membrane are given by

$$\frac{\partial^2 \omega(r, \phi, t)}{\partial r^2} + \frac{1}{r} \frac{\partial \omega(r, \phi, t)}{\partial r} + \frac{1}{r^2} \frac{\partial^2 \omega(r, \phi, t)}{\partial \phi \partial r} = \frac{\rho}{\tau} \frac{\partial^2 \omega(r, \phi, t)}{\partial t^2}$$

where r is the distance from the center point of the membrane along a radius and ϕ is the angle around the center. Calculate the natural frequencies if the membrane is clamped around its boundary at $r = R$.

6.52. Discuss the orthogonality condition for Example 6.6.1.

Section 6.7

6.53. Calculate the response of Example 6.7.1 for $l = 1$ m, $E = 2.6 \times 10^{10}$ N/m^2 and $\rho = 8.5 \times 10^3$ kg/m^3. Plot the response using the first three modes at $x = l/2$, $l/4$, and $3l/4$. How many modes are needed to represent accurately the response at the point $x = l/2$?

6.54. Repeat Example 6.7.1 for a modal damping ratio of $\zeta_n = 0.1$.

6.55. Repeat Problem 6.53 for the case of Problem 6.54. Does it take more or fewer modes to accurately represent the response at $l/2$?

6.56. Calculate the form of modal damping ratios for the clamped string of equations (6.151) and the clamped membrane of equation (6.152).

6.57. Calculate the units on γ and β in equation (6.153).

6.58. Assume that E, I, and ρ are constant in equations (6.153) and (6.154) and calculate the form of the modal damping ratio ζ_n.

6.59. Calculate the form of the solution $w(x, t)$ for the system of Problem 6.58.

6.60. For a given cantilevered composite beam, the following values have been measured for bending vibration:

$$E = 2.71 \times 10^{10} \, \text{N/m}^2 \qquad \rho = 1710 \, \text{kg/m}^3$$

$$A = 0.597 \times 10^{-3} \text{m}^3 \qquad l = 1 \, \text{m}$$

$$I = 1.64 \times 10^{-9} \text{m}^4 \qquad \gamma = 1.75 \, \text{N} \cdot \text{s/m}^2$$

$$\beta = 20,500 \, \text{N} \cdot \text{s/m}^2$$

Calculate the solution for the beam to an initial displacement of $w_t(x, 0) = 0$ and $w(x, t) = 3 \sin \pi x$.

6.61. Plot the solution of Example 6.7.2 for the case $w_t(x, 0) = 0$, $w(x, 0) = \sin(\pi x/l)$, $\gamma = 10 \, \text{N} \cdot \text{s/m}^2$, $\tau = 10^4 \text{N}$, $l = 1$ m, and $\rho = 0.01$ kg/m.

6.62. Calculate the orthogonality condition for the system of Example 6.7.2. Then calculate the form of the temporal solution.

6.63. Calculate the form of modal damping for the longitudinal vibration of the beam of Figure 6.14 with boundary conditions specified by equation (6.157).

Section 6.8

6.64. Calculate the response of the damped string of Example 6.8.1 to a disturbance force of $f(x, t) = (\sin \pi x/l) \sin 10t$.

6.65. Consider the clamped–free bar of Example 6.3.2. The bar can be used to model a truck bed frame. If the truck hits an object (at the free end) causing an impulsive force of 100 N, calculate the resulting vibration of the frame. Note here that the truck cab is so massive compared to the bed frame that the end with the cab is modeled as clamped. This is illustrated in Figure 6.19.

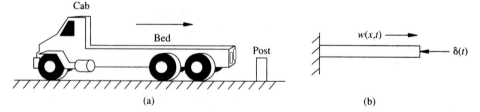

Figure 6.19   (a) Model of a truck hitting an object; (b) simplified vibration model.

6.66. A rotating machine sits on the second floor of a building just above a support column as indicated in Figure 6.20. Calculate the response of the column in terms of E, A, and ρ of the column modeled as a bar.

6.67. Recall Example 6.8.2, which models the vibration of a building due to a rotating machine imbalance on the second floor. Suppose that the floor is constructed so that the beam is clamped at one end and pinned at the other, and recalculate the response (recall Example 6.5.1). Compare your solution and that of Example 6.8.2, and discuss the difference.

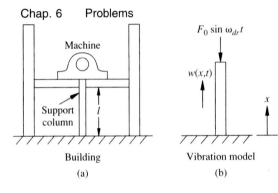

Figure 6.20 (a) Model of a rotating out-of-balance machine mounted on top of a column on the second floor of a building; (b) vibration model.

6.68. Use the modal analysis procedure suggested at the end of Section 6.8 to calculate the response of a clamped–free beam with a sinusoidal loading $F_0 \sin \omega t$ at its free end.

MATLAB VIBRATION TOOLBOX

If you have not yet used the *Vibration Toolbox*, return to the end of Chapter 1 for a brief introduction to using MATLAB files or open the file INTRO.TXT on the disk on the inside back cover. The files contained in folder/directory VTB6 may be used to help solve the problems listed above. The M-files from earlier chapters (VTBX_X, etc.) may also be useful. The exercises below are intended to help you gain some experience with the concepts introduced in this chapter for calculating vibration responses of distributed-parameter systems and to build experience with the various formulas. The MATLAB functions `plot` and `fzero` are particularly useful in solving the problems above.

TOOLBOX PROBLEMS

TB6.1. Use file VTB6_1 to investigate the effects of changing the length, density, and modulus (or tension) on the frequency and mode shapes of bars and shafts. For example, consider the clamped–free bar of Example 6.3.1. Study the effect of changing the length l on the frequencies by calculating ω_n for fixed value of E and ρ (say, for aluminum). What happens to the mode shapes?

TB6.2. Use VTB6_2 to calculate the frequencies for torsional vibration for the first three boundary conditions of Table 6.3. In particular check the results of Example 6.4.1.

TB6.3. Compare the mode shapes of a cantilevered beam with those of the fixed-pinned beam of Example 6.5.1 by using file VTB6_3 to plot the mode shapes.

TB6.4. Use VTB6_4 to compare the effects of rotary inertia and shear deformation on a pinned–pinned beam by trying several different values of the various physical parameters. Try to conclude circumstances under which the Timoshenko theory gives drastically different frequencies than the Euler–Bernoulli theory does.

7 Vibration Testing and Experimental Modal Analysis

This chapter presents methods of testing and measurement useful for obtaining experimental models of a variety of devices and structures. The analyzer pictured here receives analog signals from tranducers (accelerometers) mounted on the structure, digitizes the signals, and transforms them into the frequency domain for analysis. The entire process is controlled by a personal computer. A schematic of such a test setup is given in Figure 7.1. The picture at the bottom exhibits the use of a laser vibrometer to measure velocity without physically being mounted on the test object (car). Section 7.1 describes the measurement hardware, and the remainder of the chapter is devoted to methods of analyzing the data. In particular, the method of modal analysis is discussed.

This chapter discusses vibration measurement and focuses in particular on techniques associated with experimental modal analysis. Vibration measurements are made for a variety of reasons. As pointed out in previous chapters, especially in Chapter 5 on design, the natural frequencies of a structure or machine are extremely important in predicting and understanding a system's dynamic behavior. Hence a major reason for performing a vibration test of a system is to determine its natural frequencies. Another reason for vibration testing is to verify an analytical model proposed for the test system. For example, the analytical models proposed for the various examples of Section 4.8 and those of Chapter 6 yield a specific set of frequencies and mode shapes. A vibration test can then be performed on the same system. If the measured frequencies and mode shapes agree with those predicted by the analytical model, the model is verified and can be used in design and response prediction with some confidence.

Another important use of vibration testing is to determine experimentally the dynamic durability of a particular device. In this case a test article is driven or forced to vibrate by specified inputs for a specific length of time. When the test is over, the test piece must still perform its original task. The purpose of this type of testing is to provide experimental evidence that a machine part or structure can survive a specific dynamic environment.

Vibration testing is also used in machinery diagnostics for maintenance. The idea here is continuous monitoring of the natural frequencies of a structure or machine. A shift in frequency or some other vibration parameter may indicate a pending failure or a need for maintenance. This use of vibration measurement is part of the general topic of condition monitoring of machinery. It is similar in concept to observing the oil pressure in an automobile engine to determine if engine failure may occur or if maintenance is required.

The primary requirement of each of the aforementioned uses of vibration tests is a determination of a system's natural frequencies. Hence this chapter focuses on *experimental modal analysis* (EMA), which is the determination of natural frequencies, mode shapes, and damping ratios from experimental vibration measurements. The fundamental idea behind modal testing is that of resonance introduced in Section 2.2. If a structure is excited at resonance, its response exhibits two distinct phenomena, as indicated in Figure 2.5. As the driving frequency approaches the natural frequency of the structure, the magnitude at resonance rapidly approaches a sharp maximum value, provided that the damping ratio is less than about 0.5. The second, often neglected phenomenon of resonance is that the phase of the response shifts by $180°$ as the frequency sweeps through resonance, with the value of the phase at resonance being $90°$. This physical phenomenon is used to determine the natural frequency of a structure from measurements of the magnitude and phase of the structure's forced response as the driving frequency is swept through a wide range of values.

The vibration test methods presented in this chapter depend on several assumptions. First it is assumed that the structure or machine being tested can be described adequately by a lumped-parameter model. There are several other assumptions commonly made but not stated (or understated) in vibration testing. The most obvious of these is that the system under test is linear and is driven by the test input only in its linear range. This assumption is essential and should not be neglected.

Vibration testing and measurement for modeling purposes has grown into a large industry. This field is referred to as *modal testing, modal analysis*, or *experimental modal analysis*. Understanding modal testing requires knowledge of several areas. These include instrumentation, signal processing, parameter estimation, and vibration analysis. These topics are presented in the following sections. The first few sections of this chapter deal with the hardware considerations and digital signal analysis necessary for making a vibration measurement for any purpose.

7.1 MEASUREMENT HARDWARE

The data acquisition and signal processing hardware has changed considerably over the past decade and continues to change rapidly as the result of advances in solid-state and computer technology. Hence specific hardware capabilities change very quickly and only generic hardware is discussed here. A vibration measurement generally requires several hardware components. The basic hardware elements required consist of a source of excitation, called an *exciter*, for providing a known or controlled input force to the structure, a *transducer* to convert the mechanical motion of the structure into an electrical signal, a signal conditioning amplifier to match the characteristics of the transducer to the input electronics of the digital data acquisition system, and an analysis system (or analyzer) in which signal processing and modal analysis computer programs reside. This arrangement is illustrated in Figure 7.1; it includes a power amplifier and a signal generator for the exciter, as well as a transducer to measure, and possibly control, the driving force or other (input). Each of these devices and their functions are discussed briefly in this section.

First consider the excitation system. This system provides an input motion or, more commonly, a driving force $F_i(t)$, as in equation (4.130). The physical device may take several forms, depending on the desired input and the physical properties of the test structure. The two most commonly used exciters in modal testing are the *shaker* (electromagnetic or electrohydraulic) and the *impulse hammer*. The preferred device is often the electromagnetic exciter, which has the ability, when properly sized, to provide inputs large enough to result in easily measured responses. Also, the output is easily controlled electronically, sometimes using force feedback. The excitation signal, which

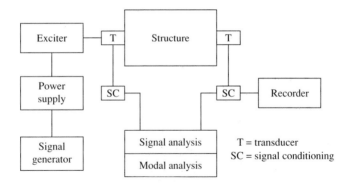

Figure 7.1 Schematic of hardware used in performing a vibration test.

can be tailored to match the requirements of the structure being tested, can be a swept sinusoidal, random, or other appropriate signal. A swept sine input consists of applying a harmonic force of constant magnitude f_i at a variety of different frequencies, ranging from a small value to larger values covering a frequency range of interest. At each incremental value of the driving frequency, the structure is allowed to reach steady state before the response magnitude and phase are measured. The electromagnetic shaker is basically a linear electric motor consisting of coils of wire surrounding a shaft in a magnetic field. An alternating current applied to the coil causes a force to be applied to the shaft, which, in turn, transfers force to the structure. The input electrical signal to the shaker is usually a voltage that causes a proportional force to be applied to the test structure. Thus a signal generator can be used to impart a variety of different input signals to the structure.

Since shakers are attached to the test structure and since they have significant mass, care should be taken by the experimenter in choosing the size of shaker and method of attachment to minimize the effect of the shaker on the structure. The shaker and its attachment can add mass to the structure under test (called *mass loading*) as well as otherwise constraining the structure. Mass loading will lower the apparent measured frequency since $\omega = \sqrt{k/m}$. Mass loading and other effects can be minimized by attaching the shaker to the structure through a *stinger*. A stinger consists of a short thin rod (usually made of steel or nylon) running from the driving point of the shaker to a force transducer mounted directly on the structure. The stinger serves to isolate the shaker from the structure, reduces the added mass, and causes the force to be transmitted axially along the stinger, controlling the direction of the applied force more precisely.

In recent years the *impact hammer* has become a popular excitation device. The use of an impact hammer avoids the mass loading problem and is much faster to use than a shaker. An impact hammer consists of a hammer with a force transducer built into the head of the hammer. The hammer is then used to hit (impact) the test structure and thus excite a broad range of frequencies. The impact hammer is intended to apply an impulse to the structure as modeled and analyzed in Section 3.1. As indicated in Section 4.6, the impulse response contains excitations at each of the system's natural frequencies. The peak impact force is nearly proportional to the hammerhead mass and the impact velocity. The load cell (force transducer) in the head of the hammer provides a measure of the impact force.

Figure 7.2 illustrates both time history and the corresponding frequency response of a typical hammer hit. Note that the time history is not a perfect delta function (as in Figure 3.1) but rather has a finite time duration, T. Hence the frequency response is not a flat straight line as indicated by the transform of an exact impulse, but rather, has the periodic form given in Figure 7.2. The duration of the pulse and hence the shape of the frequency response is controlled by the mass and stiffness of both the hammer and the structure. In the case of a small hammer mass used on a hard structure (such as metal), the stiffness of the hammer tip determines the shape of the spectrum and in particular the cutoff frequencies ω_c. The *cutoff frequency* is the largest value of frequency reasonably well excited by the hammer hit. As illustrated in the figure, ω_c corresponds roughly to the point where the magnitude of the frequency response falls more than 10 to 20 dB from its maximum value. This means that at frequency higher than ω_c, the test structure does not receive enough energy to excite modes above ω_c. Thus ω_c determines the useful range of frequency excitation.

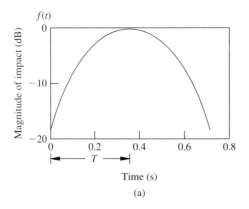

(a)

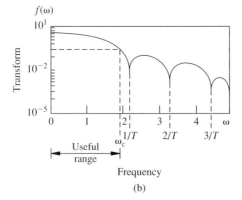

(b)

Figure 7.2 Time (a) and frequency response (b) of a hammer hit, indicating the useful range of excitation and its dependence on the pulse duration, T.

The upper frequency limit excited by the hammer is decreased by increasing the hammerhead mass and is increased with increasing stiffness of the tip of the hammer. The hammer hit is less effective in exciting the modes of the structure with frequencies larger than ω_c than it is for those less than ω_c. The built-in force transducer in impact hammers should be dynamically calibrated for each tip used, as this will affect the sensitivity. Although the impact hammer is simple and does not add mass loading to the structure, it is often incapable of transforming sufficient energy to the structure to obtain adequate response signals in the frequency range of interest. Also, peak impact loads are potentially damaging, and the direction of the applied load is difficult to control. Nonetheless, impact hammers remain a popular and useful excitation device, as they generally are much faster to use than shakers, are portable, and are relatively inexpensive.

Next consider the transducers required to measure the response of the structure as well as the impact force. The most popular and widely used transducers are made from piezoelectric crystals. Piezoelectric materials generate electrical charge when strained. By various designs, transducers incorporating these materials can be built to produce signals proportional to either force or local acceleration. *Accelerometers*, as they are called, actually consist of two masses, one of which is attached to the structure, separated by a piezoelectric material. The piezoelectric material acts like a very stiff spring. This causes the transducer to have a resonant frequency. The maximum measurable frequency is usually a fraction of the accelerometer's resonance frequency (recall Figure 2.17). In fact, the upper frequency limit is usually determined by the so-called mounted resonance, since the connection of the transducer to the structure is always somewhat compliant. The dynamics of accelerometers are discussed in some detail in Section 2.6.

Strain gauges can also be used to pick up vibrational responses. A *strain gauge* is a metallic or semiconductor material that exhibits a change in electrical resistance when subjected to a strain. Strain gauges are constructed by bending a conducting wire back and forth in a serpentine fashion over a very small surface which is then bonded to the device or structure to be measured. As the structure strains, the resistance of the wire serpentine changes. The gauge is made part of a Wheatstone bridge circuit which is used to measure the resistance change of the gauge and hence the strain in the test specimen (see Figliola and Beasley, 1991). Strain gauges are also used to form load cells.

The output impedance of most transducers is not well suited for direct input into signal analysis equipment. Hence *signal conditioners*, which may be charge amplifiers or voltage amplifiers, match and often amplify signals prior to analyzing the signal. It is very important that each set of transducers along with signal conditioning are properly calibrated in terms of both magnitude and phase over the frequency range of interest. While accelerometers are convenient for many applications, they provide weak signals if one is interested in lower-frequency vibrations incurred in terms of velocity or displacement. Even substantial low-frequency vibration displacements may result in only small accelerations, since a harmonic displacement of amplitude X has acceleration of amplitude $-\omega^2 X$. Strain gauges and potentiometers as well as various optical, capacitive, and inductive transducers are often more suitable than accelerometers for low-frequency motion measurement.

Once the response signal has been properly conditioned, it is routed to an analyzer for signal processing. There are several types of analyzers in use. The type that has become the standard is called a digital Fourier analyzer, also called the fast Fourier transform (often abbreviated FFT) analyzer; it is introduced briefly here. Basically, the analyzer accepts analog voltage signals that represent the acceleration (force, velocity, displacement, or strain) from a signal conditioning amplifier. This signal is filtered and digitized for computation. Discrete frequency spectra of individual signals and cross-spectra between the input and various outputs are computed. The analyzed signals can then be manipulated in a variety of ways to produce such information as natural frequencies, damping ratios, and mode shapes in numerical or graphic displays.

While almost all of the commercially available analyzers are marketed as turnkey devices, it is important to understand a few details of the signal processing performed by these analysis units in order to carry out valid experiments. This forms the topic of the next two sections.

7.2 DIGITAL SIGNAL PROCESSING

Much of the analysis done in modal testing is performed in the frequency domain, inside the analyzer. The analyzer's task is to convert analog time-domain signals into digital frequency-domain information compatible with digital computing and then to perform the required computations with these signals. The method used to change an analog signal, $x(t)$, into frequency-domain information is the Fourier transform [defined by equations (3.43) and (3.44)], or a *Fourier series* as defined by equations (3.20) to (3.23). The Fourier series is used here to introduce the digital Fourier transform (DFT).

As pointed out in Section 3.3, a periodic time signal of period T can be represented by a Fourier series in time of the form given by equation (3.20) with Fourier coefficients, or spectral coefficients as defined by equations (3.21) to (3.23). Essentially the spectral coefficients represent frequency-domain information about a given time signal. These equations are repeated in Window 7.1.

The Fourier coefficients a_n and b_n given by equations (3.21) to (3.23) also represent the connection between Fourier analysis and vibration experiments. The analog output signals of accelerometers and force transducers, represented by $x(t)$, are inputs to the analyzer. The analyzer, in turn, calculates the spectral coefficients of these signals, thus setting the stage for a frequency-domain analysis of the signals. Some signals and their Fourier spectrum are illustrated in Figure 7.3. The analyzer first converts the analog signals into digital records. It samples the signals $x(t)$ at many different equally spaced values and produces a digital record, or version, of the signal in the form of a set of numbers $\{x(t_k)\}$. Here, $k = 1, 2, \ldots, N$, where the digit N denotes the number of samples and t_k indicates a discrete value of the time.

This process is performed by an analog-to-digital (A/D) converter. This conversion from an analog to a digital signal can be thought of in two ways. First, one can imagine a gate that samples the signal every Δt seconds and passes through the signal $x(t_k)$. The

---

Window 7.1
Review of the Fourier Series of a Periodic Signal $F(t)$ of Period T

$$F(t) = \frac{a_0}{2} + \sum_{n=1}^{\infty} (a_n \cos n\omega_T t + b_n \sin n\omega_T t) \tag{3.20}$$

where

$$\omega_T = \frac{2\pi}{T}$$

$$a_0 = \frac{2}{T} \int_0^T F(t)\, dt \tag{3.21}$$

$$a_n = \frac{2}{T} \int_0^T F(t) \cos n\omega_T t\, dt \qquad n = 1, 2 \dots \tag{3.22}$$

$$b_n = \frac{2}{T} \int_0^T F(t) \sin n\omega_T t\, dt \qquad n = 1, 2 \dots \tag{3.23}$$

---

process of A/D conversion can also be considered as multiplying the signal $x(t)$ by a square-wave function, which is zero over alternate values of t_k and has the value of 1 at each t_k for a short time. Some signals and their digital representation are illustrated in Figure 7.3.

In calculating digital Fourier transforms care must be taken in choosing the sampling time (i.e., the time elapsed between successive t_k's). A common error introduced in digital signal analysis caused by improper sampling time is called *aliasing*. Aliasing results from A/D conversion and refers to the misrepresentation of the analog signal by the digital record. Basically, if the sampling rate is too slow to catch the details of the analog signal, the digital representation will cause high frequencies to appear as low frequencies. The following example illustrates two periodic analog signals of different frequency and phase that produce the same digital record.

Example 7.2.1

Consider the signals $x_1(t) = \sin[(\pi/4)t]$ and $x_2(t) = -\sin[(7\pi/4)t]$, and suppose that these signals are both sampled at 1-s intervals. The digital record of each signal is given in the following table.

t_k	0	1	2	3	4	5	6	7	8	ω_i
x_1	0	0.707	1	0.707	0	-0.707	-1	-0.707	0	$\frac{1}{8}\pi$
x_2	0	0.707	1	0.707	0	-0.707	-1	-0.707	0	$\frac{7}{8}\pi$

As is easily seen from the table, the digital sample records of x_1 and x_2 are the same [i.e., $x_1(t_k) = x_2(t_k)$ for each value of k]. Hence no matter what analysis is performed on the digital record, x_1 and x_2 will appear the same. Here the sampling frequency, $\Delta\omega$, is one. Note that the difference between the frequency of the first signal, $x_1(t)$, and the sampling frequency is $\frac{1}{8} - 1 = -\frac{7}{8}$, which is the frequency of the second signal $x_2(t)$.

To avoid aliasing, the sampling interval, denoted by Δt, must be chosen small enough to provide at least two samples per cycle of the highest frequency to be calculated. That is, to recover a signal from its digital samples, the signal must be sampled at a rate at least twice the highest frequency in the signal. In fact, experience (see Otnes and

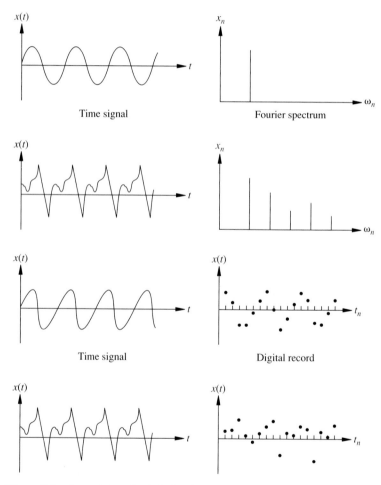

Figure 7.3   Several signals, their Fourier representation, and their digital representations.

Encochson, 1972) indicates that 2.5 samples per cycle is a better choice. This is referred to as the *sampling theorem*, or *Shannon's sampling theorem*.

Aliasing can be avoided in signals containing many frequencies by subjecting the analog signal $x(t)$ to an *antialiasing* filter. An antialiasing filter is a low-pass (i.e., only allows low frequencies through) sharp-cutoff filter. The filter effectively cuts off frequencies higher than about half the maximum frequency of interest, denoted by ω_{max}, and also called the *Nyquist frequency*. Most modern digital analyzers provide built-in antialiasing filters.

Once the digital record of the signal is available, the discrete version of the Fourier transform is performed. This transform provides a series representation of a discrete-time history value. This is accomplished by a digital Fourier transform or series defined by

$$x_k = x(t_k) = \frac{a_0}{2} + \sum_{i=1}^{N/2} \left(a_i \cos \frac{2\pi i t_k}{T} + b_i \sin \frac{2\pi i t_k}{T} \right) \qquad k = 1, 2, \ldots N \qquad (7.1)$$

where the *digital spectral coefficients* are given by

$$a_0 = \frac{1}{N} \sum_{k=1}^{N} x_k \qquad (7.2)$$

$$a_i = \frac{1}{N} \sum_{k=1}^{N} x_k \cos \frac{2\pi i k}{N} \qquad (7.3)$$

$$b_i = \frac{1}{N} \sum_{k=1}^{N} x_k \sin \frac{2\pi i k}{N} \qquad (7.4)$$

These are the digital versions of equation (3.21), (3.22), and (3.23), respectively. The task of the analyzer is to calculate equations (7.2) to (7.4) given the digital record $x(t_k)$, also denoted by x_k, for the measured signals. The transform size or number of samples, N, is usually fixed for a given analyzer and is a power of 2. Some common sizes are 512 and 1024.

Writing out equations (7.2) to (7.4) for each of the N samples yields N linear equations in the N spectral coefficients ($a_0, \ldots, a_{N/2}, b_0, \ldots, b_{N/2}$). These equations can also be written in the form of matrix equations. In matrix form they become

$$\mathbf{x} = C\mathbf{a} \qquad (7.5)$$

where $\mathbf{x}$ is the vector of samples with elements x_k and $\mathbf{a}$ is the vector of the spectral coefficients, a_0, a_i and b_i. The matrix C consists of elements containing the coefficients $\cos(2\pi i t_k/T)$ and $\sin(2\pi i t_k/T)$ as indicated in equation (7.1). The solution of equation (7.5) for the spectral coefficients is then given simply by

$$\mathbf{a} = C^{-1}\mathbf{x} \qquad (7.6)$$

The task of the analyzer is to compute the matrix C^{-1} and hence the coefficient $\mathbf{a}$. The most widely used method of computing the inverse of this matrix C is called the fast

Fourier transform (FFT), developed by Cooley and Tukey (1965). Note that while **x** represents the digital version, the spectral coefficient **a** represents the frequency content of the response (or input) signal.

To make digital analysis feasible, the periodic signal must be sampled over a finite time (*N* must obviously be finite). This can give rise to another problem referred to as *leakage*. To make the signal finite, one could simply cut the signal at any integral multiple of its period. Unfortunately, there is no convenient way to do this for complicated signals containing many different frequencies. Hence, if no further steps are taken the signal may be cut off midperiod. This causes erroneous frequencies to appear in the digital representation because the digital Fourier transform of the finite-length signal assumes that the signal is periodic *within* the sample record length. Thus the actual frequency will "leak" into a number of fictitious frequencies. This is illustrated in Figure 7.4.

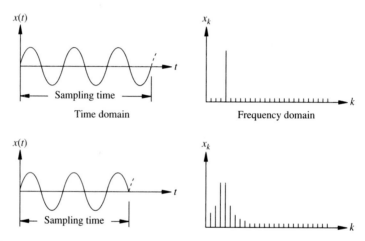

Figure 7.4   Example of leakage (i.e., frequencies caused by not sampling over an integer multiple of frequencies).

Leakage can be corrected to some degree by the use of a *window function*. Windowing, as it is called, involves multiplying the original analog signal by a weighting function, or window function, $w(t)$, which forces the signal to be zero outside the sampling period. A common window function, called the *Hanning window*, is illustrated in Figure 7.5, along with the effect it has on a periodic signal. A properly windowed signal will yield a spectral plot with much less leakage. This is also illustrated in the figures.

As noted in this section, if the signal's properties are precisely known (i.e., the frequency content), the choice of sampling rate and *N* would be obvious and correct. However, the reason a signal is being measured in the first place is to determine its frequency content; hence part of the art of modal analysis is choosing the sampling frequency and data size, *N*.

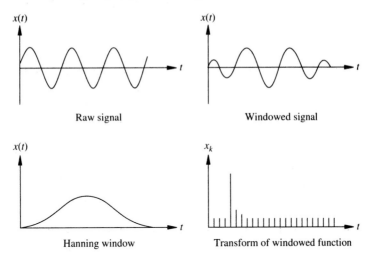

Figure 7.5 Use of a window function, in this case a Hanning window, to reduce leakage in calculating the frequency content of a signal.

7.3 RANDOM SIGNAL ANALYSIS IN TESTING

The transducer used to measure both the input and output during a vibration test usually contains noise (i.e., random components that make it difficult to analyze the measured data in a deterministic fashion). In addition, confidence in a measured quantity is increased by performing a number of identical tests and averaging the results. This is fairly common practice when measuring almost anything. In fact, the stiffness of a single structure is determined by multiple measurements, not just one, as indicated in Figure 1.3 of Section 1.1. Thus it is important to consider the random input vibration response developed in Section 3.5.

Recall the definition of the autocorrelation function of a signal and the associated power spectral density (PSD). These are reviewed in Window 7.2. Also recall that the PSD of the input or driving force can be related to the PSD of the response and the frequency response function of the system by

$$S_{xx}(\omega) = |H(\omega)|^2 S_{ff}(\omega) \tag{7.7}$$

as indicated by equation (3.60) and reviewed in Window 7.3. Equation (7.7) relates the dynamics of the test structure contained in $H(\omega)$ to measurable quantities (i.e., the PSDs). As pointed out at the end of Section 3.7, the common approach to measuring the frequency response function is to average several matched sets of input force time histories and output response time histories. These averages are used to produce correlation functions which are transformed to yield the corresponding PSDs. Equation (7.7) is then used to calculate the magnitude of the frequency response function $|H(\omega)|$. The experimental vibration data are then taken from the plot of $|H(\omega)|$ as indicated in Figure 3.15, or by means to be discussed in Section 7.4.

Window 7.2
Review of Some of the Definitions Used in Random Vibration Analysis

The autocorrelation function of the random signal $x(t)$ is given by

$$R_{xx}(\tau) = \lim_{T \to \infty} \frac{1}{T} \int_0^T x(t) x(t + \tau) \, dt \tag{3.48}$$

The power spectral density (PSD) of a signal is the Fourier transform of the signal's autocorrelation:

$$S_{xx}(\omega) = \frac{1}{2\pi} \int_{-\infty}^{\infty} R_{xx}(\tau) e^{-j\omega\tau} \, d\tau \tag{3.49}$$

The frequency response function can also be related to the cross correlation between the two signals $x(t)$ and $f(t)$. The *cross correlation function*, denoted $R_{xf}(\tau)$, for the two signals $x(t)$ and $f(t)$ is defined by

$$R_{xf}(\tau) = \lim_{T \to \infty} \frac{1}{T} \int_0^T x(t) \, f(t + \tau) \, dt \tag{7.8}$$

Here $x(t)$ is considered to be the response of the structure to the driving force $f(t)$. Similarly, the *cross-spectral density* is defined as the Fourier transform of the cross correlation:

$$S_{xf}(\omega) = \frac{1}{2\pi} \int_{-\infty}^{\infty} R_{xf}(\tau) \, e^{-j\omega\tau} \, d\tau \tag{7.9}$$

These correlation and density functions also allow calculation of the transfer functions of test structures. The frequency response function, $H(j\omega)$, can be shown (see e.g., Ewins, 1984) to be related to the spectral density functions by the two equations

$$S_{fx}(\omega) = H(j\omega) S_{ff}(\omega) \tag{7.10}$$

and

$$S_{xx}(\omega) = H(j\omega) S_{xf}(\omega) \tag{7.11}$$

These hold if the structure is excited by a random input $f(t)$ resulting in the response $x(t)$. Note that the cross-correlation functions include information about the phase and magnitude of the structure's transfer function and not just the magnitude as in the case of the correlation function of equation (7.7) repeated on the bottom right corner of Window 7.3.

Window 7.3
Comparison between Calculations for the Response of a
Spring–Mass–Damper System to Deterministic and Random Excitations

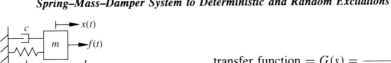

$$\text{transfer function} = G(s) = \frac{1}{ms^2 + cs + k}$$

Frequency response function:

$$G(j\omega) = H(\omega) = \frac{1}{k - m\omega^2 + c\omega j}$$

Impulse response function:

$$h(t) = \frac{1}{m\omega_d}e^{-\zeta\omega t}\sin \omega_d t$$

which has Laplace transform

$$L[h(t)] = \frac{1}{ms^2 + cs + k} = G(s)$$

and the Fourier transform of the impulse response function is just the frequency response function $H(\omega)$. These quantities relate to the input and response by

For deterministic $f(t)$:

$$X(s) = G(s)F(s)$$

$\leftrightarrow$

For random $f(t)$:

$$S_{xx}(\omega) = |H(\omega)|^2 S_{ff}(\omega)$$

$$x(t) = \int_0^t h(t - \tau)f(\tau)\,d\tau$$

$\leftrightarrow$

$$E[\bar{x}^2] = \int_{-\infty}^{\infty} |H(\omega)|^2 S_{ff}(\omega)\,d\omega$$

The spectrum analyzer calculates (or estimates) the various spectral density functions from the transducer outputs. Then, using equation (7.10) or (7.11), the analyzer can calculate the desired frequency response function $H(j\omega)$. Note that equations (7.10) and (7.11) use different power spectral densities to calculate the same quantity. This fact can be used to check the consistency of $H(j\omega)$. The *coherence function*, denoted by γ^2, is defined to be the ratio of the two values of $H(j\omega)$ calculated from equations (7.10) and (7.11). In particular, the coherence function is defined to be

$$\gamma^2 = \frac{|S_{xf}(\omega)|^2}{S_{xx}(\omega)S_{ff}(\omega)} \tag{7.12}$$

which always lies between 0 and 1. In fact, if the measurements are consistent, $H(j\omega)$ should be the same value, independent of how it is calculated and the coherence should be 1 ($\gamma^2 = 1$). The coherence is a measurement of the noise in the signal. If it is zero, the measurement is of a pure noise; if the value of the coherence is 1, the signals x and f are not contaminated with noise. In practice, coherence versus frequency is plotted versus frequency (see Figure 7.6) and is taken as an indication of how accurate the measurement process is over a given range of frequencies. Generally, the values of $\gamma^2 = 1$ should occur at values of ω near the structure's resonant frequencies. Near

Coherence

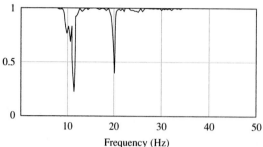

Figure 7.6 Sample plot of a coherence function.

resonance the signals are large and hence less affected by noise. In practice, data with a coherence of less than 0.75 are not used and indicate that the test should be done over.

7.4 MODAL DATA EXTRACTION

Once the frequency response of a test structure is calculated from equation (7.10) or (7.11), the analyzer is used to construct various vibration parameter information from the processed measurements. This is what is referred to as *experimental modal analysis*. In what follows it is assumed that the frequency response function $H(j\omega)$ has been measured via equation (7.10) or (7.11) or their equivalents.

The task of interest is to compute the natural frequencies, damping ratios, and modal amplitudes associated with each resonant peak of the measured frequency response function. There are several ways to examine the measured frequency response function to extract these data. To examine all of them is beyond the scope of this book and the interested reader should consult Ewins (1984). To illustrate the basic method, consider the somewhat idealized (compliance) frequency response function record of Figure 7.7, resulting from measurements taken between two points on a simple structure. Here it is assumed that a sinusoidal force of adjustable frequency is applied to one point on the structure and that the displacement response is measured at a second point. The response is measured for many values of the driving frequency to produce the plot of Figure 7.7. The procedure is examined for a single-degree-of-freedom system as illustrated in Section 3.7. and Figure 3.15.

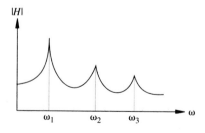

Figure 7.7 Sample magnitude plot of the frequency response function from a test article excited at one point and measured at another point.

One of the gray areas in modal testing is deciding on the number of degrees of freedom to assign to a test structure. In many cases, simply counting the number of clearly defined peaks or resonances, three in Figure 7.7, determines the order, and the procedure continues with a three-mode model. However, this procedure is not accurate if the structure has closely spaced natural frequencies, or repeated natural frequencies.

The easiest method to use on these data is the so-called *single-degree-of-freedom curve fit* (often called the SDOF method) approach. In this method the frequency response function for the compliance is sectioned off into frequency ranges bracketing each successive peak. Each peak is then analyzed by assuming that it is the frequency response of a single-degree-of-freedom system. This assumes that in the vicinity of the resonance, the frequency response function is dominated by that single mode.

In other words, in the frequency range around the first resonant peak, it is assumed that the plot is due to the response of a damped single-degree-of-freedom system due to a harmonic input at, and near, the first natural frequency. Recall from Section 3.7 that the point of resonance corresponds to that value of the frequency for which the magnification curve has its maximum or peak value and the phase shift is 90°. Hence each of the frequencies ω_1, ω_2, and ω_3 of the plot of Figure 7.7 is determined simply by noting where the three peaks lie on the horizontal (frequency) axis and confirmed by examining the value of the phase at each of these frequencies (should be 90°).

The damping ratio associated with each peak is assumed to be the modal damping ratio, ζ_i, defined in Sections 4.5 and 6.7 in the modal coordinate system. To obtain the modal damping ratios, consider the frequency response functions (compliance) magnitude plot illustrated in Figure 7.8. For systems with light enough damping, so that the peak

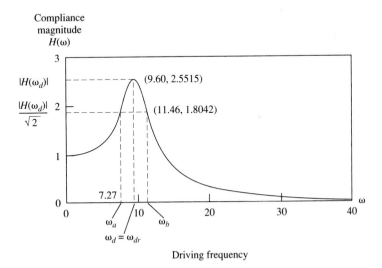

Figure 7.8 Magnitude of the compliance frequency response function of a single-degree-of-freedom system, illustrating the calculation of the modal damping ratio by using the *quadrature peak picking* method for lightly damped systems.

of $|H(\omega)|$ at resonance is well defined, the modal damping ratio ζ is related to the frequencies corresponding to the two points on the magnitude plot, where

$$|H(\omega_a)| = |H(\omega_b)| = \frac{|H(\omega_d)|}{\sqrt{2}} \tag{7.13}$$

by $\omega_b - \omega_a = 2\zeta\omega_d$, so that

$$\zeta = \frac{\omega_b - \omega_a}{2\omega_d} \tag{7.14}$$

Here ω_d is the damped natural frequency at resonance and ω_a and ω_b satisfy condition (7.13). The condition of equation (7.13) is also called the *3-dB down point*, since $H(\omega_d)/\sqrt{2}$ corresponds to $H(\omega_d)$ minus 3 dB when the magnitude is plotted on a log scale.

Equation (7.14) can be shown to apply for inertance (acceleration) transfer functions as well. The peak of the inertance frequency response magnitude plot also occurs at ω_d. Often, because the damping is small, ω, the natural frequency, and ω_d, the damped natural frequency, are taken to be the same. In fact, for $\zeta = 0.01$, $\omega_d = \omega\sqrt{1 - \zeta^2} = 0.999949998\omega$ and for $\zeta = 0.1$, $\omega_d = 0.99498\omega$, so that they are very nearly the same for an order-of-magnitude spread of damping ratios. Both the natural frequency and the damping ratio can be determined directly for accelerometer measurements and force input measurements, by plotting the magnitude of the inertance frequency response function and applying the method of Figure 7.8.

In the case of a multiple-degree-of-freedom system, as indicated by the three peaks of Figure 7.7, the number of peaks indicates the number of degrees of freedom. Each peak is then treated as if it resulted from a single-degree-of-freedom system. For example, the three natural frequencies of Figure 7.7 are determined by the positions of the peaks on the frequency axis, and the three damping ratios are determined by treating each peak as if it were from a single-degree-of-freedom system, computing the 3-db down frequencies and using equation (7.14) three times. This yields the three modal damping ratios ζ_1, ζ_2, and ζ_3.

Example 7.4.1

Consider the experimentally determined compliance transfer function plotted in Figure 7.9 and calculate the number of degrees of freedom, modal damping ratios, and natural frequencies.

Solution Since the magnitude plot indicates two distinct peaks, the test system is assumed to have two degrees of freedom. This is confirmed by examining the phase plot at the peaks. Since the phase is $\pm 90°$ at each peak, each of the peaks corresponds to a natural frequency. Reading the vertical axis yields

$$\omega_1 = 10 \text{ Hz} \qquad \omega_2 = 20 \text{ Hz}$$

Next since $|H(\omega_1)| = 0.0017$, the 3-dB down points are those two values of ω where $H(\omega_a) = H(\omega_b) = 0.0017/\sqrt{2} = 0.0012$. From the plot, these values yield $\omega_a = 9.75 \text{ Hz}$

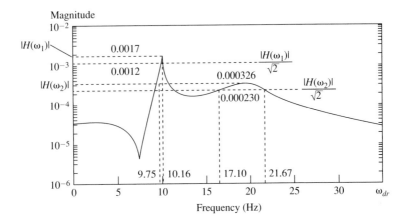

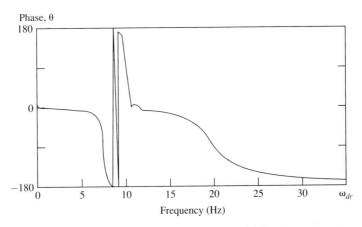

Figure 7.9   Plot of the magnitude and phase versus driving frequency of a test specimen, illustrating the peak amplitude method of determining modal damping ratios and natural frequencies.

and $\omega_b = 10.16$ Hz. Using equation (7.14) yields

$$\zeta_1 = \frac{\omega_b - \omega_a}{2\omega_1} = \frac{10.16 - 9.75}{20} = 0.02$$

Repeating this procedure for the second peak yields

$$\zeta_2 = \frac{\omega_b - \omega_a}{2\omega_2} = \frac{21.67 - 17.10}{40} = 0.11$$

7.5 MODAL PARAMETERS BY CIRCLE FITTING

In Section 7.4, the frequency and damping ratios are essentially determined by visually examining the frequency response function. In this section a more systematic method is examined which can be programmed so that a analyzer can calculate the frequencies and damping ratios in a more automated fashion. This method also assumes that a single mode dominates the behavior of the mobility transfer function in a frequency range around the natural frequency. If the real part of the mobility frequency response function is plotted versus the imaginary part of the frequency response function for a range of frequencies, a circle results. Plots of $\text{Re}[(H(\omega)]$ versus $\text{Im}[H(\omega)]$ are called *Nyquist plots*, or *Nyquist circles*, or Argand plane plots.

The mobility transfer function and corresponding frequency response function are presented in Section 7.3 and reviewed in Window 7.4. The real and imaginary parts of the mobility frequency response function can be calculated to be (the subscript "*dr*" is dropped here for convenience).

$$\text{Re}(\alpha) = \frac{\omega^2 c^2}{(k - \omega^2 m)^2 + (\omega c)^2} \tag{7.15}$$

and

$$\text{Im}(\alpha) = \frac{\omega(k - \omega^2 m)}{(k - \omega^2 m)^2 + (\omega c)^2} \tag{7.16}$$

respectively. The imaginary part is plotted versus the real part in Figure 7.10 for increasing values of ω. The plots of Figure 7.10 are formed by computing values for the pairs $[\text{Im}(\alpha), \text{Re}(\alpha)]$ for each value of ω. This triple of values—ω, $\text{Im}(\alpha)$, $\text{Re}(\alpha)$—correspond to information available in digital forms in the analyzer used to manipulate measured data.

Window 7.4
Mobility Frequency Response Function

Recall from Table 3.2 that the mobility transfer function is the ratio of the Laplace transform of the response velocity to the force input. From equation (3.84) this is

$$\frac{sX(s)}{F(s)} = sH(s) = \frac{s}{ms^2 + cs + k}$$

for a single-degree-of-freedom system. Substitution of $s = j\omega$ yields the mobility frequency response function defined by

$$\alpha(\omega) = j\omega H(\omega) = \frac{j\omega}{(k - \omega^2 m) + j\omega c}$$

which is a complex-valued function usually denoted $\alpha(\omega)$. However, $\alpha(\omega)$ is used to denote the frequency response function associated with any transfer function.

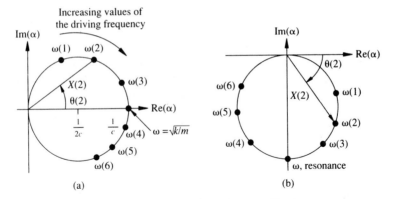

Figure 7.10   (a) Plot of the imaginary part of the mobility frequency response function versus the real part for increasing values of ω, starting at ω = 0. (b) Receptance transfer function. The values in parentheses correspond to numbered data points. These are called Nyquist plots.

The imaginary part is plotted versus the real part for equally spaced increments of frequency. That is, the frequency range of interest is divided up into equally spaced values of the driving frequency ω_{dr} (say, every 1 Hz or every 0.1 Hz). At each of the values of ω_{dr}, the quantities $\text{Re}(\alpha)$ and $\text{Im}(\alpha)$ are calculated from the measured data in the analyzer, and these are plotted as indicated in Figure 7.10(a). For a single-degree-of-freedom system, equations (7.15) and (7.16) predict that these points will all lie on a circle tangent to the origin centered on the $\text{Re}(\alpha)$ axis. The equally spaced driving frequencies are labeled ω(1), ω(2), and so on, in the figure. Note that these points are not equally spaced around the circle, but are at equal increments of frequency. Note also on this plot that the input force magnitude is lined up along the $\text{Re}(\alpha)$ axis and that the response of magnitude X, lags the force by the phase angle θ. These quantities are also marked on the plot of Figure 7.10 for the second value of the driving frequency [these are labeled $X(2)$, $\theta(2)$]. Thus the distance from the origin to any point on the circle drawn through the data points is the magnitude of the response. Since resonance is defined as the driving frequency at which the response magnitude X is the largest, the point on the circle farthest from the origin corresponds to the resonant condition. Resonance is also defined, for small damping, as the condition when the driving frequency and the system's natural frequency coincide. Hence the point labeled ω corresponds to the system's natural frequency, which is calculated from equations (7.15) and (7.16) evaluated at the point of intersection of the circle and the $\text{Re}(\alpha)$ axis.

The fact that the Nyquist plot of the mobility frequency response function is a circle can be seen by defining $A = \text{Re}(\alpha) - (1/2c)$, $B = \text{Im}(\alpha)$ and using equations (7.15) and (7.16), so that

$$A^2 + B^2 = \left[\text{Re}(\alpha) - \frac{1}{2c}\right]^2 + (\text{Im}(\alpha))^2 = \left[\frac{1}{2c}\right]^2$$

which is the equation of a circle of radius $(1/2c)$ centered at the point $[\text{Im}(\alpha) = 0,$ $\text{Re}(\alpha) = 1/2c]$. Note from equation (7.16) that at the point where the circle intersects the $\text{Re}(\alpha)$ axis, $\text{Im}(\alpha) = 0$, so that $\omega_{dr} = \sqrt{k/m}$, which is the condition for resonance. If the frequency at the point where the circle crosses the axis is not necessarily a point that was measured or plotted, the value of the resonant frequency can be determined by fitting a circle to the points $\omega(i)$ and numerically determining the value of ω corresponding to the intersection of the axis and the circle. This also gives the value of the damping coefficient from the simple relationship $\text{Re}(\alpha) = 1/c$ at the value of ω corresponding to resonance.

Figure 7.10(b) shows the Nyquist circle for the receptance transfer function (i.e., displacement measurements) for the same system. Most analyzers allow the user to plot any of the transfer functions given in Table 3.1, and their corresponding Nyquist plots. The point corresponding to resonance on the receptance plot of a single-degree-of-freedom system given in Figure 7.10(b) can be characterized in four different ways. The point corresponding to the natural frequency can be thought of as:

1. The point on the circle corresponding to a maximum distance from the origin.
2. The point lying halfway between the two adjacent frequencies forming the largest arc length on the circle. The arc between $\omega(3)$ and $\omega(4)$ is the largest in the example given in Figure 7.10(b).
3. The point on the circle a maximum distance away from the $\text{Re}(\alpha)$ axis.
4. The point on the circle intersecting the line $\text{Im}(\alpha)$ axis.

For single-degree-of-freedom systems, each of these points is, of course, the same. However, for multiple-degree-of-freedom systems several changes occur. First the Nyquist circle becomes many circles (roughly one for each mode) and the circles drift from being tangent to the origin centered on the $\text{Im}(\alpha)$ axis. As this happens, the four points listed above no longer coincide. This is illustrated in Figure 7.11. The single point labeled ω in Figure 7.10(b) becomes the four points labeled 1, 2, 3, and 4 in Figure 7.11(a) as the circle moves away from the origin. These labels correspond numerically to the list characterizing the resonance point given above. Any one of these four points could be used to define the natural frequency for the mode described by the circle. The most common choice, and the choice least affected by the presence of the other modes, is to use point 2 (i.e., the point halfway between adjacent frequencies defining the largest arc length).

Referring to Figure 7.11(b), the procedure for using the circle is as follows. First, the analyzer computes the points marked x in the figure from $\text{Re}(\alpha)$ and $\text{Im}(\alpha)$ evaluated at equal intervals of the driving frequency. A numerical curve-fit procedure is then used to calculate the best circle through these points, the center of the circle (O), and the arc lengths between each point x. This determines point 2 as defined by the largest arc length. The equations for $\text{Re}(\alpha)$ and $\text{Im}(\alpha)$ are then used to calculate the value for the natural frequency, denoted ω_3 here, since it corresponds to the third mode. The points ω_a and ω_b and O are also calculated and used to derive the angle α.

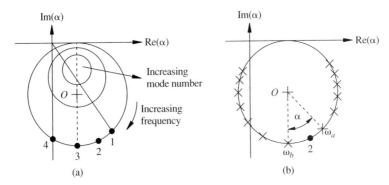

Figure 7.11 Nyquist plot (receptance or compliance) for a three-degree-of-freedom lightly damped system. (a) Four points that could be used to define the frequency of the third mode, defined by the circle centered at O. (b) Third mode plotted without the other modes and the geometry label that is used to determine the modal damping ratio. Point 2 corresponds to the natural frequency for the third mode. The points labeled ω_a and ω_b are adjacent frequencies that form the largest arc length, and α is the angle between the radii defined by ω_a and ω_b. The **x**'s denoted measured points of equally spaced frequencies.

As long as the angle $\alpha/2 < 45°$, which it will be for any reasonable amount of data, the angle α is related to the modal damping ratio ζ_3, and natural frequency ω_3 by

$$\tan\frac{\alpha}{2} = \frac{1 - (\omega/\omega_3)^2}{2\zeta_3\omega/\omega_3} \tag{7.17}$$

Applying equation (7.17) to ω_a yields

$$\tan\frac{\alpha}{2} = \frac{(\omega_a/\omega_3)^2 - 1}{2\zeta_3\omega_a/\omega_3} \tag{7.18}$$

and applying equation (7.17) to ω_b yields

$$\tan\frac{\alpha}{2} = \frac{1 - (\omega_b/\omega_3)^2}{2\zeta_3\omega_b/\omega_3} \tag{7.19}$$

Equations (7.18) and (7.19) can be added to yield (after some manipulation)

$$\zeta_3 = \frac{\omega_b^2 - \omega_a^2}{2\omega_3[\omega_a\tan(\alpha/2) + \omega_b\tan(\alpha/2)]} \tag{7.20}$$

which yields an expression for the damping ratio for the mode under study.

To check this result, note that if the half-power points used in Section 7.4 are taken to be those frequencies ω_a and ω_b where $\alpha = 90°$, equation (7.20) reduces to

$$\zeta_i = \frac{\omega_b - \omega_a}{2\omega_i} \tag{7.21}$$

which matches the quadrature formula given to equation (7.14) of Section 7.4.

The method of using the Nyquist circle for determining natural frequencies and damping ratios is first to divide the frequency response function into segments by looking at the magnitude plot to determine regions of ω for which the frequency response looks approximately like that of a single-degree-of-freedom system (i.e., take a frequency range around each peak). The data points comprising each peak are then chosen to use in generating a Nyquist circle for that mode. Each frequency range must contain at least three points.

The circle generated by these points will contain noise, and so on, and will not be perfect circles. To rectify this, the data points are curve fit to a circle using a simple least squares method. This yields the equation of the ("best") circle through the data points. This circle is then used to calculate ω_i and ζ_i from the formulas above. Since a curve-fit procedure is used, this method is called the *circle fit* method of extracting modal parameters. The method was formulated by Kennedy and Pancu (1947).

7.6 MODE SHAPE MEASUREMENT

Determining the mode shapes from experimentally measured transfer functions is slightly more complicated and involves the measurement of several transfer functions. First, the concept of a transfer function matrix, or *receptance matrix*, needs to be established. To this end, consider the response of the multiple-degree-of-freedom system as described by equation (4.122) to a harmonic force input represented in a complex form by $\mathbf{f}e^{j\omega_{dr}t}$. The equation of motion becomes

$$M\ddot{\mathbf{x}} + C\dot{\mathbf{x}} + K\mathbf{x} = \mathbf{f}e^{j\omega_{dr}t} \qquad (7.22)$$

The forced response can be constructed by assuming that the solution $\mathbf{x}(t)$ is harmonic, of the form $\mathbf{x}(t) = \mathbf{u}e^{j\omega_{dr}t}$. Substitution of this form into equation (7.22) yields

$$(K - \omega_{dr}^2 M + j\omega_{dr}C)\mathbf{u} = \mathbf{f} \qquad (7.23)$$

after rearranging terms and factoring out the nonzero scalar $e^{j\omega_{dr}t}$. Equation (7.23) relates the magnitude of the response vector (i.e., the vector $\mathbf{u}$), to the magnitude of the input vector $\mathbf{f}$, both of which are constants. Solving equation (7.23) yields

$$\mathbf{u} = (K - \omega_{dr}^2 M + j\omega_{dr}C)^{-1}\mathbf{f} \qquad (7.24)$$

The inverse of the complex matrix coefficient above is the *receptance matrix*, denoted $\alpha(\omega_{dr})$ and defined by

$$\alpha(\omega_{dr}) = (K - \omega_{dr}^2 M + j\omega_{dr}C)^{-1} \qquad (7.25)$$

so that equation (7.24) becomes simply $\mathbf{u} = \alpha(\omega_{dr})\mathbf{f}$. A two-degree-of-freedom example of the receptance matrix is used in Section 5.4 on damped absorbers to calculate equation (5.35). An undamped version of $\alpha(\omega_{dr})$ for a two-degree-of-freedom system is given by equation (5.18).

The receptance matrix can be further analyzed by recalling the transformation used in Chapter 4 to derive modal coordinates. In particular, recall that the modal stiffness matrix can be represented in diagonal form by

$$\Lambda_K = \text{diag}[\omega_i^2] = P^T M^{-1/2} K M^{-1/2} P \tag{7.26}$$

where P is the matrix of normalized eigenvectors of the matrix $M^{-1/2} K M^{-1/2}$. Similarly, the modal damping matrix can be written as

$$\Lambda_c = \text{diag}[2\zeta_i \omega_i] = P^T M^{-1/2} C M^{-1/2} P \tag{7.27}$$

if the damping is assumed to be proportional. Multiplying equation (7.26) by $M^{1/2} P$ from the left and $P^T M^{1/2}$ from the right yields

$$K = M^{1/2} P \Lambda_K P^T M^{1/2} \tag{7.28}$$

since $P^T P = I$. Similarly, the damping matrix can be written from equation (7.27) as

$$C = M^{1/2} P \Lambda_C P^T M^{1/2} \tag{7.29}$$

Substitution of equations (7.28) and (7.29) for C and K into equation (7.25) for the receptance matrix yields

$$\alpha(\omega_{dr}) = [M^{1/2} P (\Lambda_K - \omega_{dr}^2 I + j\omega_{dr} \Lambda_C) P^T M^{1/2}]^{-1} \tag{7.30}$$

$$= [S (\Lambda_K - \omega_{dr}^2 I + j\omega_{dr} \Lambda_C) S^T]^{-1} \tag{7.31}$$

$$= [S \,\text{diag}[\omega_i^2 - \omega_{dr}^2 + 2\zeta_i \omega_i \omega_{dr} j] S^T]^{-1} \tag{7.32}$$

where $S = M^{1/2} P$. Here the inside matrix is a combination of diagonal matrices and hence is diagonal. Recall from matrix theory that $(AB)^{-1} = B^{-1} A^{-1}$ (see Appendix C), so that equation (7.32) can be written as

$$\alpha(\omega_{dr}) = S^{-T} \text{diag} \left[\frac{1}{\omega_i^2 - \omega_{dr}^2 + 2\zeta_i \omega_i \omega_{dr} j} \right] S^{-1} \tag{7.33}$$

since the inverse of a diagonal matrix is obtained simply by inverting its nonzero diagonal elements. Note this formulation assumes proportional damping.

Equation (7.33) for the receptance matrix can be expressed as a summation of n matrices rather than the product of three matrices by realizing that the columns of S^{-T} are the mode shape vectors of the undamped system, denoted $\mathbf{u}_i$, of equation (4.111). Equation (7.33) can thus be written as

$$\alpha(\omega_{dr}) = \sum_{i=1}^{n} \left[\frac{\mathbf{u}_i \mathbf{u}_i^T}{(\omega_i^2 - \omega_{dr}^2) + (2\zeta_i \omega_i \omega_{dr})j} \right] \tag{7.34}$$

where $\mathbf{u}_i \mathbf{u}_i^T$ is the outer product of two $n \times 1$ mode shape vectors. This outer product results in an $n \times n$ matrix. This representation provides a connection between the receptance matrix and the system's mode shapes, which can be exploited in testing to provide a measurement of the test article's mode shapes.

The elements of the receptance matrix located at the intersection of the sth row and rth column of $\alpha(\omega_{dr})$ is essentially the transfer function between the response at point s, u_s, and the input at point r, f_r, when all other inputs are held at zero. The srth element of $\alpha(\omega_{dr})$ is

$$\alpha_{sr}(\omega_{dr}) = \sum_{i=1}^{n} \frac{[\mathbf{u}_i \mathbf{u}_i^T]_{sr}}{\omega_i^2 - \omega_{dr}^2 + (2\zeta_i \omega_i \omega_{dr})j} \tag{7.35}$$

which relates the transfer function between a given input and output, $\alpha_{sr}(\omega_{dr})$, to elements of the mode shapes $\mathbf{u}_i$. This interpretation of $\alpha_{sr}(\omega_{dr})$ is a generalization of the single-degree-of-freedom concept of a transfer function. Since $\alpha(\omega)$ is a matrix, it cannot be written as the ratio of an output to an input. However, each element of $\alpha(\omega_{dr})$ is a transfer function:

$$\frac{u_s}{f_r} = [\alpha(\omega_{dr})]_{sr} = H_{sr}(\omega_{dr}) \tag{7.36}$$

where $H_{sr}(\omega_{dr})$ is the transfer function between an input at point r and an output at point s. An example of equation (7.36) is the ratio x_1/F_0k used in Sections 5.3 and 5.4 to discuss absorbers.

If it is assumed that the modes, or peaks, of the system are well spaced, the summation in equation (7.35) evaluated at a natural frequency will be dominated by one term, the term corresponding to that frequency. This can be seen by substituting $\omega_{dr} = \omega_i$ into equation (7.35) and taking the magnitude. This yields the approximation

$$|\alpha_{sr}(\omega_i)| = \frac{|\mathbf{u}_i \mathbf{u}_i^T|_{sr}}{|(\omega_i^2 - \omega_i^2) - 2\zeta_i \omega_i \omega_i j|} = \frac{|\mathbf{u}_i \mathbf{u}_i^T|_{sr}}{2\zeta_i \omega_r^2} \tag{7.37}$$

where it is assumed that the contributions from the other terms in the summation are all much smaller because of the nonzero term $(\omega_i^2 - \omega_{dr}^2)$ in their denominators. Equation (7.37) can be rearranged to yield

$$|\mathbf{u}_i \mathbf{u}_i^T|_{sr} = |2\zeta_i \omega_i^2||H_{sr}(\omega_i)| \tag{7.38}$$

where $|H_{sr}(\omega_i)| = |\alpha_{sr}(\omega_i)|$ is the magnitude of the frequency response function measured between points s and r and evaluated at the ith natural frequency. Equation (7.38) holds for proportionally damped systems with underdamped, widely spaced modes. This equation relates the measured damping ratio, ζ_i, measured natural frequency, ω_i, and the measured magnitude of the transfer function, $|H_{sr}(\omega_i)|$, to the ith mode shape, ω_i, and hence provides a measure of the mode shape of the test structure.

Equation (7.38) only provides a measurement of the magnitude of one element of the matrix $\left[\mathbf{u}_i \mathbf{u}_i^T\right]_{sr}$. The phase plot of $H(\omega_i)$ is used to determine the sign of the element $|\mathbf{u}_i \mathbf{u}_i^T|_{sr}$. Equation (7.37) is a mathematical statement that the ith peak in the transfer function plot of Figure 7.8 results from only a single-degree-of-freedom system. Note that the matrix $\mathbf{u}_i \mathbf{u}_i^T$ has n^2 elements but that only n of them are unique where n is the number of measured natural frequencies. Hence n measurements of $|H_{sr}(\omega_i)|$ must be made. This is accomplished by stepping through the n elements of the vector $\mathbf{f}$ one at a time so that r ranges from 1 to n. This amounts to measuring the response at point s with first an input at point 1, then at point 2, and so on, until all n input positions have been used. This provides a measurement of one row of the matrix $\mathbf{u}_i \mathbf{u}_i^T$. From this one row, the entire vector $\mathbf{u}_i$ can be determined as the following example illustrates. Note that this process could be interchanged so that the driving point is fixed and the measurement point is moved. The following example illustrates this procedure.

Example 7.6.1

Consider a simple beam of Figure 7.12. A transfer function measurement made by applying a force at point 1 and measuring the response at point 1 (called the *driving point* frequency response) yields three distinct peaks, indicating that the system has three natural frequencies and can be modeled by a three-degree-of-freedom system. This initial measurement suggests that the beam be measured at two other points in order to establish enough data to determine the mode shapes. These other two points are marked on the beam of Figure 7.12. Since a shaker is used, it is easier to move the accelerometer to obtain the required two additional transfer functions. Alternatively, a multichannel frequency analyzer can be used with two additional accelerometers to obtain simultaneously the required three transfer functions: $H_{11}(\omega_{dr})$, $H_{21}(\omega_{dr})$, and $H_{31}(\omega_{dr})$. This is the procedure illustrated in Figure 7.12. Plots of the three transfer functions are given in Figure 7.13.

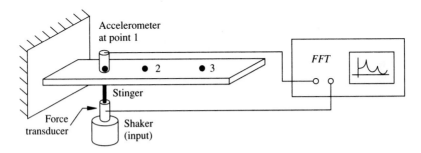

Figure 7.12 Cantilevered beam labeled with three measurement and driving points.

From the driving point transfer function $H_{11}(\omega_{dr})$ the values of the modal damping ratio and natural frequencies are obtained using the peak method illustrated in Example 7.4.1. They are

$$\omega_1 = 10 \text{ rad/s} \qquad \zeta_1 = 0.01$$

$$\omega_2 = 20 \text{ rad/s} \qquad \zeta_2 = 0.01 \qquad (7.39)$$

$$\omega_3 = 32 \text{ rad/s} \qquad \zeta_3 = 0.05$$

These values can be checked against similar calculations of the remaining two transfer functions $H_{21}(\omega_{dr})$ and $H_{31}(\omega_{dr})$. To determine the mode shape vectors, the value of $|H_{11}(\omega_1)|$ is measured to be $|H_{11}(\omega_1)| = 0.423$ and the phase of $H_{11}(\omega_1)$ is noted to be phase $[H_{11}(\omega_1)] = -90°$. In addition the magnitude and phase of the remaining two transfer functions yields

$$|H_{21}(\omega_1)| = 0.917 \qquad \text{phase}[H_{21}(\omega_1)] = -90°$$

$$|H_{31}(\omega_1)| = 2.317 \qquad \text{phase}[H_{31}(\omega_1)] = -90° \qquad (7.40)$$

From equation (7.38) and the measured values of $\zeta_1, \omega_1, H_{11}(\omega_1), H_{21}(\omega_1)$ and $H_{31}(\omega_1)$, the first row of the matrix $\mathbf{u}_1 \mathbf{u}_1^T$ is known:

$$|\mathbf{u}_1 \mathbf{u}_1^T|_{11} = 0.846 \qquad |\mathbf{u}_1 \mathbf{u}_1^T|_{21} = 1.834 \qquad |\mathbf{u}_1 \mathbf{u}_1^T|_{31} = 4.633 \qquad (7.41)$$

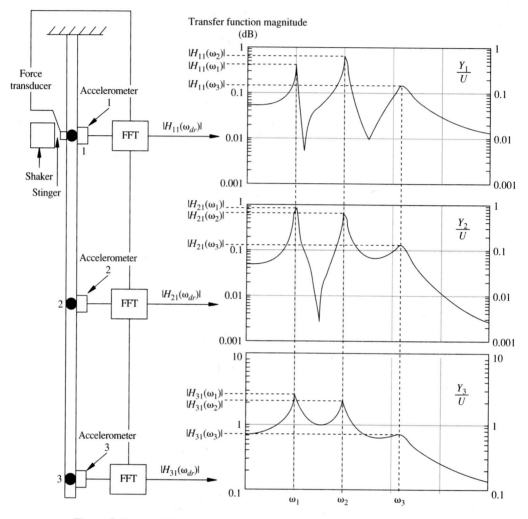

Figure 7.13 (cont'd on next page) Instrumentation required to construct the necessary frequency response curves to allow computation of the mode shapes for a three-degree-of-freedom model of the test specimen. The peaks of the three transfer functions determine the values of the receptance matrix and hence the system mode shapes.

This along with the phase information allows determination of the vector $\mathbf{u}_1$. To see this, let $\mathbf{u}_1 = [a_1 \quad a_2 \quad a_3]^T$, so that

$$\mathbf{u}_i \mathbf{u}_i^T = \begin{bmatrix} a_1^2 & a_1 a_2 & a_1 a_3 \\ a_2 a_1 & a_2^2 & a_2 a_3 \\ a_3 a_1 & a_3 a_2 & a_3^2 \end{bmatrix}$$

Note that this matrix is symmetric. Hence from the values given in equation (7.41), the elements of the matrix $|\mathbf{u}_1 \mathbf{u}_1^T|$ must satisfy

$$a_1^2 = 0.846 \qquad a_1 a_2 = 1.834 \qquad a_1 a_3 = 4.633 \qquad (7.42)$$

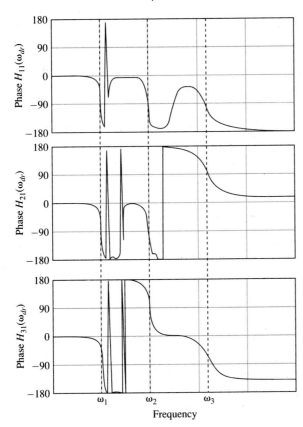

Figure 7.13 (cont'd)

This forms a set of three equations in three unknowns which is readily solved to yield

$$a_1 = 0.920 \qquad a_2 = 1.993 \qquad a_3 = 5.036 \tag{7.43}$$

Using the phase as either a $+$ or $-$ sign [i.e., $H(\omega_1)$ is either in phase or out of phase], the phase is either $-90°$ or $+90°$ at resonance. If phase $[H_{ij}(\omega_1)] = +90°$, the element associated with $[u_1 u_1^T]_{ij}$ is assigned a positive value. If the phase is $-90°$, the element is assigned a negative value. Examining the phase plots of Figure 7.13 and the numerical values given in equation (7.43), the vector $\mathbf{u}_1$ becomes

$$\mathbf{u}_1 = \begin{bmatrix} a_1 \\ a_2 \\ a_3 \end{bmatrix} = \begin{bmatrix} -0.920 \\ -1.993 \\ -5.036 \end{bmatrix}$$

Next consider the $\omega_2 = 20\text{rad/s}$ peak in each of the three transfer functions of Figure 7.13. These along with equation (7.38) yield $|(\mathbf{u}_2\mathbf{u}_2^T)|_{11} = 7.12$, $|(\mathbf{u}_2\mathbf{u}_2^T)|_{21} = 7.72$, and $|(\mathbf{u}_2\mathbf{u}_2^T)|_{31} = 15.681$. Again using these values and the phase information yields

$$\mathbf{u}_2 = \begin{bmatrix} -2.67 \\ -2.89 \\ 5.873 \end{bmatrix}$$

Similarly, the measurement of the $\omega_3 = 32\text{rad/s}$ peak, the corresponding phase values, and the modal data in equation (7.38) yields

$$\mathbf{u}_3 = \begin{bmatrix} -4.22 \\ 2.99 \\ -15.311 \end{bmatrix}$$

Hence the three mode shapes $\mathbf{u}_1$, $\mathbf{u}_2$, and $\mathbf{u}_3$ are determined.

☐

The method of determining the mode shapes, natural frequencies and damping ratios illustrated in Example 7.6.1 is only one of many methods available for extracting modal data from frequency response functions constructed from test data. These are discussed in Ewins (1984). The notation used here is consistent with that used in Chapters 4 and 5. However, the modal analysis and testing community has begun to try to standardize the notation, and it will likely differ from that used here.

7.7 VIBRATION TESTING FOR ENDURANCE AND DIAGNOSTICS

A part manufactured for use in a machine or structure must obviously be able to function in its operating environment. In particular, the device under consideration must be able to withstand all the dynamic loads that might be applied to it and it must continue to function. Often, the behavior of a device over time cannot be predicted analytically with 100% accuracy. Hence sample devices are subjected to dynamic loads in various controlled testing environments. For example, an electronic module might be dropped from a height of 10 feet 20 times and still be expected to function. A valve might be mounted on a shaker and driven with a random input for 12 hours, after which it must still open and close. The idea here is very simple: Characterize the dynamic environment that a given device will experience during its normal use. Condense this into a worst-case series of laboratory tests. Apply the loads to the device and then see if it still works.

The most difficult part of this procedure is estimating a reasonable description of the dynamic loads that a given device is likely to experience during its useful service life. The next problem is to devise a test procedure that faithfully reproduces the dynamic input data specified for the devices. The basic elements are similar to those used in experimental modal analysis: shakers, accelerometers, and some kind of recording device. In addition, a feedback device is often applied to the shaker to make sure that it produces the desired force and frequency. The entire test system is then coupled to a computer control system which both records data and controls the input to the shaker.

The control computer can be used to program the shaker to perform long hours of testing at a variety of loads and frequencies and signal inputs. For example, a load may call for a device to be driven at 10g for 3 hours at 10 Hz, then 3 hours at 50 Hz followed by 3 hours of random vibration of a specified strength. A schematic of a computer-controlled vibration testing device is given in Figure 7.14. In the figure the test profile data containing timing and signal information is fed to the control computer

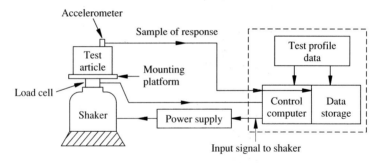

Figure 7.14 Schematic of a computer-controlled vibration endurance test.

which assigns the appropriate signal to the power supply, which in turn drives the shaker. The force output of the shaker is monitored by a load cell. This signal is returned to the computer, which matches the signal to the required test profile. If there are some differences, the computer adjusts its output to the shaker accordingly. The acceleration of the test article is also measured if transmissibility values are needed and to provide a record of the test. All signals are stored in the computer data storage section to provide evidence that the tests were performed and exactly what the responses and inputs were.

Another use of vibration testing is diagnostics or machine health monitoring. The basic idea here is that periodic measurements of frequency and damping may yield information regarding changes in the integrity of a structure or predict the pending failure of a machine. If a system's frequencies are measured and monitored over a period of time and a change is observed, then since $\omega = \sqrt{k/m}$, some change in the system's mass or stiffness must be the cause. A change in stiffness could imply a cracked or malfunctioning part and a change in mass might reflect excessive wear.

Example 7.7.1

A static deflection test is performed on a cantilevered aluminum beam, both with and without a small cut in the aluminum. The results indicate that with the cut the modulus is measured to be 10% lower than its nominal value of 71×10^{10} N/m^2. Based on this information, the fundamental frequency of aircraft wings are measured after each flight to see if any fatigue cracks are present. What sort of frequency shift would be expected to detect a crack?

Solution   From Table 6.4 the first natural frequency of a cantilevered Euler–Bernoulli beam is

$$\omega_1 = \frac{1.8751}{l^2} \sqrt{\frac{EI}{\rho A}}.$$

For a 2-m-long wing with estimated parameters

$$I = 5 \times 10^{-5} \text{m}^4$$

$$A = 0.05 \text{m}^2$$

the nominal value of the wing frequency will be ($\rho = 2.7 \times 10^3 \text{kg/m}^3$)

$$\omega_1 = \frac{1.8751}{(2)^2} \sqrt{\frac{(71 \times 10^{10})(5 \times 10^{-5})}{(2.7 \times 10^3)(0.05)}} \approx 240 \text{ rad/s}$$

If E changes by 10% (i.e., is reduced to 63.9×10^{10}) the new frequency becomes

$$\omega_1 = \frac{1.8751}{(2)^2} \sqrt{\frac{(63.9 \times 10^{10})(5 \times 10^{-5})}{(2.7 \times 10^3)(0.05)}} \approx 228 \text{ rad/s}$$

Thus the change in frequency is both noticeable and possible to measure, forming a reasonable diagnostic.

□

In some cases the vibration response is examined as a signature of the device. If the time history of the response changes over time, it is possible that the change has been caused by some deterioration of the part. The use of these ideas is an emerging technology and engineers are involved in trying to make strong connections between certain types of changes (such as frequency) and the condition or health of the device.

As an example of health monitoring, consider the plot of Figure 7.15. The plot consists of a record of displacement measurements of a bearing housing of a rotating shaft on a machine made over several months. The measurements over time indicate a trend. The increase (changes in normal operating deflections) in later months is thought to show that something is changing in the bearing system so that maintenance or repair is required. The other way to examine this is to stop the machine and physically look for damage or wear. If the machine is required for production, stopping the machine to dismantle it could be very expensive. Using vibration monitoring techniques, such as indicated in Figure 7.15, the routine of stopping the machine periodically to check it or waiting for it to fail can be avoided. A more complete discussion of machine health monitoring can be found in Wowk (1991).

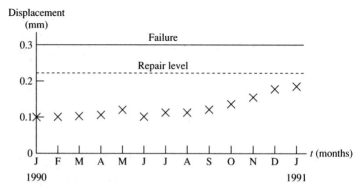

Figure 7.15 Record of average displacement of a housing for the bearing of a rotating shaft over a period of months.

PROBLEMS

Section 7.1

7.1. A low-frequency signal is to be measured by using an accelerometer. The signal is physically a displacement of the form $5 \sin(0.2t)$. The noise floor of the accelerometer (i.e., the smallest magnitude signal it can detect) is 0.4 volt/g. Can the accelerometer measure this signal?

7.2. Referring to Chapter 2, calculate the response of a single-degree-of-freedom system to a unit impulse and then to a unit triangle input lasting T seconds. Compare the two responses. The differences correspond to the differences between a "perfect" hammer hit and a more realistic hammer hit, as indicated in Figure 7.2. Use $\zeta = 0.01$ and $\omega = 4$ rad/s for your model.

7.3. Compare the Laplace transform of $\delta(t)$ with the Laplace transform of the triangle input of Figure 7.2 and Problem 7.2.

7.4. Plot the error in measuring the natural frequency of a single-degree-of-freedom system of mass 10 kg and stiffness 350 N/m if the mass of the excitation device (shaker) is included and varies from 0.5 to 5 kg.

7.5. Calculate the Fourier transform of $f(t) = 3 \sin 2t - 2 \sin t - \cos t$ and plot the spectral coefficients.

Section 7.2

7.6. Represent $5 \sin 3t$ as a digital signal by sampling the signal at $\pi/3, \pi/6$, and $\pi/12$ seconds. Compare these three digital representations.

7.7. Compute the Fourier coefficient of the signal $|120 \sin 120 \, \pi t|$.

7.8. Consider the periodic function

$$x(t) = \begin{cases} -5 & 0 < t < \pi \\ 5 & \pi < t < 2\pi \end{cases}$$

and $x(t) = (t + 2\pi)$. Calculate the Fourier coefficients. Next plot $x(t)$: $x(t)$ represented by the first term in the Fourier series, $x(t)$ represented by the first two terms of the series, and $x(t)$ represented by the first three terms of the series. Discuss your results.

7.9. Consider a signal $x(t)$ with maximum frequency of 500 Hz. Discuss the choice of record length and sampling interval.

Section 7.4

7.10. Consider the magnitude plot of Figure 7.16. How many natural frequencies does this system have, and what are their approximate values?

7.11. Consider the experimental transfer function plot of Figure 7.17. Use the methods of Example 7.4.1 to determine ζ_i and ω_i.

7.12. Consider a two-degree-of-freedom system with frequency $\omega_1 = 10$ rad/s, $\omega_2 = 15$ rad/s, and damping ratios $\zeta_1 = \zeta_2 = 0.01$. With modal matrix $\frac{1}{\sqrt{2}} \begin{bmatrix} 1 & -1 \\ 1 & 1 \end{bmatrix}$ calculate the transfer function of this system for an input at x_1 and a response measurement at x_2.

7.13. Plot the magnitude and phase of the transfer function of Problem 7.12 and see if you can reconstruct the modal data ($\omega_1, \omega_2, \zeta_1$ and ζ_2) from your plot.

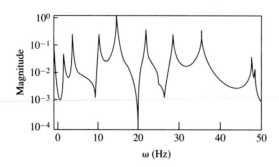

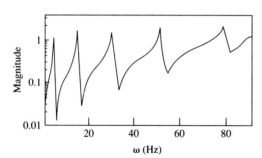

Figure 7.16 Sample magnitude plot of $H(\omega)$ for a simple structure.

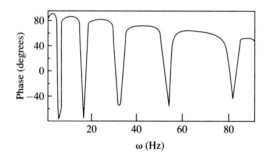

Figure 7.17 Experimental plot of phase and magnitude of a simple laboratory structure.

7.14. Consider equation (7.14) for determining the damping ratio of a single mode. If the measurement in frequency varies by 1%, how much will the value of ζ change?

7.15. Discuss the problems of using equation (7.14) if the natural frequencies of the structure are very close together.

7.16. Discuss the limitation of using equation (7.15) if ζ is very small. What happens if ζ is very large?

7.17. Consider the two-degree-of-freedom system described by

$$\begin{bmatrix} 1 & 0 \\ 0 & 1 \end{bmatrix}\begin{bmatrix} \ddot{x}_1 \\ \ddot{x}_2 \end{bmatrix} + \begin{bmatrix} 0 & 0 \\ 0 & c \end{bmatrix}\begin{bmatrix} \dot{x}_1 \\ \dot{x}_2 \end{bmatrix} + \begin{bmatrix} 2 & -1 \\ -1 & 2 \end{bmatrix}\begin{bmatrix} x_1 \\ x_2 \end{bmatrix} = \begin{bmatrix} f_0 \sin \omega t \\ 0 \end{bmatrix}$$

and calculate the transfer function $|X/F|$ as a function of the damping parameter c.

7.18. Plot the transfer function of Problem 7.17 for the four cases: $c = 0.01$, $c = 0.2$, $c = 1$, and $c = 10$. Discuss the difficulty in using these plots to measure ζ_i and ω_i for each value of c.

7.19. Use a numerical procedure to calculate the natural frequencies and damping ratios of the system of Problem 7.18. Label these on your plots from Problem 7.18 and discuss the possibility of measuring these values using the methods of Section 7.4.

Section 7.5

7.20. Using the definition of the mobility transfer function of Window 7.4, calculate the Re and Im parts of the frequency response function and hence verify equations (7.15) and (7.16).

7.21. Using equations (7.15) and (7.16), verify that the Nyquist plot of the mobility frequency response function does in fact form a circle.

7.22. Consider a single-degree-of-freedom system of mass 10 kg, stiffness 1000 N/m, and damping ratio of 0.01. Pick five values of ω between 0 and 20 rad/s and plot five points of the Nyquist circle using equations (7.15) and (7.16). Do these form a circle?

7.23. Derive equation (7.20) for the damping ratio from equations (7.18) and (7.19). Then verify that equation (7.20) reduces to equation (7.21) at the half-power points.

7.24. Consider the experimental curve fit Nyquist circle of Figure 7.18. Determine the modal damping ratio for this mode.

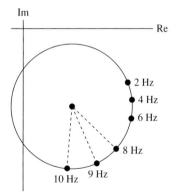

Figure 7.18 Experimentally determined Nyquist circle consisting of five data points. The 9-Hz point is constructed as halfway along the longest arc.

Section 7.6

7.25. Referring to Section 5.4 and Window 5.3, calculate the receptance matrix of equation (7.25) for the following two-degree-of-freedom system, without using the system's mode shapes.

$$\begin{bmatrix} 2 & 0 \\ 0 & 1 \end{bmatrix} \begin{bmatrix} \ddot{x}_1 \\ \ddot{x}_2 \end{bmatrix} + \begin{bmatrix} 3 & -1 \\ -1 & 1 \end{bmatrix} \begin{bmatrix} \dot{x}_1 \\ \dot{x}_2 \end{bmatrix} + \begin{bmatrix} 6 & -2 \\ -2 & 2 \end{bmatrix} \begin{bmatrix} x_1 \\ x_2 \end{bmatrix} = \begin{bmatrix} f_0 \\ 0 \end{bmatrix} \sin \omega_{dr} t$$

7.26. Repeat Problem 7.25 using the undamped mode shapes. Note that the system has proportional damping since $C = \alpha M + \beta K$, where $\alpha = 0$, $\beta = 1/2$. Use this result, and the result of Problem 7.25 to verify equation (7.33).

7.27. Compare equation (7.36) to equations (5.19) and (5.20) for the undamped vibration absorber problem and with equation (5.29) for the damped vibration absorber.

7.28. Consider the damped vibration absorber equation given by (5.29) and write out the four terms of the matrix $H_{sr}(\omega_{dr})$ given in equation (7.36). Physically interpret each term of H_{sr} by relating the input and output points to Figure 5.15.

7.29. Consider the transfer function of Figure 7.19 and determine the natural frequencies.

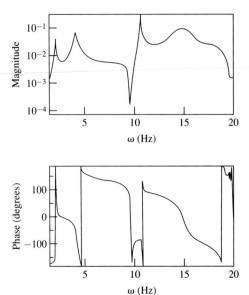

Figure 7.19 Sample magnitude and phase plot of a simple structure.

7.30. Try to determine the modal damping ratios from the plot of Figure 7.19.

7.31. The first row of the matrix $[\mathbf{u}_i \mathbf{u}_i^T]$ is measured to be

$$[1 \quad -1 \quad 3 \quad -0.25 \quad 4]$$

Construct the entire matrix.

MATLAB VIBRATION TOOLBOX

You may use the files contained in the *Vibration Toolbox*, first discussed at the end of Chapter 1 immediately following the problems, to help solve some of the exercises above. If you have not yet used the *Vibration Toolbox*, return to Chapter 1 for a brief introduction or open the file titled INTRO.TXT on the disc on the inside back cover. The files contained in folder/directory VTB7 are specifically for this chapter and contain several data sets and calculations to augment the text examples presented in this chapter. In particular, the subdirectory/folder VTB7_3.M contains data from actual experiments that can be plotted and used to perform a modal analysis. This is not a substitute for a good hands-on session with an analyzer but will at least provide the reader some experience with modal test results. The following exercises are suggested to help build experience with the material presented in this chapter.

TOOLBOX PROBLEMS

TB7.1. Open file VTB7_1. This is a demonstration program that inputs a periodic signal, digitizes it, calculates its digital Fourier transform, and plots out both the digital record and its PSD. You may also use this to try some DFTs of your own. The basic MATLAB function used here is `fft(x)`, which performs a digital Fourier transform of the data vector x.

TB7.2. Open file VTB7_1. This is a demonstration file that performs a crude power spectral density calculation (S_{xx}). For a given function $x(t)$, the digital Fourier transform is computed followed by the computation of S_{xx}. Both $x(t)$ and $S_{xx}(\omega)$ are plotted.

TB7.3. Open file VTB7_2. This file calculates $H(j\omega)$ from input data are $f(t)$ and the output data $x(t)$. A plot of $H(\omega)$ versus ω as well as a phase plot is given. A sample time history data file named "TIME_DAT.MAT" can be loaded to run this program.

TB7.4. Subfolder/directory VTB7_3 contains several data files containing force transducer and accelerometer data from actual laboratory experiments on a clamped-free beam as illustrated in Figure 7.12. The data are manipulated to produce frequency response information (both magnitude and phase) like those of Figure 7.13. Plot these data and use the techniques of Section 7.6 to extract the modal parameters. Use the command `abs` to get magnitude of $H(j\omega)$ or type "`help vtb7`" for help.

TB7.5. Subfolder/directory VTB7_3 contains experimental data presented as Nyquist plots. Plot these data and use the techniques of Section 7.5 to determine the natural frequencies and damping ratios. Type "`help vtb7`" for help.

8

Finite Element Method

This chapter introduces the finite element method for vibration analysis. The finite element method is extremely useful for modeling complicated structures and machines such as the aircraft and bridge pictured here. In finite element analysis, the vibration of the structure is approximated by the motion of the points of connection of the lines illustrated in the photograph of the airplane. Such animations are valuable in locating troublesome vibration, in design, and in predicting the response of such devices before they are actually constructed and tested. The finite element methods presented here are often used in conjunction with the modal analysis test methods of Chapter 7 and the analysis methods of Chapter 4 for multiple-degree-of-freedom systems.

The finite element method is a powerful numerical technique that uses variational and interpolation methods for modeling and solving boundary value problems such as described in Chapter 6, associated with distributed parameter vibration problems (e.g., bars, beams, and plates). The method is also extremely useful for complicated devices and structures with unusual geometric shapes (e.g., trusses, frames, and machine parts). The finite element method is very systematic and modular. Hence the finite element method may easily be implemented on a digital computer to solve a wide range of practical vibration problems simply by changing the input to a computer program. In fact, several large commercial finite element codes are available. These commercial codes can be run on almost every type of computer, ranging from "laptops" to large super computers.

The finite element method approximates a structure in two distinct ways. The first approximation made in finite element modeling is to divide the structure up into a number of small simple parts. These small parts are called *finite elements* and the procedure of dividing up the structure is called *discretization*. Each element is usually very simple, such as a bar, beam, or plate, which has an equation of motion that can easily be solved or approximated. Each element has endpoints called *nodes*, which connect it to the next element. The collection of finite elements and nodes is called a *finite element mesh* or *finite element grid*.

The equation of vibration for each individual finite element is then determined and solved. This forms the second level of approximation in the finite element method. The solutions of the element equations are approximated by a linear combination of low-order polynomials. Each of these individual polynomial solutions is made compatible with the adjacent solution (called continuity conditions) at nodes common to two elements. These solutions are then brought together in an assembly procedure, resulting in global mass and stiffness matrices, which describe the vibration of the structure as a whole. This global mass and stiffness model represents a lumped-parameter approximation of the structure which can be analyzed and solved using the methods of Chapter 4. The vector $\mathbf{x}(t)$ of displacements associated with the solution of the global finite element model corresponds to the motion of the nodes of the finite element mesh.

The finite element procedure is best illustrated by examining some of the simple beams discussed in Chapter 6. Because simple structures have closed-form solutions, developing the finite element approximation on such structures provides an easy comparison with a more exact solution. However, the power and usefulness of the finite element method is not found in examining simple structures with closed-form solutions, but rather in modeling and solving complicated parts and structures that do not have closed-form solutions.

As a final introductory comment, note that the word *node* in finite element analysis means something completely different from a *node* in vibration analysis. This is an unfortunate situation that must be kept in mind. A *node* in vibration analysis refers to a node of a mode shape (i.e., a place where no motion occurs). In finite element analysis, a node is a point on the structure representing the boundary between two elements, corresponding to the coordinate or point on the structure which represents the motion of the structure as a whole. The nodes in finite element methods are used to capture the global motion of the structure as it vibrates.

The phrase *finite element method* is often abbreviated "FEM." This abbreviation can also denote the phrase *finite element model*. Another often used abbreviation is "FEA," which denotes *finite element analysis*. Sometimes the abbreviation "FE" is used to abbreviate *finite elements*.

8.1 EXAMPLE: THE BAR

The longitudinal vibration of a bar provides a simple example of how a finite element model is constructed as well as how it is used to approximate the vibration of a distributed-parameter system with that of a lumped-parameter system (FEM). Recall the longitudinal vibration of a bar introduced in Section 6.3, which is reviewed in Window 8.1. Two finite element models of such a bar in a cantilevered configuration are illustrated in Figure 8.1. Note that the figure illustrates two *different* finite element grids of the *same* beam. Consider first the single-element model of Figure 8.1(a). The static (time independent) displacement of this bar element must satisfy the equation

$$EA\frac{d^2u(x)}{dx^2} = 0 \tag{8.1}$$

for each value of x in the interval from 0 to l. Equation (8.1) can be integrated directly to yield

$$u(x) = c_1 x + c_2 \tag{8.2}$$

where c_1 and c_2 are constants of integration with respect to x. Hence although c_1 and c_2 are called constants, they could be functions of another variable, such as t. Equation (8.2) for the time-dependent displacement follows from the static deflection equation and is known because the structure being modeled is so simple. For more complicated structures the functional form of expression (8.2) must be guessed, usually as some low-order polynomial. As pointed out in the introduction, the finite element method proceeds with two levels of approximation. One approximation involves deciding which model of Figure 8.1 to use (i.e., which mesh and size of mesh, where to put elements and nodes, etc.). The second level of approximation is the choice of polynomials to use in equation (8.1).

Window 8.1
Review of the Vibration of an Undamped Cantilevered
Bar in Longitudinal Vibration from Example 6.3.1

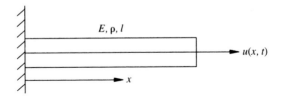

Displacement: $u(x,t)$ Velocity: $u_t(x,t)$

Acceleration: $u_{tt}(x,t)$ Elastic modulus: E

Cross sectional area: A Length: l

Equation of motion: $\rho u_{tt}(x,t) - E u_{xx}(x,t) = 0$

Boundary conditions: $u(0,t) = 0$ at $x = 0$, $u_x(l,t) = 0$ at $x = l$

Initial conditions: $u(x,0) = u_o(x)$ at $t = 0$, $u_t(x,0) = \dot{u}_o(x)$ at $t = 0$

Solution: $u(x,t) = \sum\limits_{n=1}^{\infty} A_n \sin(\omega_n t + \phi_n) \sin \dfrac{n\pi}{l} x$

where A_n and are ϕ_n constants determined by the initial
conditions: $u(x,0)$ and $u_t(x,0)$

Natural frequencies: $\omega_n = \dfrac{(2n-1)\pi}{2l}\sqrt{\dfrac{E}{\rho}}$, $n = 1,2,3,\ldots$

Mode shapes: $\sin\left(\dfrac{2n-1}{2l}\pi x\right)$, $n = 1,2,3,\ldots$

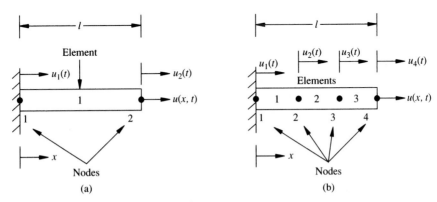

Figure 8.1 Two different finite element grids of the same cantilevered bar
of length l in longitudinal vibration. (a) Single-element, two-node mesh.
(b) Three-element, four-node mesh.

At each node, the value of u is allowed to be time dependent, hence the use of the labels $u_1(t)$ and $u_2(t)$ in Figure 8.1(a). The time-variable functions $u_1(t)$ and $u_2(t)$ are called the *nodal* displacements of the model and will eventually be solved for and used to describe the longitudinal vibration of the bar. The spatial function $u(x)$ and the temporal functions $u_1(t)$ and $u_2(t)$ are related through using the nodes as boundaries to evaluate the spatial constants in equation (8.2). At $x = 0$, equation (8.2) becomes

$$u(0) = u_1(t) = c_1(0) + c_2 \qquad t \geq 0$$

so that $c_2 = u_1(t)$. Similarly, at $x = l$, equation (8.2) yields

$$u(l) = u_2(t) = c_1 l + c_2$$

so that $c_1 = [u_2(t) - u_1(t)]/l$. Substitution of these (time-dependent) values of c_1 and c_2 into the expression for $u(x)$ given by equation (8.2) yields the approximation of the displacement $u(x, t)$ given by

$$u(x, t) = \left(1 - \frac{x}{l}\right) u_1(t) + \frac{x}{l} u_2(t) \tag{8.3}$$

If $u_1(t)$ and $u_2(t)$ were known at this stage, equation (8.3) would provide an approximate solution to the bar equation. The coefficient functions $(1 - x/l)$ and (x/l) are called *shape functions*, because they determine the spatial distribution or shape of the solution $u(x, t)$.

Next consider the energy associated with the approximation given by equation (8.3). The strain energy of a bar is given by the integral (see, e.g., Shames, 1989)

$$V(t) = \frac{1}{2} \int_0^l EA \left[\frac{\partial u(x, t)}{\partial x}\right]^2 dx \tag{8.4}$$

Substitution of the approximate solution for $u(x, t)$ given by equation (8.3), equation (8.4) yields

$$V(t) = \frac{1}{2} \int_0^l \frac{EA}{l^2} [-u_1(t) + u_2(t)]^2 dx = \frac{EA}{2l} \left(u_1^2 - 2u_1 u_2 + u_2^2\right) \tag{8.5}$$

The last expression can be recognized (recall Window 4.1) as proportioned to the product of the transpose of the vector $\mathbf{u}(t)$ defined by

$$\mathbf{u}(t) = \begin{bmatrix} u_1(t) \\ u_2(t) \end{bmatrix} \tag{8.6}$$

with the matrix K and the vector $\mathbf{u}(t)$, where

$$K = \frac{EA}{l} \begin{bmatrix} 1 & -1 \\ -1 & 1 \end{bmatrix} \tag{8.7}$$

i.e., $V(t) = \frac{1}{2}\mathbf{u}^T K \mathbf{u}$. Equation (8.7) defines the stiffness matrix associated with the single element of Figure 8.1(a).

The kinetic energy of the element can be calculated from the integral

$$T(t) = \frac{1}{2} \int_0^l A\rho(x) \left[\frac{\partial u(x, t)}{\partial t}\right]^2 dx \tag{8.8}$$

where $\rho(x)$ is the density of the bar as discussed in Section 6.3 and reviewed in Window 8.1. Using the approximation for $u(x, t)$ given by equation (8.3), the approximate velocity becomes

$$\frac{\partial u(x, t)}{\partial t} = \left(1 - \frac{x}{l}\right)\dot{u}_1(t) + \frac{x}{l}\dot{u}_2(t) \tag{8.9}$$

Assuming a constant density [i.e., $\rho(x) = \rho$] and substituting equation (8.9) into equation (8.8) yields

$$T(t) = \frac{1}{2}\frac{\rho A l}{3}(\dot{u}_1^2 + \dot{u}_1\dot{u}_2 + \dot{u}_2^2) \tag{8.10}$$

Looking at this expression as a matrix-based quadratic form yields that equation (8.10) can be factored into the velocity vector $\dot{\mathbf{u}}(t) = [\dot{u}_1(t) \ \dot{u}_2(t)]^T$, the matrix M defined by

$$M = \frac{\rho A l}{6}\begin{bmatrix} 2 & 1 \\ 1 & 2 \end{bmatrix} \tag{8.11}$$

and the vector $\dot{\mathbf{u}}^T$,

$$T(t) = \tfrac{1}{2}\dot{\mathbf{u}}^T M \dot{\mathbf{u}} \tag{8.12}$$

The matrix M defined by equation (8.11) is the mass matrix associated with the single finite element of Figure 8.1(a).

The equations of vibration can be obtained from the expressions above for the kinetic energy $T(t)$ and strain energy $V(t)$ by using the variational or Lagrangian approach introduced in Section 4.7. Recall that the equations of motion can be calculated from the energy in the structure from

$$\frac{\partial}{\partial t}\left(\frac{\partial T}{\partial \dot{u}_i}\right) - \frac{\partial T}{\partial u_i} + \frac{\partial V}{\partial u_i} = f_i(t) \qquad i = 1, 2, \ldots, n \tag{8.13}$$

where u_i is the ith coordinate of the system, which is assumed to have n degrees of freedom, and where f_i is the external force applied at coordinates u_i (for the problem at hand, $f_i = 0$). Here the energies T and V are the total kinetic energy and strain energy, respectively, in the structure.

Examination of the boundary conditions in Figure 8.1(a) indicates that the time response at the clamped end must be zero. Hence the total kinetic energy becomes

$$T(t) = \frac{1}{2}\frac{\rho A l}{3}(\dot{u}_2^2)$$

and the total strain energy becomes

$$V(t) = \frac{1}{2}\frac{EA}{l}u_2^2$$

Substitution of these two energy expressions into the Lagrange equation given in (8.13) yields

$$\frac{\rho A l}{3}\ddot{u}_2(t) + \frac{EA}{l}u_2(t) = 0 \tag{8.14}$$

This becomes

$$\ddot{u}_2(t) + \frac{3E}{\rho l^2} u_2(t) = 0 \tag{8.15}$$

which constitutes a simple finite element model of the cantilevered bar using only one element.

This finite element model of the bar can now be solved, given a set of initial conditions, for the nodal displacement $u_2(t)$. The solution to equation (8.15) is (see Window 8.2)

$$u_2(t) = \sqrt{u_0^2 + \left(\frac{\dot{u}_0}{\omega}\right)^2} \sin\left(\omega t + \tan^{-1}\frac{\omega u_0}{\dot{u}_0}\right) \tag{8.16}$$

where u_0 is the initial nodal displacement, $\dot{u}_0$ is the initial nodal velocity, and from the coefficient of $u_2(t)$ in equation (8.15) the natural frequency ω is

$$\omega = \frac{1}{l}\sqrt{\frac{3E}{\rho}}$$

This solution for $u_2(t)$ can be combined with equation (8.3) to yield the approximate solution for the transient displacement of the bar. The bar displacement becomes

$$u(x, t) = \sqrt{u_0^2 + \left(\frac{\dot{u}_0}{\omega}\right)^2} \frac{x}{l} \sin\left[\left(\frac{1}{l}\sqrt{\frac{3E}{\rho}}\right) t + \tan^{-1}\frac{\omega u_0}{\dot{u}_0}\right] \tag{8.17}$$

This describes a vibration of frequency $(1/l)\sqrt{3E/\rho}$, in contrast to the exact solution given in Window 8.1, which describes vibration at an infinite number of frequencies.

Window 8.2
Review of the Solution for the Free Response of
an Undamped Single-Degree-of-Freedom System

From Section 1.1, equation (1.10), recall that the solution of $\ddot{x}(t) + \omega^2 x(t) = 0$ subject to initial conditions $x(0) = x_0$, $\dot{x}(0) = v_0$ is

$$x(t) = \sqrt{x_0^2 + \left(\frac{v_0}{\omega}\right)^2} \sin\left(\omega t + \tan^{-1}\frac{\omega x_0}{v_0}\right)$$

Example 8.1.1

Compare the solution of the clamped–free bar of Window 8.1 obtained by the methods of Chapter 6 with the solution obtained by the finite element approach as given by equation (8.17).

Solution The solution given by the finite element approach contains a time oscillation at only one frequency, oscillating in only one spatial mode shape, whereas the exact solution consists of an infinite number of mode shapes oscillating at an infinite number of frequencies superimposed on one another (depending, of course, on the initial conditions). The single undamped natural frequency of the finite element model is

$$\omega_{FEM} = \frac{1}{l}\sqrt{\frac{3E}{\rho}} = \frac{1.732}{l}\sqrt{\frac{E}{\rho}}$$

whereas the first natural frequency of the exact solution is $\omega_1 = (\pi/2l)\sqrt{E/\rho}$ or approximately

$$\omega_1 = \frac{1.57}{l}\sqrt{\frac{E}{\rho}}$$

This is smaller than the frequency predicted by the FEM. The second (exact) natural frequency is approximately

$$\omega_2 = \frac{4.712}{l}\sqrt{\frac{E}{\rho}}$$

which is larger than ω_{FEM}. In addition, the exact first mode shape is different, yet has some of the same character, as the FEM shape function. This is illustrated in Figure 8.2.

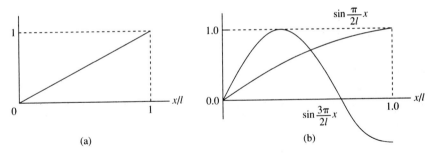

Figure 8.2   Plot of (a) the FEM shape function and (b) the first mode shape $[\sin(\pi x/2l)]$ of the exact solution, both for a clamped-free bar. Part (b) also shows the second mode shape $[\sin(3\pi/2l)]$.

If the bar is excited only in its first mode, the shape of vibration for the finite element model is a fairly reasonable approximation to the exact mode shape in the sense that it captures the basic behavior of the first mode. However, examination of the second exact mode shape in Figure 8.2 illustrates that the FEM is a terrible representation of the exact response if the bar is excited in the second mode.   □

Example 8.1.1 indicates that the finite element approximation is useful only in a certain context. However, the next section illustrates that the FEM can often be increased in size to produce a more accurate description of the structure under consideration.

8.2 THREE-ELEMENT BAR

Consider increasing the size of the finite element model of the bar in Section 8.1 to four nodes and three elements as indicated in Figure 8.1(b). Each element of the bar will again have a strain energy relationship as calculated in equation (8.5), with two differences. The first is that the length of the element becomes $l/3$, instead of l, and the second is that the node coordinates are different in each of the three elements. Taking these changes into consideration and using matrix notation, the strain energy for element 1 is

$$V_l(t) = \frac{3EA}{2l} \begin{bmatrix} 0 \\ u_2 \end{bmatrix}^T \begin{bmatrix} 1 & -1 \\ -1 & 1 \end{bmatrix} \begin{bmatrix} 0 \\ u_2 \end{bmatrix} \tag{8.18}$$

For element 2 the strain energy becomes

$$V_2(t) = \frac{3EA}{2l} \begin{bmatrix} u_2 \\ u_3 \end{bmatrix}^T \begin{bmatrix} 1 & -1 \\ -1 & 1 \end{bmatrix} \begin{bmatrix} u_2 \\ u_3 \end{bmatrix} \tag{8.19}$$

and for element 3 it becomes

$$V_3(t) = \frac{3EA}{2l} \begin{bmatrix} u_3 \\ u_4 \end{bmatrix}^T \begin{bmatrix} 1 & -1 \\ -1 & 1 \end{bmatrix} \begin{bmatrix} u_3 \\ u_4 \end{bmatrix} \tag{8.20}$$

The total strain energy is the sum

$$V(t) = V_1(t) + V_2(t) + V_3(t)$$

$$= \frac{3EA}{2l} \left\{ \begin{bmatrix} 0 \\ u_2 \end{bmatrix}^T \begin{bmatrix} 1 & -1 \\ -1 & 1 \end{bmatrix} \begin{bmatrix} 0 \\ u_2 \end{bmatrix} \right.$$

$$\left. + \begin{bmatrix} u_2 \\ u_3 \end{bmatrix}^T \begin{bmatrix} 1 & -1 \\ -1 & 1 \end{bmatrix} \begin{bmatrix} u_2 \\ u_3 \end{bmatrix} + \begin{bmatrix} u_3 \\ u_4 \end{bmatrix}^T \begin{bmatrix} 1 & -1 \\ -1 & 1 \end{bmatrix} \begin{bmatrix} u_3 \\ u_4 \end{bmatrix} \right\}$$

$$= \frac{3EA}{2l} (2u_2^2 - 2u_2u_3 + 2u_3^2 - 2u_3u_4 + u_4^2)$$

The vector of derivatives of this total strain energy indicated in the Lagrangian given by equation (8.13) becomes

$$\begin{bmatrix} \dfrac{\partial V}{\partial u_2} \\[1mm] \dfrac{\partial V}{\partial u_3} \\[1mm] \dfrac{\partial V}{\partial u_4} \end{bmatrix} = \frac{3EA}{l} \begin{bmatrix} 2u_2 - u_3 \\ -u_2 + 2u_3 - u_4 \\ -u_3 + u_4 \end{bmatrix} = \frac{3EA}{l} \begin{bmatrix} 2 & -1 & 0 \\ -1 & 2 & -1 \\ 0 & -1 & 1 \end{bmatrix} \begin{bmatrix} u_2(t) \\ u_3(t) \\ u_4(t) \end{bmatrix} \tag{8.22}$$

In a similar fashion, the total kinetic energy can be written as

$$T(t) = \frac{\rho A l}{36} \left\{ \begin{bmatrix} 0 \\ \dot{u}_2 \end{bmatrix}^T \begin{bmatrix} 2 & 1 \\ 1 & 2 \end{bmatrix} \begin{bmatrix} 0 \\ \dot{u}_2 \end{bmatrix} + \begin{bmatrix} \dot{u}_2 \\ \dot{u}_3 \end{bmatrix}^T \begin{bmatrix} 2 & 1 \\ 1 & 2 \end{bmatrix} \begin{bmatrix} \dot{u}_2 \\ \dot{u}_3 \end{bmatrix} + \begin{bmatrix} \dot{u}_3 \\ \dot{u}_4 \end{bmatrix}^T \begin{bmatrix} 2 & 1 \\ 1 & 2 \end{bmatrix} \begin{bmatrix} \dot{u}_3 \\ \dot{u}_4 \end{bmatrix} \right\}$$

(8.23)

The first term in the Lagrange equation (8.13) then becomes

$$\frac{d}{dt} \begin{bmatrix} \dfrac{\partial T}{\partial \dot{u}_2} \\[6pt] \dfrac{\partial T}{\partial \dot{u}_3} \\[6pt] \dfrac{\partial T}{\partial \dot{u}_4} \end{bmatrix} = \frac{\rho A l}{18} \begin{bmatrix} 4 & 1 & 0 \\ 1 & 4 & 1 \\ 0 & 1 & 2 \end{bmatrix} \begin{bmatrix} \ddot{u}_2(t) \\ \ddot{u}_3(t) \\ \ddot{u}_4(t) \end{bmatrix}$$

(8.24)

Combining equations (8.24) and (8.22) with equation (8.13) yields the familiar form

$$M\ddot{\mathbf{u}}(t) + K\mathbf{u}(t) = \mathbf{0}$$

(8.25)

where $\mathbf{u}(t) = [u_2 \quad u_3 \quad u_4]^T$ is the vector of nodal displacements. Here the coefficient

$$M = \frac{\rho A l}{18} \begin{bmatrix} 4 & 1 & 0 \\ 1 & 4 & 1 \\ 0 & 1 & 2 \end{bmatrix}$$

(8.26)

is the *global mass matrix* and the coefficient

$$K = \frac{3EA}{l} \begin{bmatrix} 2 & -1 & 0 \\ -1 & 2 & -1 \\ 0 & -1 & 1 \end{bmatrix}$$

(8.27)

is the *global stiffness matrix* defining the dynamic finite element model of the bar. Equation (8.25) can be solved and analyzed by the methods of Section 4.4. Note that equation (8.25) is both dynamically and statically coupled.

Example 8.2.1

Compare the natural frequencies of the finite element model of Figure 8.1(b) with those of the distributed mass model given in Window 8.1.

Solution   The natural frequencies of the three element finite element model of the clamped–free bar are determined by substituting the global stiffness matrix of equation (8.27) and the global mass matrix of equation (8.26) into equation (8.25). Following the procedures of Section 4.4 yields the natural frequency equation

$$\det \left\{ \frac{3EA}{l} \begin{bmatrix} 2 & -1 & 0 \\ -1 & 2 & -1 \\ 0 & -1 & 1 \end{bmatrix} - \omega^2 \frac{\rho A l}{18} \begin{bmatrix} 4 & 1 & 0 \\ 1 & 4 & 1 \\ 0 & 1 & 2 \end{bmatrix} \right\} = 0$$

(8.28)

Assuming that the beam is made of aluminum and is 1 m in length, then $l = 1$ m, $\rho = 2700$ kg/m^3 and $E = 7.0 \times 10^{10}$ N/m^2. Note that the value of A is not required. Equation (8.28) can be solved for the values of ω^2 using MATLAB to yield the natural frequencies

$$\omega_1 = 8092 \text{ rad/s}$$

$$\omega_2 = 26{,}458 \text{ rad/s}$$

$$\omega_3 = 47{,}997 \text{ rad/s}$$

On the other hand, the first three natural frequencies for the beam from the distributed parameter solution given in Window 8.1 are

$$\omega_1 = 7998 \text{ rad/s} \ (0.55\%)$$

$$\omega_2 = 23{,}994 \text{ rad/s} \ (9.64\%) \tag{8.29}$$

$$\omega_3 = 39{,}900 \text{ rad/s} \ (19.3\%)$$

The numbers in parentheses in equation (8.29) are percents of the difference between the actual natural frequencies of the aluminum beam as calculated from the distributed model and the three corresponding natural frequencies of the three-element finite element model of the same beam. Note that the first natural frequency of the three-element model calculated here is much closer to the actual value than the natural frequency calculated in Example 8.1.1 from a one-element model. The value for ω from a one-element bar is 8819 rad/s. It is a rule of thumb in finite element analysis that many more (usually at least twice as many) elements must be used than number of accurate frequencies required.

☐

In the preceding analysis, each element is of the same length. However, the finite element method does not require the sizes of the various elements to be uniform. In fact, for complicated structures it is often necessary to choose a nonuniform size. The following example illustrates the use of a nonuniform element size.

Example 8.2.2

Consider the longitudinal vibration of a clamped bar and calculate two natural frequencies using the two-element mesh arrangement suggested in Figure 8.3. Compare these frequencies to the exact frequencies given in Window 8.1.

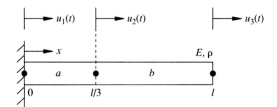

Figure 8.3 Bar of length l, area A, modulus E, and density ρ divided into two elements of dissimilar length.

Solution The energy expressions for the first element are identical to the energy expressions calculated for the three-element model of a cantilevered bar analyzed above. In particular, equation (8.18) yields that the potential energy in element 1 is

$$V_1(t) = \frac{3EA}{2l} u_2^2(t) \tag{8.30}$$

and equation (8.23) yields that the kinetic energy in element 1 is

$$T_1(t) = \frac{\rho A l}{18} \dot{u}_2^2(t) \tag{8.31}$$

Following the example of Section 8.1, the energy in the second element can be calculated from the assumed form of the solution given by

$$u(x, t) = c_3(t)x + c_4(t) \tag{8.32}$$

This displacement equation must satisfy $u(l/3, t) = u_2(t)$ and $u(l, t) = u_3(t)$, which yields the coupled algebraic equations

$$u_2(t) = \frac{l}{3}c_3(t) + c_4(t)$$

$$u_3(t) = lc_3(t) + c_4(t) \tag{8.33}$$

Equations (8.33) can be solved for $c_3(t)$ and $c_4(t)$ in terms of $u_2(t)$ and $u_3(t)$ to yield

$$c_3(t) = -\frac{3}{2l}[u_2(t) - u_3(t)] \quad \text{and} \quad c_4(t) = \frac{1}{2}[3u_2(t) - u_3(t)] \tag{8.34}$$

Substitution of these values of $c_3(t)$ and $c_4(t)$ into equation (8.32) yields

$$u(x, t) = \frac{3}{2}\left(1 - \frac{x}{l}\right)u_2(t) + \frac{1}{2}\left(\frac{3x}{l} - 1\right)u_3(t) \tag{8.35}$$

which can be used to calculate the energy expressions for the second element.
The potential energy in the second element becomes

$$V_2(t) = \frac{EA}{2}\int_{l/3}^{l}[u_x(x,t)]^2 dx = \frac{EA}{2}\int_{l/3}^{l}\left(\frac{3}{2l}\right)^2[u_3 - u_2]^2 dx$$

$$= \frac{EA}{2l}\left(\frac{3}{2}\right)\left[u_2^2(t) - 2u_2(t)u_3(t) + u_3^2(t)\right] \tag{8.36}$$

Adding the potential energy in each element as given by equations (8.30) and (8.36) yields the total potential energy in the bar:

$$V(t) = \frac{EA}{2l}\left(\frac{3}{2}\right)\left[3u_2^2(t) - 2u_2(t)u_3(t) + u_3^2(t)\right] \tag{8.37}$$

This can be written in matrix form as

$$V(t) = \frac{EA}{2l}\left(\frac{3}{2}\right)\begin{bmatrix} u_2 \\ u_3 \end{bmatrix}^T \begin{bmatrix} 3 & -1 \\ -1 & 1 \end{bmatrix}\begin{bmatrix} u_2 \\ u_3 \end{bmatrix} \tag{8.38}$$

which implies that the stiffness matrix has the value

$$K = \frac{EA}{l}\left(\frac{3}{2}\right)\begin{bmatrix} 3 & -1 \\ -1 & 1 \end{bmatrix} \tag{8.39}$$

The kinetic energy in the second element becomes

$$T_2(t) = \frac{A\rho}{2} \int_{l/3}^{l} [u_t(x, t)]^2 \, dx$$

$$= \frac{A\rho}{8l^2} \int_{l/3}^{l} \left[9(l - x)^2 \dot{u}_2^2 + 6(3x - l)(l - x)\dot{u}_2\dot{u}_3 + (3x - l)^2 \dot{u}_3^2 \right] dx \tag{8.40}$$

$$= \frac{A\rho l}{9} \left[\dot{u}_2^2 + \dot{u}_2\dot{u}_3 + \dot{u}_3^2 \right]$$

where $u_t(x, t)$ is calculated from equation (8.35). Adding equations (8.31) and (8.40), the total kinetic energy in the bar becomes

$$T(t) = T_1(t) + T_2(t) = \frac{A\rho l}{18} \left[3\dot{u}_2^2(t) + 2\dot{u}_2(t)\dot{u}_3(t) + 2\dot{u}_3^2(t) \right] \tag{8.41}$$

In matrix form this becomes simply

$$T(t) = \frac{1}{2} \frac{A\rho l}{18} \begin{bmatrix} \dot{u}_2 \\ \dot{u}_3 \end{bmatrix}^T \begin{bmatrix} 6 & 2 \\ 2 & 4 \end{bmatrix} \begin{bmatrix} \dot{u}_2 \\ \dot{u}_3 \end{bmatrix} \tag{8.42}$$

so that the system mass matrix is identified as

$$M = \frac{A\rho l}{18} \begin{bmatrix} 6 & 2 \\ 2 & 4 \end{bmatrix} \tag{8.43}$$

Using the kinetic energy and potential energy in the Euler–Lagrange equation yields that the equations of motion are

$$\frac{\rho l}{9} \begin{bmatrix} 3 & 1 \\ 1 & 2 \end{bmatrix} \begin{bmatrix} \ddot{u}_2 \\ \ddot{u}_3 \end{bmatrix} + \frac{3E}{2l} \begin{bmatrix} 3 & -1 \\ -1 & 1 \end{bmatrix} \begin{bmatrix} u_2 \\ u_3 \end{bmatrix} = \begin{bmatrix} 0 \\ 0 \end{bmatrix} \tag{8.44}$$

Using the methods of Section 4.3, the natural frequencies of equation (8.44) are

$$\omega_1 = (1.6432)\frac{1}{l}\sqrt{\frac{E}{\rho}} \quad \left(1.5708\frac{1}{l}\sqrt{\frac{E}{\rho}} \right)$$

$$\omega_2 = (5.1962)\frac{1}{l}\sqrt{\frac{E}{\rho}} \quad \left(4.7124\frac{1}{l}\sqrt{\frac{E}{\rho}} \right)$$

where the number in parentheses indicates the exact value from the solution given in Window 8.1. Although not illustrated here, the best use of nonuniform elements is for those applications where the geometry or physical parameters are variable along x.

□

8.3 BEAM ELEMENTS

Many parts and structures cannot be modeled by axial vibration only. Hence a finite element is needed to describe transverse vibration as well. Window 8.3 summarizes the vibration analysis for an Euler–Bernoulli beam as discussed in Section 6.5. Figure 8.4 indicates the coordinate system and variables used in the finite element analysis of a free–

Window 8.3
Review of the Transverse Vibration Analysis of an Undamped Pinned–Pinned Beam

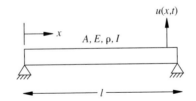

Displacement:	$u(x, t)$	Velocity:	$u_t(x, t)$
Acceleration:	$u_{tt}(x, t)$	Elastic modulus:	E
Cross sectional area:	A	Length:	l

Equation of motion: $\rho A u_{tt}(x, t) + E I u_{xxxx}(x, t) = 0$

Boundary conditions: $u(0, t) = u_{xx}(0, t) = 0$, pinned end at $x = 0$

$u(l, t) = u_{xx}(l, t) = 0$, pinned end at $x = l$

Solution: $u(x, t) = \sum_{n=1}^{\infty} A_n \sin(\omega_n t + \phi_n) \sin \dfrac{n\pi x}{l}$

where A_n and are ϕ_n are constants determined
by the initial conditions: $u(x, 0)$ and $u_t(x, 0)$

Natural frequencies: $\omega_n = \left(\dfrac{n\pi}{l}\right)^2 \sqrt{\dfrac{EI}{\rho A}}, \quad n = 1, 2, \ldots$

Mode shapes: $X_n(x) = \sin \dfrac{n\pi x}{l}, \quad n = 1, 2, \ldots$

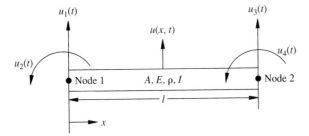

Figure 8.4 Single-finite-element model of a beam illustrating the two transverse coordinates $u_1(t)$ and $u_3(t)$ and the two slope coordinates $u_2(t)$ and $u_4(t)$ required to describe the vibration of this element.

free beam for transverse vibration. The coordinates used in the finite element model of the beam are the two linear coordinates $u_1(t)$ and $u_3(t)$ and the two rotational coordinates $u_2(t)$ and $u_4(t)$. One of each type is required to describe the motion of each node. That is, each node is modeled as having two degrees of freedom. One of these accounts for the slope and the other for the lateral motion. The transverse static displacement must satisfy

$$\frac{\partial^2}{\partial x^2}\left[EI\frac{\partial^2 u(x, t)}{\partial x^2}\right] = 0 \tag{8.45}$$

For constant values of EI this becomes $u_{xxxx} = 0$, which may be integrated to yield

$$u(x, t) = c_1(t)x^3 + c_2(t)x^2 + c_3(t)x + c_4(t) \tag{8.46}$$

where, as before, the $c_i(t)$ are constants of integration with respect to the spatial variable x [recall equations (8.1) and (8.2) for the bar]. Equation (8.46) is used to approximate the transverse displacement within the element.

Proceeding as in Section 8.1 for the bar, the unknown nodal displacements $u_i(t)$ must satisfy the boundary conditions

$$u(0, t) = u_1(t) \qquad u_x(0, t) = u_2(t)$$
$$u(l, t) = u_3(t) \qquad u_x(l, t) = u_4(t) \tag{8.47}$$

These relationships along with equation (8.46) are combined and solved for the constants of integration in equation (8.46). This yields

$$c_4(t) = u_1(t) \qquad c_3(t) = u_2(t)$$

$$c_2(t) = \frac{1}{l^2}[(3(u_3 - u_1) - l(2u_2 + u_4)] \tag{8.48}$$

$$c_1(t) = \frac{1}{l^3}[2(u_1 - u_3) + l(u_2 + u_4)]$$

Substitution of equations (8.48) into equation (8.46), and rearranging terms as coefficients of the unknown nodal displacement (again recall Section 8.1), yields the result that the approximate displacement $u(x, t)$ for the element can be expressed as

$$u(x, t) = \left[1 - 3\frac{x^2}{l^2} + 2\frac{x^3}{l^3}\right]u_1(t) + l\left[\frac{x}{l} - 2\frac{x^2}{l^2} + \frac{x^3}{l^3}\right]u_2(t)$$
$$+ \left[3\frac{x^2}{l^2} - 2\frac{x^3}{l^3}\right]u_3(t) + l\left[-\frac{x^2}{l^2} + \frac{x^3}{l^3}\right]u_4(t) \tag{8.49}$$

As before, the coefficient of each $u_i(t)$ defines the shape functions for the transverse beam element.

The mass and stiffness matrices can be calculated following exactly the same procedure as followed for the bar. In the case of the beam, the mass matrix calculation results from substituting equation (8.49) into the formula for kinetic energy:

$$T(t) = \frac{1}{2} \int_0^l \rho A[u_t(x, t)]^2 dx \tag{8.50}$$

and recognizing that this can be written in the form

$$T(t) = \frac{1}{2}\dot{\mathbf{u}}^T M \dot{\mathbf{u}} \tag{8.51}$$

where M is the desired mass matrix. The vector $\dot{\mathbf{u}}$ is the time derivative of the 4×1 vector $\mathbf{u}(t)$ of nodal displacements defined by

$$\mathbf{u}(t) = \begin{bmatrix} u_1(t) \\ u_2(t) \\ u_3(t) \\ u_4(t) \end{bmatrix} \tag{8.52}$$

After performing the integrations and factoring out the nodal displacement vector, the mass matrix for the beam of Figure 8.4 becomes

$$M = \frac{\rho Al}{420} \begin{bmatrix} 156 & 22l & 54 & -13l \\ 22l & 4l^2 & 13l & -3l^2 \\ 54 & 13l & 156 & -22l \\ -13l & -3l^2 & -22l & 4l^2 \end{bmatrix} \tag{8.53}$$

The stiffness matrix calculation proceeds in a similar fashion again following the example of the bar in Section 8.1. The strain energy for the beam results from substituting the assumed solution form given by equation (8.49) into the formula for the strain energy given by

$$V(t) = \tfrac{1}{2} \int_0^l EI\,[u_{xx}(x,t)]^2 dx \tag{8.54}$$

The result can be factored into the form

$$V(t) = \tfrac{1}{2}\mathbf{u}^T K \mathbf{u} \tag{8.55}$$

where **u** is as defined above. This defines the stiffness matrix K to be

$$K = \frac{EI}{l^3} \begin{bmatrix} 12 & 6l & -12 & 6l \\ 6l & 4l^2 & -6l & 2l^2 \\ -12 & -6l & 12 & -6l \\ 6l & 2l^2 & -6l & 4l^2 \end{bmatrix} \tag{8.56}$$

The mass and stiffness matrices of equations (8.53) and (8.56) define the beam finite element for transverse vibration.

Example 8.3.1

Use the beam finite element mass and stiffness matrices defined above to calculate the natural frequencies of a simply supported beam. Use a single element. Compare these with the frequencies obtained for a simply supported beam in Chapter 6 and reviewed in Window 8.3.

Solution The simply supported boundary condition allows the slope at the boundary to vary but fixes the displacement to be zero at the boundary. Examining Figure 8.5 the pinned boundary condition requires that both $u_1(t)$ and $u_3(t)$ be zero. This is accomplished by setting these terms to zero in the kinetic and potential energy expression. Setting $u_1(t)$ and $u_3(t)$ to zero has the effect of deleting the rows and columns of the mass and stiffness matrices corresponding to $u_1(t)$ and $u_3(t)$. Deleting the first and third row and column results in the dynamic equation

$$\frac{\rho Al}{420} \begin{bmatrix} 4l^2 & -3l^2 \\ -3l^2 & 4l^2 \end{bmatrix} \begin{bmatrix} \ddot{u}_2(t) \\ \ddot{u}_4(t) \end{bmatrix} + \frac{EI}{l^3} \begin{bmatrix} 4l^2 & 2l^2 \\ 2l^2 & 4l^2 \end{bmatrix} \begin{bmatrix} u_2(t) \\ u_4(t) \end{bmatrix} = \begin{bmatrix} 0 \\ 0 \end{bmatrix} \tag{8.57}$$

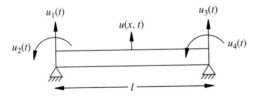

Figure 8.5 Coordinates of a simply supported beam modeled as a single finite element.

The above can be written as

$$\begin{bmatrix} 4 & -3 \\ -3 & 4 \end{bmatrix} \ddot{\mathbf{u}} + \frac{840EI}{\rho A l^4} \begin{bmatrix} 2 & 1 \\ 1 & 2 \end{bmatrix} \mathbf{u} = 0 \tag{8.58}$$

where $\mathbf{u}$ is now interpreted as $\mathbf{u} = [u_2(t) \quad u_4(t)]^T$. Following the procedures of Section 4.2 (i.e., let $\mathbf{u} = \mathbf{x}e^{\omega t}$ and solve for ω), equation (8.58) yields the two natural frequencies

$$\omega_1 = 10.95445 \sqrt{\frac{EI}{\rho A l^4}} \qquad \left(9.86965 \sqrt{\frac{EI}{\rho A l^4}}\right)$$

$$\omega_2 = 50.199605 \sqrt{\frac{EI}{\rho A l^4}} \qquad \left(39.47860 \sqrt{\frac{EI}{\rho A l^4}}\right) \tag{8.59}$$

The numbers in parentheses are the actual values calculated using the distributed-mass models and methods of Chapter 6. The first frequency is within 11% of the actual value, while the second frequency is only within 28% of the actual value. As stated previously, it is generally thought that it is required to take at least twice as many degrees of freedom in the finite element model as the number of frequencies required.

□

The next step in applying the finite element method is to incorporate more than one element in the analysis of the beam. In the case of the bar, Section 8.2, the equations of motion for multielement structures were obtained by returning to the energy computation and the Euler–Lagrange formula. Such equations become too lengthy for the more complicated beam element, so an alternative but equivalent method of superimposing individual element matrices is used to obtain the global mass and stiffness matrices. The procedure is presented by repeating the three element bar example of Section 8.2.

Example 8.3.2

Consider again the three-element model of the longitudinal vibration of the bar of Figure 8.1(b) repeated in Figure 8.6. Here the local mass and stiffness matrix for each element is obtained from the basic element matrices given by equations (8.7) and (8.11) with l replaced by $l/3$. The resulting mass and stiffness matrices are

$$M_1 = \frac{\rho A l}{18} \begin{bmatrix} 2 & 1 \\ 1 & 2 \end{bmatrix} \qquad K_1 = \frac{3EA}{l} \begin{bmatrix} 1 & -1 \\ -1 & 1 \end{bmatrix} \tag{8.60}$$

corresponding to local coordinates $u_1(t)$ and $u_2(t)$. The local equation of motion then becomes

$$\frac{\rho A l}{18} \begin{bmatrix} 2 & 1 \\ 1 & 2 \end{bmatrix} \begin{bmatrix} \ddot{u}_1 \\ \ddot{u}_2 \end{bmatrix} + \frac{3EA}{l} \begin{bmatrix} 1 & -1 \\ -1 & 1 \end{bmatrix} \begin{bmatrix} u_1 \\ u_2 \end{bmatrix} = \begin{bmatrix} 0 \\ 0 \end{bmatrix} \tag{8.61}$$

Similary, for element 2,

$$M_2 = \frac{\rho A l}{18} \begin{bmatrix} 2 & 1 \\ 1 & 2 \end{bmatrix} \qquad K_2 = \frac{3EA}{l} \begin{bmatrix} 1 & -1 \\ -1 & 1 \end{bmatrix} \tag{8.62}$$

with local coordinates u_2 and u_3. Element 3 has mass and stiffness matrices

$$M_3 = \frac{\rho A l}{18} \begin{bmatrix} 2 & 1 \\ 1 & 2 \end{bmatrix} \qquad K_3 = \frac{3EA}{l} \begin{bmatrix} 1 & -1 \\ -1 & 1 \end{bmatrix} \tag{8.63}$$

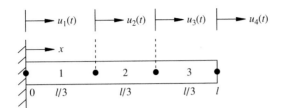

Figure 8.6 Cantilevered bar modeled with three finite elements and four nodes.

and local coordinates u_3 and u_4. These three sets of matrices and their corresponding equations [i.e., three sets of equations identical to equation (8.61) with different sets of modal displacements $u_i(t)$] can be assembled together by superimposing them to yield

$$\frac{\rho Al}{18}\begin{bmatrix} 2 & 1 & 0 & 0 \\ 1 & 2+2 & 1 & 0 \\ 0 & 1 & 2+2 & 1 \\ 0 & 0 & 1 & 2 \end{bmatrix}\begin{bmatrix} \ddot{u}_1 \\ \ddot{u}_2 \\ \ddot{u}_3 \\ \ddot{u}_4 \end{bmatrix} + \frac{3EA}{l}\begin{bmatrix} 1 & -1 & 0 & 0 \\ -1 & 1+1 & -1 & 0 \\ 0 & -1 & 1+1 & -1 \\ 0 & 0 & -1 & 1 \end{bmatrix}\begin{bmatrix} u_1 \\ u_2 \\ u_3 \\ u_4 \end{bmatrix} = \begin{bmatrix} 0 \\ 0 \\ 0 \\ 0 \end{bmatrix}$$

(8.64)

Here each of the local matrices from each of the individual elements is overlapped with that of adjacent elements. Equation (8.64) represents the global finite element model of the bar with global coordinates vector $\mathbf{u}(t) = [u_1(t) \quad u_2(t) \quad u_3(t) \quad u_4(t)]^T$ and defines the global mass and stiffness matrices, whereas M_1, M_2, M_3, and K_1, K_2, and K_3 are called local mass and stiffness matrices. The coordinates $u_i(t)$ taken separately in pairs are called local coordinates, while the vector $\mathbf{u}(t)$ defines the global coordinate system.

To finish this modeling exercise, the boundary conditions must be accounted for. Since the bar is clamped at $u_1(t)$, this coordinate is eliminated in equation (8.64) by striking out the row and column associated with it, resulting in

$$M = \frac{\rho AL}{18}\begin{bmatrix} 4 & 1 & 0 \\ 1 & 4 & 1 \\ 0 & 1 & 2 \end{bmatrix} \qquad K = \frac{3EA}{l}\begin{bmatrix} 2 & -1 & 0 \\ -1 & 2 & -1 \\ 0 & -1 & 1 \end{bmatrix}$$

(8.65)

These are in perfect agreement with the global mass and stiffness matrices derived using a complete energy calculation for a three-element model of the clamped bar in Section 8.2 as given in equations (8.26) and (8.27), respectively.

□

It is important to note that the method of superposition used in the example to assemble the global mass and stiffness matrices yields the same results as the more rigorous variational approach used in Section 8.2. Since the superposition method is much easier to use, this approach of assembling global mass and stiffness matrices will be used to examine multielement beam problems. An example is again used to illustrate the procedure.

Example 8.3.3

Derive the global mass and stiffness matrices for a clamped–free beam of Figure 8.7 using two elements and three nodes.

Solution The equations for the first element are obtained directly from equations (8.53) and (8.56) for a general beam element of length l, by replacing l with $l/2$ in these equations.

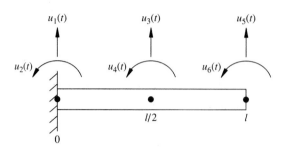

Figure 8.7 Two-element three-node mesh of a cantilevered beam illustrating the nodal coordinates.

Thus the equation for the finite element in Figure 8.7 becomes

$$\frac{\rho A l}{840}\begin{bmatrix} 156 & 11l & 54 & -\frac{13}{2}l \\ 11l & l^2 & \frac{13}{2}l & -\frac{3}{4}l^2 \\ 54 & \frac{13}{2}l & 156 & -11l \\ -\frac{13}{2}l & -\frac{3}{4}l^2 & -11l & l^2 \end{bmatrix}\begin{bmatrix} \ddot{u}_1 \\ \ddot{u}_2 \\ \ddot{u}_3 \\ \ddot{u}_4 \end{bmatrix}$$
$$+ \frac{8EI}{l^3}\begin{bmatrix} 12 & 3l & -12 & 3l \\ 3l & l^2 & -3l & 0.5l^2 \\ -12 & -3l & 12 & -3l \\ 3l & 0.5l^2 & -3l & l^2 \end{bmatrix}\begin{bmatrix} u_1 \\ u_2 \\ u_3 \\ u_4 \end{bmatrix} = \begin{bmatrix} 0 \\ 0 \\ 0 \\ 0 \end{bmatrix} \qquad (8.66)$$

At this point it is possible to apply the clamped boundary condition, as it affects only this first element. The clamped end requires that both the deflection and slope at $x = 0$ must vanish so that $u_1 = u_2 = 0$. Striking out the rows and columns associated with these two coordinates yields

$$\frac{\rho A l}{840}\begin{bmatrix} 156 & -11l \\ -11l & l^2 \end{bmatrix}\begin{bmatrix} \ddot{u}_3 \\ \ddot{u}_4 \end{bmatrix} + \frac{8EI}{l^3}\begin{bmatrix} 12 & -3l \\ -3l & l^2 \end{bmatrix}\begin{bmatrix} u_3 \\ u_4 \end{bmatrix} = \begin{bmatrix} 0 \\ 0 \end{bmatrix} \qquad (8.67)$$

The equations for the second element are identical to equation (8.66) with the vector $[u_1 \ u_2 \ u_3 \ u_4]^T$ replaced with $[u_3 \ u_4 \ u_5 \ u_6]^T$ or

$$\frac{\rho A l}{840}\begin{bmatrix} 156 & 11l & 54 & -\frac{13}{2}l \\ 11l & l^2 & \frac{13}{2}l & \frac{-3l^2}{4} \\ 54 & \frac{13}{2}l & 156 & -11l \\ -\frac{13}{2}l & -\frac{3}{4}l^2 & -11l & l^2 \end{bmatrix}\begin{bmatrix} \ddot{u}_3 \\ \ddot{u}_4 \\ \ddot{u}_5 \\ \ddot{u}_6 \end{bmatrix}$$
$$+ \frac{8EI}{l^3}\begin{bmatrix} 12 & 3l & -12 & 3l \\ 3l & l^2 & -3l & 0.5l^2 \\ -12 & -3l & 12 & -3l \\ 3l & 0.5l^2 & -3l & l^2 \end{bmatrix}\begin{bmatrix} u_3 \\ u_4 \\ u_5 \\ u_6 \end{bmatrix} = 0 \qquad (8.68)$$

Combining equations (8.67) and (8.68) using the superposition of like coordinates yields the global equation

$$\frac{\rho A l}{840}\begin{bmatrix} 312 & 0 & 54 & -6.5l \\ 0 & 2l^2 & 6.5l & -0.75l^2 \\ 54 & 6.5l & 156 & -11l \\ -6.5l & -0.75l^2 & -11l & l^2 \end{bmatrix}\begin{bmatrix} \ddot{u}_3 \\ \ddot{u}_4 \\ \ddot{u}_5 \\ \ddot{u}_6 \end{bmatrix}$$

$$+ \frac{8EI}{l^3} \begin{bmatrix} 24 & 0 & -12 & 3l \\ 0 & 2l^2 & -3l & \frac{1}{2}l^2 \\ -12 & -3l & 12 & -3l \\ 3l & \frac{1}{2}l^2 & -3l & l^2 \end{bmatrix} \begin{bmatrix} u_3 \\ u_4 \\ u_5 \\ u_6 \end{bmatrix} = \mathbf{0} \qquad (8.69)$$

which constitutes the two-element, finite element model of a cantilevered beam. Note that the matrix element in the 1–1 position in equation (8.69) is the sum of the element in the 1–1 position of equation (8.67) and the element in the 1–1 position of equation (8.68). This is also true for the elements in the mass and stiffness matrices in the 2–1, 1–2, and 2–2 positions. These four positions correspond to the common coordinates, u_3 and u_4, between the two finite elements. □

8.4 LUMPED MASS MATRICES

In this section an alternative procedure for constructing the mass matrix is considered. In Section 8.3, the mass matrix was constructed by using the shape functions derived from the static displacement of a given element along with the definition of kinetic energy. Mass matrices constructed in this fashion are called *consistent-mass matrices*, because they are derived from a set of shape functions and displacement functions consistent with the stiffness matrix calculation.

Recall from the exercises and examples that the integrations required for the mass matrix involve higher-order polynomials than those required for the stiffness matrix. Hence an alternative to performing these calculations is to use a lumped-mass approximation. This involves a simple lumping of the mass of the structure at the nodes of the finite element model in proportion to the number of elements in the model. Such mass matrices are called *inconsistent-mass matrices*.

The lumped-mass approach has an advantage in that it generally produces lower-frequency estimates and is very easy to calculate. The lumped-mass matrices are diagonal, making computation easier. However, the lumped-mass method has several disadvantages. First, it can cause errors through a loss of accuracy. If the element under consideration has a rotational coordinate, such as the slope coordinates of the beam element, this coordinate has no mass assigned to it and the resulting mass matrix becomes singular (i.e., M^{-1} does not exist). Such systems require special methods to solve.

The lumped-mass matrix is obtained by simply placing a lumped mass at each node equal to the appropriate proportions of the total mass of the system. For example, consider the bar element of Section 8.1. The total mass of the bar element of length l is ρAl. Placing one-half of this at each of the two nodes yields

$$M = \frac{\rho Al}{2} \begin{bmatrix} 1 & 0 \\ 0 & 1 \end{bmatrix} \qquad (8.70)$$

This is the lumped-mass matrix for the bar element.

Next consider the beam element of Section 8.3. The mass of an element of length l is $\rho A l$. If this mass is divided evenly among the two transverse coordinates (u_1 and u_3), the mass matrix becomes

$$M = \frac{\rho A l}{2} \begin{bmatrix} 1 & 0 & 0 & 0 \\ 0 & 0 & 0 & 0 \\ 0 & 0 & 1 & 0 \\ 0 & 0 & 0 & 0 \end{bmatrix} \tag{8.71}$$

Note that since the rotational coordinates (u_2 and u_4) are not assigned any mass, the diagonal mass matrix has two zeros along its diagonal and hence is singular. The singularity of the mass matrix can cause great difficulties in computing and interpreting the eigenvalues and hence the corresponding natural frequencies. The singular nature of the beam mass matrix can be removed by assigning some inertia to the rotational coordinates u_2 and u_4. This is done by computing the mass moment of inertia of half of the beam element about each of its ends. For a uniform beam this becomes

$$I = \frac{1}{3} \left(\frac{\rho A l}{2} \right) \left(\frac{l}{2} \right)^2 = \frac{\rho A l^3}{24} \tag{8.72}$$

Assigning this inertia to u_2 and u_4, the beam element lumped-mass matrix becomes

$$M = \frac{\rho A l}{2} \text{diag} \left[(1) \quad \left(\frac{l^2}{12} \right) \quad (1) \quad \left(\frac{l^2}{12} \right) \right] \tag{8.73}$$

This diagonal lumped-mass matrix is nonsingular and when combined with the beam stiffness matrix can easily be solved for the system's natural frequencies using the methods of Chapter 4.

Example 8.4.1

Compute the frequencies of a clamped–clamped bar of length l, modulus E, cross section A, and density ρ using two elements and both a consistent-mass and lumped-mass matrix. Compare the natural frequencies between the two systems and to the exact frequencies obtained by the methods of Chapter 6. The bar is illustrated in Figure 8.8.

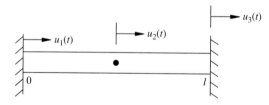

Figure 8.8   Coordinates for a two-element model of a clamped–clamped bar.

Solution   The stiffness and consistent mass matrices for a bar element are given by equations (8.7) and (8.11), respectively. Substituting $l/2$ in for l yields the following equation for each element:

element 1: $$\frac{\rho A l}{12} \begin{bmatrix} 2 & 1 \\ 1 & 2 \end{bmatrix} \begin{bmatrix} \ddot{u}_1 \\ \ddot{u}_2 \end{bmatrix} + \frac{2EA}{l} \begin{bmatrix} 1 & -1 \\ -1 & 1 \end{bmatrix} \begin{bmatrix} u_1 \\ u_2 \end{bmatrix} = \begin{bmatrix} 0 \\ 0 \end{bmatrix} \tag{8.74}$$

element 2: $$\frac{\rho A l}{12} \begin{bmatrix} 2 & 1 \\ 1 & 2 \end{bmatrix} \begin{bmatrix} \ddot{u}_2 \\ \ddot{u}_3 \end{bmatrix} + \frac{2EA}{l} \begin{bmatrix} 1 & -1 \\ -1 & 1 \end{bmatrix} \begin{bmatrix} u_2 \\ u_3 \end{bmatrix} = \begin{bmatrix} 0 \\ 0 \end{bmatrix} \tag{8.75}$$

These two equations are combined to form the global equation by striking out the first column and row of (8.74), because of the clamped boundary (see Figure 8.8) at $u_1(t)$, and the last row and column of equation (8.75) because of the clamped boundary at $u_3(t)$ (i.e., at $x = l$). This yields the single-degree-of-freedom system

$$\frac{\rho A l}{12}[2 + 2]\ddot{u}_2 + \frac{2EA}{l}[1 + 1]u_2 = 0 \tag{8.76}$$

Solving for ω^2 yields

$$\omega = 2\sqrt{3}\sqrt{\frac{E}{\rho l^2}} \cong 3.464\sqrt{\frac{E}{\rho l^2}} \tag{8.77}$$

Next consider making the same calculation with the lumped-mass matrix of equation (8.70). Substitution of $l/2$ for l in equation (8.70) and repeating equations (8.74) and (8.75) yields

element 1:
$$\frac{\rho A l}{4}\begin{bmatrix} 1 & 0 \\ 0 & 1 \end{bmatrix}\begin{bmatrix} \ddot{u}_1 \\ \ddot{u}_2 \end{bmatrix} + \frac{2EA}{l}\begin{bmatrix} 1 & -1 \\ -1 & 1 \end{bmatrix}\begin{bmatrix} u_1 \\ u_2 \end{bmatrix} = \begin{bmatrix} 0 \\ 0 \end{bmatrix} \tag{8.78}$$

element 2:
$$\frac{\rho A l}{4}\begin{bmatrix} 1 & 0 \\ 0 & 1 \end{bmatrix}\begin{bmatrix} \ddot{u}_2 \\ \ddot{u}_3 \end{bmatrix} + \frac{2EA}{l}\begin{bmatrix} 1 & -1 \\ -1 & 1 \end{bmatrix}\begin{bmatrix} u_2 \\ u_3 \end{bmatrix} = \begin{bmatrix} 0 \\ 0 \end{bmatrix} \tag{8.79}$$

Here the lumped-mass matrix is equation (8.70) with l replaced by $l/2$.

Again using the assembly procedure and applying the boundary conditions yields the single-degree-of-freedom model

$$\frac{\rho A l}{4}[1 + 1]\ddot{u}_2 + \frac{2EA}{l}[1 + 1]u_2 = 0 \tag{8.80}$$

Solving this single-degree-of-freedom system for ω yields

$$\omega = 2\sqrt{2}\sqrt{\frac{E}{\rho l^2}} \cong 2.8284\sqrt{\frac{E}{\rho l^2}} \tag{8.81}$$

The first natural frequency of a clamped–clamped bar is given in Chapter 6 as

$$\omega_1 = \pi\sqrt{\frac{E}{\rho l^2}} \cong 3.14159\sqrt{\frac{E}{\rho l^2}}$$

Note that the frequency estimate with the two different mass models are each about 10% away from the actual value as determined by the distributed-parameter model. The inconsistent-mass matrix yields an approximate value that is 10% lower and the consistent-mass matrix yields an estimate that is about 10% higher than the actual value. $\square$

8.5 TRUSSES

The power of finite element analysis is its ability to model complicated structures of odd geometry using simple elements such as bar, beam, and torsional rod elements. While the analysis of such structures is well beyond the scope of this introduction, an analysis of a truss structure is presented here to illustrate the main features of using finite element analysis on a more complicated structure.

Consider the simple truss structure of Figure 8.9. Note in particular that the coordinate system for each of the two elements (u_1, u_2, u_3, and u_4) is pointing in different directions. The truss element model describes vibration only along its axis, while the combined structure can vibrate in both the X and Y directions. To accommodate this situation, a *global coordinate system* aligned with the X–Y coordinate direction is defined. The final model for the full structure will be defined relative to the global X–Y coordinate. This is accomplished by defining global coordinates for each of the structure's nodes. These are denoted by capital U_i in the figure and are called global joint displacements.

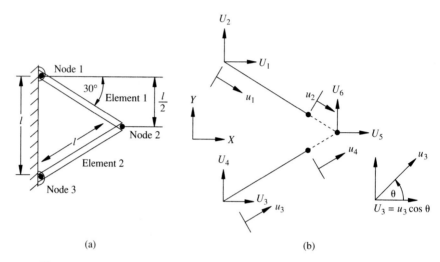

(a)   (b)

Figure 8.9   (a) Two-member framed structure mounted to a wall through pinned connections. Each of the two members is modeled as a bar. The two bars are pinned together. (b) Coordinate system.

The geometric configuration of the frame can be used to establish a relationship between the *local* nodal displacement u_i and the global joint displacements U_i. From the figure, u_3 and u_4 can be related to U_3, U_4, U_5, and U_6 by examining the projections of the global coordinates along the *local coordinate* direction (e.g., u_3 and u_4). This yields the relationships

$$u_3(t) = U_3 \cos \theta + U_4 \sin \theta$$
$$u_4(t) = U_5 \cos \theta + U_6 \sin \theta \tag{8.82}$$

where θ is the angle between the global system, X–Y, and the local coordinate system which is aligned along each of the two bars. Equation (8.82) can be written as the product of a matrix and vector:

$$\begin{bmatrix} u_3(t) \\ u_4(t) \end{bmatrix} = \begin{bmatrix} \cos \theta & \sin \theta & 0 & 0 \\ 0 & 0 & \cos \theta & \sin \theta \end{bmatrix} \begin{bmatrix} U_3(t) \\ U_4(t) \\ U_5(t) \\ U_6(t) \end{bmatrix} \tag{8.83}$$

or written symbolically as

$$\mathbf{u}_2(t) = \Gamma \mathbf{U}_2(t) \tag{8.84}$$

where Γ denotes the coordinate transformation matrix between the local and global coordinate systems and $\mathbf{U}_2$ is that part of the global coordinate vector $\mathbf{U}$ containing those coordinates associated with the second element (i.e., $\mathbf{U}_2 = [U_3 \quad U_4 \quad U_5 \quad U_6]^T$). The vector $\mathbf{u}_2$ is the collection of local coordinates (i.e., $\mathbf{u}_2 = [u_3 \quad u_4]^T$).

The kinetic and potential energy of the element 2 in the figure can now be written in two ways, which must be equivalent. That is, the energy written in terms of either coordinate system must be the same. Equating the strain energy in the local coordinate system with that of the global coordinate system yields

$$V(t) = \tfrac{1}{2}\mathbf{u}^T K_e \mathbf{u} = \tfrac{1}{2}\mathbf{U}^T \Gamma^T K_e \Gamma \mathbf{U} \tag{8.85}$$

where K_e is the element stiffness matrix of the local coordinate system. In this case the element stiffness matrix in the local coordinate system is that of a bar. Hence the stiffness matrix in the global coordinate system for element 2 becomes

$$K_{(2)} = \Gamma^T K_e \Gamma \tag{8.86}$$

Note that the stiffness matrix in the global coordinate systems is the product of the three matrices: the element stiffness matrix, the transformation matrix, and its transpose.

For the example of element 2 of Figure 8.9 the element stiffness matrix in global coordinates becomes

$$K_{(2)} = \frac{EA}{l} \begin{bmatrix} \cos\theta & 0 \\ \sin\theta & 0 \\ 0 & \cos\theta \\ 0 & \sin\theta \end{bmatrix} \begin{bmatrix} 1 & -1 \\ -1 & 1 \end{bmatrix} \begin{bmatrix} \cos\theta & \sin\theta & 0 & 0 \\ 0 & 0 & \cos\theta & \sin\theta \end{bmatrix} \tag{8.87}$$

where K_e, the general bar element stiffness matrix in the local coordinate system, is given by equation (8.7). Performing the indicated matrix products yields

$$K_{(2)} = \frac{EA}{l} \begin{bmatrix} \cos^2\theta & \sin\theta\cos\theta & -\cos^2\theta & -\sin\theta\cos\theta \\ \sin\theta\cos\theta & \sin^2\theta & -\sin\theta\cos\theta & -\sin^2\theta \\ -\cos^2\theta & -\sin\theta\cos\theta & \cos^2\theta & \sin\theta\cos\theta \\ -\sin\theta\cos\theta & -\sin^2\theta & \sin\theta\cos\theta & \sin^2\theta \end{bmatrix} \tag{8.88}$$

which is a 4×4 matrix corresponding to the global coordinates $\mathbf{U}_2$. Following this procedure for the other member of the truss, the global stiffness matrix for element 1 becomes

$$K_{(1)} = \frac{EA}{l} \begin{bmatrix} \cos^2\theta & \sin\theta\cos\theta & -\cos^2\theta & \sin\theta\cos\theta \\ -\sin\theta\cos\theta & \sin^2\theta & \sin\theta\cos\theta & -\sin^2\theta \\ -\cos^2\theta & \sin\theta\cos\theta & \cos^2\theta & -\sin\theta\cos\theta \\ \sin\theta\cos\theta & -\sin^2\theta & -\sin\theta\cos\theta & \sin^2\theta \end{bmatrix} \tag{8.89}$$

which corresponds to the global coordinate vector.

$$\mathbf{U}_{(1)} = \begin{bmatrix} U_1(t) \\ U_2(t) \\ U_5(t) \\ U_6(t) \end{bmatrix} \tag{8.90}$$

To combine the two element matrices in the global coordinates [i.e., $K_{(1)}$ and $K_{(2)}$] into a global matrix in the full global coordinate $\mathbf{U} = [U_1 \quad U_2 \quad U_3 \quad U_4 \quad U_5 \quad U_6]^T$, $K_{(1)}$ is expanded to

$$K'_{(1)} = \frac{EA}{l} \begin{bmatrix} \cos^2\theta & -\sin\theta\cos\theta & 0 & 0 & -\cos^2\theta & \sin\theta\cos\theta \\ -\sin\theta\cos\theta & \sin^2\theta & 0 & 0 & \sin\theta\cos\theta & -\sin^2\theta \\ 0 & 0 & 0 & 0 & 0 & 0 \\ 0 & 0 & 0 & 0 & 0 & 0 \\ -\cos^2\theta & \sin\theta\cos\theta & 0 & 0 & \cos^2\theta & -\sin\theta\cos\theta \\ \sin\theta\cos\theta & -\sin^2\theta & 0 & 0 & -\sin\theta\cos\theta & \sin^2\theta \end{bmatrix} \tag{8.91}$$

Here zeros have been added to those positions corresponding to the missing coordinates U_3 and U_4. Similarly, $K_{(2)}$ is expanded to become compatible with the size of the full global vector $\mathbf{U}$. This yields

$$K'_{(2)} = \frac{EA}{l} \begin{bmatrix} 0 & 0 & 0 & 0 & 0 & 0 \\ 0 & 0 & 0 & 0 & 0 & 0 \\ 0 & 0 & \cos^2\theta & \sin\theta\cos\theta & -\cos^2\theta & -\sin\theta\cos\theta \\ 0 & 0 & \sin\theta\cos\theta & \sin^2\theta & -\sin\theta\cos\theta & -\sin^2\theta \\ 0 & 0 & -\cos^2\theta & -\sin\theta\cos\theta & \cos^2\theta & \sin\theta\cos\theta \\ 0 & 0 & -\sin\theta\cos\theta & -\sin^2\theta & \sin\theta\cos\theta & \sin^2\theta \end{bmatrix} \tag{8.92}$$

The two terms $K'_{(1)}\mathbf{U}$ and $K'_{(2)}\mathbf{U}$ can be added to yield the full global stiffness matrix in the full global coordinate system $\mathbf{U}$.

For the example of Figure 8.9, the sum of equations (8.91) and (8.92) yields the global stiffness matrix (for the case that each rod of the frame has the same physical parameters) defined by

$$K\mathbf{U} = (K'_{(1)} + K'_{(2)})\mathbf{U}$$

$$= \frac{EA}{l} \begin{bmatrix} 0.75 & -0.4330 & 0 & 0 & -0.75 & 0.4330 \\ -0.4330 & 0.25 & 0 & 0 & 0.4330 & -0.25 \\ 0 & 0 & 0.75 & 0.4330 & -0.75 & -0.4330 \\ 0 & 0 & 0.4330 & 0.25 & -0.4330 & -0.25 \\ -0.75 & 0.4330 & -0.75 & -0.4330 & 1.5 & 0 \\ 0.4330 & -0.25 & -0.4330 & -0.25 & 0 & 0.5 \end{bmatrix} \begin{bmatrix} U_1 \\ U_2 \\ U_3 \\ U_4 \\ U_5 \\ U_6 \end{bmatrix} \tag{8.93}$$

Note that the effective stiffness corresponding to coordinates U_5 and U_6 has increased. This corresponds to the point where the two beams join together at a common node. Before equation (8.93) can be used as part of a vibration analysis, the boundary conditions must be applied. Examining Figure 8.9, it is clear that the pinned boundary condition at the connection of the two elements to ground requires that $U_1 = U_2 = U_3 = U_4 = 0$.

Hence after applying the boundary conditions, the global stiffness matrix reduces to

$$K = \frac{EA}{l} \begin{bmatrix} 1.5 & 0 \\ 0 & 0.5 \end{bmatrix} \tag{8.94}$$

which is obtained by deleting those rows and columns of the coefficient matrix in equation (8.93) corresponding to U_1, U_2, U_3, and U_4. Similarly, the global displacement vector reduces to $\mathbf{U} = [U_5 \quad U_6]^T$.

Next consider assembling a consistent mass matrix for this truss from the local mass matrix of the bar element given by equation (8.11) and from the frame geometry captured by the transformation Γ. Following the steps used in determining the global stiffness matrix, the kinetic energy of bar element number i (here $i = 1, 2$) as stated in global coordinates is equated to the kinetic energy of the same element stated in the local coordinate system. Using the notation defined above, this yields

$$T_{(i)} = \tfrac{1}{2}\mathbf{u}^T M_i \mathbf{u} = \tfrac{1}{2}\mathbf{U}^T M_{(i)}\mathbf{U}$$

where M_i is the element mass matrix given by equation (8.11) and Γ is the coordinate transformation between the local coordinate $\mathbf{u}$ and the global coordinates $\mathbf{U}$, as given by equation (8.83). Thus the mass matrix in the global coordinates for the elements is of the form

$$M_{(i)} = \Gamma^T M_i \Gamma$$

Each of these matrices, $M_{(1)}$ and $M_{(2)}$ in the case of the example Figure 8.9, is then expanded by adding zeros as indicated in equations (8.91) and (8.92) for the corresponding stiffness matrices. The expanded matrices are combined (added) as in equation (8.93) to produce a 6×6 matrix corresponding to the full global coordinate system. This 6×6 mass matrix is then collapsed by applying the boundary conditions to the global coordinates to produce a 2×2 mass matrix compatible with the stiffness matrix of equation (8.94).

Alternatively, a lumped-mass matrix, as defined in Section 8.4, can be defined. In this case a reasonable choice for a lumped-mass matrix is to assign the mass of each element to each of the two remaining coordinates. Since the mass of each bar (of length l) is ρAl, a reasonable lumped-mass matrix is

$$M = \rho Al \begin{bmatrix} 1 & 0 \\ 0 & 1 \end{bmatrix} \tag{8.95}$$

Problem TB8.6 at the end of the chapter addresses the difference between the lumped mass matrix as defined by equation (8.95) and the consistent mass matrix suggested above.

The vibration problem using the finite element approach for the simple truss of Figure 8.9 becomes

$$\rho Al \begin{bmatrix} 1 & 0 \\ 0 & 1 \end{bmatrix} \begin{bmatrix} \ddot{U}_5 \\ \ddot{U}_6 \end{bmatrix} + \frac{EA}{l} \begin{bmatrix} 1.5 & 0 \\ 0 & 0.5 \end{bmatrix} \begin{bmatrix} U_5 \\ U_6 \end{bmatrix} = \begin{bmatrix} 0 \\ 0 \end{bmatrix} \tag{8.96}$$

Thus in this case with a lumped-mass matrix the vibration of the truss system of Figure 8.9 is modeled as the independent motion of the tip in the x and y directions of frequency.

$$\omega_1 = \frac{1}{l}\sqrt{\frac{1.5E}{\rho}}$$

$$\omega_2 = \frac{1}{l}\sqrt{\frac{0.5E}{\rho}}$$

The model in Equation 8.96 above provides a crude result in the sense that the motion of U_6 and U_5 are decoupled. The model can be improved by choosing to model each of the two bars with more elements. This is addressed in the toolbox problems at the end of the chapter.

8.6 MODEL REDUCTION

A difficulty with many design and analysis methods is that they work best for systems with a small number of degrees of freedom. Unfortunately, many interesting problems have a large number of degrees of freedom. In fact, to obtain accurate results with finite element models, the number of elements and hence the order of the vibration is increased. Thus finite element models of practical structures and machines are often very large. One approach to this dilemma is to reduce the size of the original model by essentially removing those parts of the model that affect its dynamic response the least. This process is called *model reduction* or *reduced-order modeling*. It is an attempt to reduce the size of an FEM but still retain the dynamic character of the system.

Quite often the mass matrix of a system may be singular or nearly singular, due to some elements being much smaller than others. In fact, in the case of finite element modeling the mass matrix may contain zeros along a portion of the diagonal. Coordinates associated with zero or relatively small mass are likely candidates for being removed from the model. Another set of coordinates that are likely choices for removal from the model are those that do not respond when the structure is excited. Stated another way, some coordinates may have more significant responses than others. The distinction between significant and insignificant coordinates leads to a convenient formulation of the model reduction problem due to Guyan (1965).

Consider the undamped, forced-vibration, finite element model and partition the mass and stiffness matrices according to significant displacements denoted by $\mathbf{u}_1$ and insignificant displacements $\mathbf{u}_2$. This yields

$$\begin{bmatrix} M_{11} & M_{12} \\ M_{21} & M_{22} \end{bmatrix} \begin{bmatrix} \ddot{\mathbf{u}}_1 \\ \ddot{\mathbf{u}}_2 \end{bmatrix} + \begin{bmatrix} K_{11} & K_{12} \\ K_{21} & K_{22} \end{bmatrix} \begin{bmatrix} \mathbf{u}_1 \\ \mathbf{u}_2 \end{bmatrix} = \begin{bmatrix} \mathbf{f}_1 \\ \mathbf{f}_2 \end{bmatrix} \tag{8.97}$$

Note here that the coordinates $\mathbf{u}(t)$ have been rearranged so that those having the least significant displacements associated with them appear last in the displacement vector,

$\mathbf{u} = [\mathbf{u}_1^T \quad \mathbf{u}_2^T]$. Next consider the potential energy of the sysem defined by the scalar $V = \frac{1}{2}\mathbf{u}^T K \mathbf{u}$ or, in partitioned form,

$$V = \frac{1}{2}\begin{bmatrix} \mathbf{u}_1 \\ \mathbf{u}_2 \end{bmatrix}^T \begin{bmatrix} K_{11} & K_{12} \\ K_{21} & K_{22} \end{bmatrix} \begin{bmatrix} \mathbf{u}_1 \\ \mathbf{u}_2 \end{bmatrix} \tag{8.98}$$

Similarly, the kinetic energy of the system can be written as the scalar $T = \frac{1}{2}\dot{\mathbf{u}}^T M \dot{\mathbf{u}}$, which becomes

$$T = \frac{1}{2}\begin{bmatrix} \dot{\mathbf{u}}_1 \\ \dot{\mathbf{u}}_2 \end{bmatrix}^T \begin{bmatrix} M_{11} & M_{12} \\ M_{21} & M_{22} \end{bmatrix} \begin{bmatrix} \dot{\mathbf{u}}_1 \\ \dot{\mathbf{u}}_2 \end{bmatrix} \tag{8.99}$$

in partioned form. Since each coordinate u_i is acted on by a force f_i, the condition that there is no force in the direction of the insignificant coordinates $\mathbf{u}_2$ requires that $\mathbf{f}_2 = 0$ and that $\partial V / \partial \mathbf{u}_2 = 0$. This yields

$$\frac{\partial}{\partial \mathbf{u}_2}(\mathbf{u}_1^T K_{11}\mathbf{u}_1 + \mathbf{u}_1^T K_{12}\mathbf{u}_2 + \mathbf{u}_2^T K_{21}\mathbf{u}_1 + \mathbf{u}_2^T K_{22}\mathbf{u}_2) = \mathbf{0} \tag{8.100}$$

Hence the constraint relation between $\mathbf{u}_1$ and $\mathbf{u}_2$ must be (since $K_{12} = K_{21}^T$)

$$\mathbf{u}_2 = -K_{22}^{-1} K_{21}\mathbf{u}_1 \tag{8.101}$$

Expression (8.101) suggests a coordinate transformation (which is not a similarity transformation) from the full coordinate system to the reduced coordinate system $\mathbf{u}_1$. If the transformation matrix Q is defined by

$$Q = \begin{bmatrix} I \\ -K_{22}^{-1} K_{21} \end{bmatrix} \tag{8.102}$$

then if $\mathbf{u} = Q\mathbf{u}_1$ is substituted into equation (8.97) and this expression is premultiplied by Q^T, a new reduced-order system of the form

$$Q^T MQ \ddot{\mathbf{u}}_1 + Q^T KQ \mathbf{u}_1 = Q^T \mathbf{f} \tag{8.103}$$

results. The vector $Q^T \mathbf{f}$ now has the dimension of $\mathbf{u}_1$. Equation (8.103) represents the reduced-order form of equation (8.97), where

$$Q^T MQ = M_{11} - K_{21}^T K_{22}^{-1} M_{21} - M_{12} K_{22}^{-1} K_{21} + K_{21}^T K_{22}^{-1} M_{22} K_{22}^{-1} K_{21} \tag{8.104}$$

and

$$Q^T KQ = K_{11} - K_{12} K_{22}^{-1} K_{21} \tag{8.105}$$

These last expressions are frequently used to reduce the order of finite element vibration models in a systematic and consistent manner. Such model reduction schemes are used when a finite element model has coordinates (represented by $\mathbf{u}_2$) that do not contribute substantially to the response of the system. Model reduction can greatly simplify design and analysis problems under certain circumstances.

   If some of the masses in the system are negligible or zero, the preceding formulas can be used to reduce the order of the vibration problem simply by setting $M_{22} = 0$ in equation (8.104). This is essentially the model reduction technique referred to as *mass condensation*.

Example 8.6.1

Consider a four-degree-of-freedom system finite element model with mass matrix

$$M = \frac{1}{420} \begin{bmatrix} 312 & 54 & 0 & -13 \\ 54 & 156 & 13 & -22 \\ 0 & 13 & 8 & -3 \\ -13 & -22 & -3 & 4 \end{bmatrix}$$

and stiffness matrix

$$K = \begin{bmatrix} 24 & -12 & 0 & 6 \\ -12 & 12 & -6 & -6 \\ 0 & -6 & 2 & 4 \\ 6 & -6 & 4 & 4 \end{bmatrix}$$

Note that this system is both dynamically and statically coupled. To remove the effect of the last two coordinates, the submatrices of equation (8.97) are easily identified as

$$M_{11} = \frac{1}{420} \begin{bmatrix} 312 & 54 \\ 54 & 156 \end{bmatrix} \qquad M_{12} = \frac{1}{420} \begin{bmatrix} 0 & -13 \\ 13 & -22 \end{bmatrix} = M_{21}^T$$

$$M_{22} = \frac{1}{420} \begin{bmatrix} 8 & -3 \\ -3 & 4 \end{bmatrix} \qquad K_{22} = \begin{bmatrix} 2 & 4 \\ 4 & 4 \end{bmatrix}$$

$$K_{11} = \begin{bmatrix} 24 & -12 \\ -12 & 12 \end{bmatrix} \qquad K_{12} = \begin{bmatrix} 0 & -6 \\ -6 & -6 \end{bmatrix} = K_{21}^T$$

Using equations (8.104) and (8.105) yields

$$Q^T M Q = \begin{bmatrix} 0.9071 & -0.0357 \\ -0.0357 & 0.2357 \end{bmatrix}$$

$$Q^T K Q = \begin{bmatrix} 33 & -3 \\ -3 & 3 \end{bmatrix}$$

These matrices form the resulting reduced-order model of the structure.

$\square$

PROBLEMS

Section 8.1

8.1. Consider the one-element model of a bar discussed in Section 8.1. Calculate the finite element of the bar for the case that it is free at both ends rather than clamped.

8.2. Calculate the natural frequencies of the free–free bar of Problem 8.1. To what does the first natural frequency correspond? How do these values compare with the exact values obtained form the methods of Chapter 6?

8.3. Consider the system of Figure 8.10, consisting of a spring connected to a clamped–free bar. Calculate the finite element model and discuss the accuracy of the frequency prediction of this model by comparing it with the method of Chapter 6.

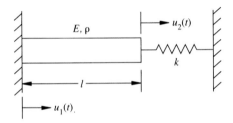

Figure 8.10   One-element model of a cantilevered bar connected to a spring.

8.4. Consider a clamped–free bar with a force $f(t)$ applied in the axial direction at the free end as illustrated in Figure 8.11. Calculate the equations of motion using a single-element finite element model.

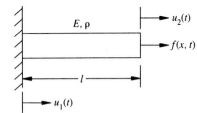

Figure 8.11   A cantilevered bar with an externally applied axial force.

8.5. Compare the solution of a cantilevered bar modeled as a single finite element with that of the distributed-parameter method summarized in Figure 8.1 truncated at three modes by calculating (a) $u(x, t)$ and (b) $u(l/2, t)$ for a 1-m aluminum beam at $t = 0.1, 1$, and 10 s using both methods. Use the initial condition $u(x, 0) = 0.1x$ m and $u_t(x, 0) = 0$.

8.6. Repeat Problem 8.5 using a five-mode model. Can you draw any conclusions?

8.7. Repeat Problem 8.5 using only the first mode in the series solution and the initial condition $u(x, 0) = 0.1 \sin(\pi x / 2l)$, $u_t(x, 0) = 0$. For this initial condition, the first mode is exact. Why?

Section 8.2

8.8. Consider the bar of Figure 8.10 and model the bar with two elements. Calculate the frequencies and compare them with the solution obtained in Problem 8.3. Assume material properties of aluminum, a cross-sectional area of 1 m and a spring stiffness of $1 \times 10^6 \text{N/m}$.

8.9. Repeat Problem 8.8 with a three-element model. Calculate the frequencies and compare them with those of Problem 8.8.

8.10. Consider Example 8.2.2. Repeat this example with node 2 moved to $l/2$ so that the mesh is uniform. Calculate the natural frequencies and compare them to those obtained in the example. What happens to the mass matrix?

8.11. Compare the frequencies obtained in Problem 8.10 with those obtained in Section 8.2 using three elements.

8.12. As mentioned in the text, the usefulness of the finite element method rests in problems that cannot readily be solved in closed form. To this end, consider a section of an air frame sketched in Figure 8.12 and calculate a two-element finite model of this structure (i.e., find M and K) for a bar with

$$A(x) = \frac{\pi}{4}\left[h_1^2 + \left(\frac{h_2 - h_1}{l}\right)^2 x^2 + 2h_1\left(\frac{h_2 - h_1}{l}\right)x\right]$$

8.13. Let the bar in Figure 8.12 be made of aluminum 1 m in length with $h_1 = 20$ cm and $h_2 = 10$ cm. Calculate the natural frequencies using the finite element model of Problem 8.12.

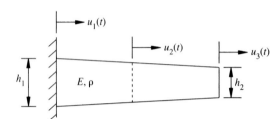

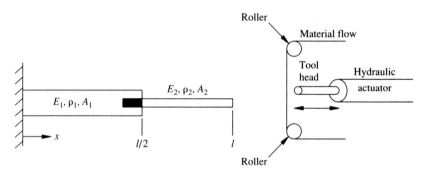

Figure 8.12 Tapered bar model of a wing section in longitudinal vibration.

8.14. Repeat Problems 8.12 and 8.13 using a three-element four-node finite element model.

8.15. Consider the machine punch of Figure 8.13. This punch is made of two materials and is subject to an impact in the axial direction. Use the finite element with two elements to model this system and estimate (calculate) the first two natural frequencies. Assume $E_1 = 8 \times 10^{10}$ Pa, $E_2 = 2.0 \times 10^{11}$ Pa, $\rho_1 = 7200$ kg/m^3, $\rho_2 = 7800$ kg/m^3, $l = 0.2$ m, $A_1 = 0.009$ m^2 and $A_2 = 0.0009$ m^2.

Figure 8.13 Bar made of two materials of two sizes used for a punch. Material passes around the rollers. At appropriate points on the material, the roller stops and the toolhead is actuated in the x direction and impacts the bar, causing a hole to be punched in the material. The punch itself (E_2) is made of hardened steel while the base is made of cast iron (E_1).

8.16. Recalculate the frequencies of Problem 8.15 assuming that it is made entirely of one material and size (i.e., $E_1 = E_2$, $\rho_1 = \rho_2$, and $A_1 = A_2$), say steel, and compare your results to those of Problem 8.15.

8.17. A bridge support column is illustrated in Figure 8.14. The column is made of concrete with a cross-sectioned area defined by $A(x) = A_0 e^{-x/l}$, where A_0 is the area of the column at ground. Consider this pillar to be cantilevered (i.e., fixed) at ground level and to be excited

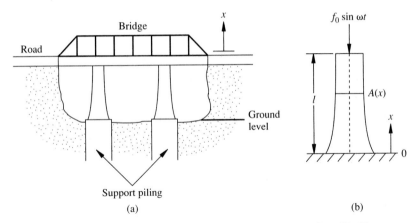

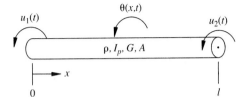

Figure 8.14   (a) Schematic of a highway bridge over a ravine. Traffic over the bridge causes a harmonic motion in the up-and-down direction (labeled x here). (b) Vibration of the column holding up the bridge. The pillars are modeled as bars of odd cross section.

sinusoidally at its tip in the longitudinal direction due to traffic over the bridge. Calculate a single element finite element model of this system and compute its approximate natural frequency.

8.18. Redo Problem 8.17 using two elements. What would happen if the "traffic" frequency corresponds with one of the natural frequencies of the support column?

8.19. Problems 8.17 and 8.18 represent approximations. As pointed out in Problem 8.18, it is important to know the natural frequencies of this column as precisely as possible. Hence consider modeling this column as a uniform bar of average cross section, calculate the first few natural frequencies, and compare them to the results in Problems 8.17 and 8.18. Which model do you think is closest to reality?

8.20. Torsional vibration can also be modeled by finite elements. Referring to Figure 8.15, calculate a single-element mass and stiffness matrix for the torsional vibration following the steps of Section 8.1. (*Hint:* $\theta(x, t) = c_1(t)\theta + c_2(t)$, $T(t) = \frac{1}{2}\int_0^l \rho I_p[\theta_t(x, t)]^2 dx$ and $V(t) = \frac{1}{2}\int_0^l GI_p[\theta_x(x, t)]^2 dx$.)

Figure 8.15   Coordinate system used for finite element analysis of torsional vibration.

Section 8.3

8.21. Use equations (8.47) and (8.46) to derive equation (8.48) and hence make sure that the author and reviewer have not cheated you.

8.22. It is instructive, though tedious, to derive the beam element deflection given by equation (8.49). Hence derive the beam shape functions.

8.23. Using the shape functions of Problem 8.22, calculate the mass and stiffness matrices given by equations (8.53) and (8.56). Although tedious, this involves only simple integration of polynominals in x.

8.24. Calculate the natural frequencies of the cantilevered beam given in equation (8.69) using $l = 1$ m and compare your results with those listed in Table 6.1.

8.25. Calculate the finite element model of a cantilevered beam one meter in length using three elements. Calculate the natural frequencies and compare them to those obtained in Problem 8.23 and with the exact values listed in Table 6.4.

8.26. Consider the cantilevered beam of Figure 8.16 attached to a lumped spring–mass system. Model this system using a single finite element and calculate the natural frequencies. Assume $m = (\rho A l)/420$.

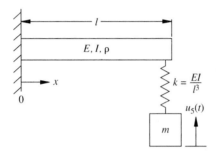

Figure 8.16 Cantilevered beam with a spring–mass system attached to its end.

8.27. Repeat Exercise 8.26 using two finite elements for the beam and compare the frequencies.

8.28. Calculate the natural frequencies of a clamped–clamped beam for the physical parameters: $l = 1$ m, $E = 2 \times 10^{11}$ N/m^2, $\rho = 7800$ kg/m^3, $I = 10^{-6}$ m^4, and $A = 10^{-2}$ m^2, using beam theory of Chapter 6 and a four-element finite element model of the beam.

8.29. Repeat Problem 8.28 with two elements and compare the frequencies with the four-element model. Calculate the frequencies of a clamped–clamped beam using one element. Any comment?

8.30. Estimate the first natural frequency of a clamped–simply supported beam. Use a single finite element.

8.31. Consider the stepped beam of Figure 8.17 clamped at each end. Both pieces are made of aluminum. Use two elements, one for each step, and calculate the natural frequencies.

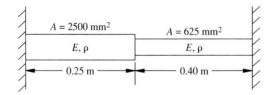

Figure 8.17 Clamped two-step aluminum beam.

8.32. Use a two-element model of nonuniform length to estimate the first few natural frequencies of a clamped–clamped beam. Use the spacing indicated in Figure 8.18. Compare the result to the actual frequencies and to those of Problems 8.28 and 8.29.

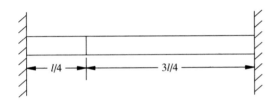

Figure 8.18 Clamped beam modeled with two nonuniform finite elements.

8.33. Calculate the first natural frequency of a clamped–pinned beam using first one, then the two elements.

Section 8.4

8.34. Refer to the tapered bar of Figure 8.12. Calculate a lumped-mass matrix for this system and compare it to the solution of Problem 8.13. Since the beam is tapered, be careful how you divide up the mass.

8.35. Calculate and compare the natural frequencies obtained for a tapered bar by using first, the consistent mass matrix (Problem 8.12), and second, the lumped-mass matrix (Problem 8.34).

8.36. Consider again the machine punch of Problem 8.16 and Figure 8.13. Calculate the natural frequencies of this system using a lumped matrix and compare the results to those obtained with the consistent-mass matrix.

8.37. Consider again the bridge support of Figure 8.14 discussed in connection with Problem 8.17. Develop a four-element finite element model of this structure using a lumped-mass approximation and calculate the natural frequencies. Use constant area elements.

8.38. Consider the torsional vibration problem illustrated in Figure 8.15 and discussed in Problem 8.20. Calculate a lumped-mass matrix for the single element.

8.39. Estimate the first three natural frequencies of a clamped–free bar of length l in torsional vibration by using a lumped-mass model and four elements.

8.40. Calculate the natural frequencies of pinned–pinned beam of length l using one element and the consistent mass matrix of equation (8.73).

8.41. Calculate the natural frequencies of pinned–pinned beam of length l using one element and the lumped-mass matrix of equation (8.73). Compare your results to those obtained with a consistent-mass matrix of Problem 8.40.

8.42. Calculate a three-element finite element model of a cantilevered beam (see Problem 8.25) using a lumped mass that includes rotational inertia. Also calculate the system's natural frequencies and compare them with those obtained with a consistent-mass matrix of Problem 8.25 and with the values obtained by the methods of Chapter 6.

8.43. Repeat Problem 8.42 using a lumped-mass matrix that neglects the rotational degree of freedom. Discuss any problem you encounter when trying to solve the related eigenvalue problem.

Section 8.5

8.44. Derive a consistent-mass matrix for the system of Figure 8.9. Compare the natural frequencies of this system with those calculated with the lumped-mass matrix computed in Section 8.5.

8.45. Consider the two beam system of Figure 8.19. Use VTB8_1 to create a two-element, rod/beam element model and compute the first three natural frequencies. Use $A = 0.0004 \text{ m}^2$, $I = 1.33 \times 10^{-8} \text{ m}^4$ and the properties of aluminum. Assume that nodes 1 and 3 are clamped.

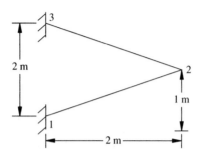

Figure 8.19 Clamped two-element beam system.

8.46. Follow the procedure of Problem 8.45 using two elements for each beam. Compare the natural frequencies and mode shapes of the four element model produced here to those of the two-element model of Problem 8.45. State which model is better and why.

8.47. Determine a finite element model of the three-bar truss of Figure 8.20 using a lumped-mass matrix.

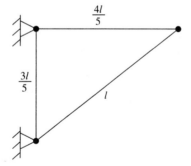

Figure 8.20 Pinned three-element truss.

8.48. Determine a finite element model for the three-bar truss of Figure 8.20 using a consistent-mass matrix.

8.49. Compare the frequencies obtained for the system of Problem 8.48 with those of Problem 8.47.

Section 8.6

8.50. Consider the machine punch of Figure 8.13. Recalculate the fundamental natural frequency by reducing the model obtained in Problem 8.16 to a single degree of freedom using Guyan reduction.

8.51. Compute a reduced-order model of the three-element model of a cantilevered bar given in Example 8.3.2 by eliminating $\mathbf{u}_2$ and $\mathbf{u}_3$ using Guyan reduction. Compare the frequencies of each model to those of the distributed model given in Window 8.1.

8.52. Consider the system defined by the matrices

$$M = \begin{bmatrix} 2 & 0 & 0 & 0 \\ 0 & 0 & 0 & 0 \\ 0 & 0 & 2 & 0 \\ 0 & 0 & 0 & 0 \end{bmatrix} \qquad K = \begin{bmatrix} 20 & -1 & 0 & 0 \\ -1 & 20 & -3 & 0 \\ 0 & -3 & 20 & -17 \\ 0 & 0 & -17 & 17 \end{bmatrix}$$

Use mass condensation to reduce this to a two-degree-of-freedom system with a nonsingular mass matrix.

8.53. Recall the punch press problem modeled in Figure 4.27 and treated in Example 4.8.3. The mass and stiffness matrices are given by

$$M = \begin{bmatrix} 0.4 \times 10^3 & 0 & 0 \\ 0 & 2.0 \times 10^3 & 0 \\ 0 & 0 & 8.0 \times 10^3 \end{bmatrix} \qquad K = \begin{bmatrix} 30 \times 10^4 & -30 \times 10^4 & 0 \\ -30 \times 10^4 & 38 \times 10^4 & -8 \times 10^4 \\ 0 & -8 \times 10^4 & 88 \times 10^4 \end{bmatrix}$$

Recalling that the only external force acting on the machine is at the $x_1(t)$ coordinate, reduce this to a single-degree-of-freedom system using Guyan reduction to remove x_2 and x_3. Compare this single frequency with those of Example 4.8.3.

MATLAB VIBRATION TOOLBOX

If you have not yet used the *Vibration Toolbox* program, return to the end of Chapter 1 for a brief introduction to MATLAB files or open the file "INTRO.TXT" on the disk on the inside back cover. The finite element modeling method introduced in this chapter is ideally suited for computer implementation. Toolbox folder/directory VTB_8 contains a finite element program. This interactive program calls for the user to input node locations, followed by the physical parameters between the nodes (i.e., E, A, I, G, and ρ). The code is based on a two-dimensional Timoshenko beam. The file contains an example of how to use the program for an aluminum beam with five elements. The following problems should help clarify the finite element procedure.

TOOLBOX PROBLEMS

TB8.1. Use the file VTB_8 to solve for the frequencies of a cantilevered aluminum bar using five elements. Compare the frequencies to those of the analytical solution of the distributed parameter model.

TB8.2. Use file VTB_8 to solve for the natural frequencies of a pinned–pinned beam of Window 8.3 using 10 elements. Compare your results with the analytical solution.

TB8.3. Recalculate the frequencies of the machine punch of Figure 8.13 using four elements for each bar. How does your result compare with those obtained in Problem 8.16?

TB8.4. Write a MATLAB file to perform mass condensation.

TB8.5. Use file VTB_8 to calculate the consistent-mass matrix of the two-bar truss system of Section 8.5. Compute the frequencies of this system and compare them to those calculated in Section 8.5.

TB8.6. Use VTB_8 to create a finite element model of the bridge in Figure 8.21. The bridge is made of steel, and the beam cross sections are 0.1 m wide and 0.15 m high. The x and y displacements of nodes 1 and 5 are constrained to be zero. Find the first seven natural frequencies and mode shapes. Removing the center diagonal members, find the first seven natural frequencies and mode shapes again. What is the most significant result of removing the center diagonal members? Can you explain why the seventh natural frequency and mode shape do not change?

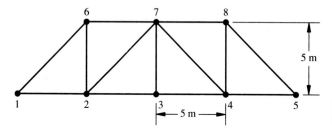

Figure 8.21   A 13 element bridge model.

9 Computational Considerations

This chapter focuses on the numerical aspects of computing vibration solutions. Advanced methods of computing and the efficiency of computations become extremely important for complicated structures such as the Hubble telescope pictured here. The telescope's solar panels initially experienced unacceptable levels of vibration. NASA engineers were quickly able to find a solution, but the computations required were too large to fit on the Hubble's on-board computer. Once efficient numerical computations were considered, the problem was solved and the Hubble performed as desired. On a more down to earth scale, parts such as the football helmet pictured here are numerically analyzed for shock and vibration performance. Various numerical methods of computing the natural frequencies of vibrating systems are discussed in this chapter. Some are more efficient than others.

In most of the preceding chapters, the vibration analysis was performed for the purpose of predicting the response of a structure or machine given input forces and/or initial conditions. This chapter discusses methods of obtaining the response of a vibrating system using a computing device (i.e., digital computer). Such methods are referred to as numerical methods and address both the calculation of the time response of a system (simulation) and the calculation of a system's natural frequencies, damping ratios, and mode shapes. Thus the numerical methods discussed here are those that address numerical integration and those that address the computational algebraic eigenvalue problem.

It is assumed here that the reader has access to a computer and probably a favorite programming language available. Many, if not all, of the methods discussed here are available in public-domain software packages as "canned" programs. Hence no large computer programs, per se, are presented. However, for those without a preferred software package, a disk of files using the MATLAB language is located on the inside back cover. The files are keyed to this book as noted at the end of each of the preceding chapters. In addition, an inexpensive software/text package of MATLAB is available from the publishers (*Student Edition of* MATLAB, 1992). Thus rather than focusing on extensive programs, the material that follows addresses how these numerical approaches are used in vibration analysis and design, as well as on using MATLAB to solve vibration problems. This chapter uses the MATLAB code in examples and problems. If you have not yet used the *Vibration Toolbox* program, return to the end of Chapter 1 for a brief introduction to using MATLAB files or open the file/directory INTRO.TXT on the disk on the inside back cover.

In Section 4.2, Example 4.2.5 illustrates the results of computing natural frequencies using a cheap, solar-powered calculator. By using a limited computational device, the example illustrates just how numerical error can show up in vibration computations. In the case of Example 4.2.4, numerical error causes the matrix product $P^T P$ to be not quite equal to the identity matrix as the theory predicts, and the matrix $\Lambda = P^T K P$ to have diagonal elements not quite equal to the squares of the natural frequencies. This same problem was computed again in Example 4.3.2 using a more reasonable computing device (a PC with MATLAB), and the results for the same problem clearly cause $P^T P$ to be the identity matrix and Λ to be a diagonal matrix of the squares of the natural frequencies as they should. Hence computational power can have a significant effect on vibration analysis.

The first section in this chapter discusses some general ideas on numerical errors. Sections 9.2 to 9.6 examine the computation of natural frequencies and mode shapes.

The last section concentrates on methods of simulating the solution, or time response, of the structure or device of interest. Sections 9.5, 9.6 and 9.7 may be studied independently of the first four sections by readers interested only in computing modal information and time responses.

9.1 NUMERICAL ERRORS

The range of vibration problems that can be solved exactly is very limited. In the early chapters of this book (and in most books) the values of the system's parameters are chosen so that the required arithmetic operations result in frequencies and amplitudes that are computed by integer arithmetic or other simplifications, such as taking the square root of a perfect square. Most interesting problems are likely to involve evaluating numbers such as $\sqrt{1.57/\pi}$ and $e^{\pi/2}$, which must be computed using digital computations on a calculator or computer. Such numbers can be computed only approximately. In the "good old days" of slide rules, these types of errors were more obvious. Modern computational devices tend to generate a false feeling of exactness. While it is well beyond the scope of this book to introduce binary arithmetic and error analysis, some simple ideas are presented to help the reader understand potential sources of error when using computer codes and calculators to solve vibration problems.

Any computing machine from calculators to large supercomputers has a finite amount of space representing a number. Unfortunately, many of the numbers of interest in engineering are irrational numbers having infinite representations (e.g., π, e, $\sqrt{2}$). The designer of each computing device must decide on the most efficient method of representing numbers in the finite record length of the computer so that an acceptable compromise is obtained between the range of numbers available (how long or small) and the accuracy of their representation. This is accomplished following the idea of scientific notation (e.g., 1.3×10^2) using a floating-point number system.

In general, a number x can be represented in terms of an arbitrary base β as

$$a_n\beta^n + a_{n-1}\beta^{n-1} + \cdots + a_1\beta + a_0 + b_1\beta^{-1} + b_2\beta^{-2} + \cdots + b_m\beta^{-m}$$

where each of the coefficients a_i and b_j is an integer between 0 and $\beta - 1$. In floating-point form this number x is expressed as

$$x = f\beta^N \tag{9.1}$$

where f, called the *mantissa*, satisfies $\beta^{-1} \le f < 1$ and where N is an integer. Such a floating-point representation is called normalized since $f < 1$. In this case equation (9.1) is replaced by an approximation in which the mantissa is represented by a fixed number of significant figures (in the base β system). This approximation is the first source of error in computing numbers using a digital computer.

The fitting of an irrational number into the finite representation as a fixed number of "significant" β-based figures requires a scheme for eliminating part of the number. In one scheme, called *chopping*, a real (positive) number is represented as the largest machine number that does not exceed it. Another rule for abbreviating a real number, called *symmetric rounding*, represents a number by the nearest machine number. As an example, the number e (the exponential $e = 0.2718281828\ldots \times 10^1$), with symmetric rounding in a machine with five significant digits would be 0.27183×10^1, whereas with chopping e would be represented as 0.27182×10^1. It should be noted that exceptions are made in most current computing machines and programming languages so that quantities known to be integers become represented in their exact binary form. This usually diminishes the introduction of additional errors in subsequent calculations.

Example 9.1.1

Recalling the error introduced in the natural frequency calculation of Example 4.2.5, by using only four significant figures, consider performing some simple arithmetic using four significant digits. Compute the number

$$x = \frac{1.234}{0.1234} - \frac{1.234}{0.1233} \tag{9.2}$$

using four significant figures, first by computing the indicated quotients and then subtracting, and second, by factoring out the common factor, that is, by

$$x = 1.234 \left(\frac{1}{0.1234} - \frac{1}{0.1233} \right) \tag{9.3}$$

and calculating the difference first, then multiplying.

Solution Preceding by following the formula suggested in equation (9.2) yields (here $\approx$ implies rounded to four digits)

$$\frac{1.234}{0.1234} \approx 0.1000 \times 10^2 \tag{9.4}$$

and

$$\frac{1.234}{0.1233} \approx 0.1001 \times 10^2 \tag{9.5}$$

Subtracting equation (9.5) yields

$$x = -0.0001 \times 10^2 = -0.1000 \times 10^{-1} \tag{9.6}$$

where the rightmost number is in normalized form. The correct result, however, is $x = -0.8110 \times 10^{-2}$!

Next consider performing the same results using the formula of equation (9.3) to calculate x. This results in

$$\frac{1}{0.1234} - \frac{1}{0.1233} \approx 8.104 - 8.110 \approx -0.6000 \times 10^{-2}$$

Multiplying by 1.234 yields

$$x = -0.7404 \times 10^{-2}$$

which is significantly closer to the correct result of -0.8110×10^{-2}.

$\square$

Example 9.1.1 illustrates two major points. First, there is error associated with storing a number in a computer that has more significant digits than that available in the machine and that this error is increased during arithmetic operations. Second, the size of the error resulting in a machine calculation depends a great deal on the order in which the computations are made. This is intended to underscore the idea that the algorithm used to compute a number (such as a natural frequency) can have a very important impact on the accuracy of the result as algorithms dictate the order in which computations are made. While algorithm design is not discussed in this book, it is important for the user of vibration calculations on computers to understand the errors that may result from the use of different algorithms to compute the same frequencies, mode shapes, or simulations.

Another important concept in computer calculations is that of *overflow* and *underflow*. This can best be understood by considering a machine that effectively assigns one decimal digit for the exponent of a floating-point number. Thus, if 0.3000×10^5 is multiplying by 0.1234×10^6, the number 0.3702×10^{10} results, which is too large to fit in the machine. This is called *overflow*. Similarly, if a number such as 0.1234×10^{-12} results from a computation, this is too small to fit in the machine (it would read this as zero) and *underflow* results.

The errors discussed above are usually referred to as *rounding errors*. Another source of error, called *truncation error*, results from truncating a numerical process. This usually occurs in simulation work, where a finite number of terms are used to represent an infinite series (e.g., Section 3.3). For example, the divergent infinite sum

$$\sum_{n=1}^{\infty} \frac{1}{n}$$

appears to converge to a finite value, for a machine with finite accuracy. For example, suppose that $1/200 = 0.005$ is the smallest number in the machine. Then to two decimal places, every number smaller than 0.005 is zero and the infinite sum becomes

$$\sum_{n=1}^{\infty} \frac{1}{n} \approx \sum_{n=1}^{200} \frac{1}{n} \approx 6.16$$

which is not a very good approximation of infinity!

The above is a brief introduction to some of the sources of error in using computers to calculate vibration properties and in solving design problems. The reader interested in a more serious treatment should consult a proper treatment of numerical methods such as Forsyth et al. (1977).

The following sections encourage the use of preprogrammed or "canned" computer routines for making vibration calculations. On the other hand, the preceding examples might tend to destroy the reader's confidence in using canned programs. The purpose here is not to destroy confidence in such programs but rather to encourage wise and reasonable use of commercially available software for calculating vibration quantities and performing simulations. Whenever you use the results from calculations, they should be checked to make sure that they make sense.

9.2 INFLUENCE COEFFICIENTS AND DUNKERLEY'S FORMULA

This section presents a simple method of approximating the lowest natural frequency of a complicated system without solving the eigenvalue problem. The method uses influence coefficients and a result from matrix theory. The influence coefficients of an n-degree-of-freedom system are defined in terms of the system's stiffness matrix. The relationship between force and displacement for a single-degree-of-freedom spring system is simply $f = kx$. Solving this for x yields $x = (1/k)f$. The reciprocal of the stiffness $a = 1/k$ is defined as the spring's *influence coefficient* for a single-degree-of-freedom system. If more than one spring is present, as in Figure 1.21, the influence coefficient becomes $a = 1/k_{eq}$.

The idea of an influence coefficient can be extended to a multiple-degree-of-freedom system with static deflection equation

$$\mathbf{f} = K\mathbf{x} \tag{9.7}$$

where K is the $n \times n$ stiffness matrix and $\mathbf{x}$ is the $n \times 1$ vector of static deflections resulting from application of the $n \times 1$ (static) force vector $\mathbf{f}$. If the matrix K in equation (9.7) is nonsingular (i.e., no rigid-body motion is present), both sides of equation (9.7) can be multiplied by K^{-1} to yield

$$\mathbf{x} = K^{-1}\mathbf{f} = A\mathbf{f} \tag{9.8}$$

The coefficient matrix $A = K^{-1}$ is defined to be the *flexibility influence matrix* and its elements a_{ij} are called the flexibility influence coefficients. The elements of the matrix A can be determined without calculating the inverse of the stiffness matrix K and hence avoiding some of the numerical difficulties suggested in Section 9.1. The coefficients a_{ij} are determined by considering one of the n equations represented by the matrix equation (9.8). The deflection of the ith coordinate, $\mathbf{x}$ denoted x_i (see Window 9.1), is

$$x_i = \sum_{j=1}^{n} a_{ij} f_j \tag{9.9}$$

If the vector of force $\mathbf{f}$ is chosen such that only a unit force is applied at location j, then $\mathbf{f} = [0 \quad 0 \quad \cdots \quad 1 \quad \cdots \quad 0 \quad 0]^T$, where the 1 is in the jth position, then equation (9.9) reduces to

$$x_i = a_{ij} \tag{9.10}$$

Window 9.1

Do not be confused between $\mathbf{x}_i$ and x_i. The bold character $\mathbf{x}_i$ denotes the ith vector of a set (i.e., one of $\{\mathbf{x}_1, \mathbf{x}_2, \mathbf{x}_3, \ldots \mathbf{x}_n\}$), whereas the italic character x_i denotes the ith element of a vector $\mathbf{x}$.

Thus the flexibility influence coefficient a_{ij} is the deflection at point i due to a unit force applied at point j. The following example illustrates the calculation of the flexibility influence A.

Example 9.2.1

Consider the three-degree-of-freedom system of Figure 9.1 and determine the flexibility influence coefficients using the concept of static deflection and equation (9.10).

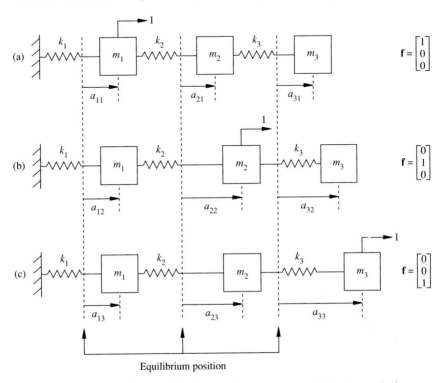

Figure 9.1   Three-degree-of-freedom spring–mass system indicating static deflections by unit forces applied at each mass in turn.

Solution   In the figure, the displacement under static application of the unit force on mass m_1 is denoted by x_1, that of m_2 is x_2, and that of m_3 is denoted by x_3. The dashed line indicates the equilibrium position with no force applied. From Figure 9.1, a unit force applied at m_1 causes the spring k_1 to deflect a distance a_{11} as indicated. Hence

$$a_{11} = \frac{1}{k_1}$$

Since the force $\mathbf{f} = [\,1 \quad 0 \quad 0\,]^T$ is applied statically, the springs k_2 and k_3 do not compress or extend, so that the masses m_2 and m_3 must move the same amount as m_1 does. Hence $a_{21} = a_{31} = a_{11} = 1/k_1$, as indicated in the figure.

Next consider applying a unit force to mass m_2. The resulting deflection is indicated in Figure 9.1b. In this configuration the two springs k_1 and k_2 act on m_2 to provide resistance to the static unit force as if mass m_1 were not there. Thus the stiffness experienced at m_2 is that of two springs in series, so that (recall Figure 1.27)

$$a_{22} = \frac{1}{k_{eq}} = \frac{1}{k_1} + \frac{1}{k_2}$$

The mass m_3 must undergo the same translation so that $a_{32} = a_{22}$. The mass m_1 will undergo a smaller displacement determined by reciprocity. The principle of reciprocity requires that the deflection at point i due to a load a point j is the same as the deflection at point j due to a load at point i for a linear system. Hence $a_{12} = a_{21} = 1/k_1$.

Next consider a unit force applied to mass m_3 as indicated in Figure 9.1(c). Mass m_3 is acted on by springs k_1, k_2, and k_3 as if m_1 and m_2 were not there because a static deflection is considered. Hence

$$a_{33} = \frac{1}{k_{eq}} = \frac{1}{k_1} + \frac{1}{k_2} + \frac{1}{k_3}$$

From reciprocity, $a_{13} = a_{31} = 1/k_1$ and

$$a_{23} = a_{32} = \frac{1}{k_1} + \frac{1}{k_2}$$

Thus the flexibility influence matrix becomes

$$A = \begin{bmatrix} \dfrac{1}{k_1} & \dfrac{1}{k_1} & \dfrac{1}{k_1} \\[2mm] \dfrac{1}{k_1} & \left(\dfrac{1}{k_1} + \dfrac{1}{k_2}\right) & \left(\dfrac{1}{k_1} + \dfrac{1}{k_2}\right) \\[2mm] \dfrac{1}{k_1} & \left(\dfrac{1}{k_1} + \dfrac{1}{k_2}\right) & \left(\dfrac{1}{k_1} + \dfrac{1}{k_2} + \dfrac{1}{k_3}\right) \end{bmatrix}$$

Note that the stiffness matrix K can be calculated from the equation of motion as discussed in Chapter 4. Equation (4.83) for $n = 3$ yields that the stiffness matrix for the system is

$$K = \begin{bmatrix} k_1 + k_2 & -k_2 & 0 \\ -k_2 & k_2 + k_3 & -k_3 \\ 0 & -k_3 & k_3 \end{bmatrix}$$

The reader can easily verify that $AK = I$ (see Problem 9.9). Hence, the flexibility influence matrix calculated above is correct and is the inverse of the stiffness matrix.

□

It should be obvious that the calculation of K^{-1} using the influence coefficient method suggested in Example 9.2.1 is easier and more accurate than writing down the equations of motion to find the stiffness matrix K and actually computing the inverse of the matrix K. The numerical evaluation of the matrix A involves very few computations compared to taking the inverse of K and hence is less subject to computational error. This is discussed in Problem 9.11.

The influence coefficient matrix can be used to estimate the lowest natural frequency of a multiple-degree-of-freedom system. The method, called *Dunkerley's formula*, results from a straightforward consideration of the characteristic equation for an undamped system. From Chapter 4, equation (4.19), the natural frequencies are determined from the characteristic equation given by

$$\det(-\omega^2 M + K) = 0 \tag{9.11}$$

From linear algebra (see Appendix C) the determinant is unchanged if multiplied by the determinant of a nonsingular matrix [i.e., $\det A \det B = \det (AB)$]. Hence equation (9.11) can also be written as

$$\det(K^{-1})\det\left(M - \frac{1}{\omega^2}k\right) = \det\left(K^{-1}M - \frac{1}{\omega^2}I\right) = \det\left(AM - \frac{1}{\omega^2}I\right) = 0 \tag{9.12}$$

where A is the flexibility influence matrix. Equation (9.12) expresses the eigenvalue problem for the matrix AM with eigenvalues $\lambda = 1/\omega^2$.

If the mass matrix M is diagonal, the matrix product AM becomes simple to calculate:

$$
AM = \begin{bmatrix} a_{11} & a_{12} & \cdots & a_{1n} \\ a_{21} & a_{22} & \cdots & a_{2n} \\ \cdots & \cdots & \cdots & \cdots \\ \cdots & \cdots & \cdots & \cdots \\ a_{n1} & a_{n2} & \cdots & a_{nn} \end{bmatrix} \begin{bmatrix} m_1 & 0 & \cdots & 0 \\ 0 & m_2 & \cdots & 0 \\ \cdots & \cdots & \cdots & \cdots \\ \cdots & \cdots & \cdots & \cdots \\ 0 & 0 & \cdots & m_n \end{bmatrix}
$$

$$
= \begin{bmatrix} a_{11}m_1 & a_{12}m_2 & \cdots & a_{1n}m_n \\ a_{21}m_1 & a_{22}m_2 & \cdots & a_{2n}m_n \\ \cdots & \cdots & \cdots & \cdots \\ \cdots & \cdots & \cdots & \cdots \\ a_{n1}m_1 & a_{n2}m_2 & \cdots & a_{nn}m_n \end{bmatrix} \tag{9.13}
$$

A simple result from linear algebra states that the trace of a matrix, defined as the sum of its diagonal elements, is equal to the sum of its eigenvalues (see Appendix C). Applied to the matrix AM of equation (9.13), the trace condition yields

$$\text{trace}(AM) = \sum_{i=1}^{n} a_{ii}m_i = \sum_{i=1}^{n} \lambda_i = \sum_{i=1}^{n} \frac{1}{\omega_i^2} \tag{9.14}$$

where the ω_i are the system's natural frequencies. Writing out the summation in equation (9.14) yields

$$a_{11}m_1 + a_{22}m_2 + \cdots + a_{nn}m_n = \frac{1}{\omega_1^2} + \frac{1}{\omega_2^2} + \cdots + \frac{1}{\omega_n^2} \tag{9.15}$$

Equation (9.15) can be used to make a fairly effective and easy-to-calculate approximation of the first or lowest frequency of a system under certain assumptions. If the lowest frequencies are assumed to be reasonably spaced, so that $\omega_1^2 < \omega_2^2$, then

$$\frac{1}{\omega_1^2} > \frac{1}{\omega_2^2} > \frac{1}{\omega_3^2} \cdots > \frac{1}{\omega_n^2}$$

and it is reasonable to approximate the sum on the right side of equation (9.15) with the single term $1/\omega_1^2$. Equation (9.15) then becomes the inequality

$$a_{11}m_1 + a_{22}m_2 + \cdots + a_{nn}m_n > \frac{1}{\omega_1^2} > 0 \tag{9.16}$$

This inequality can be rewritten

$$\omega_1 > \frac{1}{\sqrt{\sum_{i=1}^n a_{ii}m_i}} \tag{9.17}$$

since $a > b > 0$ implies that $1/b > 1/a > 0$ for any positive real numbers a and b.

The approximation

$$\frac{1}{\omega_1^2} < a_{11}m_1 + a_{22}m_2 + \cdots + a_{nn}m_n \tag{9.18}$$

is called *Dunkerley's formula* for the fundamental frequency. The right-hand side of equation (9.18) is easy to evaluate in the sense that if the value of each component's mass is known, each m_i is known. In addition, the diagonal coefficients in the flexibility matrix a_{ii} are the easiest ones to evaluate for structures with diagonal mass matrix, as this essentially becomes a series of equivalent spring constants for the system (recall Section 1.5). Inequality (9.17) illustrates that a fundamental natural frequency estimated by using Dunkerley's formula will be smaller than the actual value [i.e., the expression $(\sum a_{ii}m_i)^{-1/2}$ represents a lower bound to the exact fundamental natural frequency]. The following example illustrates the use of Dunkerley's formula for estimating the fundamental natural frequency.

Example 9.2.2

Consider the two-degree-of-freedom system of Example 4.1.5 repeated in Figure 9.2. For $k_1 = 24$, $k_2 = 3$, $m_1 = 9$, and $m_2 = 1$, estimate the fundamental natural frequency and compare it with the exact value of $\omega_1 = \sqrt{2}$ given in Example 4.1.5.

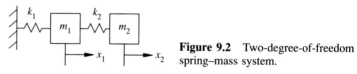

Figure 9.2   Two-degree-of-freedom spring–mass system.

Solution   Following the method of Example 9.2.1, the two diagonal influence coefficients become

$$a_{11} = \frac{1}{k_1} = \frac{1}{24}, \qquad a_{22} = \frac{1}{k_1} + \frac{1}{k_2} = \frac{1}{24} + \frac{1}{3} = \frac{27}{72}$$

so that

$$\sum_{i=1}^{2} a_{ii}m_i = \frac{1}{24}9 + \frac{27}{72}(1) = 0.375 + 0.375 = 0.75$$

Hence using Dunkerley's formula the approximate value of ω_1 is

$$\omega_1 \approx \frac{1}{\sqrt{\sum a_{ii} m_i}} = \frac{1}{\sqrt{0.75}} = 1.1547 \text{ rad/s}$$

The exact value is $\omega_1 = \sqrt{2} = 1.4142 > 1.1547$. Hence the estimated value is lower than the exact value as predicted.

□

Example 9.2.2 is simple enough to calculate exactly. The approximate value is of more interest in cases where it is difficult to obtain the exact values (see Problem 9.20). In addition, Dunkerley's formula is useful in the design setting to predict the results of changes in the design of a system. It is very important to note that Dunkerley's formula assumes (1) diagonal mass matrix, (2) nonsingular stiffness matrix (i.e., no rigid motion), and (3) separation of the first and second natural frequencies. The accuracy of the estimate depends on the size of separation between the numbers $1/\omega_1^2$ and $1/\omega_2^2$.

Example 9.2.3

Consider the three-degree-of-freedom system of Figure 9.1 and Example 9.2.1. Let $k_1 = 10$, $k_2 = 100$, and $k_3 = 10$ (N/m). Also let $m_1 = m_2 = m_3 = 1$ kg and estimate the fundamental natural frequency using Dunkerley's formula.

Solution   From Example 9.2.1, the diagonal coefficients are

$$a_{11} = \frac{1}{k_1} = 0.1$$

$$a_{22} = \frac{1}{k_1} + \frac{1}{k_2} = \frac{k_2 + k_1}{k_1 k_2} = \frac{110}{1000} = 0.11$$

$$a_{33} = \frac{1}{k_1} + \frac{1}{k_2} + \frac{1}{k_3} = 0.1 + 0.01 + 0.1 = 0.21$$

The estimate of ω_1 is then

$$\omega_1 \approx \frac{1}{\sqrt{(0.1)(1) + (0.11)(1) + (0.21)(1)}} = 1.543 \text{ rad/s}$$

Using the methods of Chapter 4 yields $\omega_1 = 1.675$ rad/s, which is slightly larger than the estimate provided by Dunkerley's formula.

□

As mentioned, Dunkerley's formula is more useful as a design indicator than as a computational device. As the following sections illustrate, frequencies of most systems are easily obtained by sophisticated computer codes not available 30 years ago. However, an important design question can be answered by the use of influence coefficients and Dunkerley's formula. In particular it is often the lowest natural frequency of a structure that is of interest in vibration suppression (recall Chapter 5). If a device or mass is added to an existing system, the fundamental frequency will change. Dunkerley's formula can be used to predict the change by treating the terms $a_{ii}m_i$ as reciprocal frequencies. Then equation (9.18) becomes

$$\frac{1}{\omega_1^2} \approx \frac{m_1}{k_{eq}(1)} + \frac{m_2}{k_{eq}(2)} + \cdots + \frac{m_n}{k_{eq}(n)} \tag{9.19}$$

$$\frac{1}{\omega_1^2} \approx \frac{1}{\omega_{11}^2} + \frac{1}{\omega_{22}^2} + \cdots + \frac{1}{\omega_{nn}^2} \tag{9.20}$$

where $k_{eq}(1)$ is the equivalent stiffness of mass m_1, without consideration for the presence of the other masses. Similarly, the notation ω_{ii} denotes a frequency of a single-degree-of-freedom system with stiffness $k_{eq}(i)$ and mass m_i. The following example illustrates the use of equation (9.20) in the design of an aircraft wing.

Example 9.2.4

Consider the airplane wing illustrated in Figure 9.3. The frequency of the system as it vibrates in torsion is measured to be 170 rad/s. In the airplane's initial design, it was determined that no harmonic forces would be applied to the wing section with frequency content greater than 100 rad/s. At a later date, the aircraft's flight range is to be extended by adding external fuel pods as indicated in Figure 9.3. Use Dunkerley's formula to decide if the addition of a fuel pod will cause the aircraft to be unsafe. The torsional stiffness is determined to be (from Figure 1.17) $k_{eq} = 3.3 \times 10^7$ N·m/rad, and the added inertia of the tank is approximately $J = 10^3$ kg · m^2.

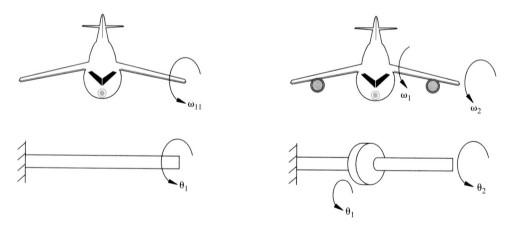

Figure 9.3 Simplified torsional vibration model of a fighter plane wing with and without an external fuel tank (pod), used to determine if adding the fuel pods will cause unsafe wing vibration or not.

Solution The frequency of the torsional vibration of the pod about the center of the wing treated as a massless shaft is about

$$\omega_{22} = \sqrt{\frac{k_{eq}}{J}} = \sqrt{\frac{3.3 \times 10^7}{10^3}} = \sqrt{3.3 \times 10^4} = 182 \text{ rad/s}$$

From Dunkerley's formula, the lowest frequency of the new system with the tank added is estimated from

$$\omega_1 \approx \sqrt{\frac{1}{1/\omega_{11}^2 + 1/\omega_{22}^2}} = \frac{1}{\sqrt{1/170^2 + 1/182^2}} = 124 \text{ rad/s}$$

Since this is 25% higher than the nearest driving frequency of 100 rad/s, the aircraft should be able to fly with the extra fuel tank.

$\square$

9.3 RAYLEIGH'S METHOD

An examination of the energy in a conservative system can be used to provide another estimate of the fundamental natural frequency of a vibrating system without solving an eigenvalue problem. Rayleigh noted that the frequency of vibration is the minimum value of an expression for the energy in a system. This is called *Rayleigh's principle*. This approach follows the energy method expressed in equation (1.62)—that for conservative systems the maximum kinetic energy and the maximum potential energy must be the same. Expressions for the potential and kinetic energy of a *n*-degree-of-freedom spring–mass system are given in Section 4.7 and again in Section 8.1 following equation (8.5) and in equation (8.12). The kinetic energy can be expressed as the vector product

$$T(t) = \tfrac{1}{2}\dot{\mathbf{x}}^T M \dot{\mathbf{x}} \tag{9.21}$$

where $\dot{\mathbf{x}}$ is the vector of velocities and M is the mass matrix. Similarly, the potential energy $U(t)$ can be expressed as the quadratic form

$$U(t) = \tfrac{1}{2}\mathbf{x}^T K \mathbf{x} \tag{9.22}$$

If the system is oscillating at one frequency, the solution is of the form $\mathbf{x}(t) = (\sin \omega t)\mathbf{u}$, where $\mathbf{u}$ is a constant vector (recall Section 4.3). The velocity is just $\dot{\mathbf{x}}(t) = (\omega \cos \omega t)\mathbf{u}$. The maximum value of the potential energy occurs when the displacement $\mathbf{x}(t)$ is at maximum, or $\mathbf{x}_{\max}(t) = \mathbf{u}$. Hence

$$U_{\max} = \tfrac{1}{2}\mathbf{u}^T K \mathbf{u} \tag{9.23}$$

Similarly, the maximum value of the kinetic energy occurs at the maximum velocity, which is $\dot{\mathbf{x}}_{\max}(t) = \omega\mathbf{u}$. Hence

$$T_{\max} = \tfrac{1}{2}\omega^2\mathbf{u}^T M \mathbf{u} \tag{9.24}$$

is the maximum kinetic energy. For conservative systems the maximum potential energy must be the same as the maximum kinetic energy. Equating equations (9.23) and (9.24) and solving for ω^2 yields

$$R(\mathbf{u}) = \frac{\mathbf{u}^T K \mathbf{u}}{\mathbf{u}^T M \mathbf{u}} = \omega^2 \tag{9.25}$$

This quantity is defined as the *Rayleigh quotient*. It can be shown (see, e.g., Inman, 1989) that the value $R(\mathbf{u})$ of the Rayleigh quotient approaches the lowest natural frequency ω_1 as the vector $\mathbf{u}$ is changed. In particular, the smallest value of $R(\mathbf{u})$ is ω_1^2, so that

$$R(\mathbf{u}) \geq \omega_1^2 \tag{9.26}$$

for all vectors $\mathbf{u}$. This provides an estimate of the fundamental (first) natural frequency that is larger than ω_1, while Dunkerley's formula provided an estimate of ω_1 that is smaller than ω_1:

$$R(\mathbf{u}) = \frac{\mathbf{u}^T K \mathbf{u}}{\mathbf{u}^T M \mathbf{u}} \geq \omega_1^2 > \frac{1}{\sum_{i=1}^{n} a_{ii} m_i} \tag{9.27}$$

In comparing Dunkerley's formula to the Rayleigh quotient, note that $R(\mathbf{u})$ is defined for systems with rigid-body modes, while Dunkerley's formula is not. In addition, at the value $\mathbf{u} = \mathbf{u}_1$, the first mode shape, $R(\mathbf{u}_1)$ is exact:

$$R(\mathbf{u}_1) = \omega_1^2 \tag{9.28}$$

whereas Dunkerley's formula always yields an approximate value.

Example 9.3.1

Use Rayleigh's quotient to estimate the lowest natural frequency for the three-degree-of-freedom model of a machine given in Example 4.8.3. The mass and stiffness matrices are given as

$$M = (10^3) \begin{bmatrix} 0.4 & 0 & 0 \\ 0 & 2 & 0 \\ 0 & 0 & 8 \end{bmatrix} \qquad K = (10^4) \begin{bmatrix} 30 & -30 & 0 \\ -30 & 38 & -8 \\ 0 & -8 & 88 \end{bmatrix}$$

respectively.

Solution For the sake of simple arithmetic, choose a trial vector of $\mathbf{u} = [1 \quad 0 \quad 0]^T$. Then

$$\omega^2 \approx R(\mathbf{u}) = \frac{[1 \quad 0 \quad 0] \begin{bmatrix} 30 & -30 & 0 \\ -30 & 38 & -8 \\ 0 & -8 & 88 \end{bmatrix} \begin{bmatrix} 1 \\ 0 \\ 0 \end{bmatrix} (10^4)}{[1 \quad 0 \quad 0] \begin{bmatrix} 0.4 & 0 & 0 \\ 0 & 2 & 0 \\ 0 & 0 & 8 \end{bmatrix} \begin{bmatrix} 1 \\ 0 \\ 0 \end{bmatrix} (10^3)} = \frac{30 \times 10^4}{0.4 \times 10^3} = 750$$

This yields an estimate of $\omega_1 = \sqrt{750} = 27.3$. Next consider $\mathbf{u} = [1 \quad 1 \quad 0]^T$. This yields

$$R(\mathbf{u}) = 33.33$$

so that $\omega_1 \approx \sqrt{33.33} = 5.77$ rad/s. Since this value of $R(\mathbf{u})$ yields a value lower than the previous value, the number 5.77 rad/s is a better estimate than the previously estimated value of 27.3 rad/s. Various vectors can be tried; the smallest value of $R(\mathbf{u})$ will be the closest to the actual value. In this case the actual value of the fundamental natural frequency is $\omega_1 = 5.3872$ rad/s.

□

As Example 9.3.1 illustrates, Rayleigh's quotient yields an easy estimate of the lowest frequency. It is also useful in design and in analytical development. However, with readily available computation devices, the natural frequencies of a device, or structure, are usually calculated using a numerical procedure of the type discussed in the following sections. Inexpensive calculators often can be found with programs for eigenvalue analysis and hence can be used to compute natural frequencies and mode shapes of low-order systems (up to four or five degrees of freedom). Larger-order systems, say up to 20 degrees of freedom, can be solved on inexpensive or public-domain software on personal computers. Professional-quality software on high-end PCs can be used to solve systems with up to 100 to 200 degrees of freedom very accurately. Most commercial finite element codes that generate models on the order of 1000 degrees of freedom include a sophisticated program for computing natural frequencies. Hence it is unlikely that modern vibration engineers would need to write software for computing frequencies based on Rayleigh's quotient. Rather, it is important to have a feeling for the nature of such computations, their accuracy, and their limitations. The following sections are intended to provide this.

9.4 MATRIX ITERATION

From Chapter 4 it is clear that the eigenvalues and eigenvectors of the matrix $M^{-1}K$ or $M^{-1/2}KM^{-1/2}$ are related to the natural frequencies and mode shapes of vibration. Hence a solution to the matrix eigenvalue problem yields a solution to the vibration problem. One basic but restricted method of computing eigenvalues and eigenvectors of a matrix is a procedure called *matrix iteration*. This method is introduced here.

Let A be an $n \times n$ matrix with distinct real eigenvalues, denoted μ_i, ordered such that

$$\mu_1 > \mu_2 > \mu_3 > \cdots > \mu_n \tag{9.29}$$

These are fairly restrictive assumptions but often met in rotating systems. The matrix iteration procedure is started by assuming a trial vector $\mathbf{x}$ and multiplying it by the matrix A. This produces the column vector $A\mathbf{x}$, which is then normalized. Let the normalized version of $A\mathbf{x}$ be denoted $A\bar{\mathbf{x}}$. The normalized vector $A\bar{\mathbf{x}}$ is again multiplied by the matrix A, to yield $AA\bar{\mathbf{x}}$, which is again normalized. This process is repeated until these successive normalized column vectors converge (i.e., until multiplying by A and normalizing does not produce a noticeably different vector). The constant used to normalize the last step yields the largest eigenvalue. The following delineates the procedure.

Recall that any n-dimensional column vector can be represented as a linear combination of n orthogonal eigenvectors of a symmetric matrix. This was used in Section 4.4 to calculate the time response of a multiple-degree-of-freedom system to a set of initial conditions. Let $\{\mathbf{u}_i\}$ denote the n, $n \times 1$ eigenvectors of the matrix A, which are assumed to be linearly independent and orthogonal. Then an arbitrary $n \times 1$ trial vector $\mathbf{x}$ can be written as

$$\mathbf{x} = c_1 \mathbf{u}_1 + c_2 \mathbf{u}_2 + c_3 \mathbf{u}_3 + \cdots + c_n \mathbf{u}_n \qquad (9.30)$$

where c_i are constants that at this point are unknown, as are $\mathbf{u}_i$. Here only the trial vector $\mathbf{x}_1$ and the matrix A are known. Multiplying equation (9.30) by the matrix A produces the first step in the iteration or

$$\mathbf{x}_1 = A\mathbf{x} = c_1 A\mathbf{u}_1 + c_2 A\mathbf{u}_2 + \cdots + c_n A\mathbf{u}_n \qquad (9.31)$$

which becomes

$$\mathbf{x}_1 = A\mathbf{x} = c_1 \mu_1 \mathbf{u}_1 + c_2 \mu_2 \mathbf{u}_2 + \cdots + c_n \mu_n \mathbf{u}_n \qquad (9.32)$$

where the fact that μ_i, $\mathbf{u}_i$ are eigensolutions for A is used (i.e., $A\mathbf{u}_i = \mu_i \mathbf{u}_i$). Next let $\mathbf{x}_2 = A(A\mathbf{x})$, $\mathbf{x}_3 = A(\mathbf{x}_2)$ and so on. Each time the multiplication of A on the right side of equation (9.32) increases the exponent on the eigenvalue μ_i, so that

$$\mathbf{x}_r = c_1 \mu_1^r \mathbf{u}_1 + c_2 \mu_2^r \mathbf{u}_2 + \cdots + c_n \mu_n^r \mathbf{u}_n \qquad (9.33)$$

If it is further assumed that μ_1^r is much larger than μ_2^r, the only term on the right side of equation (9.33) with a significant value will be $c_1 \mu_1^r \mathbf{u}_1$. A specific case where this last assumption is true is where the matrix A is the stiffness influence coefficient matrix, so that $\mu_1 = 1/\omega_1^2$. Note that once $\mathbf{x}_r$ does not change when multiplied by the matrix A, corresponding elements of $\mathbf{x}_r$ and $\mathbf{x}_{r+1}$ can be used to calculate μ_1. The vector $\mathbf{x}_{r+1}$ becomes the first eigenvector (i.e., $\mathbf{u}_1 = \mathbf{x}_{r+1}$) and equation (9.33) reduces to

$$\mathbf{x}_r = c_1 \mu_1^r \mathbf{u}_1 \qquad \text{and} \qquad \mathbf{x}_{r+1} = c_1 \mu_1^{r+1} \mathbf{u}_1 \qquad (9.34)$$

Comparing the ith element of each expression yields

$$(\mathbf{x}_r)_i = c_1 \mu_1^r (\mathbf{u}_1)_i \quad \text{and} \quad (\mathbf{x}_{r+1})_i = c_1 \mu_1^{r+1} (\mathbf{u}_1)_i$$

and dividing yields

$$\frac{(\mathbf{x}_{r+1})_i}{(\mathbf{x}_r)_i} = \mu_1 = \frac{1}{\omega_1^2} \qquad (9.35)$$

Expression 9.35 holds for any value of the index i from 1 to n.

The above describes how the matrix iteration method calculates the first eigenvalue and eigenvector of a matrix with real eigenvalues and with well-spaced eigenvalues. The remaining eigenvalues and eigenvectors can be calculated in a similar fashion using the concept of *matrix deflation*. Matrix deflation is a technique for eliminating the known eigenvalue from the characteristic equation. The eigenvalue μ_1 can be removed from the characteristic equation, which has the form

$$\det(A - \mu I) = 0 \qquad (9.36)$$

First the eigenvector $\mathbf{u}_1$, corresponding to μ_1, is normalized. Then a deflated matrix, D_1, is defined by

$$D_1 = A - \mu_1 \mathbf{u}_1 \mathbf{u}_1^T \qquad (9.37)$$

The new matrix D_1 does not have μ_1 as an eigenvalue. Applying the matrix iteration method to the matrix D_1 will produce the eigenvalue μ_2 and its corresponding eigenvector $\mathbf{u}_2$. The matrix D_1 is then deflated using equation (9.37) (i.e., $D_2 = D_1 - \mu_2 \mathbf{u}_2 \mathbf{u}_2^T$). This procedure is followed until all the eigenvalues and eigenvectors have been constructed.

The matrix iteration method or *power iteration method* is able to calculate eigenvalues and eigenvectors for matrices with well-spaced real eigenvalues and real eigenvectors. There are, of course, more general, faster, and numerically efficient methods of computing eigenvalues and eigenvectors of matrices which do not satisfy these criteria (i.e., systems with complex and/or repeated eigenvalues and complex eigenvectors). Some of these methods are discussed in Section 9.6.

9.5 EIGENVALUE PROBLEMS OF VIBRATION ANALYSIS

The vibration problem for a multiple-degree-of-freedom system, or a finite element model of a system, can be expressed as an eigenvalue problem in several ways. In Section 4.2 the undamped vibration problem is related to the symmetric eigenvalue problem by calculating the inverse-square root of the positive definite mass matrix M. In particular, if the equations of motion are stated in matrix/vector form as

$$M\ddot{\mathbf{x}} + K\mathbf{x} = \mathbf{0} \tag{9.38}$$

where M is the mass matrix and K is the $n \times n$, symmetric positive semidefinite stiffness matrix (see Window 9.2), there are four different eigenvalue problems that can be used to solve for the natural frequencies and mode shapes. As before, $\mathbf{x}$ is an $n \times 1$ vector of displacements and $\ddot{\mathbf{x}}$ is the second time derivative of $\mathbf{x}$.

Window 9.2

A matrix A is symmetric if $A = A^T$ and positive definite if for every real, nonzero vector $\mathbf{x}$, the scalar $\mathbf{x}^T A \mathbf{x} > 0$. If $\mathbf{x}^T A \mathbf{x} \geq 0$ for every nonzero, real vector $\mathbf{x}$, the matrix A is said to be positive semidefinite.

The first approach considered is the standard approach of multiplying equation (9.30) from the left by the inverse of the matrix M. This yields

$$\ddot{\mathbf{x}} + M^{-1} K \mathbf{x} = \mathbf{0} \tag{9.39}$$

First the matrix M^{-1} must be calculated. In many cases M is a diagonal matrix (i.e., $M = \text{diag}[m_1 \ m_2 \ \cdots \ m_n]$ and the inverse becomes simply

$$
M^{-1} =
\begin{bmatrix}
\dfrac{1}{m_1} & 0 & 0 & \cdots & \cdots & 0 \\[2mm]
0 & \dfrac{1}{m_2} & 0 & \cdots & \cdots & 0 \\[2mm]
\cdots & 0 & \dfrac{1}{m_3} & 0 & \cdots & \cdots \\[2mm]
\cdots & \cdots & \cdots & \cdots & \cdots & \cdots \\[2mm]
\cdots & \cdots & \cdots & \cdots & \cdots & \cdots \\[2mm]
0 & \cdots & \cdots & \cdots & \cdots & \dfrac{1}{m_n}
\end{bmatrix}
\tag{9.40}
$$

In this case calculating the matrix $M^{-1}K$ becomes a simple matrix multiplication. If K is a banded matrix, $M^{-1}K$ will also be banded. However, the product is usually not symmetric, even though both K and M are symmetric matrices, and computing the eigenvalues of a symmetric matrix is more efficient than calculating the eigenvalue problem of an asymmetric matrix.

If the matrix M is not diagonal (i.e., if the system is dynamically coupled), the inverse can be computed using the eigenvalue problem (solved in Section 9.4) for the matrix M. Let $\mathbf{v}_i$ denote the eigenvector of the matrix M and let μ_i denote the associated eigenvalues. Then the statement of the eigenvalue problem for the matrix M becomes

$$
M\mathbf{v}_i = \mu_i \mathbf{v}_i \qquad \mathbf{v}_i \neq \mathbf{0}
\tag{9.41}
$$

Since M is symmetric and positive definite, each of the μ_i is greater than zero and each $\mathbf{v}_i$ can be normalized such that they form an orthonormal set (see Window 4.2). The $n \times n$ matrix V formed by taking the vectors $\mathbf{v}_i$ as its columns [similar to equation (4.49)], i.e., $V = [\mathbf{v}_1 \quad \mathbf{v}_2 \quad \cdots \quad \mathbf{v}_n]$, is called orthogonal because $V^T = V^{-1}$, so that $V^T V = I$, the $n \times n$ identity. The matrix V also diagonalizes the matrix M, so that

$$
V^T M V = \text{diag}(\mu_1, \mu_2, \ldots, \mu_n)
\tag{9.42}
$$

Thus the matrix M can be decomposed into

$$
M = V\,\text{diag}(\mu_1, \mu_2, \ldots, \mu_n)V^T
\tag{9.43}
$$

This decomposition can be used to calculate the inverse of the matrix M using the identity $(AB)^{-1} = B^{-1}A^{-1}$, via

$$
M^{-1} = V\,\text{diag}\left(\frac{1}{\mu_1}, \frac{1}{\mu_2}, \cdots, \frac{1}{\mu_n}\right)V^T
\tag{9.44}
$$

Hence one way to compute the inverse of the matrix M is to solve the eigenvalue problem associated with M by the power method of Section 9.4 and use equation (9.44) to construct the inverse.

Example 9.5.1

Use equations (9.44) and (9.43) to show that $MM^{-1} = I$, as it should.

Solution   From equations (9.44) and (9.43) MM^{-1} becomes

$$MM^{-1} = V \; \text{diag}(\mu_1, \cdots \mu_n) V^T V \; \text{diag} \left(\frac{1}{\mu_1} \cdots \frac{1}{\mu_n} \right) V^T$$

Since $V^T V = I$ this becomes

$$MM^{-1} = V \; \text{diag}(\mu_1 \cdots \mu_n) \; \text{diag} \left(\frac{1}{\mu_1}, \frac{1}{\mu_2}, \cdots, \frac{1}{\mu_n} \right) V^T$$

Multiplying the inner two diagonal matrices together produces the diagonal matrix of elements $\mu_i(1/\mu_i) = 1$. Hence

$$MM^{-1} = VIV^T = VV^T = I$$

which verifies that equation (9.44) is correct.

$\square$

It is also useful to note that for any symmetric matrix, the formula for a function of matrix is given in the same form as equation (9.44). That is, if $f(\cdot)$ is any real function for which $f(\mu_i)$ is defined,

$$f(M) = V \; \text{diag}[f(\mu_1), \; f(\mu_2), \; \cdots, \; f(\mu_n)] V^T \tag{9.45}$$

defines the function f of the symmetric matrix M. In particular, the matrix $M^{-1/2}$ was used extensively in Chapter 4. Using this function of a matrix definition, the matrices $M^{-1/2}$ and $M^{1/2}$ can be calculated from the eigenvalues and eigenvectors of M by

$$M^{1/2} = V \; \text{diag} \left(\mu_1^{1/2}, \mu_2^{1/2}, \cdots, \mu_n^{1/2} \right) V^T \tag{9.46}$$

and

$$M^{-1/2} = V \; \text{diag} \left(\mu_1^{-1/2}, \mu_2^{-1/2}, \cdots, \mu_n^{-1/2} \right) V^T \tag{9.47}$$

This formulation allows the extension of the material of Chapter 4 to dynamically coupled systems. Again, in the case of only static coupling the matrix M is diagonal and computation of $M^{-1/2}$ or $M^{1/2}$ is trivial. For instance, if $M = \text{diag}(\mu_1, \mu_2, \ldots, \mu_n)$, $M^{-1/2} = \text{diag} \left(\mu_1^{-1/2}, \mu_2^{-1/2}, \ldots, \mu_n^{-1/2} \right)$, as was used in Chapter 4.

The use of equation (9.44) to calculate the inverse of a matrix is not a very numerically efficient method. A better way to calculate M^{-1} is to view the calculation as the solution of the set of linear equations

$$MA = K \tag{9.48}$$

where A and K are vectors or matrices, K is known, and A is unknown. This can be solved by a Gaussian elimination (think of the case where A and K are vectors) and back substitution. Gaussian elimination is essentially a systematic version of the method of elimination often taught in high school algebra classes and involves a series of steps, each of which eliminates one variable from the system of equations. To examine this method, consider the set of linear algebraic equations of the form

$$A\mathbf{x} = \mathbf{b} \tag{9.49}$$

where A is a known $n \times n$ matrix, $\mathbf{x}$ is an unknown $n \times 1$ vector of the form $\mathbf{x} = [\, x_1 \quad x_2 \quad \ldots \quad x_n \,]^T$, and $\mathbf{b}$ is an $n \times 1$ vector of known elements. The method of Gaussian elimination uses the fact that if P is any $n \times n$ nonsingular matrix, equation (9.49) and

$$P A \mathbf{x} = P \mathbf{b} \tag{9.50}$$

have the same solution $\mathbf{x}$. The approach of the Gaussian elimination algorithm is to find a matrix P such that the matrix PA is upper triangular (i.e., PA has zeros as every element below the diagonal elements). If such a matrix P can be determined, the last equation in the relation specified by equation (9.50) will be of the form

$$x_n = (P\mathbf{b})_n \tag{9.51}$$

where $(P\mathbf{b})_n$ denotes the nth element of the vector $P\mathbf{b}$. The second-to-last equation will be only in x_{n-1} and x_n. Hence each element x_i is calculated from the last equation backward (back substitution) until the entire vector is known.

The required matrix P in equations (9.51) and (9.50) can be determined from a series of elementary matrices. Elementary matrices are nonsingular matrices which when postmultiplied by a given matrix (A, in this case) results in the subtraction of multiples of one column of the matrix A from each of the other columns of A. An example of an elementary matrix is the matrix P_i defined to be the $n \times n$ identity matrix with the ith column replaced by

$$\begin{bmatrix} 0 \\ 0 \\ \vdots \\ 1 \\ -p_{i,i+1} \\ \vdots \\ -p_{i,n} \end{bmatrix} \tag{9.52}$$

where the element $p_{i,j}$ are to be determined by the elements of the matrix A in such a way as to reduce A to a triangular form.

The algorithm for Gaussian elimination then proceeds in a step-by-step manner by writing the initial set of linear equations given in equation (9.49) as

$$A_0 \mathbf{x} = \mathbf{b}_0 \tag{9.53}$$

Multiplying this expression by P_1 to get $P_1 A_0 \mathbf{x} = P_1 \mathbf{b}_0$ produces the next step. The last expression is renamed $A_1 \mathbf{x} = \mathbf{b}_1$, where $A_1 = P_1 A_0$ and $\mathbf{b}_1 = P_1 \mathbf{b}_0$. Then P_2 is multiplied times $A_1 \mathbf{x} = \mathbf{b}_1$ to yield $A_2 \mathbf{x} = \mathbf{b}_2$, where $A_2 = P_2 A_1$ and $\mathbf{b}_2 = P_2 \mathbf{b}_1$. This is repeated so that the rth step is

$$A_r \mathbf{x} = \mathbf{b}_r \tag{9.54}$$

Here multiplying by P_r has made A_{r-1} upper triangular in the first r column. For example, for $n = 6$ and $r = 3$, equation (9.54) has the form

$$
\begin{bmatrix}
\times & \times & \times & \times & \times & \times \\
0 & \times & \times & \times & \times & \times \\
0 & 0 & \times & \times & \times & \times \\
0 & 0 & 0 & \times & \times & \times \\
0 & 0 & 0 & \times & \times & \times \\
0 & 0 & 0 & \times & \times & \times
\end{bmatrix}
\begin{bmatrix}
x_1 \\ x_2 \\ x_3 \\ x_4 \\ x_5 \\ x_6
\end{bmatrix}
=
\begin{bmatrix}
\times \\ \times \\ \times \\ \times \\ \times \\ \times
\end{bmatrix}
\tag{9.55}
$$

where $\times$ represents any nonzero number. Equation (9.55) illustrates how the equation $A\mathbf{x} = \mathbf{b}$ becomes more "triangular" at each step.

Examination of each step illustrates that the matrix $P_r A_{r-1}$ has as its rth column,

$$
P_r
\begin{bmatrix}
a_{1r}^{(r-1)} \\
a_{2r}^{(r-1)} \\
\vdots \\
a_{nr}^{(r-1)}
\end{bmatrix}
\tag{9.56}
$$

where $a_{ij}^{(r-1)}$ denotes the ijth element of the matrix A_{r-1}. The choice of the matrix P_r that performs the desired triangularization is then given by equation (9.52) with

$$
p_{ir} = \frac{a_{ir}^{(r-1)}}{a_{rr}^{(n-1)}}
\tag{9.57}
$$

This formula results in a matrix P_r such that $P_r A_{r-1}$ has one more upper triangular column, eliminating x_r from the equation in the $r - 1$ position. Successive multiplication of the matrices P_r defined by equations (9.57) and (9.52) yields a triangular system of equations which are easily solved for the elements x_r of the vector $\mathbf{x}$.

The solution of the linear system of equations $A\mathbf{x} = \mathbf{b}$ is then solved by computing $P_n P_{n-1} P_{n-2} \cdots P_1 A\mathbf{x} = P_n P_{n-1} P_{n-2} \cdots P_1 \mathbf{b}$. The method works well and is numerically superior to computing the solution via $\mathbf{x} = A^{-1}\mathbf{b}$. However, the Gaussian elimination method fails if any of the pivot elements $a_{rr} = 0$. The matrix equation $MA = K$ given in equation (9.48) can also be solved by a Gaussian elimination procedure (applied to each column of A) or by more sophisticated triangularization procedures. This produces a numerically more reliable calculation of the matrix $A = M^{-1}K$ than is obtained by first calculating the matrix M^{-1} via equation (9.44) and computing the matrix product of M^{-1} and K. The procedure of computing the matrix A from the equation $MA = K$ to form the matrix $M^{-1}K$ without actually computing an inverse is called *matrix division*.

The ability to effectively compute the inverse matrix M^{-1} and the inverse square root matrix $M^{-1/2}$ allows the formulation of five eigenvalue problems associated with vibration analysis. Each of these eigenvalue problems has computational advantages and disadvantages depending on the nature of both the structures and values of the elements of both the mass and stiffness matrices. These various forms of the eigenvalue problems are discussed next.

Recall the undamped vibration problem of equation (9.38), repeated here:

$$M\ddot{x} + Kx = 0 \tag{9.58}$$

where M and K are symmetric $n \times n$ positive definite matrices. The solution of this problem proceeds by (recall Section 4.1) assuming a solution of the form $x = e^{j\omega t}u$, where u is a nonzero vector of constants (and ω will become the natural frequency) and substituting this into equation (9.58). This yields

$$-\omega^2 Mu + Ku = 0 \tag{9.59}$$

or

$$Ku = \lambda Mu \tag{9.60}$$

where $\lambda = \omega^2$. Equation (9.60) is a statement of the *generalized eigenvalue problem*. The important feature of this formulation is that M and K are generally sparse, banded matrices that are symmetric and positive definite. These features are very useful in solving for the n eigenvalues λ_i and n eigenvectors u_i. The eigenvalues λ_i are just the squares of the natural frequencies ω_i^2. The solution of equation (9.58) is of the form

$$x(t) = \sum_{i=1}^{n} c_i \sin(\omega_i t + \phi_i)u_i \tag{9.61}$$

so that the eigenvectors u_i are also the vibration mode shapes (recall Section 4.3). Here c_i and ϕ_i are constants determined by initial conditions. Because M and K are symmetric, it is known from matrix theory that λ_i and u_i are real valued.

Next consider equation (9.39), which is obtained simply by multiplying equation (9.58) by the matrix M^{-1}. Again assuming a solution of the form $x = e^{j\omega t}u$ yields

$$-\omega^2 u + M^{-1}Ku = 0 \tag{9.62}$$

or

$$(M^{-1}K)u = \lambda u \tag{9.63}$$

This is the standard *algebraic eigenvalue problem*. The matrix $M^{-1}K$ is neither symmetric nor banded. Again there are n eigenvalues λ_i, which are the squares of the natural frequencies ω_i^2, and n eigenvectors u_i. The solution of equation (9.39), $x(t)$, is again of the form

$$x(t) = \sum_{i=1}^{n} c_i \sin(\omega_i t + \phi_i)u_i \tag{9.64}$$

where c_i and ϕ_i are constants to be determined by the initial conditions. Hence the eigenvectors u_i are also the mode shapes. Since $M^{-1}K$ is not symmetric, the solution of the algebraic eigenvalue problem could yield complex values for the eigenvalues and eigenvectors. However, they are known to be real valued because of the generalized eigenvalue problem formulation, equation (9.60), which has the same eigenvalues and eigenvectors.

Next consider the vibration problem (of Section 4.2) obtained by substitution of the coordinate transformation $\mathbf{x} = M^{-1/2}\mathbf{q}$ into equation (9.58) and multiplying by $M^{-1/2}$. This yields the form

$$\ddot{\mathbf{q}}(t) + M^{-1/2}KM^{-1/2}\mathbf{q}(t) = \mathbf{0} \tag{9.65}$$

where the matrix $M^{-1/2}KM^{-1/2}$ is symmetric but not necessarily sparse or banded unless M is diagonal. The solution of equation (9.65) is obtained by assuming a solution of the form $\mathbf{q} = e^{j\omega t}\mathbf{v}$ where $\mathbf{v}$ is a nonzero vector of constants. Substituting this form into equation (9.65) yields

$$-\omega^2\mathbf{v} + M^{-1/2}KM^{-1/2}\mathbf{v} = \mathbf{0} \tag{9.66}$$

or

$$(M^{-1/2}KM^{-1/2})\mathbf{v} = \lambda\mathbf{v} \tag{9.67}$$

where again $\lambda = \omega^2$. This is the *symmetric eigenvalue problem* and again results in n eigenvalues λ_i, which are the squares of the natural frequencies ω_i^2, and n eigenvectors $\mathbf{v}_i$. The solution of equation (9.65) becomes

$$\mathbf{q}(t) = \sum_{i=1}^{n} c_i \sin(\omega_i t_i + \phi_i)\mathbf{v}_i \tag{9.68}$$

where c_i and ϕ_i are again constants of integration. The solution in the original coordinate system $\mathbf{x}$ is obtained from this last expression by multiplying by the matrix $M^{-1/2}$:

$$\mathbf{x} = M^{-1/2}\mathbf{q} = \sum_{i=1}^{n} c_i \sin(\omega_i t_i + \phi_i)M^{-1/2}\mathbf{v}_i \tag{9.69}$$

Hence the mode shapes are the vectors $M^{-1/2}\mathbf{v}_i$, where $\mathbf{v}_i$ are the eigenvectors of the symmetric matrix $M^{-1/2}KM^{-1/2}$. Since the eigenvalue problem here is symmetric, it is known that the eigenvalues and eigenvectors are real valued, as are the mode shapes. The numerical advantage here is that the eigenvalue problem is symmetric, so more efficient numerical algorithms can be used to solve it. The numerical disadvantage is in calculating the matrix $M^{-1/2}KM^{-1/2}$.

Again consider the vibration problem defined in equation (9.39), which is

$$\ddot{\mathbf{x}} + M^{-1}K\mathbf{x} = \mathbf{0} \tag{9.70}$$

This equation can be transformed to a first-order vector differential equation by defining two new $n \times 1$ vectors, $\mathbf{y}_1$ and $\mathbf{y}_2$, by

$$\begin{aligned} \mathbf{y}_1 &= \mathbf{x} \\ \mathbf{y}_2 &= \dot{\mathbf{x}} \end{aligned} \tag{9.71}$$

Note that $\mathbf{y}_1$ is the vector of displacements and $\mathbf{y}_2$ is a vector of velocities. Differentiating these two vectors yields

$$\begin{aligned} \dot{\mathbf{y}}_1 &= \dot{\mathbf{x}} = \mathbf{y}_2 \\ \dot{\mathbf{y}}_2 &= \ddot{\mathbf{x}} = -M^{-1}K\mathbf{y}_1 \end{aligned} \tag{9.72}$$

where the equation for $\dot{\mathbf{y}}_2$ has been expanded by solving equation (9.70) for $\ddot{\mathbf{x}}$. Equations (9.72) can be recognized as the first-order vector differential equation

$$\dot{\mathbf{y}} = A\mathbf{y} \tag{9.73}$$

Here

$$A = \begin{bmatrix} 0 & I \\ -M^{-1}K & 0 \end{bmatrix} \tag{9.74}$$

is called the *state-matrix*. The 0 denotes an $n \times n$ matrix of zeros, I denotes the $n \times n$ identity matrix, and the *state vector* $\mathbf{y}$ is defined by the $2n \times 1$ vector

$$\mathbf{y} = \begin{bmatrix} \mathbf{y}_1 \\ \mathbf{y}_2 \end{bmatrix} = \begin{bmatrix} \mathbf{x} \\ \dot{\mathbf{x}} \end{bmatrix} \tag{9.75}$$

The solution of equation (9.73) proceeds by assuming the exponential form $\mathbf{y} = \mathbf{z}e^{\lambda t}$, where $\mathbf{z}$ is nonzero vector of constants and λ is a scalar. Upon substitution into equation (9.73) yields $\lambda \mathbf{z} = A\mathbf{z}$ or

$$A\mathbf{z} = \lambda \mathbf{z} \qquad \mathbf{z} \neq 0 \tag{9.76}$$

This is again the standard algebraic eigenvalue problem. While the matrix A has many zero elements, it is now a $2n \times 2n$ eigenvalue problem. It can be shown that the $2n$ eigenvalues λ_i again corresponds to the n natural frequencies ω_i by the relation $\lambda_i = \omega_i j$, where $j = \sqrt{-1}$. The extra n eigenvalues are $\lambda_i = -\omega_i j$, so that there are still only n natural frequencies, ω_i. The $2n$ eigenvectors, $\mathbf{z}$ of the matrix A, however, are of the form

$$\mathbf{z}_i = \begin{bmatrix} \mathbf{u}_i \\ \lambda_i \mathbf{u}_i \end{bmatrix} \tag{9.77}$$

where $\mathbf{u}_i$ are the mode shapes of the corresponding vibration problem. The matrix A (see Window 9.3) is not symmetric and the eigenvalues λ_i and eigenvectors $\mathbf{z}_i$ would therefore be complex. In fact, the eigenvalues λ_i in this case are imaginary numbers of the form $\omega_i j$.

Window 9.3

Do not be confused by the matrix A. The symbol A is used to denote any matrix. In this chapter, A is used to denote

$$A = K^{-1} \qquad \text{the influence matrix}$$

$$A = M^{-1}K$$

$$A = \begin{bmatrix} 0 & I \\ M^{-1}K & M^{-1}C \end{bmatrix} \qquad A = \begin{bmatrix} 0 & I \\ M^{-1}K & 0 \end{bmatrix}$$

to name a few. Which matrix A is being discussed should be clear from the text.

For large-order systems, computing the eigenvalues using equation (9.76) becomes numerically more difficult because it is of order $2n$ rather than n. The main advantage of the state-space form is in numerical simulations and in solving the damped multiple-degree-of-freedom vibration problem, which is discussed next. Now consider the damped vibration problem of the form

$$M\ddot{\mathbf{x}} + C\dot{\mathbf{x}} + K\mathbf{x} = \mathbf{0} \qquad (9.78)$$

where C represents the viscous damping in the system (see Section 4.5) and is assumed only to be symmetric and positive semidefinite. The state-matrix approach and related standard eigenvalue problem of equation (9.76) can also be used to describe the non-conservative vibration problem of equation (9.78). Multiplying equation (9.78) by the matrix M^{-1} yields

$$\ddot{\mathbf{x}} + M^{-1}C\dot{\mathbf{x}} + M^{-1}K\mathbf{x} = \mathbf{0} \qquad (9.79)$$

Again it is useful to rewrite this expression in a first-order or state-space form by defining the two $n \times 1$ vectors $\mathbf{y}_1 = \mathbf{x}$ and $\mathbf{y}_2 = \dot{\mathbf{x}}$ as indicated in equation (9.71). Then equation (9.72) becomes

$$\dot{\mathbf{y}}_1 = \dot{\mathbf{x}} = \mathbf{y}_2$$
$$\dot{\mathbf{y}}_2 = \ddot{\mathbf{x}} = -M^{-1}K\mathbf{x} - M^{-1}C\dot{\mathbf{x}} \qquad (9.80)$$

where the expression for $\ddot{\mathbf{x}}$ in equation (9.80) is taken from equation (9.79) for the damped system by moving the terms $M^{-1}C\dot{\mathbf{x}}$ and $M^{-1}K\mathbf{x}$ to the right of the equal sign. Renaming $\mathbf{x} = \mathbf{y}_1$ and $\dot{\mathbf{x}} = \mathbf{y}_2$ in equation (9.80) and using matrix notation yields

$$\dot{\mathbf{y}} = \begin{bmatrix} \dot{\mathbf{y}}_1 \\ \dot{\mathbf{y}}_2 \end{bmatrix} = \begin{bmatrix} 0\mathbf{y}_1 + I\mathbf{y}_2 \\ -M^{-1}K\mathbf{y}_1 - M^{-1}C\mathbf{y}_2 \end{bmatrix} = \begin{bmatrix} 0 & I \\ -M^{-1}K & -M^{-1}C \end{bmatrix} \begin{bmatrix} \mathbf{y}_1 \\ \mathbf{y}_2 \end{bmatrix} = A\mathbf{y} \quad (9.81)$$

where the vector $\mathbf{y}$ is defined as the state vector of equation (9.75) and the state matrix A for the damped case is defined as the partitioned form

$$A = \begin{bmatrix} 0 & I \\ -M^{-1}K & -M^{-1}C \end{bmatrix} \qquad (9.82)$$

The eigenvalue analysis for the system of equation (9.81) proceeds directly as for the undamped state-matrix system of equation (9.73).

A solution of equation (9.82) is again assumed of the form $\mathbf{y} = \mathbf{z}e^{\lambda t}$ and substituted into equation (9.82) to yield the eigenvalue problem of equation (9.76) (i.e., $A\mathbf{z} = \lambda \mathbf{z}$). This again defines the standard eigenvalue problem of dimension $2n \times 2n$. In this case, the solution again yields $2n$ values λ_i which may be complex valued. The $2n$ eigenvectors $\mathbf{z}_i$ described in equation (9.77) may also be complex valued (if the corresponding λ_i is complex). This, in turn, causes the physical mode shape $\mathbf{u}_i$ to be complex valued as well as the free-response vector $\mathbf{x}(t)$.

Fortunately, there is a rational physical interpretation of the complex eigenvalue, modes, and the resulting solution determined by the state-space formulation of the eigenvalue problem given in equation (9.82). The physical time response $\mathbf{x}(t)$ is simply taken to be the real part of the first n coordinates of the vector $\mathbf{y}(t)$ computed from

$$\mathbf{x}(t) = \sum_{i=1}^{2n} c_i \mathbf{u}_i e^{\lambda_i t} \tag{9.83}$$

The time response is discussed in more detail in Section 9.7 and was introduced in equation (4.115). The physical interpretation of the complex eigenvalues λ_i is taken directly from the complex numbers arising from the solution of an underdamped single-degree-of-freedom system given in equations (1.32) and (1.33) of Section 1.3. In particular, the complex eigenvalues λ_i will appear in complex conjugate pairs in the form

$$\lambda_i = -\zeta_i \omega_i - \omega_i \sqrt{1 - \zeta_i^2}\, j$$

$$\lambda_{i+1} = -\zeta_i \omega_i + \omega_i \sqrt{1 - \zeta_i^2}\, j \tag{9.84}$$

where $j = \sqrt{-1}$, ω_i is the undamped natural frequency of the ith mode and ζ_i is the *modal damping ratio* associated with the ith mode. The solution of the eigenvalue problem for the state matrix A of equation (9.82) produces a set of complex numbers of the form $\lambda_i = \alpha_i + \beta_i j$, where $\text{Re}(\lambda_i) = \alpha_i$ and $\text{Im}(\lambda_i) = \beta_i$. Comparing these expressions with equations (9.84) yields

$$\omega_i = \sqrt{\alpha_i^2 + \beta_i^2} = \sqrt{\text{Re}(\lambda_i)^2 + \text{Im}(\lambda_i)^2} \tag{9.85}$$

$$\zeta_i = \frac{-\alpha_i}{\sqrt{\alpha_i^2 + \beta_i^2}} = \frac{-\text{Re}(\lambda_i)}{\sqrt{\text{Re}(\lambda_i)^2 + \text{Im}(\lambda_i)^2}} \tag{9.86}$$

which provides a connection to the physical notions of natural frequency and damping ratios for the underdamped case. (See Inman, 1989, for the overdamped and critically damped cases.)

The complex-valued mode shape vectors $\mathbf{u}_i$ also appear in complex conjugate pairs and are referred to as *complex modes*. The physical interpretation of a complex mode is that each element describes the relative magnitude and phase of the motion of the degree of freedom associated with that element when the system is excited at that mode only. In the undamped real mode case, the mode shape vector is real (recall Section 4.1) and indicates the relative positions of each mass at any given instant of time at a single frequency. The difference between the real mode case and the complex mode case is that if the mode is complex, the relative position of each mass can also be out of phase by the amount indicated by the complex part of the mode shapes entry (recall that a complex number can be thought of as a magnitude and a phase rather than a real part and an imaginary part).

The state-space formulation of the eigenvalue problem for the matrix A given by equation (9.82) is related to the most general linear vibration problem. It also forms the most difficult computational eigenvalue problem of the five problems discussed above. The following example illustrates each of the various eigenvalue problems for a simple two-degree-of-freedom system. The example also serves to summarize the preceding.

The actual computations of the various eigenvalue problems are deferred to the following section.

Example 9.5.2

Consider the two-degree-of-freedom system of Figure 4.14 repeated in Figure 9.4. Consider the model of this system with the values (SI units) $m_1 = 9$, $m_2 = 1$, $k_1 = 24$, and $k_2 = 3$, For the undamped case, $c_1 = c_2 = 0$, of course. In the damped case, let $c_1 = 8$ and $c_2 = 1$.

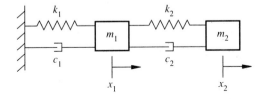

Figure 9.4 Simple two-degree-of-freedom system with damping.

Solution Using the techniques of Chapter 4, the equations of motion can be written in matrix form as

$$\begin{bmatrix} 9 & 0 \\ 0 & 1 \end{bmatrix} \ddot{\mathbf{x}} + \begin{bmatrix} 9 & -1 \\ -1 & 1 \end{bmatrix} \dot{\mathbf{x}} + \begin{bmatrix} 27 & -3 \\ -3 & 3 \end{bmatrix} \mathbf{x} = \mathbf{0} \tag{9.87}$$

so that

$$M = \begin{bmatrix} 9 & 0 \\ 0 & 1 \end{bmatrix} \qquad C = \begin{bmatrix} 9 & -1 \\ -1 & 1 \end{bmatrix} \qquad K = \begin{bmatrix} 27 & -3 \\ -3 & 3 \end{bmatrix} \tag{9.88}$$

With these values for the coefficient matrices, calculate the four different eigenvalue problems discussed above for the undamped case ($C = 0$) and the state-space eigenvalue problem for the damped case. First consider the generalized eigenvalue problem equation (9.60) (i.e., $K\mathbf{u} = \lambda M\mathbf{u}$). This yields simply

$$\begin{bmatrix} 27 & -3 \\ -3 & 3 \end{bmatrix} \mathbf{u} = \lambda \begin{bmatrix} 9 & 0 \\ 0 & 1 \end{bmatrix} \mathbf{u} \tag{9.89}$$

The characteristic equation for this form can simply be calculated to be

$$\det \left(\begin{bmatrix} 27 - 9\lambda & -3 \\ -3 & 3 - \lambda \end{bmatrix} \right) = 9\lambda^2 - 54\lambda + 72 = 9(\lambda - 2)(\lambda - 4) = 0 \tag{9.90}$$

Recall that this is not the preferred way to calculate the eigenvalues. However, the characteristic equation can be used as a convincing argument that each of the four eigenvalue problems produce the same natural frequencies. The solution of equation (9.90) yields $\lambda_1 = 2$, $\lambda_2 = 4$, so that $\omega_1 = \sqrt{2}$ and $\omega_2 = 2$.

Next consider the standard eigenvalue problems of equation (9.63) (i.e., $M^{-1}K\mathbf{u} = \lambda\mathbf{u}$). In this case the mass matrix M is diagonal so that M^{-1} is easy to compute; that is, $M^{-1} = \text{diag}\begin{bmatrix} \frac{1}{9} & 1 \end{bmatrix}$ and

$$M^{-1}K = \begin{bmatrix} \frac{1}{9} & 0 \\ 0 & 1 \end{bmatrix} \begin{bmatrix} 27 & -3 \\ -3 & 3 \end{bmatrix} = \begin{bmatrix} 3 & -\frac{1}{3} \\ -3 & 3 \end{bmatrix} \tag{9.91}$$

The standard eigenvalue problem is then

$$\begin{bmatrix} 3 & -\frac{1}{3} \\ -3 & 3 \end{bmatrix} \mathbf{u} = \lambda\mathbf{u} \tag{9.92}$$

Note that this is an asymmetric eigenvalue problem. The characteristic equation for this form becomes

$$\det \begin{bmatrix} 3-\lambda & -\frac{1}{3} \\ -3 & 3-\lambda \end{bmatrix} = \lambda^2 - 6\lambda + 8 = 0 \tag{9.93}$$

Note that this is identical to the characteristic equation generated by the previous form of the eigenvalue problem since dividing equation (9.90) by 9 does not change the value of λ. Thus the generalized eigenvalue problem and the standard eigenvalue problem results in the same eigenvalues (and hence the same natural frequencies, $\omega_1 = \sqrt{2}$ and $\omega_2 = 2$).

The third eigenvalue problem considered is the symmetric eigenvalue problem of equation (9.67). This requires calculation of the matrix $M^{-1/2}$, which is made simple in this example because M is diagonal. Hence $M^{-1/2} = \text{diag}[\,\frac{1}{3}\quad 1\,]$, so that

$$M^{-1/2}KM^{-1/2} = \begin{bmatrix} \frac{1}{3} & 0 \\ 0 & 1 \end{bmatrix}\begin{bmatrix} 27 & -3 \\ -3 & 3 \end{bmatrix}\begin{bmatrix} \frac{1}{3} & 0 \\ 0 & 1 \end{bmatrix} = \begin{bmatrix} 3 & -1 \\ -1 & 3 \end{bmatrix} \tag{9.94}$$

as was given in Example 4.2.1. The eigenvalue problem then becomes

$$\begin{bmatrix} 3 & -1 \\ -1 & 3 \end{bmatrix}\mathbf{v} = \lambda\mathbf{v} \tag{9.95}$$

Note that this eigenvalue problem is symmetric. The characteristic equation for this form becomes

$$\det \begin{bmatrix} 3-\lambda & -1 \\ -1 & 3-\lambda \end{bmatrix} = \lambda^2 - 6\lambda + 8 = 0 \tag{9.96}$$

which is identical to the other two characteristic equations given in equations (9.93) and (9.90).

Next consider the state-space formulation of the undamped system given by equation (9.74). The matrix $M^{-1}K$ is as calculated in equation (9.91), so that the state matrix of equation (9.74) becomes

$$A = \begin{bmatrix} 0 & 0 & 1 & 0 \\ 0 & 0 & 0 & 1 \\ -3 & \frac{1}{3} & 0 & 0 \\ 3 & -3 & 0 & 0 \end{bmatrix} \tag{9.97}$$

The state-space eigenvalue problem given by equation (9.76) is

$$\begin{bmatrix} 0 & 0 & 1 & 0 \\ 0 & 0 & 0 & 1 \\ -3 & \frac{1}{3} & 0 & 0 \\ 3 & -3 & 0 & 0 \end{bmatrix}\mathbf{z} = \lambda\mathbf{z} \tag{9.98}$$

which is not symmetric but contains many zero elements. It is also a 4×4 matrix eigenvalue problem rather than a 2×2 as in the previous three eigenvalue problems formulations (since $2n = 4$ in this case). The characteristic equation becomes

$$\det\left(\begin{bmatrix} -\lambda & 0 & 1 & 0 \\ 0 & -\lambda & 0 & 1 \\ -3 & \frac{1}{3} & -\lambda & 0 \\ 3 & -3 & 0 & -\lambda \end{bmatrix}\right) = (\lambda^2)^2 + 6(\lambda^2) + 8 = 0 \tag{9.99}$$

which yields the four complex eigenvalues $\lambda_1 = -\sqrt{2}j$, $\lambda_2 = \sqrt{2}j$, $\lambda_3 = -2j$, and $\lambda_4 = 2j$.

Note that the state-space formulation of the eigenvalue problem increases the number of eigenvalues from $2(n)$ to $4(2n)$, but does so by repeating each (natural frequencies). In the case of the first three eigenvalue problem formulations, the natural frequencies are $\omega_1^2 = 2$ and $\omega_2^2 = 4$, so that $\omega_1 = \pm\sqrt{2}$ and $\omega_2 = \pm 2$. In the state-space formulation for natural frequencies they are determined from $\lambda_i = \omega_i j$, so that again $\omega_1 = \pm\sqrt{2}$ and $\omega_2 = \pm 2$. Only the positive numbers are used as frequencies; hence in all cases the associated natural frequencies are $\omega_1 = \sqrt{2}$ and $\omega_2 = 2$.

Next consider the damped system and the associated state-space eigenvalue problem as defined by equations (9.81) and (9.82). The required blocks of the state matrix A are $M^{-1}K$, which is given in equation (9.91), and $M^{-1}C$. The value of $M^{-1}C$ is

$$M^{-1}C = \begin{bmatrix} \frac{1}{9} & 0 \\ 0 & 1 \end{bmatrix} \begin{bmatrix} 9 & -1 \\ -1 & 1 \end{bmatrix} = \begin{bmatrix} 1 & -\frac{1}{9} \\ -1 & 1 \end{bmatrix} \tag{9.100}$$

Forming the state matrix of equation (9.82) yields

$$A = \begin{bmatrix} 0 & 0 & 1 & 0 \\ 0 & 0 & 0 & 1 \\ -3 & \frac{1}{3} & -1 & \frac{1}{9} \\ 3 & -3 & 1 & -1 \end{bmatrix} \tag{9.101}$$

so that the state-space eigenvalue problem given by equation (9.81) is

$$\begin{bmatrix} 0 & 0 & 1 & 0 \\ 0 & 0 & 0 & 1 \\ -3 & \frac{1}{3} & -1 & \frac{1}{9} \\ 3 & -3 & 1 & 1 \end{bmatrix} \mathbf{z} = \lambda \mathbf{z} \tag{9.102}$$

Again this is an asymmetric 4×4 eigenvalue problem. The characteristic equation becomes

$$\det\left(\begin{bmatrix} 0-\lambda & 0 & 1 & 0 \\ 0 & 0-\lambda & 0 & 1 \\ -3 & \frac{1}{3} & -1-\lambda & \frac{1}{9} \\ 3 & -3 & 1 & -1-\lambda \end{bmatrix}\right) = \lambda^4 + 2\lambda^3 + \frac{62}{9}\lambda^2 + \frac{16}{3}\lambda + 8 = 0 \tag{9.103}$$

In this case there are four different roots. They are complex valued and appear in conjugate pairs, as predicted by equation (9.84). The values of λ can be determined by using a simple polynomial root finder. They are

$$\lambda_1 = -\frac{1}{3} - 1.3744j \qquad \lambda_2 = -\frac{1}{3} + 1.3744j$$

$$\tag{9.104}$$

$$\lambda_3 = -\frac{2}{3} - 1.8856j \qquad \lambda_4 = -\frac{2}{3} + 1.8856j$$

The presence of the damping term shows up in the characteristic equation by filling in the missing powers of the characteristic polynomial of the undamped case given by equation (9.99).

Equations (9.85) and (9.86) can be applied to the four roots of equation (9.104) to yield the values of the natural frequencies and damping ratios. They are (the values in parentheses are the calculated values)

$$\omega_1 = \sqrt{2} \ (1.4142) \qquad \omega_2 = 2 \ (1.9999)$$

$$\zeta_1 = \frac{1}{3\sqrt{2}} \ (0.2357) \qquad \zeta_2 = \tfrac{1}{3} \ (0.3333) \tag{9.105}$$

Note that the natural frequencies agree with those calculated by each of the four other eigenvalue problems.

Note also that the damping poses the most difficult computational problem, as it requires the evaluation of the roots of a fourth-order polynomial rather than a simple quadratic as required for the undamped system. While the method of solving the eigenvalue problem used in this example is not the preferred approach, it does illustrate the similarities and differences in the various formulations of the vibration problem.

□

A symmetric generalized eigenvalue problem can also be formulated for systems with damping in such a way as to avoid calculating the inverse of the matrix M. Note that if the mass matrix M has a very large element and a very small element, it could be computationally difficult to calculate M^{-1}. Consider substituting a new set of variables $\mathbf{y}_1 = \mathbf{x}$ and $\mathbf{y}_2 = \dot{\mathbf{x}}$ into the damped system of equation (9.78). This yields the expression

$$M\dot{\mathbf{y}}_2 = -C\mathbf{y}_2 - K\mathbf{y}_1 \tag{9.106}$$

Next consider the identity (recall that $\dot{\mathbf{y}}_1 = \mathbf{y}_2$)

$$-K\dot{\mathbf{y}}_1 = -K\mathbf{y}_2 \tag{9.107}$$

Combining equations (9.106) and (9.107) into a single equation in the vector $\mathbf{y}^T = [\,\mathbf{y}_1^T \quad \mathbf{y}_2^T\,]$ yields

$$\begin{bmatrix} -K & 0 \\ 0 & M \end{bmatrix} \begin{bmatrix} \dot{\mathbf{y}}_1 \\ \dot{\mathbf{y}}_2 \end{bmatrix} = \begin{bmatrix} 0 & -K \\ -K & -C \end{bmatrix} \begin{bmatrix} \mathbf{y}_1 \\ \mathbf{y}_2 \end{bmatrix} \tag{9.108}$$

This expression can be written as

$$A\dot{\mathbf{y}} = B\mathbf{y} \tag{9.109}$$

where

$$A = \begin{bmatrix} -K & 0 \\ 0 & M \end{bmatrix} \qquad B = \begin{bmatrix} 0 & -K \\ -K & -C \end{bmatrix} \qquad \mathbf{y} = \begin{bmatrix} \mathbf{x} \\ \dot{\mathbf{x}} \end{bmatrix} \tag{9.110}$$

Note that both matrices A and B are symmetric. Substitution of $\mathbf{y} = \mathbf{z}e^{\lambda t}$ into equation (9.109) yields the symmetric generalized eigenvalue problem

$$\lambda A\mathbf{z} = B\mathbf{z} \tag{9.111}$$

where $\lambda_i = \zeta_i\omega_i \pm (\omega_i\sqrt{1-\zeta_i^2})j$ relates the solution of the symmetric generalized eigenvalue problem to the calculation of natural frequencies and damping ratios. Note, this approach does not require the calculation of M^{-1}. This is important when the values of the various mass elements in a system differ by orders of magnitude.

The preceding examples are of low order and hence simple to calculate by solving for the characteristic equation and calculating its roots. For systems with a larger number of degrees of freedom, the matrix iteration procedure of Section 9.4 can be used to calculate the solution to the various eigenvalue problems associated with vibration analysis. However, more sophisticated methods for computing eigenvalues and eigenvectors have been developed and are readily available, as discussed in the next section.

9.6 COMPUTER CODES FOR MATRIX COMPUTATIONS (MATLAB)

The calculation of the solution of the linear algebraic equation $A\mathbf{x} = \mathbf{b}$ and the algebraic eigenvalue problem has formed the object of intensive study and algorithm development over the last 30 years. Many sophisticated methods have been developed to take advantage of special properties such as symmetry, definiteness, bandedness, and other sparse structures. Much of this work was funded by government agencies and hence is in the public domain. Perhaps the most powerful of these programs are LINPACK and EISPACK (see Dongarra et al., 1979, and Smith et al., 1976). As a result, many calculators have routines for calculating eigenvalues and eigenvectors for lower-order matrices. In addition, several commercially available and inexpensive (on the order of the cost of a textbook) software packages are available which provide numerically efficient routines for solving the eigenvalue, eigenvector, and linear equation problems. These codes are available to run on a variety of machines from small personal computers to mainframe and super computers. One such code is MATLAB, introduced at the end of Chapter 1. MATLAB (Moler, 1980) is actually a commercial computer program that packages the routines of LINPACK and EISPACK into a simple-to-manipulate high-level language (*Student Edition of* MATLAB, 1992). MATLAB is an interactive system in which it is possible to express matrix computations at a very high level. MATLAB is written in C and reduces complex algorithms such as the solution of linear equations or the solution of the algebraic eigenvalue problem to a single line of code. Such specialized languages eliminate the need for most engineers to write their own C or FORTRAN codes for solving eigenvalue problems in a majority of engineering applications.

In this section we introduce the use of the matrix subroutines for solving vibration problems in the user-friendly MATLAB format. A disk of MATLAB files for vibration analysis is included in the inside back cover of this book located in file VTB9. The disk requires the student edition of MATLAB to run. For those not having a MATLAB program, it is available for about one-half the price of a textbook (see the information on the inside back cover). The disk included here should work with any recent version of MATLAB. A discussion of MATLAB commands for solving the vibration problems of Section 9.5 is presented next. A detailed discussion of the various algorithms used in MATLAB can be found in Golub and Van Loan (1989).

Some of the material presented here also appears at the end of the preceding chapters. This material is repeated here for completeness. Recall from Chapter 4 that a matrix is entered into the MATLAB program via the simple command A = followed by a row with list of values. For example, the stiffness matrix of Example 9.5.2 is entered by typing

```
A = [27  -3 ;  -3  3]
```

where spaces separate the elements of a row and the semicolon separates the rows. Entering the matrix A in this fashion results in the output

```
A =
    27  -3
    -3   3
```

and stores the matrix A with these values. Note that no dimensioning of A is required. Carriage returns can be used to replace the colons for larger matrices. Matrices can be entered into MATLAB via other avenues, such as ASCII files (see student edition users' manual).

The elements of a matrix may also be valid MATLAB expressions. For example

```
k1 = 3
k2 = 4
e = 1000
p = 10
K = [k1+k2 -k2 0; -k2 k2+sqrt(e/p) -sqrt(e/p); 0 -sqrt(e/p) sqrt(e/p)]
```

results in the stiffness matrix

```
K =
    7   -4    0
   -4   14  -10
    0  -10   10
```

Some valid MATLAB expressions are illustrated in Table 9.1.

TABLE 9.1   USEFUL MATLAB EXPRESSIONS FOR THE SCALAR x

`abs(x)`	Absolute value of x	`log(x)`	Natural logarithm of x
`+`	Addition	`sqrt(x)`	Positive square root of x
`−`	Subtraction	`i, j`	Square root of minus one
`*`	Multiplication	`exp(x)`	e^x
`^`	Power ($5^2 \Rightarrow$ 5^2)	`sin(x)`	$\sin x$
`/`	Right division	`atan(x)`	$\tan^{-1} x$
`\`	Left division	`pi`	π, the number
`1.1e-6`	1.1×10^{-6}	`real(x)`	Real part of x
		`imag(x)`	Imaginary part of x

This value of the matrix K remains intact until it is altered, overwritten by another command, or until the program is terminated by the quit or exit commands. The value of K or any series of commands can be saved by simply typing save. Then K will be stored in a file named matlab.mat (save stiff saves K in a file named stiff.mat and load stiff returns the value of the matrix K).

The transpose of a matrix (or of a vector) is simply calculated by typing A' (apostrophe). For example, if **x** is the vector

$$\mathbf{x} = \begin{bmatrix} 1 \\ 2 \\ 3 \end{bmatrix}$$

entered as

```
X = [1 ; 2 ; 3]
```

then

```
X' =
      1   2   3
```

Matrix arithmetic operations are straightforward as long as the operations indicated are dimensionally compatible. For example, if

```
A = [1    2;   3   4]
B = [0    0;   1   2]
```

then

```
C = A + B
```

returns

```
C =
    1   2
    4   6
```

and

```
D = A * B
```

returns

```
D =
    2   4
    4   8
```

Similarly, if X = [0 2] is a row vector, then Y = D*X' returns

```
Y =
      8
     16
```

but Y = D*X is not defined (the dimensions are incompatible).

Matrix division is also possible in the sense of Gaussian elimination. If A and B are compatible, previously defined matrices, and if A is a nonsingular square matrix, then

```
X = A\B
```

produces the solution to the linear system of equations $AX = B$. Similarly,

```
X = B/A
```

produces the solution to $XA = B$. The inverse of a matrix A can also be computed by the simple command INV(A). Thus the equation $AX = B$ could also be solved by the command

```
X = inv(A)*B
```

However, this procedure is not as numerically stable as using matrix division. Table 9.2 lists other valid matrix functions.

TABLE 9.2   VALID MATRIX FUNCTIONS

sqrtm(A)	Computes the square root of the matrix A
expm(A)	Computes the matrix exponential of A
logm(A)	Computes the matrix logarithm of A
trace(A)	Computes the trace of the matrix A
det(A)	Computes the determinant of the matrix A

Example 9.6.1

Write a MATLAB program to compute the matrix $M^{-1/2}KM^{-1/2}$ used in Section 9.5.

Solution   To avoid using the inverse function it is wise to compute $\tilde{K} = M^{-1/2}KM^{-1/2}$ by solving $M^{1/2}\tilde{K}M^{1/2} = K$ for $\tilde{K}$ using left and right matrix division. The appropriate program is

```
M = [enter numerical values for mass matrix here]
K = [enter numerical values for the stiffness matrix here]
MS = sqrtm(M)
KH = (MS\K)/MS
```

Here, KH denotes K tilde, or $\tilde{K}$.

□

Powers of a matrix can also be computed using the power notation `^` so that `A^p` computes the pth power of the matrix A. If p is a positive integer, `A^p` is computed by successive multiplication. If p is not an integer then the eigensolution of A is first computed, and equation (9.45) for the function of a matrix is used to compute `A^p`. MATLAB has several special commands which are useful in vibration analysis. In particular, `eye(n)` returns an $n \times n$ identity matrix and `zeros(n)` defines an $n \times n$ matrix of zeros. The individual elements of a matrix, a single column or a single row, can be referred to by using the subscript notation for a matrix. Hence `M(1,2)` refers to the element of the matrix M in the first row, second column position. The notation `M(:,3)` refers to the vector consisting of the third column of the matrix M, and `M(1:5,3)` is the 5×1 vector consisting of the first five elements of the third column of M. These special matrices can be used to create the matrices of special structure used in vibration analysis, as the following example illustrates.

Example 9.6.2

Create the state matrix of Example 9.6.2 for the damped system of equation (9.78)

Solution

```
M = eye(2)  (or M = zeros(2))
M(1,1) = 9
M(2,2) = 1
```

creates the mass matrix of Example 9.5.2:

$$M = \begin{bmatrix} 9 & 0 \\ 0 & 1 \end{bmatrix}$$

Using the mass matrix M defined above, the commands

```
C = [9 -1; -1 1]
K = [27 -3, -3 3]
MIK = M\K
MIC = M\C
A = [zeros(2) eye(2); - MIK - MIC]
```

return

```
A = 0   0        1   0
    0   0        0   1
   -3   0.3333  -1   0.1111
    3  -3        1  -1
```

which agrees with equation (9.101).

□

The command `eig(A)`, where A is any square matrix, is used to compute the eigenvalues and right eigenvectors of the matrix A. The command

```
eig(A)
```

produces a column vector with entire entries equal to the eigenvalues of A (i.e., the n scalars λ satisfying $A\mathbf{x} = \lambda\mathbf{x}$).

Example 9.6.3

Compute the eigenvalues of the matrix A of equation (9.101).

Solution   Enter the matrix A

```
A = [0 0 1 0; 0 0 0 1; -3  1/3 -1 1/9; 3 -3 1 -1]
```

then

```
eig(A)
```

returns

```
ans =
   -0.3333   + 1.3744i
   -0.3333   - 1.3744i
   -0.6667   + 1.8856i
   -0.6667   - 1.8865i
```

☐

The normalized eigenvectors of the matrix A can be obtained by using the command

```
[V, D] = eig(A)
```

which creates a matrix V with columns equal to the eigenvectors of the matrix A and a diagonal matrix D with diagonal elements equal to the eigenvalues of the matrix A. The three matrices V, D, and A satisfy $AV = VD$. The eigenvectors are scaled so that the norm of each is 1.0. If A and B are two $n \times n$ matrices, the solution to the generalized eigenvalue problem $A\mathbf{x} = \lambda B\mathbf{x}$ is obtained by the command

```
[V, D] = eig(A,B)
```

which produces a matrix V whose columns are the eigenvectors of the system and a diagonal matrix D with diagonal elements corresponding to the eigenvalues of the system. The four matrices satisfy the equation $AV = BVD$.

The eigenvalue routines used in MATLAB make use of the Schur and Hessenburg forms and as such provide numerically efficient algorithms. One of the problems encountered in computing the eigenvalues of a matrix concerns the case where the elements of the matrix A have widely varying magnitudes. In such cases round-off error can cause significant error in the eigenvalue computations. To prevent this, the matrix A is often balanced before computing its eigenvalues. This generally leads to more accurate values of the system's eigenvalues. However, if the matrix A contains small elements, which are small because of round-off error, balancing may scale this number "up," which can lead to incorrect eigenvectors. In such cases the command eig(A,'no balance') can be used to eliminate the automatic balancing of the matrix.

Example 9.6.4

Write MATLAB programs to calculate the natural frequencies, mode shapes, and damping ratios of a general damped linear system of the form $M\ddot{\mathbf{x}} + C\dot{\mathbf{x}} + K\mathbf{x} = \mathbf{0}$.

Solution For example, with M, C, and K entered as data the program (the size command counts the number of rows and columns of the argument matrix and the max command gives the largest of these two values and the semicolon avoids printing the preceding execution)

```
n = max (size(M));
A = [zeros(n) eye(n); - M\K - M\C];
[V,D] = eig(A)
ReD = (D+D')/2;
ImD = (D'-D)*i/2;
W = (ReD^2+ImD^2).^.5
Z = -ReD/W
```

returns the eigenvectors, or complex mode shapes as the columns of the matrix V, the natural frequencies as the diagonal elements of the diagonal matrix W, and the damping ratios as the diagonal elements of the diagonal matrix Z. The matrix ReD is the diagonal matrix of the real parts of the eigenvalues and ImD is the diagonal matrix of the imaginary parts of the eigenvalues.   ☐

Example 9.6.5

Enter the matrices of Example 9.5.2, equation (9.88), and run the program of Example 9.6.4 to get the system's natural frequencies, damping ratios, and mode shapes.

Solution Enter the matrices

```
M = [9   0; 0   1];
C = [9   -1;- 1   1];
K = [27   - 3; - 3   3];
n = max (size (M));
A = [zeros(n) eye(n); - M\K - M\C];
[V,D] = eig(A)
```

returns

```
V =
    -0.1240 - 0.0679j  -0.1240 + 0.0679j   -0.1711 - 0.0637j  -0.1711 + 0.0637j
     0.3721 + 0.2038j   0.3721 - 0.2038j   -0.5131 - 0.1911j  -0.5131 + 0.1911j
     0.2108 - 0.1886j   0.2108 + 0.1886j    0.1446 - 0.2139j   0.1446 + 0.2139j
    -0.6323 + 0.5658j  -0.6323 - 0.5658j    0.4337 - 0.6418j   0.4337 + 0.6418j

D =
    -0.6667 + 1.8856j          0                   0                    0
           0         -0.6667 -1.8856j              0                    0
           0                   0          -0.3333 + 1.3744j             0
           0                   0                   0           -0.3333 - 1.3744j
```

Entering the commands

```
        ReD = (D+D')/2;
        ImD = (D'-D)*i/2;
        W = (ReD^2+ImD^2).^.5
```

returns

```
    W =
           2.0000            0               0               0
              0           2.0000             0               0
              0              0            1.4142             0
              0              0               0            1.4142
```

Entering the command

```
        z = - RealD/W
```

returns

```
    Z =
           0.3333            0               0               0
              0           0.3333             0               0
              0              0            0.2357             0
              0              0               0            0.2357
```

Note that the eigenvalues [i.e., the natural frequencies given by the matrix W (i.e., 2 and 1.4142) and the damping ratios given by the matrix Z (i.e., 0.3333 and 0.2357)] agree with the values given previously by equations (9.105). Recall that the values of ω and ζ are repeated because they join in pairs to form terms $e^{\zeta\omega t} \sin \omega t$ through the Euler formula (recall Window 1.5). Note also that the eigenvectors given by the matrix V above are complex conjugate pairs, but that the damping matrix is proportional so that real eigenvectors can be found through proper orthogonalization. In addition, the matrices V and W are not ordered according to the common practice in vibration analysis of listing the lowest frequency and its associated eigenvector first. Use the SORT command.

☐

9.7 COMPUTER SIMULATION OF THE TIME RESPONSE

The time history of the response of a vibrating system is easily calculated in closed form for the simple, free-response of the single-degree-of-freedom system of Chapter 1. In Chapter 2 the response to a harmonic input is also relatively easy to obtain in closed form. However, computing the response to an arbitrary input, as discussed in Chapter 3, does not always result in closed-form analytical solutions. The time response of the multiple-degree-of-freedom systems of Chapter 4 can be computed by first solving the eigenvalue and eigenvector problems and forming the series solutions for the time response indicated in equations (4.115) and (4.131) by modal analysis. However, this is not numerically efficient. It is more efficient to compute the time response using numerical procedures of time integration and differential equation solving.

There are many schemes for numerically solving ordinary differential equations, such as those of vibration analysis listed in Section 9.5. Several numerical solution schemes are presented here. The basis of numerical solutions of ordinary differential equations is to essentially undo calculus by representing each derivative by a small but finite difference (recall the definition of a derivative from calculus given in Window 9.4). A numerical solution of an ordinary differential equation is a procedure for constructing approximate values: $x_1, x_2, \ldots, x_n$ of the solution $x(t)$ at the discrete values of time: $t_0 < t_1 < t_2 < \cdots < t_n$. Effectively, a numerical procedure produces a list of discrete values $x_i = x(t_i)$ that approximate the solution rather than a continuous function $x(t)$, which is the exact solution. The initial conditions of the vibration problem of interest form the starting point. For a single-degree-of-freedom system of the form

$$m\ddot{x} + c\dot{x} + kx = 0 \qquad x(0) = x_0 \qquad \dot{x}(0) = \dot{x}_0 \qquad (9.112)$$

the initial values x_0 and $\dot{x}_0$ form the first two points of the numerical solution. Let T be the total length of time over which the solution is of interest (i.e., the equation is to be solved for values of t between $t = 0$ and $t = T$). The time interval $T - 0$ is then

---

Window 9.4

The definition of a derivative of $x(t)$ at $t = t_i$ is

$$\frac{dx(t_i)}{dt} = \lim_{\Delta t \to 0} \frac{x(t_{i+1}) - x(t_i)}{\Delta t}$$

where $t_{i+1} = t_i + \Delta t$ and $x(t)$ is continuous.

divided up into n intervals (so that $\Delta t = T/n$). Then equation (9.112) is calculated at the values of $t_0 = 0, t_1 = \Delta t, t_2 = 2\Delta t, \ldots, t_n = n\Delta t = T$ to produce an approximate representation, or simulation, of the solution.

The concept of a numerical solution is easiest to grasp by first examining the numerical solution of a first-order scalar differential equation. This can then be extended to a first-order vector differential equation which subsequently solves the vibration problem in the form of equation (9.81) discussed in Section 9.5. To this end consider the first-order differential equation

$$\dot{x}(t) = ax(t) \qquad x(0) = x_0 \tag{9.113}$$

The Euler method proceeds from the definition of the slope form of the derivative given in Window 9.4, before the limit is taken:

$$\frac{x_{i+1} - x_i}{\Delta t} = ax_i \tag{9.114}$$

where x_i denotes $x(t_i)$, x_{i+1} denotes $x(t_{i+1})$, and Δt indicates the time interval between t_i and t_{i+1} (i.e., $\Delta t = t_{i+1} - t_i$). This expression can be manipulated to yield

$$x_{i+1} = x_i + \Delta t(ax_i) \tag{9.115}$$

This formula computes the discrete value of the response x_{i+1} from the previous value x_i, the time step Δt, and the system's parameter a. This numerical solution is called an *Euler* or *tangent line method*. The following example illustrates the use of the Euler formula for computing a solution.

Example 9.7.1

Use the Euler formula to compute the numerical solution of $\dot{x} = -3x$, $x(0) = 1$ for various time increments in the time interval 0 to 4, and compare the results to the exact solution.

Solution   First, the exact solution can be obtained by direct integration or by assuming a solution of the form $x(t) = Ae^{at}$. Substitution of this assumed form into the equation $\dot{x} = -3x$ yields $Aae^{at} = -3Ae^{at}$ or $a = -3$, so that the solution is of the form $x(t) = Ae^{-3t}$. Applying the initial conditions $x(0) = 1$ yields $A = 1$. Hence the analytical solution is simply $x(t) = e^{-3t}$.

Next consider a numerical solution using the Euler method suggested by equation (9.115). In this case the constant $a = -3$, so that $x_{i+1} = x_i + \Delta t(-3x_i)$. Suppose that a very crude time step is taken (i.e., $\Delta t = 0.5$) and the solution is formed over the interval from $t = 0$ to $t = 4$. Then Table 9.3 illustrates the values obtained from equation (9.115):

$$x_0 = 1$$

$$x_1 = x_0 + (0.5)(-3)(x_0) = -0.5$$

$$x_2 = -0.5 - (1.5)(-0.5) = 0.25$$

$$\vdots$$

forms the column marked "Euler." The column marked "exact" is the value of e^{-3t} at the indicated elapsed time for a given index. Note that while the Euler approximate gets close

TABLE 9.3 COMPARISON OF THE EXACT SOLUTION
OF $\dot{x} = -3x, x(0) = 1$ TO THE SOLUTION OBTAINED BY
THE EULER METHOD WITH LARGE TIME STEP FOR THE
INTERVAL $t = 0$ TO 4

Index	Elapsed time	Exact	Euler	Absolute error
0	0	1.0000	1.0000	0
1	0.5000	0.2231	−0.5000	−0.7231
2	1.0000	0.0498	0.2500	0.2002
3	1.5000	0.0111	−0.1250	−0.1361
4	2.0000	0.0025	0.0625	0.0600
5	2.5000	0.0006	−0.0312	−0.0318
6	3.0000	0.0001	0.0156	0.0155
7	3.5000	0.0000	−0.0078	−0.0078
8	4.0000	0.0000	0.0039	0.0039

to the correct final value, the value oscillates around zero while the exact value does not. This points out a possible source of error in a numerical solution. On the other hand, if Δt is taken to be very small, the difference between the solution obtained by the Euler equation and the exact solution becomes hard to see as Figure 9.5 illustrates. Figure 9.5 is a plot of $x(t)$ obtained via the Euler formula for $\Delta t = 0.1$. Note that it looks very much like the exact solution e^{-3t}.   ☐

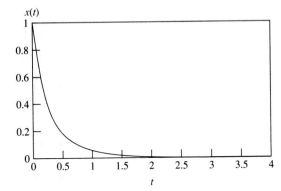

Figure 9.5 Plot of $x(t_i)$ versus t_i for $\dot{x} = -3x$ using $\Delta t = 0.1$ in equation (9.115) with $x(0) = 1$.

It is important to note from the example that two sources of error are present in computing the solution of a differential equation using a numerical scheme such as the Euler method. The first is called the *formula error*, which is the difference between the exact solution and the solution obtained by the Euler approximation. This is the error indicated in the last column of Table 9.3. Note that this error accumulates as the index increases because the value at each discrete time is determined by the previous value, which is already in error. This can be somewhat controlled by the time step and the nature of the formula. The other source of error is the *round-off error* as discussed in

Section 9.1, due to machine arithmetic. This is, of course, controlled by the computer and its architecture. Both sources of error can be significant. The successful use of a numerical method requires an awareness of both sources of errors in interpreting the results of a computer simulation of the solution of any vibration problem.

The Euler method can be improved upon by decreasing the step size, as Example 9.7.1 illustrates. Alternatively, a more accurate procedure can be used to improve the accuracy (smaller formula error) without decreasing the step size Δt. Several methods exist (such as the improved Euler method and various Taylor series methods) and are discussed in Boyce and DiPrima (1986), for instance. Only the Runge–Kutta method is presented and used here.

The Runge–Kutta method was developed by two different researchers from about 1895 to 1901 (C. Runge and M. W. Kutta). The Runge–Kutta formulas (there are several) involve a weighted average of values of the right-hand side of the differential equation taken at different points between the time intervals t_i and $t_i + \Delta t$. The derivations of various Runge–Kutta formulas are tedious but straightforward and are not presented here (see Boyce and DiPrima, 1986). One useful formulation can be stated for the first-order problem $\dot{x} = f(x, t)$, $x(0) = x_0$, where f is any scalar function as

$$x_{n+1} = x_n + \frac{\Delta t}{6}(k_{n1} + 2k_{n2} + 2k_{n3} + k_{n4}) \tag{9.116}$$

where

$$k_{n1} = f(x_n, t_n)$$

$$k_{n2} = f\left(x_n + \frac{\Delta t}{2}k_{n1}, t_n + \frac{\Delta t}{2}\right)$$

$$k_{n3} = f\left(x_n + \frac{\Delta t}{2}k_{n2}, t_n + \frac{\Delta t}{2}\right)$$

$$k_{n4} = f(x_n + \Delta t k_{n3}, t_n + \Delta t)$$

The sum in parentheses in equation (9.116) represents the average of six numbers each of which looks like a slope at a different time; for instance, the term k_{n1} is the slope of the function at the "left" end of the time interval.

Such formulas can be enhanced by treating Δt as a variable, Δt_i. At each time step t_i, the value of Δt_i is adjusted based on how rapidly the solution $x(t)$ is changing. If the solution is not changing very rapidly, a large value of Δt_i is allowed without increasing the formula error. On the other hand, if $x(t)$ is changing rapidly, a small Δt_i must be chosen to keep the formula error small. Such step sizes can be chosen automatically as part of the computer code for implementing the numerical solution. The Runge–Kutta and Euler formulas listed above can be applied to vibration problems by recalling from Section 9.5 that the most general (damped) vibration problem can be put into a first-order form following equation (9.81).

Consider the forced response of a damped linear system. From Chapter 4 the most general case can be written as

$$M\ddot{x} + C\dot{x} + Kx = F(t) \qquad x(0) = x_0, \quad \dot{x}(0) = \dot{x}_0 \qquad (9.117)$$

Following equation (9.80), let $\dot{y}_1 = x$ and $y_2 = \dot{x}$. Then multiplying equation (9.117) by M^{-1} yields the coupled first-order vector equations.

$$\dot{y}_1 = y_2$$
$$\dot{y}_2 = -M^{-1}Ky_1 - M^{-1}Cy_2 + M^{-1}F(t) \qquad (9.118)$$

with initial conditions $y_1(0) = x_0$ and $y_2(0) = \dot{x}_0$. Equation (9.118) can be written as the single first-order equation

$$\dot{y}(t) = Ay(t) + f(t) \qquad y(0) = y_0 \qquad (9.119)$$

where A is the state matrix given by equation (9.82):

$$A = \begin{bmatrix} 0 & I \\ -M^{-1}K & -M^{-1}C \end{bmatrix}$$

and

$$y(t) = \begin{bmatrix} y_1(t) \\ y_2(t) \end{bmatrix} \qquad f(t) = \begin{bmatrix} 0 \\ M^{-1}F(t) \end{bmatrix} \qquad y_0 = \begin{bmatrix} y_1(0) \\ y_2(0) \end{bmatrix}$$

Here $y(t)$ is the $(2n \times 1)$ state vector, where the first $n \times 1$ elements correspond to the displacement $x(t)$ and where the second $n \times 1$ elements correspond to the velocities $\dot{x}(t)$. Equation (9.119) is a vector form (with $f = 0$) of equation (9.113), which was solved by the Euler method.

Effectively, the Euler method of numerical solution given in equation (9.115) can be applied directly to the vector (state-space) formulation given in equation (9.119), simply by calling the scalar x_i the vector y_i and replacing the scalar a with the matrix A:

$$y(t_{i+1}) = y(t_i) + \Delta t \, Ay(t_i) \qquad (9.120)$$

which defines an Euler formula for integrating the general vibration problem described in equation (9.119) for the zero-force input case. This can be extended to the forced response case by including the term $f_i = f(t_i)$:

$$y_{i+1} = y_i + \Delta t \, Ay_i + f_i \qquad (9.121)$$

where y_{i+1} denotes $y(t_{i+1})$, and so on, using $y(0)$ as the initial value.

Equation (9.121) can be used to integrate a general linear vibration problem to compute its time response. The formulation used in Example 9.6.2 lists a numerically efficient method of computing the state-matrix A given the mass, damping, and stiffness matrices using MATLAB. The following example illustrates the procedure for a simple two-degree-of-freedom system.

Example 9.7.2

Solve the system given in Figure 9.4 with equations of motion as given by equation (9.87). From the matrix values of equation (9.88), the equation of motion becomes

$$\begin{bmatrix} 9 & 0 \\ 0 & 1 \end{bmatrix}\ddot{\mathbf{x}} + \begin{bmatrix} 9 & -1 \\ -1 & 1 \end{bmatrix}\dot{\mathbf{x}} + \begin{bmatrix} 27 & -3 \\ -3 & 3 \end{bmatrix}\mathbf{x} = \mathbf{0}$$

Choose an initial condition of $\mathbf{x}(0) = [0 \quad 0]^T$, $\dot{\mathbf{x}}(0) = [1 \quad 0]^T$ and plot the solutions [i.e., the time histories $x_1(t)$ and $x_2(t)$] for the interval $t = 0$ to $t = 5$, using MATLAB.

Solution   The heart of the problem is to program equation (9.121). This generally requires a FOR loop (or DO loop). In MATLAB a "Do loop" is written in one line. For example, to assign 0 to the first 10 elements of a vector $\mathbf{x}(i)$ requires the single line: for i = 1:10, x(i) = 0, end. However, this can be better accomplished by the single line X(1:10) = ZEROS(10,1). Program VTB9_4 in file/directory VTB9 in the *Vibration Toolbox* uses this statement applied to equation (9.121) to produce a matrix, each column of which is the time response in sequence. The input to this file consists of the three matrices M, C, and K: the initial conditions, the size of the time increment, and the final time. In particular, typing

```
n = 50;
dt = .1;
x0 = [0 0]';
xd0 = [1 0]';
m = [9 0; 0 1];
c = [9 -1;-1 1];
k = [27 -3;-3 3];
[x,v] = VTB9_4(n,dt,x0,xd0,m,c,k)
```

returns the two rectangular matrices $\mathbf{x}$ and $\mathbf{v}$. The columns of the matrix $\mathbf{x}$ are the values of displacement response $\mathbf{x}(t)$ at each time step. The value of each column of $\mathbf{v}$ is the velocity at each time step. The output x is printed in Table 9.4.   □

As suggested above, the Euler formula method can be greatly improved by using a Runge–Kutta program. MATLAB has two different Runge–Kutta-based simulations: ode23 and ode45. These are automatic step-size integration methods. The M-file ode23 uses a simple second- and third-order pair of formulas for medium accuracy and ode45 uses a fourth- and fifth-order pair for greater accuracy. Each of these corresponds to a formulation similar to that expressed in equations (9.116) with more terms and a variable step size Δt. In general, the Runge–Kutta simulations are of a higher quality than those obtained by the Euler method.

Example 9.7.3

Use the ode23 file to simulate the response to $3\ddot{x} + \dot{x} + 2x = 0$ subject to the initial conditions $x(0) = 0$, $\dot{x}(0) = 0.25$ over the time interval $0 \le t \le 20$.

TABLE 9.4 OUTPUT OF THE DISPLACEMENT FOR THE SYSTEM OF EXAMPLE 9.7.3[a]

x =
Columns 1 through 7

0	0.1000	0.1900	0.2681	0.3330	0.3838	0.4202
0	0	0.0100	0.0310	0.0631	0.1056	0.1570

Columns 8 through 14

0.4424	0.4509	0.4467	0.4311	0.4054	0.3714	0.3307
0.2153	0.2779	0.3418	0.4042	0.4619	0.5120	0.5521

Columns 15 through 21

0.2850	0.2362	0.1858	0.1353	0.0861	0.0393	-0.0042
0.5798	0.5936	0.5922	0.5753	0.5428	0.4954	0.4344

Columns 22 through 28

-0.0434	-0.0780	-0.1076	-0.1319	-0.1510	-0.1650	0.1741
0.3614	0.2787	0.1886	0.0939	-0.0027	-0.0983	0.1901

Columns 29 through 35

-0.1787	-0.1792	-0.1761	-0.1698	-0.1609	-0.1498	-0.1370
-0.2757	-0.3528	-0.4192	-0.4735	-0.5145	-0.5413	-0.5538

Columns 36 through 42

-0.1229	-0.1080	-0.0925	-0.0768	-0.0612	-0.0458	-0.0308
-0.5520	-0.5364	-0.5081	-0.4681	-0.4182	-0.3599	-0.2952

Columns 43 through 49

-0.0165	-0.0029	0.0099	0.0218	0.0327	0.0426	0.0514
-0.2260	-0.1544	-0.0824	-0.0117	0.0559	0.1189	0.1758

Columns 50 through 51

0.0590	0.0655
0.2256	0.2675

[a] The first element in column 1 is the value of $x_1(t)$ at the first time step. The second element is the value of $x_2(t)$ at the first time step. For example, where the first element of column 8 is the value of $x_1(t)$ at $t = (8 - 1) dt = (7)(0.1)$ s [i.e., $x_1(0.7) = 0.4424$]. Similarly, $x_2(0.7) = 0.2153$.

Solution The first step is to write the equation of motion in first-order form. This yields

$$\dot{x}_1 = x_2$$

$$\dot{x}_2 = -\tfrac{2}{3}x_1 - \tfrac{1}{3}x_2$$

Next an M-file is created to store the equations of motion. An M-file is created by choosing a name, say sdof.m, and entering (open New under the file window for a Macintosh system or use the save command in DOS).

```
function xdot = sdof(t,x)
xdot = zeros(2,1);
xdot(1) = x(2);
xdot(2) = -(2/3)*x(1)-(1/3)*x(2);
```

Next, go to the command mode and enter

```
t0 =0;tf = 20;
x0 = [0  0.25]';
```

```
[t,x] = ode23('sdof',t0,tf,x0);
plot(t,x)
```

The first line establishes the initial (t0) and final times (tf). The second line creates the vector containing the initial conditions x0. The third line creates the two vectors t, containing the time history and x containing the response at each time increment in t, by calling ode23 applied to the equations set up in sdof. The fourth line plots the vector x versus the vector t. This is illustrated in Figure 9.6.

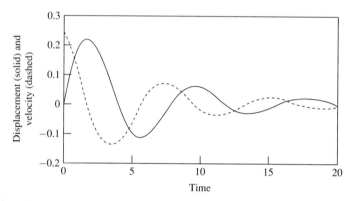

Figure 9.6   Plot of the displacement $x(t)$ of the single-degree-of-freedom system of Example 9.7.4 (solid line) and the corresponding velocity $\dot{x}(t)$ (dashed line).

Toolbox file VTB9 contains several files for numerical simulation of multiple-degree-of-freedom systems including both the Euler method and Runge–Kutta methods.

PROBLEMS

Section 9.1

9.1. Consider numbers restricted to four significant figures in base 10 ($\beta = 10$) given by π, e, and $\sqrt{2}$. Write these numbers in normalized form and in each case show that $\beta^{-1} \leq f < 1$ is satisfied.

9.2. Again consider the numbers of π, e, and $\sqrt{2}$. Represent these numbers using eight significant figures in normalized form first using chopping and then using symmetric rounding.

9.3. Determine the limits of your computing device by checking the owners manual or by trial and error (i.e., determine how many significant figures are used, etc.).

9.4. Consider multiplying, subtracting, and adding the two numbers 1.234 and 0.1234 using four significant figures. What is the round-off error?

9.5. A machine assigns two digits to the exponent in scientific notation of a floating-point number. What is the largest number it can handle before overflow? What is the smallest number it can handle before underflow? Assume four significant digits. What is the overflow value for your computing device?

9.6. Determine the number of terms needed in the Taylor series for the exponential e to estimate its value to three decimal places.

9.7. Calculate $\sum(1/n)$ using the first 500 terms. How much better is this than the value given in Section 9.1 for 200 terms?

9.8. Use a computer or programmable calculator to evaluate the sums $\sum(1/n)$ stopping when
  (a) the last term becomes zero to three decimals places, and
  (b) two adjacent sums yield the same value to three decimal places.
  Next compare the answer to (a) and (b) in terms of the number of terms in each step. Note from Problem 9.7 that neither of these is correct; but both seem sensible if the results were unknown.

Section 9.2

9.9. Referring to the flexibility influence matrix calculation of Example 9.2.1, show that $A = K^{-1}$ by calculating KA and AK and showing that $KA = AK = I$, the 3×3 identity matrix.

9.10. Consider the two-degree-of-freedom spring–mass system of Figure 9.7 and calculate the flexibility influence matrix. Also calculate the inverse of the matrix K given by Equation (4.9) using the inverse formula of Example 4.1.4 and verify that the influence matrix is correct.

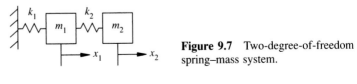

Figure 9.7   Two-degree-of-freedom spring–mass system.

9.11. Consider again the system of Figure 9.7 with matrices A and K^{-1} as calculated in Problem 9.10. Discuss the differences in using the influence matrix versus the matrix inverse formula of Example 4.1.4 to calculate the matrix A using four significant figures and chopping for the values $k_1 = 3$ and $k_2 = 4$.

9.12. Apply Dunkerley's formula to the building vibration problem of Example 4.4.3 and compare this estimate of ω_1 with the exact value given in the example of $\omega_1 = 0.3883$. Discuss the difference in the effort required for each calculation.

9.13. Consider the beam of Figure 9.8, which is simply supported and carries three equal masses. Such a configuration can be used to model the vibration of a floor of a building in a factory with heavy machinery mounted on the floor. Use flexibility coefficients and Dunkerley's formula to estimate the fundamental natural frequency. Compare your results to the exact fundamental frequency of the floor without the machine mass added.

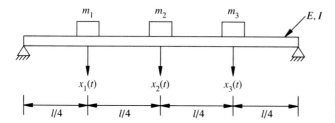

Figure 9.8 Model of the vibration of the floor of a machine room in a commercial building.

9.14. Consider the floor of Figure 9.8 again with only one mass added at point $l/3$. Without the machine mass m, the mass of the floor is M and the fundamental natural frequency is

$$\omega = \pi^2 \sqrt{\frac{EI}{Ml^3}}$$

Estimate the new natural frequency when the machine of mass m is bolted to the floor.

9.15. A three-story building is protected from mild earthquake loads by constructing it such that the foundation and building can slide or move elastically as indicated in Figure 9.9. Estimate the lowest natural frequency in terms of wall stiffness (EI), the floor masses (assumed to be equal to m), the height of the floors (assumed equal to h), and the foundation stiffness k_0.

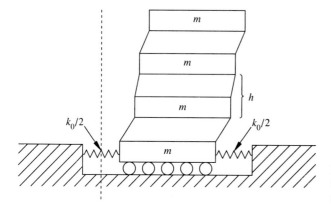

Figure 9.9 Simple model of a three-story building, indicating an earthquake protection device.

9.16. Again consider the fighter plane of Figure 9.3, this time examining the transverse vibration of a wing pod added at the end of the wing (at length l from the center) of mass M equal to the mass of the wing. Model the wing as a uniform cantilevered beam with a concentrated mass at the end. Calculate the fundamental natural frequency by considering the loaded wing and then the massless wing and apply Dunkerley's formula.

9.17. Recall from Chapter 7 that the problem of mass loading in vibration tests can produce erroneous frequency measurements. Dunkerley's formula can be used to sort out the fundamental frequency of the test structure, the measured frequency and the frequency due to the shaker mounted on the structure. Let ω_1 be the measured frequency (i.e., that of the structure and the shaker combined). Let ω_{11} be the frequency of the structure itself and let ω_{22} be the natural frequency of the exciter of mass m_2. Derive a correction equation using Dunkerley's formula with $n = 2$.

9.18. Turn to Figure 4.37 of Problem 4.46. Use Dunkerley's formula on the influence coefficient method to estimate the lowest natural frequency. If you worked Problem 4.46, compare your answer to the exact solution. Are the assumptions of Dunkerley's formula satisfied?

9.19. Solve the floor building vibration problem of Figure 4.38 for the fundamental natural frequency. Compare the results to the alternative model of Figure 9.8.

9.20. Use Dunkerley's formula to estimate ω_1 of the four-story building model of Example 4.4.3 and Figure 4.9.

Section 9.3

9.21. Consider the system of Example 4.1.3. Pick four different vectors and calculate the Rayleigh quotient for each vector. Compare these numbers to the first, or lowest, natural frequency of the system.

9.22. Again consider the system of Example 4.8.3. Calculate the stiffness influence matrix. Use this matrix with Dunkerley's formula and the Rayleigh's quotient estimates of Example 9.3.1 to find upper and lower bounds on the lowest natural frequency.

9.23. Consider Example 9.3.1. Use Rayleigh's quotient and the trial vector $\mathbf{u}^T = [\,1 \quad 1 \quad -1\,]$ to estimate ω_1. Compare these results with the trial vector $\mathbf{u}^T = 10[\,1 \quad 1 \quad -1\,] = [\,10 \quad 10 \quad -10\,]$ and the trial vector $\alpha\mathbf{u}^T$ where α is any constant. What conclusion can you draw?

9.24. Write a MATLAB program to calculate Rayleigh's quotient. Use this to try to obtain the first natural frequency and eigenvector of the system of Problem 9.21. Note that the value of the vector $\mathbf{u}$ such that $R(\mathbf{u})$ is a minimum is the first eigenvector.

9.25. Convince yourself that the minimum value of $R(\mathbf{u})$ corresponds to the first natural frequency squared by differentiating $R(\mathbf{u})$. Note that the derivative of $\mathbf{x}^T A\mathbf{x}$ with respect to the vector $\mathbf{x}$ for symmetric values of A is $2A\mathbf{x}$.

9.26. Use Rayleigh's quotient to determine an estimate of the first natural frequency of the system of Figure 9.8.

9.27. Use Dunkerley's formula and Rayleigh's quotient to estimate bounds on the first natural frequency of the machine room floor system of Figure 9.8.

9.28. Use Dunkerley's formula and Rayleigh's quotient to find bounds on the first natural frequency of the building system given in Figure 9.9.

Section 9.4

9.29. Compute the first eigenvalue of the system of Example 9.2.3 using matrix iteration. Also compute the first eigenvector.

9.30. Use the results of Problem 9.29 to compute the deflected matrix $D = A - \mu\mathbf{u}_1^T\mathbf{u}_1$ and then compute μ_2, the second eigenvalue.

Section 9.5

9.31. Consider again the system of Figure 9.7. Write the equation of motion in the form of equation (9.39).

9.32. Consider the machine floor system of Figure 9.8. Write the equation of motion in the form of equation (9.39).

9.33. Consider the building model of Figure 9.9. Write the equation of motion in the form of equation (9.39).

9.34. Consider the mass matrix

$$M = \begin{bmatrix} 10 & -1 \\ -1 & 1 \end{bmatrix}$$

and calculate M^{-1} and $M^{-1/2}$.

9.35. Following Example 9.5.1, show that

$$M^{-1/2}M^{1/2} = I$$

$$M^{1/2}M^{1/2} = M$$

9.36. Consider the problem of Figure 9.7. Write the eigenvalue problem for the system in the form of equation (9.60) and compare this to equation (9.63).

9.37. Consider the symmetric form of the conservative eigenvalue problem stated in equations (9.65) to (9.67). Use the mass and stiffness matrices of Figure 9.7 to compare the eigenvectors of equations (9.67) with those of equation (9.63). Are they related by $M^{1/2}$?

9.38. Write down the eigenvalue problem for the problem of Figures 9.8 and 9.9 for the symmetric form given by equation (9.67).

9.39. Consider the undamped machine floor vibration problem of Figure 9.8. Write the equation of motion and corresponding eigenvalue problem in the first-order state-space form given by equation (9.74).

9.40. Repeat Problem 9.39 for the system of Figure 9.9.

9.41. Again consider the undamped machine floor vibration problem of Figure 9.8. Write the eigenvalue problem of equation (9.111) for this system.

9.42. Solve the problem of Example 9.5.2 using the formulation of equation (9.111) and compare your answer with that obtained in Example 9.5.2.

9.43. Write a MATLAB program to compute the Rayleigh quotient.

9.44. Consider the matrix and vector

$$A = \begin{bmatrix} 1 & -\varepsilon \\ -\varepsilon & \varepsilon \end{bmatrix} \qquad b = \begin{bmatrix} 10 \\ 10 \end{bmatrix}$$

write a MATLAB program to solve $Ax = b$ for $\varepsilon = 0.1, 0.01, 0.001, 10^{-6}$, and 1. Solve this by using either INV(A), A\b or both.

9.45. Write a MATLAB program to calculate the natural frequencies and mode shapes of the system of Example 4.8.3. Use the undamped equation and the form given by equation (9.60).

9.46. Write a MATLAB program to compute the natural frequencies and mode shapes of the undamped version of the system of Example 4.8.3 using the formulation of equations (9.63) and (9.67). Discuss which is faster. Use the function Flops, which returns the cumulative number of floating-point operations.

9.47. Repeat Problem 9.46 by comparing the state-space formulation of equations (9.74) and (9.76) with that of the first-order form given in equation (9.111). Note that Flops(0) resets the count.

9.48. Consider the damped system of equation (9.81). Write a MATLAB program to calculate the natural frequencies, damping ratios, and mode shapes.

9.49. Write a MATLAB program to calculate the natural frequencies, mode shapes, and damping ratios using the first-order form given by equation (9.111).

9.50. Consider the damped system of Example 4.8.1. Use the programs of Problems 9.48 and 9.49 to solve for the natural frequencies, mode shapes, and damping ratios. Use `Flops` to compare your results.

Section 9.6 (Refer to the MATLAB Vibration Toolbox)

9.51. Use file VTB9_1 to solve for the modal information of Example 4.1.5.

9.52. Write a program to perform the normalization of Example 4.4.2 (i.e., calculate α such that the vector $\alpha\mathbf{x}$ is normal).

9.53. Use file VTB9_1 to calculate the natural frequencies and mode shapes obtained for the system of Example 4.2.5 and Figure 4.4.

9.54. Following the modal analysis solution of Window 4.4, write a MATLAB program to compute the time response of the system of Example 4.3.2.

9.55. Redo Problem 9.11 for Figure 9.7 using the eigenvalue approach of file VTB9_1 and compare your answer to that of Problem 9.11.

9.56. Solve Example 9.2.4 for the vibration of a wing with a fuel pod using file VTB9_1 and compare the frequencies with those obtained in the example by using Dunkerley's formula.

9.57. Use file VTB9_2 to solve the damped vibration problem of Example 4.6.1 by calculating the natural frequencies, damping ratios, and mode shapes.

9.58. Consider the vibration of the airplane of Problems 4.43 and 4.44 as given in Figure 4.36, reproduced as Figure 9.10.

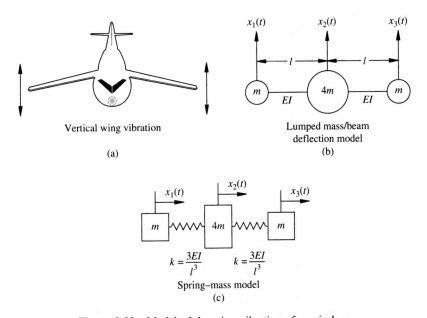

Figure 9.10   Model of the wing vibration of an airplane.

The mass and stiffness matrices are given as

$$M = m \begin{bmatrix} 1 & 0 & 0 \\ 0 & 4 & 0 \\ 0 & 0 & 1 \end{bmatrix} \qquad K = \frac{EI}{l^3} \begin{bmatrix} 3 & -3 & 0 \\ -3 & 6 & -3 \\ 0 & -3 & 3 \end{bmatrix}$$

where $m = 3000$ kg, $l = 2$ m, $I = 5.2 \times 10^{-6}$ m^4, $E = 6.9 \times 10^9$ N/m^2, and the damping matrix C is taken to be $C = (0.002)K$. Use file VBT9_1 to calculate the natural frequencies, mode shapes, and damping ratios.

9.59. Consider the proportionally damped, dynamically coupled system given by

$$M = \begin{bmatrix} 9 & -1 \\ -1 & 1 \end{bmatrix} \qquad C = \begin{bmatrix} 3 & -1 \\ -1 & 1 \end{bmatrix} \qquad K = \begin{bmatrix} 49 & -2 \\ -2 & 2 \end{bmatrix}$$

and use file VTB9_2 to calculate the mode shapes, natural frequencies, and damping ratios.

Section 9.7 (refer to the MATLAB Vibration Toolbox)

9.60. Run the program listed for Table 9.4 and check to see that you get the same results.

9.61. Solve the system of Example 1.7.3 for the vertical suspension system of a car with $m = 1361$ kg, $k = 2.668 \times 10^5$ N/m, and $c = 3.81 \times 10^4$ kg/s subject to the initial condition of $x(0) = 0$ and $\dot{x}(0) = 0.01$.

9.62. Use both the Euler approach and the ode34 file to compute the solution of Example 9.7.2. To make the comparison interesting, pick $dt = 1$ for the Euler case. Note that ode34 picks its own dt. Discuss the comparison of the two results.

9.63. Solve for the time response of Example 4.4.3 (i.e., the four-story building of Figure 4.9 using first VTB9_3 then VTB9_4. Compare the results of the two MATLAB solutions with those calculated using the modal analysis approach and given in Figure 4.11.

9.64. Reproduce the plots of Figure 4.12 for the two-degree-of-freedom system of Example 4.5.1 using file VTB9_3.

9.65. Compare the results obtained in Problem 9.54 (i.e., the time response using modal analysis) with that of computing the time response of the same system using the Euler program of VTB9_3. Which is more efficient?

9.66. Write an Euler-based program in MATLAB to compute the general forced response of a multiple-degree-of-freedom system.

10 Nonlinear Vibration

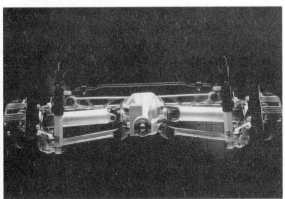

This chapter introduces the topic of nonlinear vibration. Nonlinear effects must often be considered in the design and analysis of devices and structures in order to increase operating regions and improve performance. The photograph on the left illustrates an active suspension for a Corvette. The performance of such systems is greatly improved by including the nonlinear characteristic. This is also true for the vibration isolation of the electronic components for a missile system, pictured here. Such systems must be isolated against shock and vibration loads during lift off. Chapter 5 introduces shock and vibration isolation. Nonlinear analysis can be used to improve the design of such systems.

This book is concerned primarily with time-invariant linear vibration problems. With the exception of various damping mechanisms, every mathematical model used in the preceding chapters is linear and assumes, among other things, that the physical system described by the equations behaves linearly. Namely, each system studied in the preceding chapters has a unique (single) equilibrium position (provided that no rigid-body modes are present) and the equilibrium is asymptotically stable if all of the system's eigenvalues have negative real parts, regardless of the initial conditions. A differential equation is linear if the state variables and their derivatives appear only to the first or zeroth power and are not multiplied by any other state variable. If this is not the case, the system under consideration is nonlinear. As pointed out in Section 2.1, the transient and steady-state response of a linear vibrating system can be expressed as a sum of the system's response to initial conditions with that of the response to the input force taken separately. The equations of motion are such that a linear combination of inputs to the system leads to the same linear combination of outputs (responses). This is called the *superposition principle*. In addition, asymptotic stability of a linear system implies that the system is also bounded-input, bounded-output stable. Linear vibration analysis and measurement also make use of the property that a sinusoidal input leads to a sinusoidal response of the same frequency.

All of these simple properties are violated by nonlinear systems. Consequently, much of the analysis used in previous sections for linear systems is not directly applicable to nonlinear systems. The behavior of nonlinear vibrating systems is much more complex. For instance, modal analysis does not apply to nonlinear systems because it depends on the superposition of solutions, which is not generally possible for nonlinear problems. It is very difficult, if not impossible, to find closed-form solutions to most nonlinear vibration problems. While many of the numerical methods used for linear problems may still apply to nonlinear problems, it is very important to understand the qualitative behavior of such systems before using these methods. In particular, the question of the stability of the solution to a nonlinear system is very important. Another qualitative question of interest is whether or not a given nonlinear system can be approximated successfully by a linear system. These questions are addressed in this chapter.

This chapter presents some fundamental ideas and concepts of nonlinear vibration phenomena. Only an introduction to nonlinear vibrations for single-degree-of-freedom systems is provided. The subject is best studied by consulting a book on nonlinear vibrations (see, e.g., Hagedorn, 1981; Cook, 1986; or Nayfeh and Mook, 1979). All systems considered in this chapter are modeled by nonlinear ordinary differential equations. An example of a nonlinear term is velocity-squared damping, discussed in Section 2.8. The equations discussed in Sections 2.7 and 2.8 for Coulomb damping and other forms of damping are examples of nonlinear systems. The pendulum problem of Example 1.8.1 is another example of a nonlinear system.

10.1 SINGLE-DEGREE-OF-FREEDOM PHASE PLOTS

In this section a single-degree-of-freedom nonlinear system is used to point out some of the fundamental definitions, phenomena, and behavior of nonlinear vibrating systems. Consider the simple single-degree-of-freedom system described by

$$\ddot{x} + f(x, \dot{x}) = 0 \qquad (10.1)$$

where $f(x, \dot{x})$ is a function, potentially nonlinear, of the displacement x and the velocity $\dot{x}$. Note that this is a generalization of the standard single-degree-of-freedom linear system of equations (1.2) and (1.25), where $f(x, \dot{x}) = 2\zeta\omega\dot{x} + \omega^2 x$. In equation (10.1) the function f does not contain t explicitly and hence the system is called *autonomous*.

A very common and useful method of examining the solution of equation (10.1) is to look at the problem in state-variable form and to plot the two states, velocity and position, in the same plane, called the *phase plane*. This plot, also called a *phase plot*, is used to characterize the solutions of nonlinear single-degree-of-freedom problems and to make qualitative statements about such systems. The phase plot is used to discuss the main characteristics of the solution of a nonlinear system, because the solution of a nonlinear system is usually not expressible as a simple (useful) function of time (such as $\sin \omega t$, $e^{-\zeta\omega t} \sin \omega_d t$, $te^{-\omega t}$, etc.). The phase plots of the solution can often be obtained without actually solving the equations of motion and provide a graphical representation of the main features of any motion of the system.

Equation (10.1) in state-variable form becomes

$$\begin{aligned} \dot{x}_1 &= x_2 \\ \dot{x}_2 &= -f(x_1, x_2) \end{aligned} \qquad (10.2)$$

where $x_1 = x$, the position, and $x_2 = \dot{x}$, the velocity, are the state variables for this simple single-degree-of-freedom system. A solution curve in the phase plane is a plot of x_2 versus x_1 satisfying equations (10.1) and (10.2) and is referred to as a *trajectory*, *orbit*, or *path*. Each point (x_1, x_2) on the trajectory corresponds to a specific time. A collection of trajectories that show the typical behavior of equation (10.1) throughout the phase plane is called a phase plot or *phase portrait* of the system. Arrowheads are placed on the trajectories to indicate the direction of increasing time. The following example illustrates this terminology for a simple undamped linear oscillator.

Example 10.1.1

Let $f(x_1, x_2) = (k/m)x_1$ in equation (10.2) and sketch the phase plot.

Solution   In this case the system is a linear single-degree-of-freedom undamped oscillator and the solution is well known. The solution has the form

$$x_1(t) = A \sin(\omega t + \phi)$$

$$x_2(t) = A\omega \cos(\omega t + \phi)$$

where A and ϕ are constants of integration determined by initial conditions. These two equations can be manipulated to produce a relation between x_1 and x_2 by eliminating the time dependence. Squaring the expressions for x_1 and x_2 and adding the result yields

$$x_1^2 + \frac{x_2^2}{\omega^2} = A^2$$

This, of course, is the equation of an ellipse in the x_1–x_2 plane, as plotted in Figure 10.1.

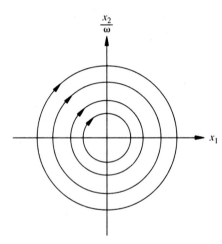

Figure 10.1   Phase plot of a linear undamped single-degree-of-freedom system. Each curve corresponds to a different value of A. For a fixed physical system (i.e., fixed ω), different initial conditions determine different values of A, and hence each curve represents a different set of initial conditions.

Alternatively, this equation for the phase plot can be derived by writing down the equation of motion directly as $\ddot{x} = -\omega^2 x$, so that

$$\ddot{x} = \frac{d\dot{x}}{dt} = \frac{d\dot{x}}{dx}\frac{dx}{dt} = \frac{d\dot{x}}{dx}\dot{x} = -\omega^2 x$$

where the chain rule has been used to expand $d\dot{x}/dt$. The last two terms in this expression yield

$$\frac{d\dot{x}}{dx}\dot{x} = -\omega^2 x \qquad \text{or} \qquad \dot{x}\, d\dot{x} = -\omega^2 x\, dx$$

This expression is integrated directly to yield

$$\frac{\dot{x}^2}{2} = -\omega^2 \frac{x^2}{2} + \text{constant}$$

Substitution of $\dot{x} = x_2$ and $x = x_1$ again yields

$$x_1^2 + \frac{x_2^2}{\omega^2} = A^2$$

which is the equation of an ellipse.

Different values of A and ω yield different trajectories in the phase plane. For a given trajectory (i.e., a particular value of A and ω), each point on the curve corresponds to the position and velocity at a particular instant of time [i.e., $x_1(t_0), x_2(t_0)$], which defines the solution at the time t_0. To determine the direction in the figure, note that if $\dot{x}_2$ is positive, then x_2 increases as t increases. Similarly, if $\dot{x}_1$ is positive, x_1 increases as t increases. For example, in the first quadrant both x_1 and x_2 are positive, so that $\dot{x}_1 = x_2 > 0$, and from the equation of motion, $\dot{x}_2 = -\omega^2 x_1 < 0$. Hence x_1 must be increasing and x_2 must be decreasing in the first quadrant. Thus the arrows are in the clockwise direction in Figure 10.1 and point in the direction of increasing time. If t_0 is the initial time, the points $x_1(t_0)$ and $x_2(t_0)$ correspond to the initial conditions and the solution of the system at some time $t_1 > t_0$ is somewhere along the trajectory in the direction of the arrow.

For a given system, ω is fixed (by the values of m and k). Each set of initial conditions determines a specific value of A, which in turn determines a single curve in Figure 10.1. Note also that the equation for a phase trajectory relates position and velocity and is, in the absence of damping, a statement of the conservation of energy. Hence each phase trajectory in Figure 10.1 represents a certain energy level of the system.   □

An important qualitative property of a nonlinear system is stability. Recall from equation (1.76) in Section 1.8 that a system is stable if the solution $x(t)$ satisfies $|x(t)| < M$ for all time $t > 0$, where M is a finite constant for all choices of initial conditions. The linear system illustrated in Example 10.1.1 is stable. Thus it is expected that phase plots of nonlinear systems that are stable will look similar to those of Figure 10.1. In fact, any trajectory in this phase plot satisfies

$$x_1^2 + x_2^2 < M$$

where M is a constant. This shows that the system satisfies the definition of stability given in Section 1.8 and reviewed in Window 10.1. Note that the literature in systems and controls refers to this situation as *marginally stable*.

Window 10.1
Review of the Definition of Stability for Section 1.8

The equilibrium state $\mathbf{x}_e$ is *stable* if for some finite constant M, $||\mathbf{x}(t)|| < M$ for all $t > 0$, where $||\mathbf{x}(t)|| = \sqrt{x_1^2 + x_2^2 + \cdots + x_n^2}$. If an equilibrium state is not stable, it is unstable. If an equilibrium state $\mathbf{x}_e$ is stable and tends toward $\mathbf{0}$ as time goes to infinity, the equilibrium state is defined to be *asymptotically stable*.

Another important definition for the analysis of nonlinear systems is that of the equilibrium point. An *equilibrium point*, denoted $\mathbf{x}_e$, is a state for which the phase velocity is zero. That is, if we represent the equations of motion as the nonlinear vector state equation,

$$\dot{\mathbf{x}} = \mathbf{F}(\mathbf{x}) \tag{10.3}$$

then the equilibrium point, $\mathbf{x}_e$ is the vector of constants that satisfies the relations

$$\mathbf{F}(\mathbf{x}_e) = \mathbf{0} \tag{10.4}$$

Here $\mathbf{F}$ is a nonlinear vector function of the state vector, $\mathbf{x} = [x_1, x_2, x_3, \ldots, x_n]^T$, and $\dot{\mathbf{x}} = [\dot{x}_1, \dot{x}_2, \dot{x}_3, \ldots, \dot{x}_n]^T$ is the vector phase velocity.

For the linear system of Example 10.1.1, the vector $\mathbf{F}$ is $\mathbf{F} = [x_2 \quad -\omega^2 x_1]^T$ and the equilibrium condition yields $x_2 = 0$ and $x_1 = 0$ (i.e., $\mathbf{x}_e = \mathbf{0}$). Hence there is a single equilibrium position at the origin for the linear undamped system of Example 10.1.1. In general, for equation (10.1) the equilibrium condition is

$$x_2 = 0, \qquad f(x_1, x_2) = 0$$

This means that equilibrium points are always on the x_1 axis in the phase plane. From equation (10.2) it follows that the phase trajectories always cut the x_1 axis at right angles (except at equilibrium points). Also note that closed trajectories in the phase plane represent periodic solutions since as time progresses the response will eventually pass through at the same (x_1, x_2) coordinates and the motion repeats itself indefinitely.

Example 10.1.2

Calculate the equilibrium position for the nonlinear system defined by $\ddot{x} + x - \beta^2 x^3 = 0$, or in state equation form, letting $x_1 = x$ as before,

$$\dot{x}_1 = x_2$$

$$\dot{x}_2 = x_1(\beta^2 x_1^2 - 1)$$

Solution These equations represent the vibration of a "soft spring" and correspond to an approximation of the pendulum problem of Example 1.4.2 where $\sin x \approx x - x^3/6$. The equations for the equilibrium position are

$$x_2 = 0$$

$$x_1(\beta^2 x_1^2 - 1) = 0$$

There are three solutions to this set of algebraic equations, corresponding to the three equilibrium positions of the soft spring. They are

$$\mathbf{x}_e = \begin{bmatrix} 0 \\ 0 \end{bmatrix}, \begin{bmatrix} \frac{1}{\beta} \\ 0 \end{bmatrix}, \begin{bmatrix} -\frac{1}{\beta} \\ 0 \end{bmatrix}$$

The phase plane for this example can be determined by solving for x_1 as a function of x_2. Eliminating dt from the state equations yields

$$\frac{dx_2}{dx_1} = \frac{-x_1 + \beta^2 x_1^3}{x_2}$$

or

$$x_2 \, dx_2 = -x_1 \, dx_1 + \beta^2 x_1^3 \, dx_1$$

Integrating this last expression yields

$$\frac{x_2^2}{2} = -\frac{x_1^2}{2} + \beta^2 \frac{x_1^4}{4} + c$$

where c is a constant of integration which depends on the initial conditions. Figure 10.2 gives a plot of x_2 versus x_1 for various values of the initial conditions. The directions are determined by the first state equation $\dot{x}_1 = x_2$. Note that for some initial conditions the motion is a stable oscillation around the equilibrium position at the origin. For other initial conditions the motion is unstable. Motions originating near the equilibrium at $x_1 = \pm 1/\beta$ move away from the equilibrium position as t increases. The important point of this example is that a nonlinear system can have multiple equilibrium points and that the qualitative behavior of the response depends strongly on the initial conditions given to the system. □

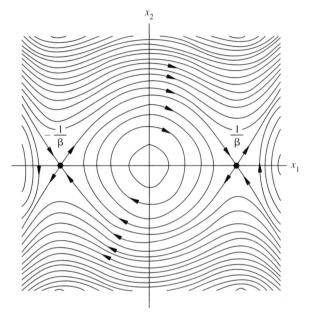

Figure 10.2   Phase plot of a soft spring illustrating multiple equilibrium points.

Another interesting phenomena of vibrating nonlinear systems is that oscillation of a fixed amplitude and constant frequency is possible without any external driving force. Such exotic behavior is called *self-excited vibration* or *limit cycle behavior*. This phenomenon is exhibited by nonlinear damping indicated by the middle term on the left side of the equation

$$m\ddot{x} + 2c(x^2 - 1)\dot{x} + kx = 0 \qquad (10.5)$$

where m, c, and k are positive real numbers. Called *Van der Pol's equation*, this represents a spring–mass system with a damping mechanism that depends on position. Note that for small displacements $x(t)$ the coefficient of $\dot{x}$ is negative and the effect of the damping term is to add energy to the system, causing the response to grow. As the response $x(t)$ grows, the damping term becomes positive and starts to subtract energy from the system, causing the response to decrease. Hence the response cannot grow without bound and cannot decrease to zero. The resulting motion is a sustained oscillation independent of initial conditions, called a *limit cycle*.

This sustained oscillation is not to be confused with the sustained vibration of an undamped linear spring–mass system. Limit cycle behavior differs from that of a linear spring–mass system without damping, because the amplitude of a limit cycle oscillation does not depend on the initial condition as does a linear undamped oscillation. In addition, the response of a linear undamped system is very sensitive to changes in m and k, while limit cycle behavior is not very sensitive to changes in m and k. This can be seen from the example of equation (10.5), where the limit cycle behavior depends on the difference $x^2 - 1$.

Example 10.1.3

Examine the equilibrium of the Van der Pol equation given by equation (10.5).

Solution   In state-space form the Van der Pol equation becomes ($x = x_1, \dot{x} = x_2$)

$$\dot{x}_1 = x_2$$

$$\dot{x}_2 = -\frac{k}{m}x_1 - \frac{2c}{m}(x_1^2 - 1)x_2$$

The equilibrium becomes simply
$$x_1 = 0 \qquad x_2 = 0$$

This nonlinear system has a single equilibrium at the origin.

$\square$

Calculations of the phase portrait of the Van der Pol equation and the associated limit cycles are defined in Section 10.2. More details on the nature of stability from the phase portraits will be given in the next section, which relates linearization to the phase portrait.

Another important nonlinear behavior is called *bifurcation*. As the parameter of a nonlinear vibrating system changes, not only does the stability of its equilibrium potentially change but the number of equilibrium points can also change. As an example, consider the equations of a hardening spring given by

$$\ddot{x} + \alpha x + \beta x^3 = 0 \tag{10.6}$$

and of similar form to the soft spring of Example 10.1.2, with $\alpha = 1$. This is called the (undamped) *Duffing equation*. As the ratio α/β varies from positive to negative, the equilibrium (at $x = 0$ for $v = 0$) splits into three at $x = (0, \sqrt{\alpha/\beta}, -\sqrt{\alpha/\beta})$ for $\alpha/\beta < 0$. This is called a *pitchfork bifurcation* because of the shape that results if the equilibrium point is plotted versus the ratio α/β. The soft spring also exhibits this bifurcation as the coefficient of x is varied.

10.2 LINEARIZED EQUATIONS

In some situations the linearized version of the equations can lead to useful information about the solution to the nonlinear case. In this section a method of linearizing the equations of motion of a nonlinear system is presented. In addition, how the behavior of the linearized system can be used to analyze the behavior of the nonlinear version is illustrated. To this end, consider the general autonomous second-order (but single-degree-of-freedom) system given by

$$\dot{x}_1 = g(x_1, x_2)$$
$$\dot{x}_2 = f(x_1, x_2)$$

(10.7)

The functions g and f are assumed to be smooth enough so that a Taylor series expansion can be written for each. A particular point (x_1, x_2) is said to be an *ordinary point* of equation (10.7) if either g and/or f do not vanish when evaluated at this point. On the other hand, the point (x_1, x_2) is defined to be a *singular point* of equation (10.7) if both g and f vanish when evaluated at this point. Note that equilibrium points are singular points.

Let the point (x_{1e}, x_{2e}) denote an equilibrium point of equations (10.7) and consider a Taylor series expansion about the equilibrium point. Then let

$$x_1 = x_{1e} + x_1' \qquad x_2 = x_{2e} + x_2'$$

(10.8)

so that x_1 and x_2 are points near an equilibrium. Substitution into equations (10.7) yields

$$\dot{x}_1 = \dot{x}_{1e} + \dot{x}_1' = g(x_{1e} + x_1', x_{2e} + x_2')$$
$$= g|_e + \frac{\partial g}{\partial x_1}\Big|_e x_1' + \frac{\partial g}{\partial x_2}\Big|_e x_2' + \text{(smaller-valued terms}$$
$$\text{of higher order)}$$

(10.9)

and

$$\dot{x}_2 = \dot{x}_{2e} + \dot{x}_2'$$
$$= f|_e + \frac{\partial f}{\partial x_1}\Big|_e x_1' + \frac{\partial f}{\partial x_2}\Big|_e x_2' + \text{(smaller-valued terms}$$
$$\text{of higher order)}$$

(10.10)

Here the subscript e denotes that the function is to be evaluated at the equilibrium point (x_{1e}, x_{2e}). Within the region near the equilibrium point (i.e., for small values of x_1' and x_2'), the "smaller valued terms" remaining in the expansion are negligible and equations (10.9) and (10.10) become of the form

$$\dot{x}_1 = a_{11}x_1 + a_{12}x_2$$
$$\dot{x}_2 = a_{21}x_1 + a_{22}x_2$$

(10.11)

where the primes have been suppressed. Note that $\dot{x}_{1e}$ and $\dot{x}_{2e}$ are both zero because the equilibrium points x_{1e} and x_{2e} are constants. Also, $g|_e$ and $f|_e$ are zero by definition. Equations (10.11) can be written in linear state-space form as

$$\dot{\mathbf{x}} = A\mathbf{x}$$

(10.12)

where the state vector $\mathbf{x} = [x_1 \quad x_2]^T$ and where A is the time invariant matrix given by

$$A = \begin{bmatrix} \dfrac{\partial g}{\partial x_1}\Big|_e & \dfrac{\partial g}{\partial x_2}\Big|_e \\ \dfrac{\partial f}{\partial x_1}\Big|_e & \dfrac{\partial f}{\partial x_2}\Big|_e \end{bmatrix} \qquad (10.13)$$

which defines the constants a_{ij} in equation (10.11). Equation (10.12) is the *linearized* version of the nonlinear equations (10.7) in the state-space region near the equilibrium position given by $\mathbf{x}_e$.

Example 10.2.1

Calculate a linearization of the Van der Pol equation for $m = 1$.

Solution   The state-variable form of equation (10.5) is given in example 10.1.3 so that $g(x_1, x_2) = x_2$ and $f(x_1, x_2) = -2c(x_1^2 - 1)x_2 - kx_1$. From equation (10.13), the linearized state matrix is formed by differentiating g and f to yield

$$g_{x_1} = 0 \qquad g_{x_2} = 1 \qquad f_{x_1} = -4cx_1x_2 - k \qquad f_{x_2} = -2cx_1^2 + 2c$$

Evaluating these derivatives at the single equilibrium point $x_1 = x_2 = 0$ yields

$$g_{x_1} = 0 \qquad g_{x_2} = 1 \qquad f_{x_1} = -k \qquad f_{x_1} = 2c$$

Upon substitution into equation (10.13) the linear state matrix becomes

$$A = \begin{bmatrix} 0 & 1 \\ -k & 2c \end{bmatrix}$$

Thus the linearized version of Van der Pol's equation near $\mathbf{x} = \mathbf{0}$ becomes

$$\begin{bmatrix} \dot{x}_1 \\ \dot{x}_2 \end{bmatrix} = \begin{bmatrix} 0 & 1 \\ -k & 2c \end{bmatrix} \begin{bmatrix} x_1 \\ x_2 \end{bmatrix}$$

in state-space form.

$\square$

The stability of equation (10.12) is determined by the nature of the eigenvalues of the state matrix A. The various combinations of eigenvalues for the matrix A give various characterizations of the equilibrium of the related nonlinear trajectories in the phase plane. The solution of the linearized system given by equation (10.12) can be represented as

$$\mathbf{x}(t) = e^{\lambda_1 t}\mathbf{x}_1 + e^{\lambda_2 t}\mathbf{x}_2$$

where $\mathbf{x}_1$ and $\mathbf{x}_2$ are the eigenvectors of the matrix A and where λ_1 and λ_2 are the corresponding eigenvalues of the matrix A. If each of the real parts of the eigenvalues is negative, each term $e^{\lambda_i t}\mathbf{x}_i$ will decay to zero as t grows. Hence solutions near this equilibrium point are asymptotically stable (recall the definitions of Section 1.8). The following characterization of the eigenvalues of the state matrix relates to the stability of the equilibrium of the linearized system.

Consider the situation where the eigenvalues of A are repeated, but simple real roots (i.e., where $\mathbf{x}_1$ and $\mathbf{x}_2$ are independent). In this case the equilibrium is called a *proper node*. If the eigenvalues of A are positive, the equilibrium is unstable as all trajectories move away from the node. On the other hand, if the eigenvalues are negative, the trajectories move toward the node, which is then referred to as an *attractor*, and the equilibrium is stable. If the eigenvalues of A are repeated and degenerate (x_1 and x_2 are dependent), the equilibrium is called an *improper node*. This node is stable if the eigenvalues are negative and unstable if they are positive. These cases are illustrated in Figure 10.3.

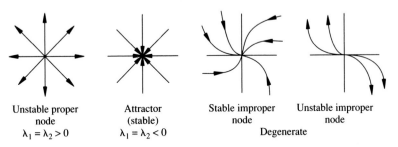

Unstable proper node	Attractor (stable)	Stable improper node	Unstable improper node
$\lambda_1 = \lambda_2 > 0$	$\lambda_1 = \lambda_2 < 0$		

Degenerate

Figure 10.3   Four possible phase plots for real eigenvalues for a two-state system, illustrating the behavior of the linearized equilibrium position.

If the eigenvalues of A are real but distinct, the node is still an improper node but with a more parabolic slope. Again the equilibrium is stable or unstable, depending on the sign of the eigenvalues. If they are both positive (negative), the equilibrium is unstable (stable). If the one eigenvalue is positive and one is negative, a so-called *saddle point* occurs. This is illustrated in Figure 10.4 and represents an unstable equilibrium.

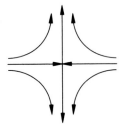

Figure 10.4   Phase plane diagram for one negative eigenvalue and one positive eigenvalue called an unstable saddle point equilibrium.

If the two eigenvalues of the matrix A are complex, three possibilities occur. If the real part is zero, periodic solutions result. The phase plane diagram is an ellipse and is referred to as a *center* or *vortex*. This is illustrated in Figure 10.5. If the eigenvalues form a complex conjugate pair, the trajectories are spirals into or out of the equilibrium point. If the real part is positive, the equilibrium is unstable and the spiral moves away from the equilibrium. If the real part is negative, the spiral moves toward the equilibrium point and the trajectory is stable. This is illustrated in Figure 10.5.

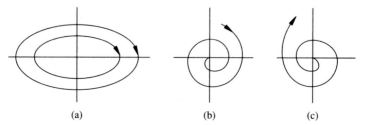

         (a)   (b)   (c)

Figure 10.5 Possible phase plane diagrams for complex eigenvalues: (*a*) stable vortex; (*b*) spiral (stable); (*c*) spiral (unstable).

Example 10.2.2

Consider the soft spring of Example 10.1.2 and use the linearization method discussed above to examine the behavior of the trajectories near the equilibrium points. Recall that there are three equilibrium points.

Solution   Here $g(x_1, x_2) = x_2$ and $f(x_1, x_2) = \beta^2 x_1^3 - x_1$. Calculating the derivatives of equation (10.13) yields

$$g_{x_1} = a_{11} = 0 \qquad\qquad g_{x_2} = a_{12} = 1$$

$$f_{x_1} = a_{21} = -1 + 3\beta^2 x_1^2 \qquad f_{x_2} = a_{22} = 0$$

Hence the linearized state matrix A becomes

$$\begin{bmatrix} 0 & 1 \\ 3\beta^2 x_1^2 - 1 & 0 \end{bmatrix}$$

Thus for the equilibrium at (0, 0) the matrix A is given by

$$A = \begin{bmatrix} 0 & 1 \\ -1 & 0 \end{bmatrix}$$

The eigenvalues of A are $\pm(-1)^{1/2}$ (i.e., purely imaginary and hence the phase plot is a stable vortex). At the equilibrium position given by $(-1/\beta, 0)$, the linearized state matrix becomes

$$A = \begin{bmatrix} 0 & 1 \\ 2 & 0 \end{bmatrix}$$

Thus the eigenvalues of the matrix A are $\pm(2)^{1/2}$. Hence this is a saddle point. The equilibrium at $(1/\beta, 0)$ is also a saddle point. These are plotted in Figure 10.2. The linear analysis in this case is in excellent agreement with the full nonlinear plot.   □

As illustrated by the Van der Pol equation, some nonlinear systems exhibit closed-curve trajectories in the phase plane. Isolated closed curves are known as *limit cycles*. Limit cycles correspond to periodic solutions of the system. They are isolated in the sense that nearby trajectories are not limit cycles. The stable vortex curves of Figure 10.5 are not limit cycles because they are immediately surrounded by other closed curves and hence are not isolated. In contrast, the trajectories near a limit cycle either approach the limit cycle as *t* grows, called a *stable limit cycle*, or diverge from it, called an *unstable limit cycle*.

Example 10.2.3

Consider a single-degree-of-freedom system with nonlinear damping of the form

$$\ddot{x} + (x^2 + \dot{x}^2 - 1)\dot{x} + x = 0 \qquad (10.14)$$

Solution   One solution to this is $x(t) = \cos t$, which results in the closed curve $x^2 + \dot{x}^2 = 1$ in the phase plane. The curve is a limit cycle because trajectories outside and inside the limit cycle approach it asymptotically as illustrated in Figure 10.6. Methods for calculating limit cycles are presented in Hagedorn (1981), for instance.

Figure 10.6   Stable limit cycle of Example 10.2.3.

☐

The linearization of a nonlinear system can yield some insight into the stability behavior of the nonlinear system. However, care must be used in interpreting the results. For instance, the system in Example 10.2.2 has three equilibriums. If the system is operated near the equilibrium point at (0,0), stable vibration results. However, if the initial conditions or inputs are such that the system operates about one of the other two equilibriums, the result is an unstable response. This points out that the stability properties are much more complicated for nonlinear systems than for linear systems. The stability properties for a linear system are *global* in the sense that they apply for any set of operating conditions. However, the stability properties listed in Example 10.2.2 are only *local* because they hold only in regions of operation *near* a particular equilibrium point. As indicated in the remaining sections, nonlinear systems have other unusual properties that make it difficult to calculate and analyze the response.

If all the eigenvalues of the linearized version of a system have nonzero real parts, small changes in the physical parameters of the system do not alter the qualitative behavior (stability, etc.) of the phase portrait. Such systems are said to be *structurally stable* and the nonlinear system and its linearization behave similarly in the neighborhood of the associated equilibrium point.

10.3 THE PENDULUM

A very common and intuitive nonlinear vibration problem is that of a pendulum. In this section the pendulum is used to illustrate some of the ideas in Section 10.2. An inverted pendulum is presented in Example 1.8.1 and a compound pendulum is given in

Example 1.4.6. In Example 1.4.6 the equation for a hanging pendulum with gravitational restoring force is given as

$$ml^2\ddot{\theta}(t) + mgl \sin \theta(t) = 0 \qquad (10.15)$$

where l is the length of the pendulum, m the mass of the pendulum tip (for a simple pendulum, the pendulum arm is assumed to have negligible mass), g the acceleration due to gravity, and θ the angle between the pendulum arm and the normal gravity force vector (i.e., the vertical). This equation is nonlinear because one of the state variables, $\theta(t)$, appears in the argument of a nonlinear function (sine). The common approach to solving the simple pendulum problem is, of course, to linearize the $\sin \theta$ term by expanding $\sin \theta$ in a Taylor series and arguing that for small θ, $\sin \theta$ is approximately equal to θ. This, of course, results in a simple conservative oscillator as discussed in Example 10.1.1. The phase portrait for the linearized version of equation (10.15) about the equilibrium at the origin is given in Figure 10.1.

The phase portrait for the full nonlinear model of a pendulum is more complex. First, the equation for the nonlinear pendulum predicts more equilibrium points than the single equilibrium for the linear case. The state-space equations for the pendulum become

$$\dot{x}_1 = x_2$$
$$\dot{x}_2 = -\frac{g}{l} \sin x_1 \qquad (10.16)$$

where $x_1 = \theta(t)$. Solving for the equilibrium positions yields the infinite number of solutions generated by $x_2 = 0$ and $x_1 = n\pi$. Physically, this requires the pendulum to be pivoted so that it is free to swing all the way around. Equation (10.15) may be rewritten as

$$\dot{\theta}\frac{d\dot{\theta}}{d\theta} + \frac{g}{l} \sin \theta = 0 \qquad (10.17)$$

since $\dot{\theta} = f(\theta(t))$ and

$$\frac{d^2\theta}{dt^2} = \frac{d\dot{\theta}}{dt} = \frac{d\dot{\theta}}{d\theta}\frac{d\theta}{dt} = \dot{\theta}\frac{d\dot{\theta}}{d\theta} \qquad (10.18)$$

In state-variable notation (i.e., $\theta = x_1$, $\dot{\theta} = x_2$) equation (10.17) becomes

$$x_2 dx_2 + \frac{g}{l} \sin x_1 dx_1 = 0 \qquad (10.19)$$

Integrating this last expression yields

$$\frac{1}{2}x_2^2 - \frac{g}{l} \cos x_1 = C \qquad (10.20)$$

where C is a constant of integration determined by initial conditions. Equation (10.20) illustrates that each value of C, or each set of initial conditions, generates a different trajectory in the phase portrait.

Equation (10.20) expresses the conservation of energy for the pendulum. The first term on the left is the kinetic energy of the pendulum:

$$\frac{1}{2}x_2^2 = \frac{1}{2}\dot{\theta}^2$$

and the second term is the pendulum's potential energy:

$$-\frac{g}{l}\cos x_1 = -\frac{g}{l}\cos\theta$$

The value of the constant of integration, C, is determined by the initial values of the position $x_1 = \theta$ and velocity $x_2 = \dot{\theta}$. Each different value of the initial conditions corresponds to a value of C. Each value of C then results in a different relationship between x_1 and x_2 through equation (10.20). Hence each value of C corresponds to a different trajectory or curve in the phase plot. The phase portrait is given in Figure 10.7.

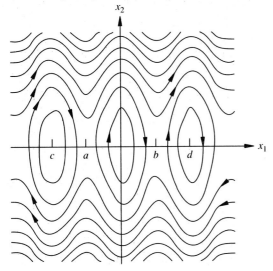

Figure 10.7 Phase portrait for the pendulum problem illustrating multiple equilibrium, at $a = -\pi, b = \pi, c = -2\pi, d = 2\pi$ and so on.

Note that the equilibrium points a and b in Figure 10.7 are unstable saddle points, whereas the origin is a stable center. The phase portrait continues this periodic pattern through the range of x_1. The saddle points correspond to initial conditions for the pendulum at the top of its swing (i.e., 180°) and is also called an inverted pendulum. Experience illustrates that if the pendulum is let go from this position, it will move (either way depending on the precise initial conditions) or fall through a large angle, and not return to this equilibrium (say, $x_1 = \pi$, $x_2 = 0$). However, values of the initial condition around the stable centers (say $x_1 = x_2 = 0$) yield curves which indicate that the motion oscillates in a stable fashion around the equilibrium position. Note that despite the absence of the time variable in the phase plane, the phase portrait yields a nice summary of the pendulum time response to various initial conditions.

The wavy lines at the top and bottom of Figure 10.7, which move from left to right across the top of the figure and from right to left along the bottom of the figure, correspond to the motion of the pendulum spinning (whirling) around its point of attachment either clockwise or counterclockwise, depending on the sign of the initial conditions. The fluctuations or wavy nature of these trajectories are due to the effect of gravity. As the value of $x_2(\theta)$ becomes very large, the lines become almost straight, parallel to the x_1 axis, because the gravitational influence is much less perceptible for high-speed whirling than for low-speed spinning.

The pendulum has two distinct types of equilibrium: one at $x_2 = 0, x_1 = n\pi, n = 0, 2, 4, \ldots$, corresponding to hanging straight down, and one at $x_2 = 0, x_1 = n\pi$, $n = 1, 3, 5, \ldots$, corresponding to standing straight up. One equilibrium is a stable center and the other is an unstable saddle point. These correspond to the physical phenomena of a pendulum. If the pendulum is moved slightly from the equilibrium point $x_2 = 0, x_1 = 2n\pi$, $n = 0, 1, 2 \ldots$, the pendulum motion falls on one of the ellipses and oscillates about the equilibrium with small amplitude. On the other hand, if the pendulum is displaced slightly from the saddle point equilibrium at $x_1 = \pi, x_2 = 0$, the pendulum path goes into a large whirling motion. In this case, $x_1 = \theta(t)$ grows without bound and the equilibrium is unstable.

The two trajectories in Figure 10.7 passing through the saddle points are each given the special name *separatrix*, as trajectories inside or between them correspond to stable center behavior and trajectories outside the region formed by these two curves correspond to saddle behavior. Physically, the separation defines the initial conditions separating the two distinct pendulum motions of oscillating about the lowest point and that of swinging around the pivot point of the pendulum. They also correspond to the maximum and minimum values of the system's potential energy.

The time it takes for the system to move from one point on the phase trajectory to another can be determined by integrating the expression for conservation of energy given by equation (10.20). Rewriting equation (10.20) in terms of θ and $\dot\theta$ and solving for $\dot\theta$ yields

$$\dot\theta = \pm\sqrt{\frac{2g}{l}\cos\theta + 2C} \tag{10.21}$$

This can be written as

$$d\theta = \pm\sqrt{\frac{2g}{l}\cos\theta + 2C}\,dt \tag{10.22}$$

or

$$dt = \pm\frac{d\theta}{\sqrt{(2g/l)\cos\theta + 2C}} \tag{10.23}$$

Integrating equation (10.23) yields

$$t = t_0 \pm \int_{\theta(t_0)}^{\theta(t)} \frac{d\theta}{\sqrt{(2g/l)\cos\theta + 2C}} \tag{10.24}$$

Each value of C determines a curve in the phase portrait. Once C is specified, equation (10.24) yields the time it takes for the pendulum to move from $\theta(t_0)$ to $\theta(t)$.

10.4 NONLINEAR DAMPING AND AVERAGING

A major cause of nonlinear behavior in vibration problems is damping. There are a variety of nonlinear damping models, very few of which lend themselves to analytical solutions. Many of these models were introduced in Sections 2.7 and 2.8. In this section a single-degree-of-freedom model is studied of the form

$$\ddot{x} + \varepsilon F_d(x, \dot{x}) + \omega^2 x = 0 \qquad (10.25)$$

where ε is a small parameter and the function $F_d(x, \dot{x})$ represents a nonlinear energy dissipation term such as Coulomb damping and air damping, defined in Sections 2.7 and 2.8. It is further assumed that equation (10.25) models some sort of oscillation so that the phase portrait of this system contains either a limit cycle or a center.

In the previous treatment of nonlinear damping given in Sections 2.7 and 2.8, the nonlinear damping force was approximated by a linear viscous damping term chosen to dissipate the same amount of energy per cycle as the nonlinear damping force F_d. Then the energy-loss calculations assume that the solution (under harmonic excitation) is harmonic. The response is then taken to be that of a damped system with a modified linear viscous damping coefficient, c_{eq}, under steady-state oscillation. Some sample nonlinear damping terms and the values of c_{eq} for each are listed in Table 10.1. In contrast, the procedure used here to analyze equation (10.25) is to smooth out, or average, the nonlinear function over a cycle of oscillation as compared to the center behavior of the $\varepsilon = 0$ case. The $\varepsilon = 0$ case, of course, corresponds to a simple (linear) harmonic oscillator. It is further assumed that $F_d(0, 0) = 0$ [i.e., that the origin of the phase plane

TABLE 10.1   DAMPING MODELS[a]

Name	$F_d(x, \dot{x})$	c_{eq}	Source
Linear viscous damping	$c\dot{x}$	c	Slow fluid
Air damping	$a\ \text{sgn}(\dot{x})\dot{x}^2$	$\dfrac{8a\omega X}{3\pi}$	Fast fluid
Coulomb damping	$\beta\ \text{sgn}\dot{x}$	$\dfrac{4\beta}{\pi\omega X}$	Sliding friction
Displacement squared damping	$d\ \text{sgn}(\dot{x})x^2$	$\dfrac{4dX}{3\pi\omega}$	Material damping
Solid damping	$b\ \text{sgn}(\dot{x})\lvert x\rvert$	$\dfrac{2b}{\pi\omega}$	Material damping

[a]The equivalent viscous damping coefficient case is calculated assuming a steady-state solution of the form $X \sin \omega t$ to a harmonic input of frequency ω, calculated using the energy balance of Section 2.8.

is an equilibrium point for equation (10.25)]. The method of averaging used here is a free-response method rather than the forced-response method used in Section 2.8.

Equation (10.25) is further simplified by scaling the time variable to the natural frequency of the linear undamped system, by substitution of $\tau = \omega t$. Then equation (10.25) can be written as (ϵ replaced by ϵ/ω^2)

$$x'' + \varepsilon F_d(x, x') + x = 0 \tag{10.26}$$

where the primes denote differentiation with respect to the scaled time τ. Note that for $\varepsilon = 0$, equation (10.26) collapses to the linear oscillator $\ddot{x} + x = 0$, which has the solution

$$x(\tau) = A \cos(\tau + \alpha) \tag{10.27}$$

It is possible to consider the case of $\alpha = 0$ (zero phase) without loss of any important features. For small enough values of ε, it seems reasonable to expect the nonlinear system to exhibit a limit cycle close to one of the circles generated by the linearized equation (the $\varepsilon = 0$ case). These circles satisfy (with $x_1 = x$ and $x_2 = x'$)

$$x_1^2 + x_2^2 = A^2 \tag{10.28}$$

with period nearly equal to 2π. The hypothesized limit cycle can be calculated by assuming that the solution to (10.26) is given by equation (10.27) and examining the change in total energy along the limit cycle.

The total energy $E(\tau)$ in the system at a given time is given by

$$E(\tau) = \frac{1}{2}x'^2 + \frac{1}{2}x^2 \tag{10.29}$$

The time rate of change of the total energy then becomes

$$\frac{dE}{d\tau} = E' = x'x'' + x'x = x'(x'' + x) = -\varepsilon x' F_d(x, x') \tag{10.30}$$

Integrating equation (10.30) over the assumed limit cycle period of 2π yields

$$E(2\pi) - E(0) = -\varepsilon \int_0^{2\pi} [(-A \sin \tau) F_d(A \cos \tau, -A \sin \tau)] d\tau \tag{10.31}$$

where it is assumed that the solution is given by equation (10.27). Since the change in energy over a complete limit cycle is zero [i.e., $E(2\pi) = E(0)$], the following energy balance results:

$$\int_0^{2\pi} F_d(A \cos \tau, -A \sin \tau) \sin \tau \, d\tau = 0 \tag{10.32}$$

Equation (10.32) can be used to determine the approximate amplitude A of the assumed limit cycle oscillation.

Example 10.4.1

Consider the vibration of a spring–mass system subject to viscous damping of the form $c\dot{x}$ and fluid damping of the form $|\dot{x}|\dot{x}$, both of relatively small amplitude. Calculate the approximate amplitude of an assumed limit cycle.

Solution   Equation (10.26) becomes

$$x'' + \varepsilon(2\zeta + |x'|)x' + x = 0 \tag{10.33}$$

Note that $F_d(0, 0) = 0$, so that the origin is an equilibrium point as required to use the method of averaging to calculate the approximate limit cycle. The energy-difference calculation of equation (10.32) for the assumed periodic solution becomes

$$\int_0^{2\pi} (2\zeta + |x'|)x' \sin \tau \, d\tau = 0 \tag{10.34}$$

Substitution of $x' = -A \sin \tau$ yields

$$\int_0^{2\pi} (2\zeta + |-A \sin \tau|)(-A \sin^2 \tau)d\tau = -2A\zeta\pi - \frac{A^2}{3}8 = 0 \tag{10.35}$$

Solving equation (10.35) for nonzero A yields the approximate limit cycle amplitude of

$$|A| = \tfrac{3}{4}\zeta\pi \tag{10.36}$$

The calculation is reasonable for small values of ε only.

□

A common characteristic of nonlinear vibration is the dependence of the frequency of vibration on the amplitude. Again considering equation (10.26) as a small distortion of the time-scaled equation $x'' + x = 0$ allows an estimate of the behavior of the limit cycle solution of equation (10.26). The idea of averaging, this time using polar coordinates, can be used to estimate the frequency dependence of the amplitude of a limit cycle oscillation.

Let $A(\tau)$ and $\theta(\tau)$ be the polar coordinates of the assumed limit cycle trajectory as indicated in Figure 10.8. Note that since $\theta(\tau)$ increases counterclockwise and the trajectory circles clockwise, the relationship between polar coordinates and state variables becomes

$$x_1 = -A \cos \theta$$
$$x_2 = A \sin \theta \tag{10.37}$$

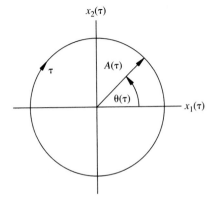

Figure 10.8   Polar coordinates used to describe a limit cycle.

The state-variable description of equation (10.26) is

$$x_1' = x_2$$
$$x_2' = -\varepsilon F_d(x_1, x_2) - x_1 \tag{10.38}$$

In terms of the state variables, differentiation of equation (10.28), which relates the amplitude to x_1 and x_2 with respect to τ yields

$$AA' = x_1 x_1' + x_2 x_2' = (x_1 + x_2')x_2$$
$$= -\varepsilon F_d(x_1, x_2)x_2 \tag{10.39}$$

Substitution of (10.37) then yields

$$A' = -\varepsilon F_d(-A\cos\theta, A\sin\theta)\sin\theta \tag{10.40}$$

A similar expression for θ is obtained by noting that $\tan\theta = -x_2/x_1$ and differentiating with respect to time. This yields

$$\theta' = \cos^2\theta \frac{-x_1 x_2' + x_2 x_1'}{x_1^2} \tag{10.41}$$

Substituting the state equations (10.38) into equation (10.41) and making use of equations (10.37) and (10.28) yields

$$\theta' = \frac{1}{A^2}(-A\cos\theta\,\varepsilon F_d + x_1^2 + x_2^2) = \frac{1}{A^2}(-A\,\varepsilon F_d\cos\theta + A^2) \tag{10.42}$$

or

$$\theta' = 1 - \frac{\varepsilon}{A}F_d\cos\theta \tag{10.43}$$

Equations (10.40) and (10.43) can be divided to obtain an approximate expression for $(dA/d\theta) = (dA/d\tau)/(d\theta/d\tau)$ by invoking the smallness of ε (e.g., $\varepsilon \ll 1$). Keeping only terms of order ε, it can be shown that

$$\frac{dA}{d\theta} \approx -\varepsilon F_d(-A\cos\theta, A\sin\theta)\sin\theta \tag{10.44}$$

The integral of $dA/d\theta$ for a limit cycle over one period is again zero, so that

$$\varepsilon \int_0^{2\pi} F_d(-A\cos\theta, \ A\sin\theta)\sin\theta\,d\theta \approx 0 \tag{10.45}$$

using the approximation of equation (10.44). This expression provides an alternative to using equation (10.32) for estimating a fixed value of the amplitude of the assumed limit cycle oscillation. From equation (10.44), the amplitude A may be calculated as a function of the polar coordinate $\theta(\tau)$. However, the expression for A arises from equation (10.45), which is accurate only for terms of order ε over one period. Thus $A(\theta)$ must be considered as only an approximation of the first element in a series, say A_0. The quantity A_0 is interpreted as the value of $A(\theta)$ at the beginning of one period of oscillation.

Equation (10.45) may also be used to calculate an approximate value for the frequency of the limit cycle oscillation. First, the period T is calculated from

$$T = \int_0^T d\tau = \int_{\theta(0)}^{\theta(T)} \frac{d\theta}{\theta'} \tag{10.46}$$

where $\theta(T) = 0$ and $\theta(0) = 2\pi$, since θ increases counterclockwise. Using equation (10.43) for θ', expanding using a simple geometric series in ε to calculate θ'^{-1}, and throwing away terms of ε^2 and higher yields

$$T \approx 2\pi + \frac{\varepsilon}{A_0} \int_{2\pi}^0 F_d(-A_0 \cos\theta, \ A_0 \sin\theta) \cos\theta \, d\theta \tag{10.47}$$

Again, using a geometric expression for calculating the algebraic inverse, equation (10.47) may be inverted to yield the frequency

$$\omega = \frac{2\pi}{T} \approx 1 - \frac{\varepsilon}{2\pi A_0} \int_{2\pi}^0 F_d(-A_0 \cos\theta, \ A_0 \sin\theta) \cos\theta \, d\theta \tag{10.48}$$

This is an approximate expression relating the amplitude and frequency of the assumed limit cycle oscillation.

Example 10.4.2

To evaluate the frequency dependence on amplitude for a simple undamped pendulum, rewrite the pendulum equation as $x'' + (\sin x - x) + x = 0$ so that for small motions,

$$F_d(x, x') = \sin x - x \tag{10.49}$$

Solution   Using a series expansion for $\sin x$, equation (10.48) becomes

$$\omega = 1 + \frac{1}{2\pi A_0} \int_0^{2\pi} [\sin(-A_0 \cos\theta) + A_0 \cos\theta] \cos\theta \, d\theta$$

$$= 1 + \frac{1}{2\pi A_0} \int_0^{2\pi} \left(-A_0 \cos\theta + \frac{A_0^3 \omega_3^2 \theta}{6} + \cdots + A_0 \cos\theta \right) \cos\theta \, d\theta \tag{10.50}$$

$$= 1 - \frac{1}{16} A_0^2$$

This illustrates that the frequency should decrease as the amplitude of oscillation increases.
□

Various forms of $F_d(x_1, x_2)$ for material damping are discussed by Bert (1973). Not all forms of nonlinear damping lend themselves to averaging. However, the discussion above indicates an example of available analysis for such systems.

10.5 FORCED RESPONSE BY PERTURBATION

Unlike the treatment of the forced response of linear systems, the forced response of a nonlinear system does not consist of a simple sum of a particular solution and the free response. Instead, new features arise because of the nonlinearities and lack of superposition. The free and forced responses interact, producing new phenomena. The effect of a nonlinearity in the forced response is easiest to discuss by examining the frequency response curves for a mildly nonlinear system. Recall from Chapters 2 and 7 that the frequency response of a simple linear single-degree-of-freedom oscillator subjected to a single harmonic input inhibits a peak at resonance as indicated in Figure 2.5. A nonlinear system, however, is likely to produce a frequency response curve that "bends" as indicated in Figure 10.9. The nonlinear frequency response is multivalued. This results in a steady-state solution having more than one possible value for a given excitation frequency. Which value of the amplitude results in a given situation depends on the initial conditions. That is, not only is the amplitude of the forced response dependent on the driving frequency (as in the linear case), but it is also a function of the initial conditions (not true in the linear case).

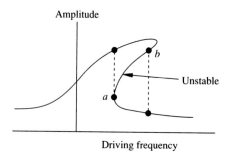

Figure 10.9 Frequency response for a nonlinear system.

Consideration of the forced response of a nonlinear system points out a very important fundamental difference between linear systems and nonlinear systems. Two important features of the forced response of a linear system are that the steady-state response is independent of both the initial conditions and the free response. In the nonlinear case the free response and steady-state response are coupled and the steady-state response depends in a critical fashion on the initial conditions. The fact that these are coupled leads to a variety of new phenomena in the nonlinear case. There are many other important nonlinear phenomena the coverage of which lie beyond the page limits of this book.

The forced response of the pendulum equation is used to present a perturbation method for analyzing periodic solutions of systems with small nonlinearities. This is called the *Lindstedt–Poincaré method*. The equation of motion for a harmonically excited pendulum is of the form ($\theta = x$)

$$\ddot{x} + \omega^2 \sin x = F \cos \omega_{dr} t \tag{10.51}$$

where ω^2 is the linear undamped natural frequency, F is the scaled amplitude of the applied force, and ω_{dr} is the driving frequency. Equation (10.51) can be approximated by replacing $\sin x$ with the first two terms of its Taylor series representation:

$$\ddot{x} + \omega^2 x - \frac{\omega^2}{6} x^3 = F \cos \omega_{dr} t \qquad (10.52)$$

Here $r^2 = \omega^2 / \omega_{dr}^2$ and consider the substitution $\tau = \omega_{dr} t$, which scales the time to the driving frequency. Then equation (10.52) becomes

$$x'' + r^2 x - \frac{r^2}{6} x^3 = \Gamma \cos \tau \qquad (10.53)$$

where $\Gamma = F / \omega_{dr}^2$ and the primes denote differentiation with respect to the scaled time τ.

It is assumed in this section that the nonlinearity is small. Hence the ratio r must be such that $r^2/6$ is a small number. To denote this, let $\varepsilon = r^2/6$, so that equation (10.53) becomes

$$x'' + r^2 x - \varepsilon x^3 = \Gamma \cos \tau \qquad (10.54)$$

Equation (10.54) provides a convenient characterization of the nonlinear problem. With $\varepsilon = 0$, equation (10.54) reduces to the linear oscillation, and with $\varepsilon \neq 0$, equation (10.54) becomes the approximate pendulum equation given in equation (10.53). The idea of the perturbation method presented here is to study the solution of equation (10.54) as a function of both τ and ε [i.e., let $x = x(\tau, \varepsilon)$]. The solution of equation (10.54) is then written as a series in the parameter ε as

$$x(\varepsilon, \tau) = x_0(\tau) + \varepsilon x_1(\tau) + \varepsilon^2 x_2(\tau) + 0(\varepsilon^3) \qquad (10.55)$$

where $0(\varepsilon^3)$ represents those terms multiplied by powers of ε greater than or equal to 3. The terms x_0, x_1, and so on, are only functions of τ. Substitution of the series given in equation (10.55) into the equation of motion (10.54) yields

$$(x_0'' + \varepsilon x_1'' + \cdots) + r^2(x_0 + \varepsilon x_1 + \cdots) - \varepsilon(x_0 + \varepsilon x_1 + \cdots)^3 = \Gamma \cos \tau \qquad (10.56)$$

By multiplying the terms of equation (10.56) and arranging the result according to powers of ε, this expression becomes

$$(x_0'' + r^2 x_0 - \Gamma \cos \tau)\varepsilon^0 + (x_1'' + r^2 x_1 - x_0^3)\varepsilon^1 + (x_2'' + r^2 x_2 - 3x_1 x_0^2)\varepsilon^2 = 0 \quad (10.57)$$

Equation (10.57) is a polynomial in ε, where ε is any arbitrary small number. For the equality to hold, each coefficient of the powers of ε must vanish. Hence from the coefficient of ε^0

$$x_0'' + r^2 x_0 = \Gamma \cos \tau \qquad (10.58)$$

from the coefficient of ε,

$$x_1'' + r^2 x_1 = x_0^3 \qquad (10.59)$$

and from the coefficient of ε^2,

$$x_2'' + r^2 x_2 = 3x_0^2 x_1 \qquad (10.60)$$

Each of these must be satisfied by the assumed solution given by equation (10.55). In addition, because the system is being forced with a period of 2π, the solution $x(\varepsilon, \tau)$ must also have period 2π. Because this system is nonlinear, there is the possibility that solutions exist at other frequencies: however, the primary concern here is to examine the nature of the response at the driving frequency. Assuming, then, that $x(\tau) = x(\tau + 2\pi)$, combined with equation (10.55), implies that

$$x_0(\tau) = x_0(\tau + 2\pi) \tag{10.61}$$

$$x_1(\tau) = x_1(\tau + 2\pi) \tag{10.62}$$

$$x_2(\tau) = x_2(\tau + 2\pi) \tag{10.63}$$

Note that equations (10.58) and (10.61), corresponding to the ε^0 term, comprise exactly the solution of the linearized equation (i.e., the case for $\varepsilon = 0$). This solution, $x_0(\tau)$, is called the *generating solution* for this perturbation method. Note also that the assumption of equations (10.61) to (10.63) means that this perturbation approach will not expose any other periodic solutions.

Assuming next that the solution of equation (10.58) occurs away from resonance, the solution to equation (10.58) is given by equation (2.8) to be

$$x_0(\tau) = A_1 \sin r\tau + A_2 \cos r\tau + \frac{\Gamma}{r^2 - 1} \cos \tau \tag{10.64}$$

Since only solutions of period 2π are of interest [i.e., since equation (10.61) must be satisfied], A_1 and A_2 are taken to be zero. The generating solution x_0 then becomes

$$x_0(\tau) = \frac{\Gamma}{r^2 - 1} \cos \tau \tag{10.65}$$

Next, the solution of equation (10.59) is computed using equation (10.65). This yields

$$x_1'' + r^2 x_1 = \left(\frac{\Gamma}{r^2 - 1}\right)^3 \cos^3 \tau \tag{10.66}$$

which can be written as

$$x_1'' + r^2 x_1 = \left(\frac{\Gamma}{r^2 - 1}\right)^3 \left(\frac{3}{4} \cos \tau + \frac{1}{4} \cos 3\tau\right) \tag{10.67}$$

by using a simple trigometric identity. Equation (10.67) is linear and can be solved by linear superposition by computing the solution using equation (2.8), neglecting the $\cos 3\tau$ forcing term and adding that to the solution obtained by neglecting the forcing term $\cos \tau$. Again, due to equation (10.62), only the particular solution is sought. This yields that

$$x_1(\tau) = \frac{3}{4} \frac{\Gamma^3}{(r^2 - 1)^4} \cos \tau + \frac{1}{4} \frac{\Gamma^3}{(r^2 - 1)^3 (r^2 - 9)} \cos 3\tau \tag{10.68}$$

where $r \neq 3$. Note that this has introduced a component in the response at a higher frequency.

The total periodic solution comes from substitution of equations (10.65) and (10.68) into equation (10.55), which yields

$$x(\varepsilon, \tau) = \left(\frac{\Gamma}{r^2 - 1} + \frac{3\varepsilon\Gamma^3}{4(r^2-1)^4} \right) \cos\tau + \frac{\varepsilon}{4} \frac{\Gamma^3}{(r^2-1)^3(r^2-9)} \cos 3\tau + 0(\varepsilon^2) \quad (10.69)$$

This solution is subject to several restrictions. One is that r cannot be an odd number because the higher harmonic terms $\cos 5\tau$, $\cos 7\tau$, and so on, will appear in the expression $0(\varepsilon^2)$. This system cannot have values of r near $1, 3, 5, \ldots$ either. If r were close to an odd integer, the corresponding coefficient would be large and the term $0(\varepsilon^2)$ could not be neglected.

Example 10.5.1

Using the perturbation techniques above, calculate the solution of the hard spring equation

$$x'' + \frac{1}{4}x + \varepsilon x^3 = \cos\tau$$

where $\varepsilon = 0.1$.

Solution   Following the procedure leading to equations (10.58) and (10.59) yields

$$x_0'' + \frac{1}{4}x_0 = \cos\tau$$

for the generating equation and

$$x_1'' + \frac{1}{4}x_1 = -x_0^3$$

from the coefficient of ε. The generating solution is

$$x_0(\tau) = -\frac{4}{3}\cos\tau$$

so that the solution of the second equation becomes

$$x_1(\tau) = +\frac{64}{27}\cos\tau + \frac{64}{945}\cos 3\tau$$

Combining $x_0(\tau)$ and $x_1(\tau)$ for the $\varepsilon = 0.1$ case yields $x(0.1, \tau) = -1.0963\cos\tau + 0.0068\cos 3\tau$.

□

The solution for the forced response of a nonlinear system computed in this section by a perturbation method is suitable only away from resonance. This solution is restricted further in the sense that it represents only one possible solution. Compared with the solution of the forced response of a linear undamped system, this solution includes a higher harmonic term (i.e., a term that oscillates at a higher frequency) than the force applied. In addition, the perturbation solution calculated here cannot be used if the frequency ratio r is near an odd integer in value. In the linear case r must not be near unity. The Lindstedt–Poincaré method presented in this section is sufficient to determine periodic solutions but does not yield the stability properties of these solutions.

10.6 FORCED RESPONSE BY DUFFING'S METHOD

In this section the undamped forced oscillation of a pendulum is again considered by the method of Duffing. This is an old method that has poor computational properties but provides an easy look at this important phenomenon of forced nonlinear systems. Again consider the solution of the approximate pendulum equation given by equation (10.52) as

$$\ddot{x} + \omega^2 x - \frac{\omega^2}{6} x^3 = F \cos \omega_{dr} t$$

Consider a harmonic generating solution of the form

$$x_0 = A \cos \omega_{dr} t \tag{10.70}$$

Substitution of this expression into equation (10.52) yields

$$\ddot{x} = -\omega^2 A \cos \omega_{dr} t + \frac{\omega^2}{6} A^3 \cos^3 \omega_{dr} t + F \cos \omega_{dr} t \tag{10.71}$$

where $\ddot{x}$ is now viewed as the second derivative of a new solution, $x_1(t)$. Using the trigonometric identity for $\cos^3$, also used in equation (10.67), yields

$$\ddot{x}_1 = -\omega^2 A \cos \omega_{dr} t + \frac{\omega^2}{24} A^3 (3 \cos \omega_{dr} t + \cos 3\omega_{dr} t) + F \cos \omega_{dr} t \tag{10.72}$$

This expression is now integrated over time twice to yield the next solution, $x_1(t)$. This yields

$$x_1(t) = a_1 + a_2 t + r^2 A \cos \omega_{dr} t - \frac{r^2 A^3}{8} \cos \omega_{dr} t - \frac{r^2 A^3}{216} \cos 3\omega_{dr} t - \frac{F}{\omega_{dr}^2} \cos \omega_{dr} t \tag{10.73}$$

where a_1 and a_2 are constants of integration. Since $x_1(t)$ must be periodic, a_1 and a_2 must be zero. Rearranging equation (10.73) yields

$$x_1(t) = \left(\frac{\omega^2}{\omega_{dr}^2} A - \frac{\omega^2 A^3}{8\omega_{dr}^2} - \frac{F}{\omega_{dr}^2} \right) \cos \omega_{dr} t - \frac{r^2 A^3}{216} \cos 3\omega_{dr} t \tag{10.74}$$

It is now argued that if the generating solution $x_0(t)$ is to be compatible with the next solution, $x_1(t)$, the coefficients of the $\cos \omega_{dr} t$ must be equal.

Equating the coefficients of the first harmonic, $\cos \omega_{dr} t$, from equations (10.70) and (10.74) yields the result that the amplitude A of the generating harmonic solution must satisfy

$$A = \frac{\omega^2}{\omega_{dr}^2} A - \frac{\omega^2 A^3}{8\omega_{dr}^2} - \frac{F}{\omega_{dr}^2} \tag{10.75}$$

Duffing's method continues by substitution of $x_1(t)$ into equation (10.52), renaming $\ddot{x}$ with $\ddot{x}_2$, integrating, comparing coefficients, and so on. This procedure iterates, producing a series solution of equation (10.52) compatible with the perturbation solution (unfortunately, Duffing's iterative scheme does not guarantee convergence).

Equation (10.75) relates the amplitude of vibration to the driving frequency for the generating harmonic response, $x_0(t)$, of the nonlinear pendulum (soft spring approximation) subject to a harmonic excitation. It is the nonlinear equivalent to the linear frequency amplitude relationship given by equation (2.6) and plotted (for the damped case) in Figure 2.5. Equation (10.75) is plotted in Figure 10.10 along with the magnitude plot for the linear case of equation (2.7). The dashed line in Figure 10.10, called a *backbone curve*, can be used to note the difference between the forced response of a linear versus nonlinear oscillator. In the linear case it is a vertical line, indicating that a sharp single resonance occurs at $r = 1$. In the nonlinear case it curves and crosses both axes. In addition, in the linear case the magnitude of the forced response at a given value of frequency ratio r, say r_0, corresponds to only *one* value of the magnitude A. On the other hand, in the nonlinear case, the frequency ratio r_0 corresponds to the three different possible values of the amplitude A (denoted a, b, and c in Figure 10.10). Note from the figure that even though damping is not present, the amplitude A remains finite for all values of r, unlike the linear case, which has a potentially infinite amplitude. Although not emphasized in the analysis presented here, it is also important to note that an essential difference between linear and nonlinear systems is the appearance of harmonic oscillations at frequencies different from the driving frequency, as indicated in equation (10.74) and Example 10.5.1.

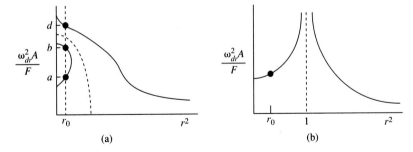

Figure 10.10 Magnitude plots for (a) an undamped nonlinear pendulum approximation and (b) an undamped linear oscillation.

PROBLEMS

Section 10.1

10.1. Sketch phase plane diagrams for a linear single-degree-of-freedom system for the three cases given by $\zeta = 1$, $\zeta < 1$, and $\zeta > 1$.

10.2. Sketch a phase plane diagram of a spring–mass damper system with (a) negative damping and positive stiffness and (b) negative stiffness and positive damping.

10.3. Locate the equilibrium points of $\ddot{x} - kx = 0$ and sketch the phase diagram. Is this system linear or nonlinear?

10.4. Air damping is often modeled as being dependent on the square of the velocity. One sample model is

$$\ddot{x} + |\dot{x}|\dot{x} + x = 0$$

Show that the origin in the phase plane is an equilibrium and the phase plot is a spiral.

10.5. Calculate the equilibrium points of equation (10.6) for several values of α and β.

10.6. Plot the function $F_1(x) = k(x - \frac{1}{6}x^3)$, which defines the nonlinear soft spring of Example 10.1.2. Compare this to a plot of $F_2(x) = k \sin x$.

10.7. Discuss why the Van der Pol equation is nonlinear by using the definition that a system is linear only if a linear combination of inputs results in a linear combination of the response to the inputs taken separately.

Section 10.2

10.8. Use the linearization method of Section 10.2 to show that both the hard and soft springs reduce to the linear oscillator equation of the form $\ddot{x} + x = 0$ near the stable equilibrium.

10.9. Derive the general form of the state matrix of the linearized version of the nonlinear single-degree-of-freedom oscillator $\ddot{x} = F(x, \dot{x})$.

10.10. Calculate the eigenvalues of the linearized Van der Pol equation of Example 10.2.1 and discuss the stability of the equilibrium as a function of k and c.

10.11. A nondimensional version of the pendulum equation given in Example 1.8.1 can be written as $(x = \theta)$

$$\ddot{x} + \sin x = 0$$

Calculate the equilibrium points and a linearization about each equilibrium.

10.12. For Problem 10.11, calculate the eigenvalues for the linearization at each equilibrium and characterize the phase portrait (i.e., vortex, saddle point, etc.) according to the classification of Section 10.2.

Section 10.3

10.13. Use equation (10.20) along with the initial conditions that $x_2 = 0$ and $x_1 = \pi/4, \pi/2, 3\pi/4$, and π radians to plot a few trajectories in the phase plane for the pendulum [i.e., numerically plot equation (10.20)].

10.14. Using the linear pendulum equation, calculate the time it takes for a pendulum to fall from $x_2 = 0, x_1 = \pi/4$ radians to the $(x_1 = 0)$ position if it has a length of 50 cm.

10.15. A simple pendulum is released with zero initial velocity at an initial displacement of $\pi/4$ radians. How long does it take to reach the downward vertical position? Assume that the pendulum has a length of 50 cm and use the nonlinear version of the pendulum equation and compare this to the result obtained in Problem 10.14.

10.16. Calculate the equation for the separatrix for the pendulum. Is the separatrix a trajectory?

10.17. The series presentation of $\sin x$ is

$$\sin x = x - \frac{1}{3!}x^3 + \frac{x^5}{5!} + \cdots$$

If the terms x^5 and higher powers are ignored and the remaining terms are used to approximate $\sin x$ in the pendulum equation, show that the soft spring equation results. Sketch the phase plot of the soft spring and compare it to that of nonlinear pendulum equations.

10.18. Numerically solve for the phase plots of the soft spring near the three equilibrium points.

10.19. If the pendulum equation is approximated by the first three terms of the series expression for $\sin x$ (see Problem 10.17), how many equilibrium points are there? Calculate them.

10.20. Show that the equilibrium of Figure 10.7 for the pendulum are unstable saddle points or stable center by using the linearization arguments of Section 10.2.

10.21. Find a set of initial conditions corresponding to the separatrix curves of the pendulum of Figure 10.7, which are not at equilibrium points.

Section 10.4

10.22. Calculate an approximate amplitude for the limit cycle of Van der Pol's equation

$$\ddot{x} + \varepsilon(x^2 - 1)\dot{x} + x = 0$$

using the energy balance method of Section 10.4.

10.23. Calculate the approximate frequency for the limit cycles of the Van der Pol equation given in Problem 10.22 to within an order of ε.

10.24. Calculate the amplitude and frequency of the limit cycle for the system

$$\ddot{x} + \varepsilon\left(\frac{\dot{x}^3}{3} - \dot{x}\right) + x = 0$$

10.25. Using energy balance, calculate the amplitude of the limit cycles of each of the following:

(a) $\ddot{x} + \varepsilon(x^2 + \dot{x}^2 - 1)\dot{x} + x = 0$

(b) $\ddot{x} + \varepsilon(x^4 - 1)\dot{x} + x = 0$

10.26. Calculate the equivalent viscous damping coefficient of Section 2.8 for the Van der Pol equation and compare the amplitude of the forced response at resonance to the limit cycle amplitude calculated in Problem 10.22.

10.27. Repeat Problem 10.26 for the system of Example 10.4.1.

Section 10.5

10.28. Consider the approximation of $\sin x \approx x - x^3/6$. For what values of x will this approximation be within 1% of the exact value of $\sin x$?

10.29. Derive equation (10.54) from equation (10.52) by performing the indicated steps. What are the units of τ?

10.30. Consider the solution given in Example 10.5.1 (for $\varepsilon = 1$) and compare a numerical integration of the exact equation to the approximate solution $x(\varepsilon, \tau)$.

10.31. Use the perturbation method of Section 10.5 to find an approximate periodic solution for the soft spring equation

$$\ddot{x} + 0.25x - 0.5x^3 = \cos t$$

and plot the response.

10.32. Use the perturbation method of Section 10.5 to find an approximate periodic solution for the system

$$\ddot{x} - 0.1x^2 - 0.5x = \cos t$$

which has velocity square damping.

10.33. Use the equivalent viscous damping approximation (see Section 2.8) to the steady-state solution for the air-damped oscillator of Problem 10.32 and compare it to the solution of that problem.

10.34. Consider the nonlinear system with quadratic displacement given by

$$\ddot{x} + \omega^2(x + \beta x^2) = F \cos \omega_{dr} t$$

where β is a small number. Use the perturbation method of Section 10.5 to find an approximate first order solution in β for $x(t)$.

Section 10.6

10.35. Obtain periodic solutions to

$$\ddot{x} + \omega^2(x - \mu^2 x^3 + \nu^4 x^5) = \sin \omega_{dr} t$$

using Duffing's method.

10.36. Plot the magnitude curve of the response in Problem 10.35.

10.37. Use the Duffing method of Section 10.6 to calculate the amplitude of the harmonic term corresponding to the driving frequency of the hard spring given by

$$\ddot{x} + 0.25x + 0.5x^3 = \sin t$$

10.38. Use the Duffing method of Section 10.6 to calculate the amplitude of the first harmonic from the soft spring equation

$$\ddot{x} + \omega^2 x - \mu x^3 = f \sin \omega_{dr} t$$

Show that if $\mu = 0$ in the solution, a linear spring results.

MATLAB VIBRATION TOOLBOX

If you have not yet used the *Vibration Toolbox* program, return to the end of Chapter 1 for a brief introduction to using MATLAB files or open the file directory "INTRO.TXT" on the disk on the inside back cover. The calculations of time responses of nonlinear systems can be accomplished using the ODE 23 and ODE 45 files discussed in Section 9.7. In addition, file VTB10_1 provides a phase plan plotting routine.

TOOLBOX PROBLEMS

TB10.1. Plot the time response of the Van der Pol equation

$$\ddot{x} + x(x^2 - 1)\dot{x} + x = 0$$

for $x(0) = 0$, $\dot{x}(0) = 0.1$ in the interval $0 \leq t \leq 20$, using either ODE 23 or VTB9_3. Try it again with $x(0) = 0.1$, $\dot{x}(0) = 0$.

TB10.2. Again consider the Van der Pol equation, this time in the form in which it appears in Example 10.1.3. Fix $m = 1$, $k = 100$, and $x(0) = 0$, $\dot{x}(0) = 0.1$ Plot the time response for several values of the damping coefficient c.

TB10.3. Compare the time response of the linearized version of Example 10.2.1 with the response generated in problem TB10.2.

TB10.4. Use file VTB10_1 to reproduce the phase portrait of Figure 10.2.

TB10.5. Use file VTB10_1 to make a phase plot for some of the systems of Problem TB10.2.

A Complex Numbers and Functions

Complex numbers occur naturally in vibration analysis from the solution of differential equations through their algebraic characteristic equations. In particular, solution of the damped single-degree of freedom system given by equation (1.36) is dependent on the values of λ satisfying the algebraic equation

$$m\lambda^2 + c\lambda + k = 0 \tag{A.1}$$

This is the familiar quadratic equation that has the solution

$$\lambda = -\frac{c}{2m} \pm \frac{1}{2m}\sqrt{c^2 - 4mk} \tag{A.2}$$

The roots given in equation (A.2) are complex valued if $c^2 - 4mk < 0$ (the underdamped case). In this case the formula given in equation (A.2) calls for the square root of a negative number. Stated algebraically, equation (A.1) in the underdamped case calls for a number j (sometimes denoted i) such that

$$j^2 = -1 \tag{A.3}$$

or symbolically, the "imaginary number" j is defined to be

$$j = \sqrt{-1} \tag{A.4}$$

This representation allows the expression of the two roots of equation (A.1) as the two pairs of real numbers

$$\left(-\frac{c}{2m}, -\frac{1}{2m}\sqrt{4mk - c^2}\right) \quad \text{and} \quad \left(-\frac{c}{2m}, \frac{1}{2m}\sqrt{4mk - c^2}\right) \tag{A.5}$$

which are written as

$$-\frac{c}{2m} - \frac{1}{2m}\sqrt{4mk - c^2}\,j \quad \text{and} \quad -\frac{c}{2m} + \frac{1}{2m}\sqrt{4mk - c^2}\,j \tag{A.6}$$

where $4mk - c^2 > 0$.

With the above as motivation, a general complex number, x, is written as $x = a + bj$. The real number a is referred to as the *real part* of the number x and the real number b is referred to as the *imaginary part* of the number x. Such complex-valued numbers are represented in a plane, called the complex plane, as illustrated in Figure A.1. The notation Re x, is used to denote the real part of the number x (i.e., Re $x = a$) and Im x is used to denote the value of the imaginary part of x (i.e., Im $x = b$).

The complex numbers $a + bj$ and $a - bj$ are called *conjugates* of each other. The notation $\bar{x}$, or x^*, is used to denote the conjugate of the complex number x. That is, if $x = a + bj$, then $\bar{x} = x^* = a - bi$. The roots of equation (A.1) appear as a complex conjugate pair in the underdamped case. Another useful property of complex numbers

520

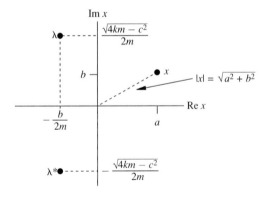

Figure A.1   Complex plane used to represent a complex number x and the roots of equation (A.1); $-(b/2m) \pm (1/2m)\sqrt{4km - c^2}\,j$. This plot is called an Argand diagram.

is their *absolute value* or *modulus*, denoted $|x|$. The modulus of a complex number is the distance from the origin in Figure A.1 to the point x:

$$|x| = |a + bj| = \sqrt{a^2 + b^2} \tag{A.7}$$

The modulus is illustrated in Figure A.1, as is the conjugate of the root λ. Note that conjugate pairs of numbers fall on the same vertical lines as they have the same real part. Also notice that a complex number and its conjugate both have the same modulus (i.e., $|x| = |x^*|$).

Complex numbers may be manipulated using real arithmetic following rules similar to those for vectors. Addition of two complex numbers is defined simply by adding the real parts and imaginary parts as separate entities. In particular, if $x = a + bj$ and $y = c + dj$, then

$$x + y = (a + c) + (b + d)j \tag{A.8}$$

In a consistent fashion, if β is a real number, then

$$\beta x = \beta a + \beta bj \tag{A.9}$$

expresses the product of a real number and a complex number.

Multiplication of two complex numbers is defined by starting with two basic definitions:

$$(a)(j) = aj$$

where a is real, and

$$(j)(j) = -1$$

then the product of two general complex numbers x and y becomes

$$(x)(y) = (a + bj)(c + dj) = (ac - bd) + (ad + bc)j \tag{A.10}$$

so that Re $(xy) = (ac - bd)$ and Im$(xy) = (ad + bc)$. The product of x and its conjugate $\bar{x}$ becomes a real number, that is,

$$x\bar{x} = (a + bj)(a - bj) = a^2 + b^2$$

This is consistent with definitions of the modulus of x defined by equation (A.7) with

$$|x|^2 = a^2 + b^2 = xx^* \tag{A.11}$$

which is a real number.

With addition defined (hence subtraction) and multiplication defined, it is important to define the division of one complex number by another. First note that the multiplicative identity of a complex number is simply the real number 1 (i.e., $1\,x = x$ for any complex number x). The inverse of the complex number $x = a + bj$ is given by

$$(a + bj)^{-1} = \frac{1}{a + bj} = \frac{1}{a^2 + b^2}(a - bj) \tag{A.12}$$

provided that $x = a + bj \neq 0$. Note that

$$(a + bj)^{-1}(a + bj) = \frac{1}{a^2 + b^2}(a - bj)(a + bj) = 1$$

the multiplicative inverse. Equation (A.12) allows the formulation of division of two complex numbers. To see this consider the division of y by x:

$$\frac{y}{x} = \frac{c + dj}{a + bj} = (c + dj)(a + bj)^{-1} \tag{A.13}$$

Invoking equation (A.12) for the inverse of x yields

$$\frac{y}{x} = (c + dj)\frac{1}{a^2 + b^2}(a - bj) = \frac{1}{a^2 + b^2}[(ac + bd) + (ad - cb)j] \tag{A.14}$$

for $x \neq 0$. Note that $x = 0$ if and only if both a and b are zero.

Note that the imaginary number j was used in equation (A.6) to manipulate the square root of a negative number. Specifically, for $c^2 - 4mk < 0$, then $4mk - c^2 > 0$ and the discriminant in equation (A.2) becomes

$$\sqrt{c^2 - 4mk} = \sqrt{(-1)(4mk - c^2)} = \left(\sqrt{4mk - c^2}\right)\left(\sqrt{-1}\right) =$$

$$\left(\sqrt{4mk - c^2}\right)j \tag{A.15}$$

which yields the complex number representation of the roots of the characteristic equation for a single-degree-of-freedom underdamped system. Note also from the definition of addition that the real and imaginary parts of a complex number are determined by simple arithmetic. If $x = a + bj$ then,

$$\text{Re } x = a = \frac{x + \bar{x}}{2}$$

$$\text{Im } x = b = \frac{x - \bar{x}}{2j} \tag{A.16}$$

Applying these formulas to the roots λ of the characteristic equation for an underdamped single degree-of-freedom system yields

$$\text{Re } \lambda = -\frac{c}{2m}$$

$$\text{Im } \lambda = -\frac{\sqrt{4mk - c^2}}{2m} \tag{A.17}$$

where λ satisfies equation (A.1).

A complex number can also be represented in terms of a polar coordinate system as illustrated in Figure A.2. Here θ is the angle between the line between the origin and the point (a, b). Note that $r = |x| = \sqrt{a^2 + b^2}$ and $\theta = \tan^{-1}(b/a)$. With this in mind x can be written as

$$x = a + bj = r(\cos\theta + j\sin\theta) \tag{A.18}$$

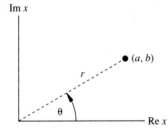

Figure A.2   Argand plot illustrating the polar representation of a complex number $x = a + bj$.

This polar representation can be used to write the exponential as

$$e^{jt} = \cos t + j\sin t \tag{A.19}$$

Equation (A.19) can be manipulated to yield

$$\cos t = \tfrac{1}{2}\left(e^{jt} + e^{-jt}\right)$$

$$\sin t = \tfrac{1}{2j}\left(e^{jt} - e^{-jt}\right) \tag{A.20}$$

These equations are referred to as Euler formulas and can be derived by writing the exponential e^{jt} as a power series and recalling the series expressions for $\sin t$ and $\cos t$. The hyperbolic functions can be written as

$$\cosh t = \tfrac{1}{2}\left(e^{t} + e^{-t}\right)$$

$$\sinh t = \tfrac{1}{2}\left(e^{t} - e^{-t}\right) \tag{A.21}$$

which are comparable to the Euler formulas for the trigonometric function.

The foregoing algebraic formulation for complex numbers can be extended to functions of a complex variable. An example of a function of a complex variable is given by equation (A.6) [i.e., $f(x) = |x| = (a^2 + b^2)^{1/2}$]. This is a real-valued function of a complex variable since x is complex and $f(x)$ is real. In general, however, a function of a complex number will also be complex. A theory of limits, differentiation, and integration can be defined, with a few cautions, following that of functions of a real variable. One major difference is that the real and imaginary parts of a complex function may have continuous derivatives of all orders at a point, yet the function itself may not be differentiable. Hence one needs to proceed with caution in examining the calculus of complex functions.

There are several graphical representations of a complex function that are useful in vibration analysis. First note that if x is a complex variable, then $f(x)$ is also potentially complex and will be of the general form

$$f(x) = u(x) + v(x)j \tag{A.22}$$

where $u(x)$ and $v(x)$ are real-valued functions. Multiplication of complex functions follows that of complex numbers as given by equation (A.10). The conjugate function is just

$$\bar{f}(x) = u(x) - v(x)j \tag{A.23}$$

and all of the formulas for arithmetic of complex numbers developed above apply to the arithmetic of complex functions. In particular,

$$|f(x)| = \sqrt{f\bar{f}} = \sqrt{u^2(x) + v^2(x)}$$

and for values of x such that $f(x) \neq 0$,

$$\frac{1}{f(x)} = \frac{1}{u^2(x) + v^2(x)}[u(x) - v(x)j] \tag{A.24}$$

which satisfies the relation $f(x)[1/f(x)] = 1$.

The graphical representation of a complex function becomes difficult because both the argument and the function require two dimensions to represent graphically. One approach is to plot $u(x)$ versus $v(x)$, as indicated in Figure A.3 for several different values of the complex variable x. These are called Nyquist plots and are used extensively in analyzing vibration measurement data. Another method of plotting complex functions is to examine the magnitude and phase separately using the polar form suggested in Figure A.2. In the case that $|f(x)|$ is plotted versus b, the imaginary part of x, the plot is called a Bode magnitude plot. Similarly, a plot of $\theta(x)$ is $\tan^{-1}[v(x)/u(x)]$ versus the imaginary part of x, called a Bode phase plot. These plots are also used extensively in analyzing vibration test data as discussed in Sections 1.6 and 7.4.

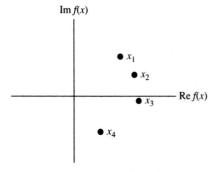

Figure A.3 Imaginary part of a complex function plotted versus the real part for several values of the variable x. Such plots are called Nyquist diagrams.

B Laplace Transforms

An integral transform is the procedure of integrating the time dependence of a function into becoming a function of an alternative variable, or parameter, which can be manipulated algebraically. A common integral transform is the Laplace transform. Laplace transforms are viewed here as a method of solving differential equations of motion by reducing the computation to that of integration and algebraic manipulation. Transforms provide both an alternative solution technique for vibration problems and an important analytical tool in the analysis and measurement of vibrating systems.

The definition of a Laplace transform of the function $f(t)$ is

$$L[f(t)] = F(s) = \int_0^\infty f(t)e^{-st}dt \tag{B.1}$$

for an integrable function $f(t)$ such that $f(t) = 0$ for $t < 0$. The variable s is complex valued. The Laplace transform changes the domain of the function from the positive real number line (t) to the complex number plane (s). The integration in the Laplace transform changes differentiation into multiplication. From the definition of the Laplace transform, it is a simple matter to see that the procedure is linear. Thus the transform of a linear combination of two functions is the same linear combination of the transform of these functions. The Laplace transform of various functions can be calculated in closed form by using equation (B.1). In addition, the Laplace transform of a derivative of an arbitrary function can easily be calculated in symbolic form. In particular, the Laplace transform of $\dot{x}(t)$ is just

$$L[\dot{x}(t)] = sX(s) - x(0) \tag{B.2}$$

where the capital X denotes a transformed version of $x(t)$. Similarly,

$$L[\ddot{x}(t)] = s^2X(s) - sx(0) - \dot{x}(0) \tag{B.3}$$

Here $x(0)$ and $\dot{x}(0)$ are the initial values of the function $x(t)$. Note that in the transformed domain, often called the s-domain, differentiating a variable $X(s)$ corresponds to simple multiplication [i.e., $sX(s)$], and integration in the time domain corresponds to dividing by s in the transform domain.

Table B.1 lists some common functions of time preceded by their Laplace transforms as calculated using equation (B.1) for the case that all of the initial conditions are set to zero. The table provides a method of quickly finding the Laplace transforms given a particular function of time by reading from right to left. However, reading the table from left to right provides the ability to determine a function, $x(t)$, from its Laplace transform $X(s)$. This gives rise to the concept of an inverse Laplace transform, denoted L^{-1}, which is formally defined by

$$L^{-1}[X(s)] = x(t) \tag{B.4}$$

which can be computed by using Table B.1.

TABLE B.1 PARTIAL LIST OF FUNCTIONS AND THEIR LAPLACE
TRANSFORMS WITH ZERO INITIAL CONDITIONS AND $t > 0$

$F(s)$	$f(t)$
(1) 1	$\delta(t_0)$ unit impulse at t_0
(2) $\dfrac{1}{s}$	1, unit step
(3) $\dfrac{1}{s+a}\left(\dfrac{1}{s-a}\right)$	$e^{-at}\left(e^{at}\right)$
(4) $\dfrac{1}{(s+a)(s+b)}$	$\dfrac{1}{b-a}(e^{-at}-e^{-bt})$
(5) $\dfrac{\omega}{s^2+\omega^2}$	$\sin\omega t$
(6) $\dfrac{s}{s^2+\omega^2}$	$\cos\omega t$
(7) $\dfrac{1}{s(s^2+\omega^2)}$	$\dfrac{1}{\omega^2}(1-\cos\omega t)$
(8) $\dfrac{1}{s^2+2\zeta\omega s+\omega^2}$	$\dfrac{1}{\omega_d}e^{-\zeta\omega t}\sin\omega_d t,\ \zeta<1,\ \omega_d=\omega\sqrt{1-\zeta^2}$
(9) $\dfrac{\omega^2}{s(s^2+2\zeta\omega s+\omega^2)}$	$1-\dfrac{\omega}{\omega_d}e^{-\zeta\omega t}\sin(\omega_d t+\phi),\ \phi=\cos^{-1}\zeta,\quad \zeta<1$
(10) $\dfrac{1}{s^n}$	$\dfrac{t^{n-1}}{(n-1)!},\ n=1,2\ldots$
(11) $\dfrac{n!}{(s-\omega)^{n+1}}$	$t^n e^{\omega t},\ n=1,2\ldots$
(12) $\dfrac{1}{s(s+\omega)}$	$\dfrac{1}{\omega}(1-e^{-\omega t})$
(13) $\dfrac{1}{s^2(s+\omega)}$	$\dfrac{1}{\omega^2}(e^{-\omega t}+\omega t-1)$
(14) $\dfrac{\omega}{s^2-\omega^2}$	$\sinh\omega t$
(15) $\dfrac{s}{s^2-\omega^2}$	$\cosh\omega t$
(16) $\dfrac{1}{s^2(s^2+\omega^2)}$	$\dfrac{1}{\omega^3}(\omega t-\sin\omega t)$
(17) $\dfrac{1}{(s^2+\omega^2)^2}$	$\dfrac{1}{2\omega^3}(\sin\omega t-\omega t\cos\omega t)$
(18) $\dfrac{s}{(s^2+\omega^2)^2}$	$\dfrac{t}{2\omega}\sin\omega t$
(19) $\dfrac{s^2-\omega^2}{(s^2+\omega^2)^2}$	$t\cos\omega t$
(20) $\dfrac{\omega_1^2-\omega_2^2}{(s^2+\omega_1^2)(s^2+\omega_2^2)}$	$\dfrac{1}{\omega_2}\sin\omega_2 t-\dfrac{1}{\omega_1}\sin\omega_1 t$

TABLE B.1   CONTINUED

	$F(s)$	$f(t)$
(21)	$\dfrac{(\omega_1^2 - \omega_2^2)s}{(s^2 + \omega_1^2)(s^2 + \omega_2^2)}$	$\cos \omega_2 t - \cos \omega_1 t$
(22)	$\dfrac{\omega}{(s + a)^2 + \omega^2}$	$e^{-at} \sin \omega t$
(23)	$\dfrac{s + a}{(s + a)^2 + \omega^2}$	$e^{-at} \cos \omega t$

The procedure for using Laplace transforms to solve equations of motion expressed as an inhomogeneous ordinary differential equation is to take the Laplace transform of both sides of the equation, treating the time derivatives symbolically using equations (B.2) and (B.3) and using Table B.1 to compute the Laplace transform of the driving force. This renders an algebraic equation in the variable $X(s)$, which is easily solved by simple manipulation. The inverse Laplace transform is applied to the resulting expression for $X(s)$ by using Table B.1 again, resulting in the time response.

As an example of using Laplace transforms to solve a homogeneous differential equation, consider the undamped single-degree-of-freedom system described by

$$\ddot{x}(t) + \omega^2 x(t) = 0 \qquad x(0) = x_0, \quad \dot{x}(0) = v_0 \tag{B.5}$$

Taking the Laplace transform of $\ddot{x} + \omega^2 x = 0$ results in

$$s^2 X(s) - s x_0 - v_0 + \omega^2 X(s) = 0 \tag{B.6}$$

by direct application of equation (B.3) and the linear nature of the Laplace transform. Algebraically solving equation (B.6) for $X(s)$ yields

$$X(s) = \frac{x_0 + s v_0}{s^2 + \omega^2} \tag{B.7}$$

Using $L^{-1}[X(s)] = x(t)$ and entries (6) and (5) of Table B.1 yields that the solution is

$$x(t) = x_0 \cos \omega t + \frac{v_0}{\omega} \sin \omega t \tag{B.8}$$

The same procedure works for calculating the forced response. However, in the forced response, calculating the algebraic solution for $X(s)$ often results in quotients of polynomials in s. These polynomial ratios may not be found in tables directly, but such quotients may be resolved into simple terms by using the method of *partial fractions*.

The method of partial fractions is one of finding unknown coefficients of terms used in combining simple fractions by computing the lowest common denominator. For instance, consider a function $X(s)$ given by

$$X(s) = \frac{s + 1}{s(s + 2)} \tag{B.9}$$

which does not appear in Table B.1, rendering it difficult to invert. This quotient of polynomials can be written as

$$\frac{s+1}{s(s+2)} = \frac{A}{s} + \frac{B}{s+2} \tag{B.10}$$

where A and B are unknown constant factors. Clearing the fractions in equation (B.10) yields

$$s + 1 = (A + B)s + 2A \tag{B.11}$$

after a little manipulation. Since the coefficients on each side of the equality must match, Equation (B.11) implies that

$$A + B = 1 \qquad \text{(from the coefficient of } s\text{)}$$

$$2A = 1 \qquad \text{(from the coefficient of } s^0\text{)}$$

so that $A = B = \frac{1}{2}$. Thus $X(s)$ may be written as

$$X(s) = \frac{1}{2s} + \frac{1}{2}\frac{1}{s+2} \tag{B.12}$$

Inverting $X(s)$ is easily performed by using entries (2) and (3) in Table B.1. This yields that

$$x(t) = \frac{1}{2}\left(1 + e^{-2t}\right) \tag{B.13}$$

providing the desired time response.

The partial fraction method requires that repeated linear factors and quadratic factors have additional coefficients. For example, if the polynomial in the denominator has a repeated linear factor such as $(s + \omega)^2$, an additional term is needed. To see this, suppose that $X(s) = \omega^2/[s(s + \omega)^2]$; then the partial fraction expansion becomes

$$\frac{\omega^2}{s(s+\omega)^2} = \frac{A}{s} + \frac{B}{s+\omega} + \frac{C}{(s+\omega)^2} \tag{B.14}$$

Multiplying both sides by $s(s + \omega)^2$ and solving for the coefficients A, B, and C by comparing coefficients of powers of s yields $A = 1, B = -1$, and $C = -\omega$, so that

$$\frac{\omega^2}{s(s+\omega)^2} = \frac{1}{s} - \frac{1}{s+\omega} - \frac{\omega}{(s+\omega)^2} \tag{B.15}$$

For quadratic factors such as $(s^2 + 2s + 5)$ the numerator of the expansion must contain a first order polynomial element of the form $As + B$. For example, consider $X(s) = (s + 3)/[(s + 1)(s^2 + 2s + 5)]$. Its partial fraction expansion is

$$\frac{s+3}{(s+1)(s^2+2s+5)} = \frac{A}{s+1} + \frac{Bs+C}{s^2+2s+5} \tag{B.16}$$

Multiplying by $(s + 1)(s^2 + 2s + 5)$ yields

$$s + 3 = (A + B)s^2 + (2A + B + C)s + (5A + C)$$

after grouping terms as coefficients of the powers of s. Comparing coefficients of s yields

$$A + B = 0 \qquad 2A + B + C = 1 \qquad 5A + C = 3$$

which has the solution $A = \frac{1}{2}, B = -\frac{1}{2}, C = \frac{1}{2}$. Thus the partial fraction expansion in equation (B.16) becomes

$$
\begin{aligned}
X(s) &= \frac{1}{2(s+1)} + \frac{1-s}{2(s^2+2s+5)} = \frac{1}{2(s+1)} - \frac{s+1-2}{2(s^2+2s+5)} \\
&= \frac{1}{2(s+1)} - \frac{1}{2}\frac{s+1}{(s+1)^2+4} + \frac{2}{2(s^2+2s+5)}
\end{aligned}
\tag{B.17}
$$

This can easily be inverted by using entries (3), (23), and (8) of Table B.1, respectively, to yield

$$x(t) = \frac{1}{2}e^{-t} - \frac{1}{2}e^{-t}\cos 2t + e^{-t}\sin 2t \tag{B.18}$$

since the denominator in the last fraction in equation (B.17) implies that $2\zeta\omega = 2$ and $\omega^2 = 5$. Hence $\omega = \sqrt{5}, \zeta = 1/\sqrt{5}$, and $\omega_d = 2$.

Matrix Basics

Some matrix definitions and manipulations useful in vibration analysis are summarized in this appendix. A matrix is an array of numbers (real or complex) arranged in rows and columns according to the following scheme:

$$A = \begin{bmatrix} a_{11} & a_{12} & \cdots & a_{1n} \\ a_{21} & a_{22} & \cdots & a_{2n} \\ \vdots & & & \\ a_{m1} & a_{m2} & & a_{mn} \end{bmatrix}$$

where A denotes the matrix as a single entity and a_{ij} denotes the element in the ij position (i.e., the element at the intersection of the ith row and jth column). Such a matrix A is said to be of order $m \times n$ and to have m rows and n columns. A majority of the matrices used in vibration analysis are square (i.e., $m = n$), having the same number of rows and columns, or are rectangular, consisting of a single row ($1 \times n$) or a single column ($n \times 1$) which is then called a row vector or column vector, respectively. However, in the analysis of vibration measurements, other size rectangular matrices occur.

Matrix arithmetic can be defined only if the matrices to be combined are of compatible dimensions. The sum of two matrices, A and B (i.e., $C = A + B$), is defined if A and B are of the same size (say $m \times n$), by

$$c_{ij} = a_{ij} + b_{ij} \tag{C.1}$$

for each value of i between 1 and m and each value of j between 1 and n. This definition of matrix addition is simply to create a new matrix C of the same size with elements formed by the sum of the corresponding elements of the two matrices A and B. Multiplication of a matrix by a scalar α is defined on a per element basis (i.e., the product of the scalar α and the matrix A, denoted αA, has elements αa_{ij}).

The product of two matrices is only defined in a specific order for matrices of a compatible size. In particular, the matrix product $C = AB$ is defined in terms of the elements of the matrix C by

$$C_{ij} = \sum_{k=1}^{p} a_{ik} b_{kj} \tag{C.2}$$

where p is the common size of each matrix. Here, if the matrix A is $m \times p$, the matrix B must be $p \times n$ in order for the definition to be consistent. Thus not all matrices can be multiplied together. The matrix C resulting from equation (C.2) will be of the size $m \times n$.

One extremely common matrix product in vibration analysis is that of a square matrix times a column vector. A matrix is square if it has the same number of rows and columns (i.e., the special case $m = n$). If A is an $n \times n$ matrix and $\mathbf{x}$ is an $n \times 1$ vector, the product $\mathbf{y} = A\mathbf{x}$ is the $n \times 1$ vector with ith element given by

$$y_i = \sum_{j=1}^{n} a_{ij} x_j \tag{C.3}$$

where $i = 1, 2, \ldots, n$. Another useful matrix manipulation in vibration analysis is the concept of the transpose of a matrix. The transpose of the matrix A is formed from A by interchanging the rows and columns of the matrix A. In particular, if A has elements a_{ij}, the transpose of A, denoted A^T, has elements a_{ji}. If A is $n \times m$, then A^T will be $m \times n$. If $\mathbf{x}$ is a column vector ($n \times 1$), then $\mathbf{x}^T$ is a $1 \times n$ row vector. The transpose operation on a matrix (or a vector) satisfies the following rules:

$$(A + B)^T = A^T + B^T \tag{C.4}$$

$$(\alpha A)^T = \alpha A^T \tag{C.5}$$

$$(AB)^T = B^T A^T \tag{C.6}$$

$$(A^T)^T = A \tag{C.7}$$

where A and B are any two matrices for which the indicated operations can be defined and α is a scalar.

The product of two vectors can be defined in two different ways, both of which are useful in vibration analysis. First consider the product $\mathbf{y}^T\mathbf{x}$, where both $\mathbf{x}$ and $\mathbf{y}$ are $n \times 1$ column vectors. Following the definition given in equation (C.2), this product becomes

$$\mathbf{y}^T\mathbf{x} = \sum_{i=1}^{n} y_i x_i \tag{C.8}$$

This corresponds to the familiar "dot product" and is also called the *inner product* or *scalar product*. The phrase scalar product arises because the product $\mathbf{y}^T\mathbf{x}$ results in a scalar. The scalar product is useful in defining the magnitude of mode shapes in vibration analysis. A second product of two column vectors can be defined by taking the product of a column vector times a row vector. This product, called an *outer product*, produces an $n \times n$ matrix:

$$\mathbf{y}\mathbf{x}^T = \begin{bmatrix} y_1 x_1 & y_1 x_2 & \cdots & y_1 x_n \\ y_2 x_1 & y_2 x_2 & \cdots & y_2 x_n \\ \vdots & \vdots & \cdots & \vdots \\ y_n x_1 & y_n x_2 & \cdots & y_n x_n \end{bmatrix} \tag{C.9}$$

The outer product is extremely useful for analyzing vibration measurements.

Returning to the inner product, if the inner product of a vector $\mathbf{x}$ and itself is formed, the result leads to the interpretation of the length of a vector. The length of a vector is an example of a vector *norm*. In particular, if $\mathbf{x}$ is an $n \times 1$ vector, the *norm* of $\mathbf{x}$ is defined by

$$||\mathbf{x}|| = (\mathbf{x}^T\mathbf{x})^{1/2} \tag{C.10}$$

If the norm of a vector is the number 1, the vector is called a *unit vector* and is said to be *normalized*. If the inner product of two vectors is zero, they are said to be *orthogonal*.

Often vectors occur in sets. If a set of $n \times 1$ vectors $\mathbf{x}_i$ satisfies the relation

$$\mathbf{x}_i^T \mathbf{x}_j = \begin{cases} 0 & i \neq j \\ 1 & i = j \end{cases} \tag{C.11}$$

for all vectors in the set, the unit vectors are each orthogonal to each other. Such a set of vectors is called *orthonormal*. Another useful property of sets of vectors is the concept of linear independence. A set of n, $n \times 1$ vectors $\mathbf{x}_i$ is *linearly dependent* if there exist n scalars $\alpha_1, \alpha_2, \dots, \alpha_n$ which are not all zero, such that

$$\alpha_1 \mathbf{x}_1 + \alpha_2 \mathbf{x}_2 + \cdots + \alpha_n \mathbf{x}_n = \mathbf{0} \tag{C.12}$$

Essentially, this statement means that there is at least one vector in the set that can be written as a linear combination of some of the other vectors in the set. If such a set of scalars cannot be found, the set of vectors $\{\mathbf{x}_i\}$ is said to be *linearly independent*. Linearly independent vectors are very useful for expressing the solution of multiple-degree-of-freedom vibration problems. A familiar example of an orthonormal, linearly independent set of vectors are the three unit vectors $(\hat{\mathbf{i}}, \hat{\mathbf{j}}, \hat{\mathbf{k}})$ used in statics and dynamics to analyze motion and forces in three dimensions.

Vectors can be differentiated by defining the derivatives in terms of each element. In this way the derivatives of a vector are simply

$$\frac{d}{dt}(\mathbf{x}) = \begin{bmatrix} \dot{x}_1 \\ \dot{x}_2 \\ \vdots \\ \dot{x}_n \end{bmatrix} \tag{C.13}$$

where the overdot denotes the usual time derivatives of each element. Such derivatives are used in representing the equations of motion for multiple-degree-of-freedom systems.

A majority of the matrices used in vibration analysis are square and have the same number of rows as columns. Some special square matrices are the *identity matrix*, I, which has its element δ_{ij}, and the zero matrix, which has a zero as each of its elements. These special matrices satisfy the following conditions for any square matrix A:

$$AI = A$$
$$IA = A$$
$$0A = 0 \tag{C.14}$$
$$A0 = 0$$

The identity matrix I is, of course, the multiplicative identity for the set of square matrices, and the zero matrix is the additive identity. A matrix that has zeros as all of its elements except those along the diagonal (i.e., a_{ii}) is called a diagonal matrix. Diagonal matrices can be manipulated almost like scalars and are often used in modal analysis.

The determinant of a matrix A is defined by the formula

$$\det(A) = \sum_{j=1}^{n} (-1)^{1+j} a_{1j} \, \det(A_{1j}) \tag{C.15}$$

where a_{1j} is the element in the $(1-j)$th position of the matrix A and A_{1j} is the $(n-1) \times (n-1)$th matrix formed from the matrix A by deleting the first row and jth column of the matrix A. For a 1×1 matrix (i.e., a scalar), the determinant of A is just

$$\det(A) = a$$

the value of the scalar. For a 2×2 matrix the determinant becomes

$$\det(A) = a_{11}a_{22} - a_{12}a_{21} \tag{C.16}$$

The expression given by equation (C.15) can be used to calculate the determinant of a 3×3 matrix in terms of the determinant formula for a 2×2 matrix given above, and so on. Note that the determinant is a scalar value.

The matrix A_{ij} formed by deleting the ith row and jth column of the matrix A is called a minor of A. Let α_{ij} denote the scalar found by taking the $\det(A_{ij})$ with a particular sign:

$$\alpha_{ij} = (-1)^{i+j} \det(A_{ij}) \tag{C.17}$$

Then the matrix defined by forming the elements α_{ij} into a matrix is called the *adjoint* of A, denoted adj A. The matrix A^{-1} such that $A^{-1}A = I$, is called the *inverse* of the matrix A and can be calculated from the adjoint by

$$A^{-1} = \frac{\text{adj } A}{\det(A)} \tag{C.18}$$

Note from this calculation that the matrix A does not have an inverse if $\det(A) = 0$. In this case A is said to be *singular*. If A^{-1} does exist [i.e., if $\det(A) \neq 0$], the matrix A is called *nonsingular*. The following is a list of properties of the matrix inverse and determinant:

$$(A^{-1})^T = (A^T)^{-1}$$

$$(AB)^{-1} = B^{-1}A^{-1}$$

$$\det(AB) = \det(A)\det(B)$$

$$\det(A^T) = \det(A)$$

$$\det(\alpha A) = \alpha \det(A) \qquad \text{where } \alpha \text{ is a scalar}$$

Each of these expressions can be derived from the definitions given above.

Some special and important types of matrices in vibration analysis are summarized in the following list; that is, a matrix A is

symmetric if $A = A^T$
skew symmetric if $A = -A^T$
positive definite if $\mathbf{x}^T A\mathbf{x} > 0$ for all $\mathbf{x} \neq \mathbf{0}$
nonnegative definite if $\mathbf{x}^T A\mathbf{x} \geq 0$ for all $\mathbf{x}$ (also called *semidefinite*)
indefinite if $(\mathbf{x}^T A\mathbf{x})(\mathbf{y}^T A\mathbf{y}) < 0$ for some $\mathbf{x}$ and $\mathbf{y}$
orthogonal if $A^T A = I$

If a matrix A is not symmetric, it can be written as the sum of a symmetric matrix, A_s, and a skew symmetric matrix, A_{ss}, defined by

$$A_s = \frac{A^T + A}{2} \qquad A_{ss} = \frac{A - A^T}{2} \tag{C.19}$$

Another useful matrix definition is that of the *trace* of a matrix. The trace of the matrix A, denoted tr(A), is simply the sum of the diagonal elements of A:

$$\text{tr}(A) = \sum_{i=1}^{n} a_{ii} \tag{C.20}$$

which is a scalar.

The frequency response methods used in steady-state vibration analysis and modal testing methods often result in complex-valued matrices. In this situation the matrix transpose used above is replaced with a conjugate transpose. In particular, let $\bar{a}_{ij}$ denote the complex conjugate of the number a_{ij}, then define A^* as the conjugate transpose by

$$A^* = \bar{A}^T \tag{C.21}$$

which has elements $\bar{a}_{ji}$. In the complex-valued case, a matrix A is said to be *Hermitian* if

$$A^* = A$$

and *unitary* if

$$A^*A = I$$

In the case of a complex-valued vector, the inner product becomes

$$\mathbf{x}^*\mathbf{y} = \sum_{i=1}^{n} \bar{x}_i y_i \tag{C.22}$$

and a complex-valued matrix A is positive definite if

$$\mathbf{x}^*A\mathbf{x} > 0 \tag{C.23}$$

for all nonzero complex vectors $\mathbf{x}$.

Next consider the matrix eigenvalue problem. Let A be a square matrix ($n \times n$). A scalar λ is an *eigenvalue* of the matrix A, with *eigenvector* $\mathbf{x}, \mathbf{x} \neq \mathbf{0}$, if

$$A\mathbf{x} = \lambda\mathbf{x} \tag{C.24}$$

is satisfied. Note that λ can be zero but $\mathbf{x}$ cannot. Also note that if $\mathbf{x}$ is an eigenvector, so is $\alpha\mathbf{x}$, where α is any scalar. As was developed in Chapter 4, the natural frequencies and mode shapes of an undamped system are calculated by solving a matrix eigenvalue problem. If the matrix A is known to be symmetric (i.e., $A = A^T$), the eigenvalues of A are real numbers and the eigenvectors are real valued. If in addition, A is positive definite, the eigenvalues must be positive numbers. If A is size $n \times n$, then A will have n eigenvalues and n eigenvectors. In particular, if A is symmetric (and real), the eigenvectors $\mathbf{x}_i$ form a linearly independent set. Furthermore, these eigenvectors can be normalized to form an orthonormal linearly independent set. These normalized eigenvectors can be used to define an orthogonal matrix P by

$$P = [\mathbf{x}_1 \quad \mathbf{x}_2 \quad \cdots \quad \mathbf{x}_n] \tag{C.25}$$

such that

$$P^T P = I \quad \text{and} \quad P^T A P = \text{diag}(\lambda_i) \tag{C.26}$$

This last expression is the foundation of the modal analysis method used so extensively in vibration theory.

The eigenvector defined in equation (C.24) is called a *right eigenvector*. A *left eigenvector* can also be defined by

$$\mathbf{y}^T A = \lambda \mathbf{y}^T \qquad \mathbf{y} \neq \mathbf{0} \tag{C.27}$$

For symmetric matrices, the left and right eigenvectors are the same. For matrices that are not symmetric, these vectors may or may not be the same. In fact the eigenvalues, and hence the eigenvectors of a nonsymmetric matrix, may or may not be complex valued. The concept of an eigenvalue and eigenvector can be generalized by introducing a *lambda matrix*, or *matrix polynomial*. In particular, the solution of the equations of motion for a lumped-parameter multiple-degree-of-freedom system with damping generates the matrix polynomial problem of calculating a scalar λ and a nonzero vector $\mathbf{x}$ satisfying

$$(M\lambda^2 + C\lambda + K)\mathbf{x} = \mathbf{0} \tag{C.28}$$

The matrix $(M\lambda^2 + C\lambda + K)$ is called a matrix polynomial (it is viewed as a polynomial in λ, with matrix coefficients). If M, C, and K are n x n matrices, there are $2n$ values of λ and $2n$ values of $\mathbf{x}$ satisfying equation (C.28). A particularly simple case to solve is that where M, C, and K are symmetric, real-valued matrices satisfying $CM^{-1}K = KM^{-1}C$ (which in general is not true). In this special case the eigenvectors of the case $C = 0$ will also satisfy (C.28). For a more complex discussion of matrix methods in vibration analysis, see Inman (1989) and Golub and Van Loan (1989).

The Vibration Literature

There are several outstanding texts and reference books on vibration that merit consultation for those who wish a second explanation for the sake of understanding or for those who wish to pursue these topics further. In addition, there are several publications devoted to presenting research results and case studies in vibration. A few such general references are listed here. A computer search in your local library should turn up the rest.

Introductory Texts

Mechanical Vibration, 4th ed., J. P. Den Hartog, McGraw-Hill, New York, 1956.

An Introduction to Mechanical Vibrations, 3rd ed., R. F. Steidel, Jr., Wiley, New York, 1989.

Theory of Vibration with Applications, 4th ed., W. T. Thomson, Prentice Hall, Englewood Cliffs, N.J., 1993.

Structural Dynamics: An Introduction to Computer Methods, R. R. Craig, Jr., Wiley, New York, 1981.

Mechanical Vibrations, 2nd ed., S. S. Rao, Addison-Wesley, Reading, Mass., 1990.

Elements of Vibration Analysis, 2nd ed., L. Meirovitch, McGraw-Hill, New York, 1986.

Shock and Vibration Handbook, 3rd ed., C. M. Harris, editor, McGraw-Hill, New York, 1988.

Advanced Texts

Vibration Problems in Engineering, 5th ed., W. Weaver, Jr., S. P. Timoshenko, and D. H. Young, Wiley, New York, 1990.

Analytical Methods in Vibration, L. Meirovitch, Macmillan, New York, 1967.

Vibration with Control Measurement and Stability, D. J. Inman, Prentice Hall, Englewood Cliffs, N.J., 1989.

Periodicals

Sound and Vibration, Acoustical Publications, Bay Village, Ohio.

Shock and Vibration Bulletin, The Vibration Institute, Willowbrook, Ill.

Journals

Journal of Sound and Vibration, Academic Press, New York.

Journal of Vibration and Acoustics, American Society of Mechanical Engineering, New York.

International Journal of Analytical and Experimental Modal Analysis (now *Modal Analysis*), Society of Experimental Mechanics by Scholarly Communication Project of Virginia Polytechnic Institute and State University, Blacksburg, Va.

Journal of Mechanical Systems and Signals, Academic Press, New York.

List of Symbols

This appendix lists the symbols used in the book. There are not enough symbols to go around so some symbols are used to denote more than one quantity. In addition, the choice of a symbol to denote a specific quantity has evolved over time often by different groups of people. Hence it is not uncommon for a particular symbol to represent several different quantities. Here, symbols have been chosen consistent with the most common used in the vibration literature.

In general, a lower case italic symbol is a scalar, an upper case italic symbol is a matrix and a bold lower case letter is used to denote a vector. However, when working with transforms, it is common to denote the Laplace Transform of a scalar by its upper case symbol. Hence, it is not possible to read an upper case italic letter and know whether the quantity is matrix or a transformed scalar. Symbols must always be clarified by the context in which it appears. Greek symbols are almost always scalars. Letters which are not italic or bold are usually units (such as N for Newtons or mm for millimeters). Units are summarized on the inside front cover.

At the writing of this book, a group of British researchers are trying to establish an international standard of symbols for modal analysis and hence, vibration. The basic problem remains, however, there are just not enough symbols to go around.

CHAPTER 1

f_k	= elastic restoring force of a spring
m	= mass
g	= acceleration of gravity
x_0	= an initial displacement
$x, x(t)$	= displacement
k	= spring constant, spring stiffness
N	= normal force
x, y	= coordinates of a two dimensional space
$\theta, \theta(t)$	= angular displacement
t	= time
ω	= $\sqrt{k/m}$ undamped natural frequency, angular frequency
ϕ	= phase angle
A	= amplitude of vibration, also a constant of integration
f	= $\omega/2\pi$ frequency in hertz, not to be confused with a force
T	= $1/f = 2\pi/\omega$, period of oscillation
v_0	= initial velocity
$a, a_1, a_2,$	= unknown constants of integration

A_1, A_2	= unknown constants of integration
λ	= constant used in solving differential equations
j	= $\sqrt{-1}$ the imaginary unit
$\bar{x}, \bar{x}^2$	= average displacement, root mean square value, respectively
$\dot{x}(t), \ddot{x}(t)$	= time derivatives of displacement, i.e., velocity and acceleration, respectively
f_c	= $c\dot{x}$ a damping force
c	= damping coefficient
ζ	= $c/2\sqrt{km}$ damping ratio
ω_d	= $\omega\sqrt{1-\zeta^2}$, for $0 < \zeta < 1$, called the damped natural frequency
c_{cr}	= $2\sqrt{km}$, critical damping coefficient so that $\zeta = c/c_{cr}$
f_{xi}	= ith force acting in the x direction
M_{0i}	= ith torque acting about the point 0
I_0, J, J_0	= moment of inertia about point 0
U_1, U_2	= potential energy at time t_1 and t_2, respectively

538

T_1, T_2	= kinetic energy at time t_1 and t_2, respectively	k_0	= radius of gyration ($k_0^2 = q_0 r$)
T_{max}, U_{max}	= maximum kinetic energy, maximum potential energy, respectively	E	= young modulus (elastic modulus)
r, l	= radius (capital R is often used as well) length, respectively	A	= cross sectional area when used with E
$\frac{d}{dt}(\ ), (\cdot)$	= always refers to a time derivative of $(\ )$	Jp	= area moment of inertia
		G	= shear modulus
m_s	= mass of a spring	ρ	= mass density
γ	= a density (specific weight)	W	= weight
q_0	= radius of the center of percussion	δ	= logarithmic decrement (see Window 4.2 for other uses of this symbol)
		x_s	= static deflection

CHAPTER 2

$F(t)$	= an external or driving force		nential)
F_0	= constant magnitude of a harmonic driving force	$H(j\omega_{dr})$	= complex frequency response function
ω_{dr}	= input frequency or driving frequency	s	= transform variable
f_0	= F_0/m	$X(s)$	= the Laplace Transform of $x(t)$
x_p	= the particular solution	θ, ϕ	= used to denote phase shifts
A_0	= amplitude of the particular solution	$H(s)$	= $X(s)/F(s)$, a transfer function
r	= frequency ratio = ω_{dr}/ω (not to be confused with a radius)	$z(t)$	= $x(t) - y(t)$, relative displacement
		Z	= amplitude of relative displacement
$y(t)$	= displacement of the base of spring-mass-damper system	ψ	= phase of relative displacement
ω_b	= frequency of base motion, i.e., the frequency of an applied harmonic displacement	z_p	= particular solution for relative displacement
		F_d	= damping force
X	= magnitude of the harmonic response	μ	= coefficient of sliding friction
Y	= magnitude of the applied harmonic displacement	A_1, B_1	= constants of integration
		ΔE	= energy dissipated
F_T	= magnitude of the force transmitted to a mass through a spring and dashpot	c_{eq}	= equivalent viscous damping coefficient
		U	= potential energy
		U_{max}	= peak potential energy
e	= radius of an out of balance mass (not to be confused with the expo-	η	= loss factor
		β	= hysteretic damping constant
		C	= drag coefficient
		α	= $C\rho A/2$

CHAPTER 3

τ	= a specific value of time, or a dummy (time) variable of integration	n, m, i	= integers (be careful not to confuse m the mass with m the index)
		ω_T	= $2\pi/T$
ϵ	= a very small positive value of time	$x_{cn}(t)$	= the nth solution if the driving force is $\cos n\omega_n t$
$\delta(t)$	= the Dirac Delta function (not a decrement)	$x_{sn}(t)$	= the nth solution if the driving force is $\sin n\omega_n t$
$\hat{F}$	= $F\Delta t$, impulse	$L[\ ]$	= the Laplace transform of [ ]
$h(t - \tau)$	= impulse response function	$\mu(t)$	= unit step function
a_0, a_n, b_n	= Fourier coefficients	x'	= $x - \bar{x}$, a zero mean variable

$S_{xx}(\omega)$ = power spectral density of the function x

$R_{xx}(\tau)$ = auto correlation of the function x

CHAPTER 4

x = displacement column vector

$x_{10}, x_{20},$ = initial conditions for the coordinates
$\dot{x}_{10} \cdots$ $x_1(t)$ and $x_2(t)$, respectively

$\mathbf{x}^T$ = the transpose of **x** or a row vector

K = the stiffness matrix

M = the mass matrix

C = the damping matrix

A^T = the transpose of the matrix A

A^{-1} = the inverse of the matrix A

$A^{1/2}$ = the matrix square root of the matrix A

u = a constant column vector also $\mathbf{u}_1, \mathbf{u}_2$, etc.

I = the identity matrix

$\tilde{K}$ = $M^{-1/2}KM^{-1/2}$

$\tilde{C}$ = $M^{-1/2}CM^{-1/2}$

$M^{-1/2}$ = the inverse of the matrix square root of the matrix M

$\mathbf{q}(t)$ = a vector of generalized coordinates

$q_i(t)$ = the ith generalized coordinate

v = an eigenvector of $\tilde{K}$

λ, λ_i = the eigenvalues

α = a scalar

CHAPTER 5

$T.R.$ = transmissibility ratio

R = $1 - T.R. =$ reduction in transmissibility

δ_s = static deflection

k_a = stiffness of the absorber

m_a = mass of the absorber

x_a = displacement coordinate of absorber mass

X_a = magnitude of the displacement of the absorber mass

μ = m_a/m ratio of absorber mass to primary mass

ω_p = natural frequency of primary system without absorber

ω_a = natural frequency of absorber without primary system

$E[x]$ = the expected value of x

M,a,b = constants

$\mathbf{w}, \mathbf{w}_i$ = mass normalized eigenvector of K

P = modal matrix

δ_{ij} = Kronecker delta (not to be confused with the Dirac delta or the log decrement)

Λ = the diagonal matrix of eigenvalues of a matrix

$\mathbf{r}(t)$ = modal coordinate vector

$r_i(t)$ = the ith modal coordinate

S = $M^{-1/2}P$, a mass weighted modal matrix

$\mathbf{0}$ = the zero vector

$\theta_1, \theta_2, \theta_3$ = rotational displacements

$\det(A)$ = the determinant of the matrix A

d_i = expansion coefficients (scalars)

η_i = modal damping ratios

β = a scalar

$\mathbf{F}(t)$ = a vector of external forces

ω_{di} = $\omega_i\sqrt{1 - \zeta_i}$ the ith damped natural frequency

$\mathbf{f}(t)$ = $P^T M^{-1/2}\mathbf{F}(t)$

Q_i = a generalized force or moment

L = $T - U =$ Lagrangian

β = ω_a/ω_p, a frequency ratio

$\mathbf{X}$ = $[X \quad X_a]^T$, where X is the magnitude of the response of the spring mass and X_a is that of the absorber mass

r = ω_{dr}/ω_p, a frequency ratio

x_m = the value of x that causes $f(x_m)$ to be an extreme (critical point)

f_x = partial derivatives of f with respect to x

f_y = partial derivatives of f with respect to y

f_{xx} = 2nd partial derivatives of f with respect to x

f_{yy} = 2nd partial derivitives of f with respect to y

γ = k_2/k_1, a stiffness ratio

$\bar{\eta}$ = loss factor

k^*	= complex stiffness	ϵ_s	= strain
E^*	= complex modulus	η_s	= shear loss factor
e_2	= ratio of modulus = E_2/E_1	G^*	= complex shear modulus
h_2	= ratio of height = H_2/H_1	E'', G''	= imaginary part of elastic and shear
g_1, g_2	= feedback control gain		modulus respectively (G'' called a
G'	= shear modulus (real part)		loss modulus)
E'	= real part of elastic modulus	k'	= shear stiffness
	= $2(1+\nu)G'$, ν Poisson's ratio	T_R	= transmissibility ratio at resonance
σ_s	= shear stress	G'_ω	= dynamic shear modulus

CHAPTER 6

		ω_n	= the natural frequency of the nth node
$w(x,t)$	= displacement	$\theta(x, t)$	= angular rotation of a shaft
$f(x,t)$	= applied force	τ	= torque when used with a rod
τ	= string tension	G	= shear modulus
$\dfrac{\partial w}{\partial x}$	= partial derivatives also denoted, w_x	J	= polar moment of area
l	= length	J_0	= polar moment of inertia (often ρJ)
c	= $\sqrt{\tau/\rho}$, wave speed also $\sqrt{E/\rho}$, de-	γ	= torsional constant
	pending on the structure	$\Theta(x)$	= spatial function of angular rotation
$X(x)$	= a spatial function of x only	$M(x,t)$	= bending moment
$T(t)$	= a temporal function of t only	$I(x)$	= cross-sectional area moment of iner-
σ	= a constant		tia
$a_1, a_2,$	= constants of integration	$f(x,t)$	= distributed force per unit area
a, b		$V(x,t)$	= shear force
$A_n, B_n,$	= constants of integration	β_n	= solution to a transcendental equation
c_n, d_n		ψ	= total angle in bending
$w_0(x),$	= the initial displacement and velocity,	κ	= shear coefficient
$\dot{w}_0(x)$	respectively	$w(x,y,t)$	= membrane displacement in the z
m,n	= indices, integers		direction
β	= separation constant	∇^2	= Laplace operator
λ	= eigenvalues	∇^4	= harmonic operator
$X_n(x)$	= eigenfunctions	D_E	= the flexural rigidity of the plate
$A(x)$	= cross sectional area of a beam	γ, β	= viscous damping coefficients

CHAPTER 7

		$S_{xf}(\omega)$	= cross spectral density function
ω_c	= cut off frequency	γ^2	= coherence function
x_k	= $x(t_k)$ = value of $x(t)$ at the discrete	$\alpha(\omega)$	= α = mobility frequency response
	time t_k		function and has a different value
a_0, a_i, b_i	= digital spectral coefficients		for the various transfer functions:
C	= matrix of digital Fourier coefficients		mobility, receptance, etc. (also used
$\mathbf{a}$	= a vector of spectral coefficients		to denote an angle in Nyquist plots)
N	= size of data set	$\alpha(\omega)$	= receptance matrix
$R_{xf}(\tau)$	= cross correlation function		= $(K - \omega_{dr}^2 M + j\omega_{dr}C)^{-1}$

CHAPTER 8

		$T(t)$	= kinetic energy
c_1, c_2 etc.	= constants of integration	$V(t)$	= potential energy
$u_i(t)$	= ith nodal displacement (local)	$U_i(t)$	= global nodal displacement

θ = angle between global and local co-ordinates

Γ = coordinate transformation matrix between local and global coordinates

K_e = element stiffness matrix (local coordinates)

$K_{()}$ = element stiffness matrix (global co-ordinates)

$K'_{()}$ = element stiffness matrix (global coordinates) expanded to fit global vector size

M_{ii}, K_{ii} = partitions of the mass as stiffness matrices, respectively

Q = reduction transformation matrix

CHAPTER 9

β = arbitrary number base

f = mantissa

a_i = floating point coefficients

A = flexibility influence matrix $= K^{-1}$

a_{ij} = elements of the flexibility influence matrix

$R(\mathbf{u})$ = Raleigh Quotient

μ_i = eigenvalues

D_i = a deflated matrix

$f(A)$ = a function of the matrix A

V = a matrix of normalized eigenvectors

A = a state matrix, or influence matrix (see Window 8.3)

CHAPTER 10

$f(x, \dot{x})$ = a nonlinear function of the displacement and velocity

x_e = equilibrium position vector with elements x_{1e}, x_{2e}, etc.

M = a constant scalar

β = a coefficient (stiffness)

$F_d(x, x')$ = a nonlinear damping function

ϵ = a small number

Γ $= F/\omega_{dr}^2$

References

BANDSTRA, J. P., 1983, "Comparison of Equivalent Viscous Damping and Nonlinear Damping in Discrete and Continuous Vibrating Systems," *Journal of Vibration, Acoustics, Stress, Reliability in Design*, Vol. 105, pp. 382–392.

BERT, C. W., 1973, "Material Damping: An Introductory Review of Mathematical Models, Measures and Experimental Techniques," *Journal of Sound and Vibration*, Vol. 29, No. 2, pp. 129–153.

BLEVINS, R. D., 1977, *Flow-Induced Vibration*, Van Nostrand Reinhold, New York.

BLEVINS, R. D., 1987, *Formulas for Natural Frequencies and Mode Shapes*, R. E. Krieger, Melbourne, Fla.

BOYCE, W. E. and DEPRIMA, P. C., 1986, *Elementary Differential Equations and Boundary Value Problems*, 4th ed., Wiley, New York.

CANNON, R. M., 1967, *Dynamics of Physical Systems*, McGraw-Hill, New York.

CAUGHEY, T. K. and O'KELLEY, M. E. J., 1965, "Classical Normal Modes in Damped Linear Dynamic Systems," ASME *Journal of Applied Mechanics*, Vol. 49, pp. 867–870.

CHURCHILL, R. V., 1972, *Operational Mathematics*, 3rd ed., McGraw-Hill, New York.

COOLEY, J. W. and TUKEY, J. W., 1965, "An Algorithm for the Machines Calcuiation of Complex Fourier Series," *Mathematics of Computation*, Vol. 19, No. 90, pp. 297–311.

COOK, P. A., 1986, *Nonlinear Dynamical Systems*, Prentice Hall, Englewood Cliffs, N.J.

COWPER, G. R., 1966, "The Shear Coefficient in Timoshenko's Beam Theory," *ASME Journal of Applied Mechanics*, Vol. 33, pp. 335–340.

CUDNEY, H. H. and INMAN, D. J., 1989, "Determining Damping Mechanisms in a Composite Beam," *International Journal of Analytical and Experimental Modal Analysis*, Vol. 4, No. 4, pp. 138–143.

DOEBELIN, E. O., 1980, *System Modeling and Response*, Wiley, New York.

DONGARRA, J., BUNCH, J. R., MOLER, C. B. and STEWART, G. W., 1978, *LINPACK Users' Guide*, SIAM Publications, Philadelphia.

D'SOUZA, F. A. and GARG, V. K., 1984, *Advanced Dynamics*, Prentice Hall, Englewood Cliffs, N.J.

EWINS, D. J., 1984, *Modal Testing: Theory and Practice*, Research Studies Press, distributed by Wiley, New York.

FIGLIOLA, R. S. and BEASLEY, D. C., 1991, *Theory and Design for Mechanical Measurement*, Wiley, New York.

FORSYTH, G. E., MALCOLM, M. A. and MOLER, C. B., 1977, *Computer Methods for Mathematical Computation*, Prentice Hall, Englewood Cliffs, N.J.

GOLUB, G. H. and VAN LOAN, C. F., 1989, *Matrix Computations*, 2nd ed., Johns Hopkins University Press, Baltimore.

GUYAN, R. I., 1965, "Reduction of Stiffness and Mass Matrices," *AIAA Journal*, Vol. 3, No. 2, p. 380.

HAGEDORN, P. 1981, *Non-linear Oscillations*, Oxford University Press, New York.

INMAN, D. J., 1989, *Vibration with Control Measurement and Stability*, Prentice Hall, Englewood Cliffs, N.J.

KENNEDY, C. C. and PANCU, C. D. P., 1947, "Use of Vectors in Vibration Measurement and Analysis," *Journal of Aeronautical Science*, Vol. 14, No. 11, pp. 603–625.

MAGRAB, E. B., 1979, *Vibration of Elastic Structural Members*, Sijthoff & Noordhoff, Winchester, Mass.

MOLER, C. B., 1980, *MATLAB Users' Guide*, Technical Report CS81-1, Department of Computer Sciences, University of New Mexico, Albuquerque, N.Mex.

NAFEH, A. H., and MOOK, D. T., 1979, *Nonlinear Oscillations*, Wiley Interscience, New York.

NASHIF, A. D., JONES, D. I. G., and HENDERSON, J. P., 1985, *Vibration Damping*, Wiley, New York.

NEWLAND, D. E., 1984, *Random Vibration and Spectral Analysis*, 2nd ed., Longman, New York.

OTNES, R. K. and ENCOCHSON, L., 1972, *Digital Time Series Analysis*, Wiley, New York.

REISMANN, H., 1988, *Elastic Plates: Theory and Application*, Wiley, New York.

REISMANN, H., and PAWLIK, P. S., 1974, *Elastokinetics*, West Publishing, St. Paul, Minn.

SHAMES, I. H., 1980, *Engineering Mechanics: Statics and Dynamics*, 3rd ed., Prentice Hall, Englewood Cliffs, N.J.

SHAMES, I. H., 1989, *Introduction to Solid Mechanics*, 2nd ed., Prentice Hall, Englewood Cliffs, N.J.

SMITH, B. T., BOYLE, J. M., DONGARRA, J., GARBOW, B., IKEBE, Y, KELMA, V. C., and MOLER, C. B., 1976, *Matrix Eigensystem Routines: EISPACK Guide*, 2nd ed., Springer-Verlag, New York.

SNOWDEN, J. C., 1968, *Vibration and Shock in Damped Mechanical Systems*, Wiley, New York.

WEAVER, W., JR., TIMOSHENKO, S. P., and YOUNG, D. H., 1990, *Vibration Problems in Engineering*, 5th ed., Wiley, New York.

WOWK, V., 1991, *Machinery Vibration: Measurement and Analysis*, McGraw-Hill, New York.

Answers to Selected Problems

CHAPTER 1

1.3. $x(t) = \cos 2t$

1.6. $B = A \cos \phi$, $C = A \sin \phi$

1.9. 2.25 Hz

1.14. 10 rad/s

1.18. $x(t) = e^{-t} \sin t$

1.22. $\zeta = 0.316$, underdamped

1.25. $x(t) = 1.155e^{-5t} \sin(0.866t - 1.047)$

1.29. $ml_2^2 \ddot{\theta} + \left(kl_1^2 + mgl_2\right)\theta = 0$, $\qquad \omega = \sqrt{\dfrac{kl_1^2 + mgl_2}{ml_2^2}}$

1.32. $c = 0.02\sqrt{Jk}$

1.38. $k = 300.98$ N/m

1.43. $E = 316 \times 10^9$ N/m

1.47. $c = 13.94$ N · s/m

CHAPTER 2

2.3. $\omega = 110$ rad/s, $\omega_{dr} = 90$ rad/s

2.8. $x(t) = 1.095e^{-.02t} \sin(2t + 1.086) + 0.03125 \cos(10t + 0.004167)$ m

2.9. $A_0 = 0.0041$ m, $\phi = -1.29$ rad

2.12. $c = 1228$ kg/s, $A_0 = 0.0023$ rad

2.19. $c = 894.4$ kg/s, $F_T = 400$ N

2.22. $c = 1331$ kg/s

2.26. $\zeta = 0.05$

2.38. $x = 1.79 \times 10^{-3}$ m

2.41. $F_0 = 1874$ N

2.44. Lubricated: $F_0 = 8.74$ N. Unlubricated: $F_0 = 37.5$ N.

2.49. $|x(0)| < 2.802$ m, $\dot{x}(0) = 0$

2.53. $F_0 = 294$ N

2.56. $\beta = 0.294, c_{eq} = 198$ kg/s

CHAPTER 3

3.1. $x(t) = 1.414e^{-t} \sin(t + 0.785) \qquad 0 < t < \pi$

$$x(t) = 1.414e^{-t} \sin(t + 0.785) - e^{-(t-\pi)} \sin t \qquad t > \pi$$

3.6. $x(t) = 1.23 \times 10^{-6}e^{-81.01t} \sin 8100.6t$ m

3.11. $$x(t) = 5(t - \sin t) \qquad 0 < t < 4$$

$$x(t) = 20 + 5[\sin(t - 4) - \sin t] \qquad t > 4$$

3.13. $x(t) = .5t - .05 \sin(10t)$ m

3.17. $\omega = 3.30$ rad/s, $\zeta = 0.303$

3.21. $F(t) = \sum_{n=1}^{\infty} b_n \sin nt$

$$\text{where } b_n = \begin{cases} 0 & n \text{ even} \\ \dfrac{4}{\pi n} & n \text{ odd} \end{cases}$$

3.25. $x(t) = -.11e^{-5t} \sin(31.22t - .0907) + .0505 \cos(3.162t - 1.57)$ m

3.29. $\bar{x}^2 = \dfrac{50\pi S_0}{3}$

3.33. $x(t) - y(t) = \dfrac{A}{m\omega^2}\left[1 - \dfrac{t}{t_0} + \dfrac{1}{t_0\omega} \sin \omega t - \cos \omega t\right] - \dfrac{A}{2}t^2 - \dfrac{A}{6t_0}t^3 \qquad 0 \le t \le 2t_0$

$$x(t) - y(t) = \dfrac{A}{m\omega^2}\left[\dfrac{1}{t_0\omega}(\sin \omega t - \sin \omega(t - 2t_0)) - \cos \omega t - \cos \omega(t - 2t_0)\right]$$

$$t > 2t_0$$

3.38. $\dfrac{X(s)}{F(s)} = \dfrac{1}{as^4 + bs^3 + cs^2 + ds + e}$

3.41. $\omega = 3$ rad/s, $\zeta = .227, c = 3.03$ kg/s, $m = 2.22$ kg, $k = 20$ N/m

3.45. $x(t) = 1 - e^{-t}$, bounded

CHAPTER 4

4.3. $\mathbf{u}_1 = \begin{bmatrix} 1 \\ .909 \end{bmatrix}$, $\mathbf{u}_2 = \begin{bmatrix} -.101 \\ 1 \end{bmatrix}$

4.7. $\omega_1 = 0$, $\omega_2 = 3.333$

4.12. $x_1(t) = 0.5 - .05 \cos 16.73t$

$x_2(t) = 0.5 + 0.5 \cos 16.73t$

4.17. $M^{1/2} = \begin{bmatrix} 3 & -2 \\ -2 & 2 \end{bmatrix}$

4.23. $\Lambda = \text{diag}(\lambda_i) = \begin{bmatrix} .454 & 0 \\ 0 & 220.05 \end{bmatrix}$

$P = [\mathbf{v}_1 \quad \mathbf{v}_2] = \begin{bmatrix} .9999 & -.0144 \\ .0144 & .9999 \end{bmatrix}$

4.28. $a = -1$

4.31. $\theta(t) = \begin{bmatrix} .2774 \cos \omega_1 t - .2774 \cos \omega_2 t \\ .3613 \cos \omega_1 t + .6387 \cos \omega_2 t \end{bmatrix}$

where $\omega_1 = .4821 \sqrt{\dfrac{k}{J_2}}$ and $\omega_2 = 1.1976 \sqrt{\dfrac{k}{J_2}}$

4.39. $\mathbf{x}(t) = \begin{bmatrix} .5(\cos 10t + \cos 20t) \\ .5(\cos 10t - \cos 20t) \end{bmatrix}$

4.46. $\mathbf{x}(t) = \begin{bmatrix} 0 \\ .001 \cos 6.3246t \\ 0 \\ 0 \end{bmatrix}$

4.53. $\mathbf{q}(t) = \begin{bmatrix} .01709 \\ -.01859 \\ .01709 \end{bmatrix} e^{-2.0771 \times 10^{-6} t} \sin\left(2.0770 \times 10^{-4} t - 1.5808\right)$

$+ \begin{bmatrix} .01744 \\ .03206 \\ .01744 \end{bmatrix} e^{-1.8142 \times 10^{-4} t} \sin\left(8.8877 \times 10^{-4} t + 1.3694\right) \text{ m}$

4.58. $\zeta_3 = .09153$

4.65. $x_2(t) = 2.2222 \times 10^4 t - 1.4351 \times 10^{-8} e^{-.5187t} \sin 5.1615t$

CHAPTER 5

5.1. $10 - 900$ Hz

5.3. $\omega = 20$ Hz, $v_{rms} = 0.95$ mm/s

5.10. $k = 807$ N/m

5.17. $A_0 = 0.062$ mm, $F_T = 86$ N

5.25. $\zeta = 0.29$

5.45. $x = 0.014$ mm

5.48. With $\dfrac{Xk}{F_0} < 1$, $\zeta = 0.767$

5.55. $\zeta_{op} = 0.422, (\omega_r)_{op} = 0.943\omega_p$

5.61. $k_2 = 8560$ N/m, $c = 9512.8$ kg/s

5.68. $E = 6.90 \times 10^{10}$ N/m^2

5.74. (a) 3573 rpm   (b) 50 cm

CHAPTER 6

6.4. $\omega_n = \dfrac{(2n-1)\pi}{2d}\sqrt{\dfrac{\tau}{\rho}}$ $X_n = \sin\dfrac{(2n-1)\pi x}{2d}$ for $n = 1, 2, 3, \cdots$.

6.5. $c_n = 0$

$$d_n = \begin{cases} \dfrac{2}{l}\dfrac{x}{\sin \sigma_n x} & 0 \le x \le \frac{l}{2} \\[2mm] \dfrac{2\left(-\frac{x}{l}+1\right)}{\sin \sigma_n x} & \frac{1}{2} < x \le 1 \end{cases}$$

6.10. For an aluminum bar 10 m long, the first natural frequency is 126 Hz.

6.19. $\omega_1 \approx \dfrac{20.29}{\rho^{1/2}}$ $\omega_2 \approx \dfrac{49.13}{\rho^{1/2}}$ $\omega_3 \approx \dfrac{79.80}{\rho^{1/2}}$

6.26. $w(x,t) = 3.58 \times 10^{-6} \displaystyle\sum_{n=1}^{\infty} \dfrac{1}{(2n-1)^2} \sin[1132(2n-1)\pi t]\sin[(2n-1)\pi x]$

6.30. $\omega_1 = 482$ Hz

$\omega_2 = 1450$ Hz

$\omega_3 = 2410$ Hz

6.34.   Torsion: $\omega_{tn} = 3162\dfrac{(2n-1)\pi}{2l}$ for $n = 1, 2, 3, \cdots$.

Longitudinal: $\omega_{ln} = 5128\dfrac{(2n-1)\pi}{2l}$ for $n = 1, 2, 3, \cdots$.

6.36. $\phi_n(x) = \cos\dfrac{n\pi x}{l}$ $\omega_n = \sqrt{\dfrac{G}{\rho}}\dfrac{n\pi}{l}$ for $n = 1, 2, 3, \cdots$.

6.40. $\omega_n = \sqrt{\dfrac{\beta_n^4 EI}{\rho A}}$ where $\cos(\beta_n l) = -\dfrac{1}{\cosh(\beta_n l)}$

$$\phi_n(x) = -\left(\dfrac{\cos(\beta_n l) + \cosh(\beta_n l)}{\sin(\beta_n l) + \sinh(\beta_n l)}\right)\sin(\beta_n x) + \cos(\beta_n x)$$

for $n = 1, 2, 3, \cdots$.

$$+ \left(\dfrac{\cos(\beta_n l) + \cosh(\beta_n l)}{\sin(\beta_n l) + \sinh(\beta_n l)}\right)\sinh(\beta_n x) - \cosh(\beta_n x)$$

6.43. $\omega_n = \left(\dfrac{n\pi}{l}\right)^2 \sqrt{\dfrac{EI}{\rho A}}$ $\quad \phi_n(x) = \sin\dfrac{n\pi x}{l}$ $\quad$ for $n = 1, 2, 3, \cdots$. and $w(x, t) = \cos(\omega_2 t)\sin\dfrac{2\pi x}{l}$

6.48. $\omega_{mn} = \sqrt{(2m-1)^2 + 4n^2}\,\sqrt{\dfrac{\tau}{\rho}\dfrac{\pi}{2}}$ $\quad$ for $m, n = 1, 2, 3, \cdots$.

6.56. (a) $\zeta_n = \dfrac{\gamma}{2\sqrt{\rho\tau}}\dfrac{n\pi}{l}$ $\quad$ for $n = 1, 2, 3, \cdots$.

$\quad$ (b) $\zeta_{mn} = \dfrac{\gamma l}{2\sqrt{\rho\tau\left(m^2 + n^2\right)}}$ $\quad$ for $m, n = 1, 2, 3, \cdots$.

6.57. $\dfrac{\text{kg}}{\text{m}\cdot\text{s}}$

6.66. $w(x, t) = \sum_{n=1}^{\infty}\left\{C_{1n}\sin\omega_n t + C_{2n}\cos\omega_n t + \left(\dfrac{F_0}{\omega_n^2 - \omega_{dr}^2}\right)\sin\omega_{dr}t\right\}\sin\dfrac{(2n-1)\pi x}{2l}$

$\quad$ where $\omega_n = \sqrt{\dfrac{E}{\rho}}\dfrac{(2n-1)\pi}{2l}$ $\quad$ for $n = 1, 2, 3, \cdots$.

CHAPTER 7

7.3. delta input: $f(t) = \delta(t)$ $\quad F(s) = 1$

$\quad$ triangle input: $f(t) = \dfrac{1}{2} - \dfrac{4}{\pi^2}\sum_{n=1,3,5,\ldots}\dfrac{1}{n^2}\cos\dfrac{2\pi n}{T}t$

$\qquad\qquad F(s) = \dfrac{1}{2s} - \dfrac{4}{\pi^2}\sum_{n=1,3,5,\ldots}\dfrac{1}{n^2}\dfrac{s}{s^2 + a^2}$ $\qquad a = \dfrac{2\pi n}{T}$

7.4. Error $= \sqrt{\dfrac{350}{(10 + m_s)}} - \sqrt{\dfrac{350}{10}}$ $\quad 0.5 < m_s < 5.0$

7.5. $\omega_T = 1$ rad/s

$\quad a_1 = -1 \qquad a_n = 0 \qquad n > 1$

$\quad b_1 = -2 \qquad b_2 = 3 \qquad b_n = 0 \qquad n > 2$

7.8. $x(t) = -\dfrac{20}{\pi}\sum_{n=1,3,5,\ldots}\dfrac{1}{n}\sin(nt)$

7.10. The system has eight modal peaks with the approximate natural frequencies (in Hz) of:

$$\omega_1 \approx 2$$

$$\omega_2 \approx 4$$

$$\omega_3 \approx 10$$

$$\omega_4 \approx 15$$

$$\omega_5 \approx 22$$

$$\omega_6 \approx 29$$

$$\omega_7 \approx 36$$

$$\omega_8 \approx 47$$

7.14. $\Delta\zeta = 0.01$

7.17. $\dfrac{X_1(s)}{F(s)} = \dfrac{s^2 + cs + 2}{s^4 + cs^3 + 4s^2 + 2cs + 3}$

$\dfrac{X_2(s)}{F(s)} = \dfrac{1}{s^4 + cs^3 + 4s^2 + 2cs + 3}$

7.24. $\zeta = 0.24$

7.25. $\alpha(\omega_{dr}) = \dfrac{1}{\det(A)} \begin{bmatrix} 2 - \omega_{dr}^2 + j\omega_{dr} & 2 + j\omega_{dr} \\ 2 + j\omega_{dr} & 6 - 2\omega_{dr}^2 + j3\omega_{dr} \end{bmatrix}$

$\det(A) = 2\omega_{dr}^4 - 12\omega_{dr}^2 + 8 + j\left(-5\omega_{dr}^3 + 8\omega_{dr}\right)$

CHAPTER 8

8.2. $\omega_1 = 0$, $\omega_2 = \sqrt{\dfrac{\pi^2 E}{\rho l^2}}$

8.13. $\omega_1 = 7092$ rad/s, $\omega_2 = 18636$ rad/s

8.15. $\omega_1 = 47556.1$ rad/s, $\omega_2 = 101975$ rad/s

8.17. $\omega_1 = \dfrac{1.984}{l}\sqrt{\dfrac{E}{\rho}}$

8.26. $\omega_1 = 3.52\dfrac{1}{l^2}\sqrt{\dfrac{EI}{\rho A}}$, $\omega_2 = 20.49\dfrac{1}{l^2}\sqrt{\dfrac{EI}{\rho A}}$, $\omega_3 = 34.98\dfrac{1}{l^2}\sqrt{\dfrac{EI}{\rho A}}$

8.28. $\omega_1 = 1134$ rad/s, $\omega_2 = 3152$ rad/s, $\omega_3 = 6253$ rad/s

8.30. $\omega = 20.5$ rad/s

8.34. $\omega_1 = 6670$ rad/s, $\omega_2 = 13106$ rad/s

8.47. $M = l\begin{bmatrix} .9 & 0 \\ 0 & .9 \end{bmatrix}$, $K = \dfrac{1}{l}\begin{bmatrix} 1.89 & .48 \\ .48 & .36 \end{bmatrix}$

8.52. $M_r = \begin{bmatrix} 2 & 0 \\ 0 & 2 \end{bmatrix}$, $K_r = \begin{bmatrix} 19.95 & -.15 \\ -.15 & 36.55 \end{bmatrix}$

CHAPTER 9

9.1. $\pi = .3142 \times 10^1, e = .2718 \times 10^1, \sqrt{2} = .1414 \times 10^1$
   $.1 \le f < 1$

9.7. 6.79, not much better than 200 terms

9.10. $A = \begin{bmatrix} \dfrac{1}{k_1} & \dfrac{1}{k_1} \\ \dfrac{1}{k_1} & \dfrac{1}{k_1} + \dfrac{1}{k_2} \end{bmatrix}$, $K^{-1} = \begin{bmatrix} \dfrac{1}{k_1} & \dfrac{1}{k_1} \\ \dfrac{1}{k_1} & \dfrac{1}{k_1} + \dfrac{1}{k_2} \end{bmatrix}$

9.12. $\omega_1 \approx .3536$ rad/s. This method is easier.

9.18. $\omega_1 \approx 3.464$ rad/s. The exact value is $\omega_1 \approx 6.325$ rad/s.
   The third assumption of Dunkerley's formula is not met since $\omega_1 = \omega_2$.

9.22. $A = \begin{bmatrix} 1.7083 \times 10^{-5} & 1.5833 \times 10^{-5} & 3.3333 \times 10^{-6} \\ 1.5833 \times 10^{-5} & 1.5833 \times 10^{-5} & 3.3333 \times 10^{-6} \\ 3.3333 \times 10^{-6} & 3.3333 \times 10^{-6} & 3.3333 \times 10^{-6} \end{bmatrix}$
   $3.92 < \omega_1 \le 5.77$

9.23. $\omega_1 \le 10.38$ rad/s for any value of α.

9.29. $\mu_1 = .3562, \mathbf{u}_1 = \begin{bmatrix} .4784 \\ .5128 \\ .7129 \end{bmatrix}$

9.30. $D_1 = \begin{bmatrix} .0185 & .0126 & -.0215 \\ .0126 & .0163 & -.0202 \\ -.0215 & -.0202 & .0290 \end{bmatrix}$, $\mu_2 = .0590$

9.31. $\ddot{\mathbf{x}} + \begin{bmatrix} \dfrac{k_1 + k_2}{m_1} & \dfrac{-k_2}{m_1} \\ \dfrac{-k_2}{m_2} & \dfrac{k_2}{m_2} \end{bmatrix} \mathbf{x} = \mathbf{0}$

9.34. $M^{-1} = \begin{bmatrix} .1111 & .1111 \\ .1111 & 1.1111 \end{bmatrix}$, $M^{1/2} = \begin{bmatrix} .3234 & .0808 \\ .0808 & 1.0510 \end{bmatrix}$

9.42. $\lambda_{1,2} = -\frac{1}{3} \pm 1.3744j$, $\lambda_{3,4} = -\frac{2}{3} \pm 1.8856j$

CHAPTER 10

10.5. $x = 0, \dot{x} = 0$

   $x = \pm \sqrt{-\dfrac{\alpha}{\beta}}, \dot{x} = 0$ where $-\dfrac{\alpha}{\beta} > 0$

10.9. $A = \begin{bmatrix} 0 & 1 \\ \dfrac{\partial F}{\partial x} \Big|_{(x_e, 0)} & \dfrac{\partial F}{\partial \dot{x}} \Big|_{(x_e, 0)} \end{bmatrix}$

10.11.

$$x = n\pi, \dot{x} = 0 \qquad \text{for } n = \ldots, -2, -1, 0, 1, 2, \ldots$$

$$\text{for } n \text{ even:} \qquad A = \begin{bmatrix} 0 & 1 \\ -1 & 0 \end{bmatrix}$$

$$\text{for } n \text{ odd:} \qquad A = \begin{bmatrix} 0 & 1 \\ 1 & 0 \end{bmatrix}$$

10.14. 0.35 s

10.16. $\dfrac{\dot{\theta}^2}{2} - \left(\dfrac{g}{l}\right) \cos\theta = \dfrac{g}{l}$, yes

10.19. One equilibrium at $\theta = 0, \dot{\theta} = 0$

10.21. $\theta = \dfrac{\pi}{2}, \dot{\theta} = \pm\sqrt{\dfrac{2g}{l}}$

10.22. $|A| = 2$

10.24. $|A| = 2, \omega \approx 1$ rad/s

10.28. $|\theta| < 0.3208\pi$

10.32. $x(t) = -\frac{4}{3}\cos t + \frac{8}{45} - \frac{8}{315}\cos 2t + \cdots$

10.37. $A = 1.85$

Photo Credits

The following people and organizations provided photographs for the chapter openings. **Chapter One:** Photo on top provided as a courtesy of Photoreporters, © DPA. Photo on bottom is by Audrey Shehyn, courtesy of AP/Wide World Photos. **Chapter Two:** Photo on top is by Spencer Grant, courtesy of FPG International. Photo on bottom provided as a courtesy of Sound and Vibration Magazine. **Chapter Three:** Photo on top provided as a courtesy of Sound and Vibration Magazine. Photo on bottom is by Rick Kopstein, courtesy of Monkmeyer Press. **Chapter Four:** Photo on top is by Jeffrey G. Russel, courtesy of Car and Driver Magazine. Photo on bottom provided as a courtesy of the American Bridge Division of USX Corp. **Chapter Five:** Photo on top provided as a courtesy of Lord Corporation. Photo on bottom provided as a courtesy of Balloon Excellsior, Inc. **Chapter Six:** Photo on top provided as a courtesy of the U.S. Air Force. Photo on bottom provided as a courtesy of Charles Gatewood. **Chapter Seven:** Photo on top provided as a courtesy of Kistler Instrument Corporation. Photo on bottom provided as a courtesy of Polytec Optronics, Inc. **Chapter Eight:** Photo on top provided as a courtesy of Triborough Bridge and Tunnel Authority. Photo on bottom provided as a courtesy of Sound and Vibration Magazine. **Chapter Nine:** Photo on top provided as a courtesy of NASA. Photo on bottom is by Dan Burns, courtesy of Monkmeyer Press. **Chapter Ten:** Photo on top provided as a courtesy of the Chevrolet Motor Division. Photo on bottom provided as a courtesy of Moog, Inc.

Index

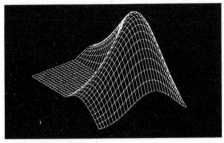

MATLAB® Technical Computing Environment

MATLAB is the companion software to **Engineering Vibration** by Daniel J. Inman (Prentice Hall, 1994). The *Engineering Vibration Toolbox*, consisting of MATLAB M-files, has been created by the author to illustrate the concepts presented in the text.

- **MATLAB Application Toolboxes** add functions for symbolic math, signal processing, control design, neural networks, and other areas.

- **Student Editions** of MATLAB and SIMULINK are available for use on students' own personal computers.

- **Educational discount plans** support classroom instruction and research.

- **Classroom Kits** provide cost-effective support for PC or Mac teaching labs.

- **MATLAB-based books** use MATLAB to illustrate basic and advanced material in a wide range of topics.

NAME

TITLE

COMPANY

DEPT. OR M/S

STREET

CITY/STATE/ZIP

PHONE

FAX   EMAIL

WHERE DID YOU PURCHASE THIS BOOK?

Computer platform – check all that apply:
☐ PC/Macintosh ☐ UNIX Workstation ☐ VAX/Supercomputer

▶ **For the fastest response, fax this card to:**
(508) 653-6284, or call us at (508) 653-1415.

I am interested in The MathWorks product information for:

☐ Simulation ☐ Control System Design ☐ Math & Visualization
☐ Signal Processing ☐ Symbolic Math ☐ Educational Discounts
☐ System Identification ☐ Chemometrics ☐ Classroom Kits
☐ Neural Networks ☐ Optimization ☐ Student Editions
☐ Statistics ☐ Image Processing ☐ MATLAB Books

Send me the free *Engineering Vibration Toolbox* disk for my:

(check one)   ☐ PC   ☐ Macintosh

The
MATH
WORKS
Inc.

BUSINESS REPLY MAIL

FIRST CLASS MAIL   PERMIT NO. 82   NATICK, MA

POSTAGE WILL BE PAID BY ADDRESSEE

The MathWorks, Inc.
24 Prime Park Way
Natick, MA 01760-9889

UNITS

QUANTITY	ENGLISH SYSTEM	S.I. SYSTEM
force	1 lb	4.448 Newtons (N)
mass	1 lb $\cdot$ sec^2/ft (slug)	14.59 kg (kilogram)
length	1 ft	0.3048 meters (m)
mass density	1 lb/ft^3	16.02 kg/m^3
torque or moment	1 lb $\cdot$ in	0.113 N $\cdot$ m
acceleration	1 ft/sec^2	0.3048 m/s^2
accel. of gravity	32.2 ft/s^2 = 386 in./sec^2	9.81 m/s^2
spring constant k	1 lb/in.	175.1 N/m
rot. spring constant k	1 lb $\cdot$ in./rad	0.113 N $\cdot$ m/rad
damping constant c	1 lb $\cdot$ sec/in.	175.1 n $\cdot$ s/m
mass moment of inertia	1 lb.in.sec^2	0.1129 kg/m^2
modulus of elasticity	10^6 lb.in.^{2}	6.895 × 10^9 N/m^2
modulus of elasticity of steel	29 × 10^{6}lb/in.^{2}	200 × 10^9 N/m^2
angle	1 degree	157.3 radian

SI UNITS PREFIXES

Multiplication Factor	Prefix	Symbol
1 000 000 000 000 = 10^{12}	terra	T
1 000 000 000 = 10^9	giga	G
1 000 000 = 10^6	mega	M
1 000 = 10^3	kilo	k
100 = 10^2	hecto	h
10 = 10	deka	da
0.1 = 10^{-1}	deci	d
0.01 = 10^{-2}	centi	c
0.001 = 10^{-3}	milli	m
0.000 001 = 10^{-6}	micro	μ
0.000 000 001 = 10^{-9}	nano	n
0.000 000 000 001 = 10^{-12}	pico	p

SOFTWARE INFORMATION:

If you have not already installed MATLAB or the Student Edition of MATLAB, do so now. You should be familiar with MATLAB before running this software. This software can only be run on a computer with a hard disk or other large storage device. To install on a PC equipped with a hard disk: 1. Boot your system, if it is not already. 2. From the DOS prompt move to the directory in which MATLAB is installed. Most likely C:\MATLAB and you can move into this directory by typing CD C:\MATLAB for instance. 3. Insert the VIBRATION TOOLBOX disk into your floppy disk drive. 4. Type *:INSTALL, where * is the label of your floppy disk drive (most likely A or B). Follow the instructions in the installation program. 5. Remove the disk and store it in a safe place. 6. Edit the file MATLAB.BAT in the MATLAB directory and add the path to the VIBRATION TOOLBOX directory to the path. Refer to Section 4.4.4 of the Student Edition of MATLAB book on setting up paths. You are now ready to run the VIBRATION TOOLBOX. On line help is available for all of the functions provided by the toolbox. For information summarizing the functions for each chapter, type help VTB?, where ? is the chapter number, from inside MATLAB. Help on specific function files is available by typing HELP VTB1_1 for instance for the function file VTB1_1. VIBRATION TOOLBOX commands can be run by typing them with the necessary arguments just as MATLAB command. For instance VTB1_1 can be run by typing VTB1_1(1,.1,1,1,0,10). Type help VTB1_1 to see what this does and what the various parameters are.